Algebraic Analysis

Volume II

Professor Mikio Sato

Algebraic Analysis

Papers Dedicated to Professor Mikio Sato on the Occasion of His Sixtieth Birthday

Volume II

Edited by

Masaki Kashiwara
Takahiro Kawai

Research Institute for Mathematical Sciences
Kyoto University
Sakyo-ku, Kyoto
Japan

ACADEMIC PRESS, INC.

Harcourt Brace Jovanovich, Publishers

Boston San Diego New York
Berkeley London Sydney
Tokyo Toronto

ACADEMIC PRESS, INC.
1250 Sixth Avenue, San Diego, CA 92101

United Kingdom Edition published by
ACADEMIC PRESS INC. (LONDON) LTD.
24-28 Oval Road, London NW1 7DX

Library of Congress Cataloging-in-Publication Data

Algebraic analysis: papers dedicated to Professor Mikio Sato on the occasion of his sixtieth birthday/edited by Masaki Kashiwara, Takahiro Kawai.
p. cm.
Includes bibliographies and index.
ISBN 0-12-400465-2 (v. 1) ISBN 0-12-400466-0 (v. 2).
1. Mathematical analysis. 2. Algebra. 3. Sato, Mikio, Date-
I. Sato, Mikio, Date- . II. Kashiwara, Masaki, Date-
III. Kawai, Takahiro.
QA300.A45 1988
515—dc19 88-6331
CIP

Printed in the United States of America

89 90 91 92 9 8 7 6 5 4 3 2 1

Dedication

We and the contributors to this collection dedicate these articles on algebraic analysis and related topics to Professor Mikio Sato.

We would like to celebrate the birthday of Professor Sato, initiator of algebraic analysis in the twentieth century.

Congratulations on your sixtieth birthday, Professor Sato!

April 18, 1988

Editors
Michio Jimbo
Masaki Kashiwara
(co-editor-in-chief)
Takahiro Kawai
(co-editor-in-chief)
Hikosaburo Komatsu
Tetsuji Miwa
Mitsuo Morimoto

VOLUME EDITORS:

Michio Jimbo

Research Institute for Mathematical Sciences
Kyoto University
Sakyo-ku, Kyoto
Japan

Masaki Kashiwara

Research Institute for Mathematical Sciences
Kyoto University
Sakyo-ku, Kyoto
Japan

Takahiro Kawai

Research Institute for Mathematical Sciences
Kyoto University
Sakyo-ku, Kyoto
Japan

Hikosaburo Komatsu

Department of Mathematics
The University of Tokyo
Bunkyo-ku, Tokyo
Japan

Tetsuji Miwa

Research Institute for Mathematical Sciences
Kyoto University
Sakyo-ku, Kyoto
Japan

Mitsuo Morimoto

Department of Mathematics
Sophia University
Chiyoda-ku, Tokyo
Japan

Contents

Contents of Volume I

Contributors

Numbers in parentheses refer to the pages on which the authors' contributions begin. Numbers smaller than 470 indicate contributions in Volume I.

Takashi Aoki (19), *Department of Mathematics and Physics, Kinki University, Kowakae 3-4-1, Higashi-Osaka-shi, Osaka 577, Japan*

K. Aomoto (25), *Department of Mathematics, Nagoya University, Furo-cho, Chikusa-ku, Nagoya 464, Japan*

Helen Au-Yang (29), *Institute for Theoretical Physics, State University of New York at Stony Brook, Stony Brook, New York 11794-3840*

Gunter Bengel (41), *Mathematisches Institut der Westfälischen Wilhelms Universität, Roxeler Strasse 64, 44 Münster, Federal Republic of Germany*

J. Bros (49), *Service de Physique Théorique, Institut de Recherche Fondamentale, CEA, CEN-Saclay, B.P. 2, 91191 Gif-sur-Yvette Cédex, France*

Etsuro Date (75), *Department of Mathematics, College of General Education, Kyoto University, Yoshida-Nihonmatsu-cho, Sakyo-ku, Kyoto 606, Japan*

Leon Ehrenpreis (85), *Department of Mathematics, Temple University, Broad St. and Montgomery Ave., Philadelphia, Pennsylvania 19122*

L. D. Faddeev (129), *Steklov Institute, Fontanka 27, Leningrad 191011, USSR*

H. Flaschka (141), *Department of Mathematics, University of Arizona, Tucson, Arizona 85721*

Victor Guillemin (155), *Department of Mathematics, Massachusetts Institute of Technology, 77 Massachusetts Avenue, Cambridge, Massachusetts 02139*

B. Helffer (171), *Department of Mathematics, Université de Nantes, 44072 Nantes Cédex, France, and Centre de Mathématiques, Ecole Polytechnique, 91128 Palaiseau Cédex, France*

C. Denson Hill (185), *Department of Mathematics, State University of New York at Stony Brook, Stony Brook, New York 11794-3840*

Ryogo Hirota (203), *Department of Applied Mathematics, Faculty of Engineering, Hiroshima University, Oaza-Shitami, Saijo-cho, Higashi-Hiroshima, Hiroshima 724, Japan*

D. Iagolnitzer (217), *Service de Physique Théorique, CEN-Saclay, B.P. 2, 91190 Gif-sur-Yvette Cédex, France*

Jun-ichi Igusa (231), *Department of Mathematics, Johns Hopkins University, Baltimore, Maryland 21218*

Mitsuru Ikawa (243), *Department of Mathematics, Osaka University, Machikaneyama-cho 1-1, Toyonaka, Osaka 560, Japan*

S. Iyanaga (9), *Otsuka 6-12-4, Bunkyo-ku, Tokyo 112, Japan*

Michio Jimbo (75, 253), *Research Institute for Mathematical Sciences, Kyoto University, Sakyo-ku, Kyoto 606, Japan*

Akiro Kaneko (267), *Department of Mathematics, University of Tokyo, Komaba 3-8-1, Meguro-ku, Tokyo 153, Japan*

Masaki Kashiwara (1, 277), *Research Institute for Mathematical Sciences, Kyoto University, Sakyo-ku, Kyoto 606, Japan*

Kiyomi Kataoka (287), *Department of Mathematics, Faculty of Science, Tokyo Metropolitan University, Fukazawa, Setagaya-ku, Tokyo 158, Japan*

Yukiyosi Kawada (305), *Nagasaki 6-37-17, Toshima, Tokyo 171, Japan*

Takahiro Kawai (1, 309, 331), *Research Institute for Mathematical Sciences, Kyoto University, Sakyo-ku, Kyoto 606, Japan*

Tatsuo Kimura (345), *Department of Mathematics, University of Tsukuba, Tennodai 1-1-1, Niihari-gun, Ibaraki 305, Japan*

Hikosaburo Komatsu (357), *Department of Mathematics, University of Tokyo, Hongo 7-3-1, Bunkyo-ku, Tokyo 113, Japan*

Michio Kuga (373), *Department of Mathematics, State University of New York at Stony Brook, Stony Brook, New York 11794–3840*

Yves Laurent (381), *Département de Mathématiques, Université de Paris-Sud, Bâtiment 425, 91405 Orsay Cédex, France*

J.-L. Lieutenant (393), *Department of Mathematics, place du 20 Août 7, 4000 Liège, Belgium*

B. Malgrange (403), *Institut Fourier, B.P. 74, 38402 St. Martin d'Hères Cédex, France*

Barry M. McCoy (29), *Institute for Theoretical Physics, State University of New York at Stony Brook, Stony Brook, New York 11794-3840*

Z. Mebkhout (413), *U.E.R. de Mathématiques, Tour 45–55 5ème étage, Université de Paris 7, 2 place Jussieu, 75251 Paris Cédex 05, France*

Tetsuji Miwa (75, 253), *Research Institute for Mathematical Sciences, Kyoto University, Sakyo-ku, Kyoto 606, Japan*

Masatake Mori (423), *Denshi-Joho-Kogaku, Institute of Information Sciences and Electronics, University of Tsukuba, Tennodai 1-1-1, Niihari-gun, Ibaraki 305, Japan*

Mitsuo Morimoto (439), *Department of Mathematics, Sophia University, Kioi-cho, Chiyoda-ku, Tokyo 102, Japan*

Yasuo Morita (457), *Mathematical Institute, Tohoku University, Aza-Aoba Aramaki, Sendai-shi, Miyagi 980, Japan*

Motohico Mulase (473), *Department of Mathematics, University of California, Los Angeles, California 90024*

Masakazu Muro (493), *Department of Mathematics, Kochi University, Akebono-cho 2-5-1, Kochi 780, Japan*

Yoshimasa Nakamura (505), *Department of Mathematics, Gifu University, Yanagito 1-1, Gifu 501-11, Japan*

Noboru Nakanishi (517), *Research Institute for Mathematical Sciences, Kyoto University, Sakyo-ku, Kyoto 606, Japan*

Isao Naruki (527), *Research Institute for Mathematical Sciences, Kyoto University, Sakyo-ku, Kyoto 606, Japan*

Masatoshi Noumi (549), *Department of Mathematics, Sophia University, Kioi-cho 7-1, Chiyoda-ku, Tokyo 102, Japan*

Toshinori Ôaku (571), *Department of Mathematics, Yokohama City University, Seto 22-2, Kanazawa-ku, Yokohama 236, Japan*

Takayuki Oda (587), *Department of Mathematics, Faculty of Science, Niigata University, Ninomachi 8050, Ikarashi 2 no chou, Niigata 950-21, Japan*

Yasunori Okabe (601), *Department of Mathematics, Hokkaido University, Kita 13-jo, Nishi 6, Kita-ku, Sapporo 060, Japan*

Masato Okado (75, 253), *Research Institute for Mathematical Sciences, Kyoto University, Sakyo-ku, Kyoto 606, Japan*

Kazuo Okamoto (647), *Department of Mathematics, University of Tokyo, Komaba 3-8-1, Meguro-ku, Tokyo 153, Japan*

Takashi Ono (659), *Department of Mathematics, Johns Hopkins University, Baltimore, Maryland 21218*

Toshio Oshima (667, 681), *Department of Mathematics, University of Tokyo, Hongo 7-3-1, Bunkyo-ku, Tokyo 113, Japan*

Jacques H. H. Perk (29), *Institute for Theoretical Physics, State University of New York at Stony Brook, Stony Brook, New York 11794-3840*

F. Pham (699), *Department of Mathematics, Université de Nice, Parc Valrose, 06034 Nice Cédex, France*

Mario Rasetti (727), *Dipartimento di Fisica, Politecnico di Torino, 10100 Torino, Italy*

Tullio Regge (727), *Dipartimento di Fisica Teorica, Università di Torino, via Pietro Giuria 1, 10125 Torino, Italy*

N. Yu. Reshetikhin (129), *Steklov Institute, Fontanka 27, Leningrad 191011, USSR*

Gian-Carlo Rota (13), *Department of Mathematics, Massachusetts Institute of Technology, 77 Massachusetts Avenue, Cambridge, Massachusetts 02139*

Yutaka Saburi (681), *Chiba Tanki College, Kuonodai 1-3-1, Ichikawa 272, Japan*

Kyoji Saito (735), *Research Institute for Mathematical Sciences, Kyoto University, Sakyo-ku, Kyoto 606, Japan*

Fumihiro Sato (789), *Mathematisches Institut der Universität, Bunsenstrasse 3–5, 3400 Göttingen, Federal Republic of Germany, and Department of Mathematics, Rikkyo University, Nishi-Ikebukuro 3-34-1, Toshima-ku, Tokyo, Japan*

Pierre Schapira (809), *Centre Scientifique et Polytéchnique, Département de Mathématiques, Université de Paris-Nord, Avenue J. B. Clément, 93430 Villetaneuse, France*

Jiro Sekiguchi (821), *Department of Mathematics, University of Electro-Communications, Dentsudai, Chofugaoka 1-5-1, Chofu 182, Tokyo, Japan*

J. Sjöstrand (171), *Department of Mathematics, Université de Paris-Sud, 91405 Orsay, France*

Henry P. Stapp (309), *Lawrence Laboratory, University of California, Berkeley, California 94720*

Hidetoshi Tahara (837), *Department of Mathematics, Sophia University, Kioi-cho 7-1, Chiyoda-ku, Tokyo 102, Japan*

Shinichi Tajima (849), *Department of Mathematics, Niigata University, Ninomachi 8050, Ikarashi, Niigata 950-21, Japan*

K. Takasaki (853), *Research Institute for Mathematical Sciences, Kyoto University, Kitashirakawa, Sakyo-ku, Kyoto 606, Japan*

Yoshitsuga Takei (331), *Department of Mathematics, Faculty of Science, Kyoto University, Sakyo-ku, Kyoto 606, Japan*

L. A. Takhtajan (129), *Steklov Institute, Fontanka 27, Leningrad 191011, USSR*

Shuang Tang (29), *Institute for Theoretical Physics, State University of New York at Stony Brook, Stony Brook, New York 11794-3840*

Nobuyuki Tose (837), *Department of Mathematics, Faculty of Science, University of Tokyo, 7-3-1 Hongo, Bunkyo, Tokyo 113, Japan*

Keisuke Uchikoshi (883), *Department of Mathematics, National Defense Academy, Boeidai, Hashirimizu 1-10-20, Yokosuka 239, Japan*

Kimio Ueno (893), *Department of Mathematics, Yokohama City University, 22-2 Seto, Kanazawa-ku, Yokohama 236, Japan*

J.-L. Verdier (901), *Ecole Normale Supérieure, 45 rue d'Ulm, 75230 Paris Cédex 05, France*

Ryoko Wada (439), *Department of Mathematics, Sophia University, Kioi-cho, Chiyoda-ku, Tokyo 102, Japan*

Masato Wakayama (681), *Sanzo 985, Higashimura-cho, Fukuyama 729–02, Japan*

Hirofumi Yamada (893), *Department of Mathematics, College of Science, University of the Ryukyus, Nishihara-cho, Okinawa 903-01, Japan*

Masao Yamazaki (911), *Mathematical Sciences Research Institute, 1000 Centennial Drive, Berkeley, California 94720, and Department of Mathematics, Faculty of Science, University of Tokyo, Hongo 7-3-1, Bunkyo-ku, Tokyo 113, Japan*

Tamaki Yano (927), *Department of Mathematics, Faculty of Science, Saitama University, Shimo-Okubo 255, Uruawa-shi, Saitama 338, Japan*

Kunio Yoshino (943), *Department of Mathematics, Jochi University, Kioi-cho 7-1, Chiyoda-ku, Tokyo 102, Japan*

Kosaku Yosida (17), *Kajiwara 3-24-4, Kamakura 247, Japan*

KP Equations, Strings, and the Schottky Problem

Motohico Mulase[1]
The Institute for Advanced Study
Princeton, New Jersey

> ... but it gives the greatest satisfaction
> to view the spirit of another age,
> to see how wise men thought before our days,
> and to rejoice how far we've come at last.

0. Historical Origin

The key point of the relationship between the KP theory and the characterization of jacobians of algebraic curves is the fact that the set A consisting of linear ordinary differential operators which commute with a given ordinary differential operator is itself a commutative algebra of transcendence degree 1 over the ground field. To prove the commutativity of A, we have to introduce fractional powers of differential operators, which are pseudo-differential operators, and a proof based on this idea was given by Gelfand–Dikii [GD] in 1975. Then Krichever [K] studied extensively the algebraic structure of commuting ordinary differential operators, and

[1] On leave from Temple University, Philadelphia, Pennsylvania.

ISBN 0-12-400466-0

obtained various exact solutions of the KP equation. When Mumford gave a lecture at UCLA about 10 years ago and mentioned the work of Krichever, Coddington pointed out that the essential part of the work had been done by Burchnall and Chaundy as early as in 1922 [BC] (not 1922 B.C. of course!), way before the study of the Russian school. Thus algebraic geometry of commuting differential operators has such a long history, although the early works have been forgotten for more than 50 years.

Krichever's work gives a method of constructing solutions of the KP equations from algebraic curves. But if we want to show, on the contrary, that every solution of the KP equations must come from an algebraic curve, then we need to prove the commutativity of the commutant A of an ordinary differential operator. For a long time, those who worked in these subjects believed that the proof was due to Gelfand–Dikii.

In the fall of 1987, I had a chance to offer a graduate course of about 30 lectures in UCLA on supersymmetry and the KP theory. One day, when I mentioned the theorem of Gelfand–Dikii, to my greatest surprise, R. Steinberg immediately commented, "*It was done by Issai Schur.*" I could not believe it, because I never knew Schur's work before. But actually it was! (I would like to thank Steinberg for drawing my attention to it.) Next day, Varadarajan told me that in his paper [Schu] which appeared in 1905, *Schur proved the commutativity of A by using fractional powers of differential operators!* Schur uses pseudo-differential operators in one variable completely freely. He took the method of pseudo-differential calculus from earlier work of S. Pincherle, Mémoire sur le calcul fonctionnel distributif, *Mathematische Annalen* **49**, 325–382 (1897), especially from its Chapitre IV. At least formal algebraic manipulation of pseudo-differential operators in one variable must have been quite familiar at the end of the 19th century. Pincherle's work could be thought of as one of the origins of algebraic analysis.

Since the fractional power Schur used is essentially a local uniformizing parameter of the curve defined as the projective scheme of the associated graded algebra of A, we could say that *essentially he had the curve too!* Thus the key point of the relationship between the KP theory and the algebraic curves was established in 1905 by Schur. Burchnall–Chaundy was not the starting point. Schur really started. Well, maybe not. Because Schur took the problem of commuting ordinary differential operators from the work of Wallenberg (Über die Vertauschbarkeit homogener linearer Differentialausdrücke, *Archiv der Mathematik und Physik*, Dritte Reihe,

Band **4**, 252–268) which appeared in 1903. Wallenberg studied the conditions for two linear ordinary differential operators to commute, stated the problem in general and studied some of the special cases such as a pair of order-2 operators, a pair of first-order and n-th order operators, and so on. He also mentioned that the case of two first-order operators had been worked out by Floquet in 1879. Therefore our theory has really a long history—more than 84 years! And Schur's work has been ignored for a long time.

I have experienced a very exciting moment. The sealed book of the past was suddenly opened in front of me, and the great works hidden in the history resurrected from the dead. I looked into them and realized that everything was there.

Of course there is another root in the theory of the KP equations. It is the study of soliton equations which goes back to J. Boussinesq (1872) and D. J. Korteweg–G. deVries (1895). Surprisingly, here again their works were forgotten for more than 50 years!

And the theory, deeply rooted in the history of mathematics, solved a problem which has another long history in mathematics—the *Schottky problem* of finding a characterization of jacobian varieties (Mulase [M1], 1983). It is also used by Shiota [Sh] (1986) in a remarkable way to solve a rather new conjecture of Novikov. And in 1987, it is realized that the theory is very closely related to the brand-new theory of theoretical physics, string theory!

The discovery of Schur's work [Schu] is very impressive to me because the modern theory of the KP equations has its origin in Sato's discovery (1981) of the fact that the *Schur polynomials solve the KP system.* By this theorem Sato was able to analyze the structure of the solution space of the KP system [Sa], and then the Kyoto school, based on Sato's work, discovered (1982) a totally unexpected relation between soliton theory and Kac–Moody algebras [DJKM]. Thus, together with his commutativity theorem mentioned above, Schur's contribution in the theory of the KP equations is really enormous.

In this article, I would like to explain the KP theory, which has many different roots in the long history of mathematics, solved a problem with another long history, and is now giving new dimensions in both mathematics and physics.

I would like to thank all of the audiences of my lectures, especially V. S. Varadarajan, for their great enthusiasm and many stimulating *interruptions!* I also thank K. T. Kim and C. Phillips for their valuable suggestions.

1. Motivations

Let me begin with the following question: *What is the most fundamental difference between classical mechanics and quantum mechanics?* One may imagine non-commutativity of physical quantities, Heisenberg's uncertainty principle, discreteness of energy, or perhaps wave-particle duality. But what was the most fundamental transition from classical physics to quantum physics? In his famous book *The Principles of Quantum Mechanics*, Dirac answered this question and said that it was the *principle of superposition*. In classical mechanics, two different states of motions never mix. But in quantum mechanics, an electron, say, can pass through two different holes on a screen to make a diffraction pattern. Namely, an actual motion in quantum mechanics is a linear combination of many different states of motion. The principle of superposition forces us to introduce the notion of Hilbert spaces of states. Physical quantities are then identified with operators acting on the Hilbert space and hence their non-commutativity follows automatically. The uncertainty principle of Heisenberg and discreteness of some physical quantities are consequences of the non-commutativity.

The principle of superposition, used by Dirac to illuminate beautifully the fundamental transition of 1925, was investigated further by Feynman. According to Feynman, a quantum path can be computed by a linear combination of all classical paths with something like Boltzmann's factors in coefficients.

What does the principle of superposition tell us when we go to string theory? In the classical picture, an orbit of interacting strings is a Riemann surface embedded in space-time together with a metric on it, which is conformally equivalent to the induced metric from space-time. Therefore, to compute quantum effects such as the vacuum–vacuum transition, we have to consider superpositions (i.e., linear combinations) of complete algebraic curves of all different genera. Physicists seem to believe that string theory is the only possible quantum theory of gravity. Therefore, the orbits of strings must not be dealt with as embedded Riemann surfaces in space-time but as abstract algebraic curves, because space-time itself must appear as a classical limit of the solution of the theory. Namely, the principle of superposition in string theory demands a mathematical framework in which we can deal with all algebraic curves of all different genera at one time as well as their infinite linear combinations. Is there any such mathematical theory? If there is one, then it must be necessarily an infinite-dimensional geometry. Therefore the usual algebraic geometry cannot provide such a framework, because by definition algebraic geometry is a science of *finitely* generated *commutative* rings.

During the year of 1987, physicists learned that the KP theory, initiated by Sato and studied mostly by Japanese mathematicians in these years, may give an example of the mathematical framework which is needed in quantum string theory. In connection with the string theory, the key point of the KP theory is that the total hierarchy of the Kadomtsev–Petviashvili equations (KP system) characterizes the Riemann theta functions of all complete algebraic curves (and their degenerated functions) from any other functions (Theorem 3). Usually moduli spaces of algebraic curves are studied in relation to moduli theory of abelian varieties. But in string theory there is no reason to talk about abelian varieties. Therefore it is much more desirable to have a theory of algebraic curves and their moduli spaces without using abelian varieties. The theorem of [M3] tells us that the jacobian varieties, and hence algebraic curves, are completely characterized by the KP system without mentioning even one word on abelian varieties. Moreover, it has been realized by the Japanese physicists Ishibashi–Matsuo–Ooguri [IMO] and Russian mathematicians Beilinson–Manin–Shechtman [BMS] that the Virasoro group acts on the space of regular solutions of the KP system and produces moduli spaces of all algebraic curves regardless of their genera. Therefore, the KP theory could give a desired mathematical framework for string theory.

The reason why string theory is interesting for a mathematician is because it tells us unexpected relations between various disciplines of mathematics itself which did not seem to be related before string theory predicted their relations. For example, the discovery of the mathematical relationship between the Virasoro algebra and the moduli theory of the algebraic curves [ADKP] has one of its motivations in string theory. For another example, Taubes's work on elliptic genera of manifolds with S^1-actions [T] was motivated by Witten's idea stating that the "*index of a Dirac operator on a loop space = a modular form*" [W2] coming from string theory. There are more examples which have been understood mathematically. Also there are interesting predictions which are not yet understood mathematically. As an example I would like to mention the relation between special three-dimensional complex manifolds called the Calabi–Yau spaces appearing in the superstring theory and the super-algebraic curves, which is expected to be understood mathematically by the super-KP theory (cf. Alvarez-Gaumé-Gomes [AG]). It seems that our task is to find, without using string theory, a string of mathematical relations between totally different subjects in mathematics which is predicted and suggested by string theory.

2. Schottky Problem

The Schottky problem, in a general sense, is a problem of finding a *good* characterization of jacobian varieties. Since jacobians form an interesting special class of abelian varieties, historically a characterization always meant a characterization *among* abelian varieties. There has been a substantial amount of work done by many great mathematicians after Schottky's original work [Scho] of 1888. A natural approach to this problem is to perform a case study for low-genus jacobians. If the genus g is less than 4, then moduli of jacobians are open dense in those of abelian varieties and there is no difficulty. The actual problem starts at $g=4$. But already genus 5 is hard enough.

Complete characterizations valid for all genera were discovered only recently. The breakthrough was made by Gunning [G], based on the trisecant relations of Fay and Mumford. He showed that if an abelian variety has one-dimensional trisecants, then it is a jacobian, under an extra condition. When Gunning's paper appeared, Welters in Barcelona immediately recognized that the extra condition was not necessary (summer of 1982). Moreover, Welters succeeded in giving a much stronger *infinitesimal criterion* [We] by March of 1983. In the very same month and year, I discovered, while at MSRI, that the KP system characterizes Riemann theta functions of jacobians *among any other functions*, based on the works of Krichever and Sato. Amazingly enough, again at the very same moment, in Utrecht, van Geemen was completing his work on the geometry of the (small) Schottky locus. His remarkable theorem [vG] says that *the jacobian locus is an irreducible component of the* (*small*) *Schottky locus for every genus.* Before him only $g=4$ and $g=5$ were known. Then Arbarello–De Concini [AD] discovered that the differential relations of the theta functions which arise from Welters's infinitesimal criterion are the consequences of the KP system, and thus obtained a characterization of jacobians among abelian varieties by the KP system, independently. Finally, Shiota (1985) realized that if one wants to characterize jacobian theta functions *only among general Riemann theta functions*, then the first equation of the KP system together with a global condition is sufficient. However, we still do not know the explicit form of the Siegel modular forms which vanish exactly at the jacobian locus.

3. Algebraic Curves

Now let me sketch how the KP system is related to algebraic curves and their jacobians. Our goal is to construct an infinite-dimensional space X of all algebraic curves of all genera on which an infinite-dimensional torus $T = \mathbf{C}^\infty = \text{ind.lim } \mathbf{C}^n$ and the Virasoro group act, and produce their jacobians and moduli spaces as orbits of their actions.

The first approximation is the following:

1st approximation of X = the set of all linear ordinary differential operators.

Let (R, ∂) be a commutative derivation algebra defined over a field κ of characteristic zero. For simplicity we take $\kappa = \mathbf{C}$, $R = \mathbf{C}[[x]]$ = the formal power series ring and $\partial = d/dx$. We denote by D the set of all linear ordinary differential operators with coefficients in R.

Now how can we associate an algebraic curve to an arbitrary operator $P \in D$? The most naive idea is to take the set of all eigenvalues of P, namely define a curve by Spec P. But since every complex number $\lambda \in \mathbf{C}$ is an eigenvalue of P, Spec $P = \mathbf{C}$ and it is not so interesting. Then how about taking multiplicities into account? Since every eigenvalue of P has multiplicity n, where n = order of P, by resolving the multiplicity we may obtain a more interesting object. Then how can we resolve the multiplicity? The natural idea is to take the maximal set of commuting operators with P and consider the simultaneous eigenspaces. So let

$$A = A_P = \{Q \in D \mid [Q, P] = 0\}.$$

To talk about the simultaneous eigenspaces of A, we have to show the commutativity of A. Since we need a little more preparation, let us postpone the proof of commutativity until a little later. Let us assume for the moment that the eigenspace corresponding to $\lambda \in \text{Spec } P$ decomposes into simultaneous eigenspaces

$$\text{Ker}(P - \lambda) = \mathbf{C}\psi_1 \oplus \mathbf{C}\psi_2 \oplus \cdots \oplus \mathbf{C}\psi_n,$$

where ψ_j is a simultaneous eigenfunction of all operators in A. Define

$$I_j = \{Q \in A \mid Q\psi_j = 0\}.$$

Since $Q_1Q_2 \in I_j$ implies $Q_1Q_2\psi_j = \alpha_1\alpha_2\psi_j = 0$ for some $\alpha_1, \alpha_2 \in \mathbf{C}$, I_j is a prime ideal of A. Therefore, for each point λ of Spec P, there are n "points" $I_1, \ldots, I_n$ sitting above it. Thus we may be able to obtain a desired curve by an n-sheeted covering of $\mathbf{C}$, which is a subset of Spec A = the set of all prime ideals of A in Grothendieck's notation.

But a curve obtained by an n-sheeted covering over $\mathbf{C}$ is a very restricted one. How can we obtain all curves? To this end, we have to get rid of n, the order. From now on let us assume that $P \in D$ is generic. Namely, we assume that P is of the following (normalized) form:

$$P = \partial^n + a_2\partial^{n-2} + a_3\partial^{n-3} + \cdots + a_n.$$

Then we can take its normalized nth root:

$$L = P^{1/n} = \partial + u_2\partial^{-1} + u_3\partial^{-2} \cdots,$$

which is a pseudo-differential operator. We define an associative algebra structure in the set E of pseudo-differential operators with coefficients in R by the commutation relation

$$[\partial^{-1}, f] = -f'\partial^{-2} + f''\partial^{-3} - \cdots,$$

where $f \in R$ and f' denotes the derivative of f. Dealing with a normalized first-order operator $L \in E$ with $L^n \in D$ is equivalent to dealing with all normalized nth-order ordinary differential operators. We can handle all normalized differential operators of all orders at one time by using an arbitrary normalized first-order pseudo-differential operator L.

So define once again

$$A_L = \{Q \in D \mid [Q, L] = 0\}.$$

If $L^n = P$, then $A_L = A_P$. Since L is a first-order operator, we can imagine that its "eigenvalues" have no multiplicity. Therefore, we have

$$\text{Spec } A_L = \text{"Spec } L\text{"}.$$

To be precise, we *define* the analytic notion of Spec L (the set of eigenvalues) by Grothendieck's notion of Spec A_L. As we noted, when $L^n = P \in D$ and eigenspaces decompose into simultaneous eigenlines, the curve Spec A_L is an n-sheeted covering over $\mathbf{C}$. This covering is given by the natural inclusion $\mathbf{C}[P] \subset A_L = A_P$. Taking their affine schemes, we have

$$\text{Spec } A_L \to \text{Spec}_{\text{Grothendieck}}\, \mathbf{C}[P] = \text{Spec}_{\text{Analytic}}\, P = \mathbf{C}.$$

On $(P-\lambda)\in \operatorname{Spec} \mathbf{C}[P]$, there are n points $I_1, I_2, \ldots, I_n$ sitting over it, since they all contain $(P-\lambda)$. Therefore the above map is an n-sheeted covering. To show Spec A_L is a curve we need the following:

Theorem 1 (Schur). *The commutant A_L of L in the set D of differential operators is itself commutative and has transcendence degree* 1 *over* $\mathbf{C}$.

This theorem follows immediately from the Lemma:

Lemma. *For any normalized first-order pseudo-differential operator L, there is an invertible monic zeroth-order pseudo-differential operator S such that*

$$S^{-1}LS=\partial.$$

The proof of this lemma is just a computation. Now since $A_0=S^{-1}A_LS$ consists of pseudo-differential operators commuting with ∂, $A_0\subset \mathbf{C}((\partial^{-1}))=\mathbf{C}[\partial]+\mathbf{C}[[\partial^{-1}]]$, and hence $A_L\cong A_0$ is commutative. By an argument in elementary number theory on orders of elements in A_0 (see [M3]), we can show that the transcendence degree of A_0 over $\mathbf{C}$ is 1. The above proof of commutativity is actually Schur's argument. In the above we transformed a differential operator to a formal Laurent series in ∂^{-1} with constant coefficients. Schur expanded it in a formal Laurent series in L^{-1} with constant coefficients. These are exactly the same arguments.

Remark. If we take the commutant $\{Q\in E \mid [Q, L]=0\}$ in E, then it is equal to $\mathbf{C}((L^{-1}))$ and hence always isomorphic (conjugate) to the maximal commutative subalgebra $\mathbf{C}((\partial^{-1}))$ of E. Thus it is not interesting at all.

Let $\operatorname{gr}(A_0)$ be the canonical graded algebra associated to the filtration of A_0. We define a complete algebraic curve C by $C=\operatorname{Proj}\operatorname{gr}(A_0)$. Now since $A_0=\{$regular functions on Spec $A_0\}$ and it has a canonical realization as a subring of $\mathbf{C}((\partial^{-1}))$, we can see how C gives a one-point completion of Spec A_0. When A_0 is of rank 1, namely if it has two elements whose orders are coprime, then we have $C=\operatorname{Spec} A_0\cup\{z=0\}$, where we identify $\partial^{-1}=z=$ a local coordinate near the point at infinity. The attached point $p=\{z=0\}$ is a smooth point of C. Since conjugation by S does not change the order of an operator, we have $A_0\cap \mathbf{C}[[z]]=\mathbf{C}$. This means that a regular function on Spec A_0 which is also regular at the point $\{z=0\}$ must be a constant. Therefore Spec $A_0\cup\{z=0\}$ is complete. In general, we cannot use z as a local parameter at infinity, but a similar argument works. Thus C is always a one-point completion of Spec A_0. Thus we have obtained a curve C, a smooth point $p \in C$ and the linear part of the local coordinate. Namely, we have a unique tangent vector $v = \partial/\partial y$, where y is the local coordinate at p.

In this way we can construct a complete algebraic curve C out of a normalized first-order ordinary pseudo-differential operator L as $C =$ "Spec L." We are already in the second approximation of the space X:

2nd approximation of X = the set of normalized first order pseudo-differential operators L.

4. KP System and Jacobian Varieties

The next natural question is this: how much do we know about L if we know its spectral data $C =$ "Spec L"? This question leads us to the notion of *isospectral deformation* of L. We define a parameter depending $L(t)$ to be an *infinitesimal isospectral deformation* of $L(0)$ if there is another parameter depending differential operator $B(t)$ such that the *Lax equation*

$$\partial L(t)/\partial t = [B(t), L(t)]$$

holds. One reason why we restrict ourselves to a *differential operator* B is because our L is pretended to be an n-th root of a differential operator P. We are looking for an isospectral deformation of P which is a family of differential operators. We do not allow P deforming to a pseudo-differential operator. The other reason is that we want to recover the commutant A_L by stationary (trivial) deformations. As we have observed that the commutant in E is not interesting, we have to restrict B to a differential operator. Since L is normalized, $\text{ord}\, \partial L(t)/\partial t$ is negative. Hence B must satisfy $\text{ord}[B, L] < 0$. Let $F = \{B \in D \mid \text{ord}[B, L] < 0\}$. Then by a simple argument we can show that F is the linear span of $1, \partial, L_+^2, L_+^3, \ldots$, where L_+^m denotes the differential operator part of L^m. (Schur seems to have known this fact, too.) Therefore, all possible infinitesimal isospectral deformations of L are given by the following system of Lax equations, called the *total hierarchy of the Kadomtsev–Petviashvili equations*, or the KP system:

$$\partial L/\partial t_m = [L_+^m, L], \qquad m = 1, 2, 3, \ldots,$$

where t_m is the deformation parameter corresponding to the conjugation by L_+^m.

Examples.

1. Let $m=1$. Then the Lax equation gives $\partial L/\partial t_1 = \partial L/\partial x$. Namely, it gives the translation of x by t_1.
2. Let $L^2 = \partial^2 + 2u$ be a differential operator. Then the even equations in the KP system are all trivial and the first nontrivial equation is the *KdV equation* $4\partial u/\partial t_3 - u_{xxx} - 12uu_x = 0$. This equation was discovered [KdV] in studies of soliton phenomena of shallow water wave motions.
3. For a general L, the first nontrivial equation among the KP system is the original *KP equation* $3\partial^2 u_2/\partial t_2^2 - (4\partial u_2/\partial t_3 - u_{2,xxx} - 12u_2u_{2,x})_x = 0$, where $L = \partial + u_2\partial^{-1} + u_3\partial^{-2} + \cdots$. This equation was introduced by physicists [KP] to study transversal stability of soliton solutions of the KdV equation. Note that the KP equation reduces to the KdV equation if we throw away the t_2 dependence.

$L(t)$ satisfies the KP system if and only if there exists $S(t) = 1 + s_1\partial^{-1} + s_2\partial^{-2} + \cdots$ such that $S(t)^{-1}L(t)S(t) = \partial$, and

$$\partial S/\partial t_m = -L_-^m \cdot S,$$

where $Q_- = Q - Q_+$. We call this equation the *KP system for S*. It is known that about half of the equations of the system are automatically integrated by introducing a new unknown function $\tau(x, t)$ called Hirota-Sato's τ-function, which is something like a potential of S. Let $p_n(t)$ be a polynomial defined by

$$1 + p_1\lambda + p_2\lambda^2 + \cdots = \exp(t_1\lambda + t_2\lambda^2 + \cdots),$$

and let $\partial_t = (\partial_1, \frac{1}{2}\partial_2, \frac{1}{3}\partial_3, \ldots)$, where $\partial_n = \partial/\partial t_n$. Then the coefficients of $S(t)$ are given by

$$s_n(t) = \frac{1}{\tau(x, t)} \cdot p_n(-\partial_t)\tau(x, t).$$

When $\tau(x, t)$ gives a solution $S(t)$ of the KP system, then it is called the *τ-function solution* of the KP system.

Krichever [K] discovered that the Riemann theta functions associated with jacobian varieties give τ-function solutions of the KP system. Namely, for any such theta function $\theta(z)$ defined on $\mathbf{C}^n$, there is a linear transformation ϕ of t-variables to z-variables and a quadratic form q in t such that $e^{q(t)}\theta(\phi(t))$ gives a τ-function solution. Then Sato [Sa] discovered that the Schur polynomials of the tensor irreducible representations of general linear groups give another class of τ-function solutions. Let $t_n = (1/n)$ trace g^n for an element g of a general linear group and $\chi^Y(t)$ be the Schur polynomial corresponding to a Young diagram Y (i.e., the character of the corresponding representation) written in t-variables. Then $\chi^Y(t)$ gives a τ-function solution. On the other hand, I established [M2] that for every initial data $L(0)$ (resp. $S(0)$) there is a unique solution $L(t)$ (resp. $S(t)$) with coefficients in $\mathbf{C}[[x, t_1, t_2, \ldots]]$. Since my theorem [M4] gives an explicit construction of the solution $S(t)$ out of its initial data, we can also establish a convergence condition of the solution: for any $\varepsilon > 0$, there is $\delta > 0$ such that if the modulus of the i-th derivative of the n-th coefficient s_n of $S(0)$ is smaller than $c^i i!\,\delta^n/n!$, then $s_n(t)$ converges absolutely for all $|t_m| < a\varepsilon^m/m!$, where a and c are some positive constants.

Let $X^0 \subset E$ be the set of all normalized first-order pseudo-differential operators. The KP system and its solvability of the initial value problem mentioned above define a vector group $T = \text{ind lim } \mathbf{C}^m$ action on X^0 as the time evolution of L. This action is an analytic action on an open subset of X^0. Every orbit of this action starting at $L \in X^0$ can be thought as a universal moduli space of isospectral deformations of L. Since the space F gives all possible deformations and A_L provides trivial deformations, the tangent space of the orbit at L is canonically isomorphic to the quotient space F/A_L. From now on let us assume that the orbit has finite dimension. (The orbit of L has finite dimension if and only if the rank of A_L is 1.) Then by the covering cohomology technique applied to a neighborhood of the point at infinity of the curve C, we can easily show that

$$F/A_L \cong H^1(C, \mathcal{O}_C),$$

where $\mathcal{O}_C$ is the structure sheaf of the curve C.

To study the global structure of an orbit, we have to define a torsion-free rank-one sheaf on C. If $L^n = P \in D$, then we attach a line $\mathbf{C}\psi_j$ to a point $I_j \in \text{Spec } A_L$. In general $A_0 \cong A_L \subset D$ defines a right A_0-module structure of D. We consider D as a left R- right A_0-bimodule. Its rank is one and $\tilde{D}$ defines a torsion-free rank-one sheaf on Spec $R \times$ Spec A_0. Restricting $\tilde{D}$ to the unique maximal ideal of Spec R and extending to C, we obtain a torsion-free rank-one sheaf $\mathcal{L}$ on C. By this construction, we can compute the cohomology of $\mathcal{L}$. It is a generic sheaf of $\deg \mathcal{L} = \dim H^1(C, \mathcal{O}_C) - 1$. (When C is nonsingular, $\mathcal{L}$ corresponds to a point in the complement of the theta divisor of the jacobian of C.) Isospectral deformations of L change the A_0-module structure of D defined by $S(t)A_0S(t)^{-1} \subset D$ and hence give deformations of the sheaf $\mathcal{L}$. Let M_L be the orbit of L under the T-action.

Then we have a covering map $H^1(C, \mathcal{O}_C) \to M_L$ and an injection $M_L \to H^1(C, \mathcal{O}_C^\times)$, and their composition coincides with the cohomology homomorphism $H^1(C, \mathcal{O}_C) \to H^1(C, \mathcal{O}_C^\times)$ given by the exponential map. Analytically the deformation of $\mathcal{L}$ is described as follows. Let $U_p \subset C$ be a small neighborhood of the point at infinity. Assume C is nonsingular. Then $\mathcal{L}$ is a line bundle and its restrictions on U_p and Spec A_0 are both trivial. Hence $\mathcal{L}$ is defined by a single transition function h on $U_p \backslash \{p\} =$ Spec $A_0 \cap U_p$. Deform h by $\exp(t_1\partial + t_2\partial^2 + t_3\partial^3 + \cdots)h$. This new transition function defines the line bundle corresponding to $L(t) \in M_L$. Note that $t_1\partial + t_2\partial^2 + t_3\partial^3 + \cdots$ gives an element of $H^1(C, \mathcal{O}_C)$ by the identification $\partial = z^{-1}$ and $\exp(t_1\partial + t_2\partial^2 + t_3\partial^3 + \cdots) \in H^1(C, \mathcal{O}_C^+)$. When we give the precise definition of X as a closure of X^0, we can see that the image of M_L fills up the connected component of $H^1(C, \mathcal{O}_C^\times)$ of degree $\dim H^1(C, \mathcal{O}_C) - 1$. Thus we have

Theorem 2 [M3]. *Every finite-dimensional orbit of L under the T-action is isomorphic to the generalized jacobian variety* $H^1(C, \mathcal{O}_C)/H^1(C, \mathbf{Z})$.

If C is nonsingular, then $H^1(C, \mathcal{O}_C)/H^1(C, \mathbf{Z})$ is nothing but the jacobian Jac(C) of C, which we identify with $\mathrm{Pic}^{g-1}(C)$, where $g = \dim H^1(C, \mathcal{O}_C)$ is the genus of C.

If we adopt the theory of τ-functions here, then our theorem states

Theorem 3. *A formal power series* $f(z_1, \ldots, z_n)$ *is a Riemann theta function associated with a jacobian variety of an algebraic curve or its degeneration if and only if there is a linear transformation* $\phi: T \to \mathbf{C}^n$ *and a quadratic form* $q(t_1, t_2, \ldots)$ *in* $t \in T$ *such that* $\exp(q(t))f(\phi(t_1, t_2, \ldots))$ *gives a* τ*-function solution of the KP system at the origin of T.*

This theorem gives a completely local characterization of the jacobian theta functions among all other formal power series. Thus we obtain a characterization of jacobian varieties without mentioning abelian varieties. It is very interesting that a *jacobian variety as a manifold appears as a solution of the KP system.* We do not have to prepare any geometric stage for a jacobian, such as complex tori and abelian varieties. The geometric structure of a jacobian simply *appears* automatically. This gives an unexpected (unwanted?) solution of the Schottky problem.

Of course if we want to characterize jacobian theta functions only among theta functions of abelian varieties, then we do not need the total hierarchy of the KP system but only one differential equation together with a global condition. Actually, Shiota proved the following

Theorem 4 (Shiota [Sh]). *A Riemann theta function θ associated with an abelian variety of dimension n is a jacobian theta function if and only if there are vectors $a_1 \neq 0$, a_2, $a_3 \in \mathbf{C}^n$ and a quadratic form $q(t_1, t_2, t_3)$ such that* $\exp(q(t))\theta(t_1 a_1 + t_2 a_2 + t_3 a_3 + b)$ *gives a global τ-function solution of the single KP equation in $t \in \mathbf{C}^3$ for all $b \in \mathbf{C}^n$.*

This theorem solved the Novikov conjecture.

Theorem 3 gives a local characterization of jacobian theta functions by a system of nonlinear partial differential equations. There is an interesting open question: *is there any set of differential equations which characterizes arbitrary Riemann theta functions of abelian varieties among any other analytic functions?* Although geometric characterization of abelian varieties among complex tori was established by the beginning of this century, any local characterization is not yet known. Since being abelian means algebraicity of a torus and by definition algebraicity is a global condition, it is not surprising that we still do not have such a characterization. It must be very hard to give a local characterization of global properties. But then why can we obtain a local characterization of jacobians? It is because of the complete integrability of the KP system. Namely, since the system is completely integrable, local information of a solution determines the global structure of the solution. Thus the true question is this: *can we find a completely integrable system which characterizes general theta functions?* The problem of understanding the relation between such a system and the KP system can be called the *adjoint Schottky problem.*

5. Virasoro Action

The solvability of the KP system in terms of $S(t)$ is established in the following way. Let $S(0)$ be an initial value. Then we define

$$U(t) = \exp(t_1\partial + t_2\partial^2 + \cdots) \cdot S(0)^{-1}.$$

Here we consider t as formal parameters and $U(t)$ as a generating function of pseudo-differential operators. We can define a rigorous mathematical framework in which the above expression makes sense as a formal pseudo-differential operator of infinite order.

Theorem 5 [M2, M4]. *There is a unique monic pseudo-differential operator $S(t)$ of order 0 and an infinite-order invertible differential operator $Y(t)$ with $Y(0)=1$ such that $U(t)=S(t)^{-1}\cdot Y(t)$.*

This theorem is a generalization of the Birkhoff decomposition of loop groups. Now, by definition, $\partial U/\partial t_n=\partial^n\cdot U$. On the other hand, it is equal to

$$-S^{-1}\cdot\frac{\partial S}{\partial t_n}\cdot S^{-1}\cdot Y+S^{-1}\cdot\frac{\partial Y}{\partial t_n}.$$

Therefore

$$-\frac{\partial S}{\partial t_n}\cdot S^{-1}+\frac{\partial Y}{\partial t_n}\cdot Y^{-1}=S\,\partial^n S^{-1}=L^n.$$

The first term of the equation is a pseudo-differential operator of negative order and the second term is an infinite order differential operator. Therefore,

$$-\frac{\partial S}{\partial t_n}\cdot S^{-1}=L^n_-$$

follows, which is the KP system we wanted to solve.

Thus the essence of the KP system is the generalized Birkhoff decomposition. The decomposition as well as the solvability holds even if we replace $R=\mathbf{C}[[x]]$ by $K=\mathbf{C}((x))[\log x]$. A normalized first order operator L with coefficients in K is said to be *quasi-regular* if there are positive integers m and n such that for every positive integer k, $x^mL^kx^n\in E$, i.e., its coefficients are in R.

Let X be the set of quasi-regular L's. This is the desired space we wanted to construct. Now X is closed under global T-action. We call a point L of X of *finite type* if the orbit of L under the KP flow (i.e., the T-action) has finite dimension. Every point of X gives the following data: $(C, p, v, \mathscr{L})$, where C is an irreducible complete algebraic curve, p is a smooth point on it, v is a non-zero tangent vector at p and $\mathscr{L}$ is a torsion-free finite-rank sheaf on C. Sato proved that the corresponding set of S, monic 0-th order operators with certain singularities at $x=0$, has a structure of an infinite Grassmannian. Let E_- be the set of pseudo-differential operators of order at most -1. Consider the quotient space $W=E/Ex$, where Ex is the left ideal generated by x in E. Following the decomposition $E=D\oplus E_-$, W decomposes $W=W_+\oplus W_-$. For every S corresponding to $L\in X$, define $\{P\in E\,|\,SP\in D\}$ and let $\Sigma\subset W=E/Ex$ be its projection image. Then the composition map $\Sigma \to W \to W_+ = W/W_-$ is Fredholm and has index 0. Therefore Σ defines an element of $\mathrm{Gr}(W_+, W)$, which consists of subspaces of W having the "same" size with W_+ in the above sense of Fredholm. We denote it by Gr and identify S with Σ. Let G_0 be the group of monic 0-th order operators with constant coefficients. Then $\pi: \mathrm{Gr} \to X$, $\pi(S) = S\,\partial S^{-1} = L$ is a principal fiber bundle over X with structure group G_0. Since G_0 is contractible, Gr and X have the same topological structure. Because of the unique solvability of the KP system in terms of S, every finite-dimensional orbit of the T-action on Gr is isomorphic to a generalized jacobian. We define the notion of *finite type* on Gr in the same way. Then every finite-type point S of Gr gives $(C, p, v, \mathscr{L}, S_0)$, where $S_0 \in G_0$ is a local trivialization of $\mathscr{L}$ near p. It is also possible to identify (v, S_0) with a full local parameter y at p. In any case our interpretation of S_0 is not canonical. Krichever's theorem tells us that if we have $(C, p, v, \mathscr{L})$ consisting of a smooth curve of genus g, a point on it, a tangent vector at p and a line bundle of degree $g-1$, then it determines a unique point $L \in X$ of finite type.

Let $G \subset E$ be the set of monic 0-th order pseudo-differential operators with coefficients in R. Then G is the big-cell of Gr. Namely, there is a "divisor" Δ at infinity such that $\mathrm{Gr} = G \cup \Delta$. Let Λ be the line bundle over Gr corresponding to the "divisor" Δ. It is called the determinant line bundle, since it coincides with the dual of the determinant bundle of the universal bundle of Gr. Let τ be a section of Λ which vanishes only on Δ. We have maps

$$T \ni t \mapsto S(t) \in \mathrm{Gr} \ni S \mapsto \tau(S) \in \mathbf{C}.$$

Let $\tau(t) = \tau(S(t))$ be the composition of the above maps. Then this is nothing but the τ-function solution of the KP system: $\tau(x, t) = \tau(x + t_1, t_2, t_3, \ldots)$.

Now let us describe the Virasoro action on Gr. The Virasoro algebra V is a special central extension of the algebra $\mathfrak{g}$ of vector fields on the circle S^1. The $\mathfrak{g}$-action on a point of Gr produces an infinitesimal deformation of the complex structure of the corresponding curve. Let C be the curve corresponding to a point S of Gr. Then $C = \operatorname{Spec} A_L \cup U_p$, where $L = S\partial S^{-1}$ and U_p is a small neighborhood of $p \in C$. Since $\operatorname{Spec} A_L$ and U_p have no deformations, deformations of the complex structure of C are given by changing the patching of $\operatorname{Spec} A_L$ and U_p. Let $\mathscr{X}(\operatorname{Spec} A_L \cap U_p)$ be the set of holomorphic vector fields on $\operatorname{Spec} A_L \cap U_p$. (Topologically this intersection is a circle.) Then its inductive limit as U_p tends to $\{p\}$ coincides with $\mathfrak{g}$. Therefore there is a natural projection $\mathfrak{g} \to H^1(C, \mathscr{T}_C)$, where $\mathscr{T}_C$ is the tangent sheaf of C. Since $H^1(C, \mathcal{T}_C)$ defines infinitesimal deformations of C by Kodaira–Spencer, we have defined a $\mathfrak{g}$-action on Gr. In other words, a vector field on S^1 determines an infinitesimal change of the patching Spec $A_L \cup U_p$.

We can extend the action to a V-action on the line bundle Λ. The one-dimensional center acts on the fiber. In the group level, the Virasoro action produces moduli spaces of the data (C, p, v).

Let Gr_f be the set of finite-type points of Gr. Then $(\mathrm{Gr}_f, T, \mathscr{V})$ gives a genus-free theory of algebraic curves, where $\mathscr{V}$ is the Virasoro group. Namely, all jacobians are obtained at one time by the T-action and all moduli spaces of curves are given by the $\mathscr{V}$-action. Actually, Gr_f is the moduli space of the data $(C, p, v, \mathscr{L}, S_0)$. To get rid of S_0, we take $X_f = \mathrm{Gr}_f / G_0$. If we also want to eliminate $\mathscr{L}$, then we have to define the quotient space X_f / T. Since an orbit of the T-action has arbitrary dimension, it is very hard to define the orbit space. I do not know how to understand this space. However, ideas of the non-commutative differential geometry might be useful here.

The section τ restricted to each of the compact orbits of the T-action gives all the Riemann theta functions associated with jacobians. On the other hand, restriction of τ to the orbits of the $\mathscr{V}$-action gives modular forms. Schur polynomials also show up as restrictions of τ to special T-orbits. In this way, these important functions can be obtained from a single function τ on the Grassmannian.

The infinite-dimensional orbits are related to infinite-genus situations as well as classification of vector bundles over algebraic curves. For example, the complete family of all different Hirzebruch surfaces appears as an infinite orbit. The geometric study of infinite-dimensional orbits has not been worked out in full generality.

6. Supersymmetrization

Supersymmetry is a new language of global analysis. For example, a supersymmetric quantum mechanics on a compact Riemannian manifold was effectively used to visualize the Morse theory and the Atiyah–Singer index theorem by Witten [W1] and Alvarez-Gaumé [A]. We now know how a harmonic p-form localizes to the critical points of Morse index p in the "strong coupling limit" with a Morse function detecting which critical points contribute to the topology of the manifold and which do not. Or we now know why the A-roof genus of a spin manifold must be of the form of product of $(x/2)/\sinh(x/2)$. These intuitive approaches of global analysis enabled mathematicians to produce new theorems such as a Morse *equality* of Helffer–Sjöstrand [HS] and an analogue of the index theorem on a loop space (cf. Taubes [T]). Since supersymmetry is a language, you do not have to speak it if you do not like. Also, any theorem proved by a supersymmetric technique can be proved by non-supersymmetric regular techniques. Perhaps an advantage of supersymmetry is to visualize global structures and to help one to find a new theorem. It also simplifies drastically the complicated arguments in index theorems and Riemann–Roch-type theorems. There is a formal resemblance between $\mathbf{Z}_2$-graded structures in supersymmetry and the K-theoretic structure in non-commutative differential geometry of Connes. I believe that the geometric structure of the *super-KP system* could be thought as an example of an algebraic version of the non-commutative differential geometry. Algebraic studies of the super-KP equations have been worked out recently by Manin–Radul [MR], Ueno–Yamada [UY] and Mulase [M4]. Geometry, especially in connection with global analysis, of the super-KP system is yet to be done.

It is natural that if we set all "odd" variables evolutions in the super-KP system to be 0, then we recover the original KP system. I discovered in [M4] that if one *eliminates* odd variables in the super-KP system, then one gets the modified KP system. The first nontrivial equation is

$$4f_{xt} - 3f_{yy} + 6f_{xx}f_y - f_{xxxx} + 6f_x^2 f_{xx} = 0.$$

The KP equation is a unification of the KdV equation and the Boussinesq equation. In a similar sense, the super-KP system is a unification of the KP system and the modified KP system. As we have observed, the KP system is a defining equation of the universal family of all isospectral deformations of all normalized differential operators. I still do not have such a simple conceptual definition of the super-KP system. In the KP case, we did not allow a differential operator to deform to a pseudo-differential operator. But in the super-KP case studied in [MR], [M4], even if we start with a super-differential operator, it deforms to a super-pseudo-differential operator. The Birkhoff-type decomposition and the solvability argument in S variable work very well in the supercategory. But the Lax equation does not work well.

If we forget about isospectral deformations and think of the KP system only by Theorem 5, then the super-KP system is the most natural generalization of the theorem to the supersymmetric case. Namely,

$$\text{KP} = \text{Birkhoff decomposition},$$

$$\text{Super-KP} = \text{Super-Birkhoff decomposition}.$$

For more detail about the super-KP system please see [M4]. Since the theory of the KP system and the infinite Grassmannian gives the critical dimension of 26 as a special case of the Riemann-Roch theorem, it is natural to ask if we can deduce 10 from the super-KP theory, which is not known.

... the days of history
make up a book with seven *Siegels.*
What you call the spirit of an age
is in reality the spirit of those men
in which their time's reflected.

References

[A] L. Alvarez-Gaumé, Supersymmetry and the Atiyah-Singer index theorem, *Comm. Math. Phys.* **90** (1983) 161-173.

[AD] E. Arbarello and C. De Concini, On a set of equations characterizing Riemann matrixes, *Ann. of Math.* **120** (1984) 119-140.

[ADKP] E. Arbarello, C. De Concini, V. Kac, and C. Procesi, Moduli space of curves and representation theory, Preprint, 1987.

[AG] L. Alvarez-Gaumé and C. Gomez, New methods in string theory, CERN Preprint CERN-TH.4775/87, 1987.

[BC] J. L. Burchnall and T. W. Chaundy, Commutative ordinary differential operators, *Proc. Lond. Math. Soc.* **21** (1922) 420-440.

[BMS] A. A. Beilinson, Yu. I. Manin and V. A. Shechtman, Localization of the Virasoro and Neveu-Schwarz algebras, Preprint, 1986.

[DJKM] E. Date, M. Jimbo, M. Kashiwara, and T. Miwa, Transformation groups for soliton equations, in *Proc. RIMS Symposium on Nonlinear Integrable Systems—Classical and Quantum Theory*, ed. by Jimbo and Miwa, World Scientific, 1983.

[G] R. C. Gunning, Some curves in abelian varieties, *Inv. Math.* **66** (1982) 377-389.

[GD] I. M. Gelfand and L. A. Dikii, Fractional powers of operators and Hamiltonian systems, *Func. Anal. Appl.* **10** (1976) 259-273.

[HS] B. Helffer and J. Sjöstrand, Puits multiples en mecanique semi-classique IV, etude du complexe de Witten, *Comm. Partial Diff. Eq.* **10** (1985) 245-340.

[IMO] N. Ishibashi, Y. Matsuo and H. Ooguri, Soliton equations and free fermions on Riemann surfaces, *Mod. Phys. Lett.* **A2** (1987) 119.

[K] I. M. Krichever, Method of algebraic geometry in the theory of nonlinear equations, *Russ. Math. Surv.* **32** (1977) 185-213.

[KdV] D. J. Korteweg and G. deVries, On the change of form of long waves advancing in a rectangular canal, and on a new type of long stationary waves, *Phil. Mag.* **39** (1895) 422-443.

[KP] B. B. Kadomtsev and V. I. Petviashvili, On the stability of solitary waves in weakly dispersing media, *Sov. Phys. Doklady* **15** (1970) 539–541.

[M1] M. Mulase, Algebraic geometry of soliton equations, *Proc. Japan Acad.* **59** (1983) 285–288.

[M2] M. Mulase, Complete integrability of the Kadomtsev-Petviashvili equation, *Advances in Math.* **54** (1984) 57–66.

[M3] M. Mulase, Cohomological structure in soliton equations and jacobian varieties, *J. Diff. Geom.* **19** (1984) 403–430.

[M4] M. Maluse, Solvability of the super-KP equation and a generalization of the Birkhoff decomposition, *Invent. Math.* **92** (1988) 1–46.

[MR] Yu. I. Manin and A. O. Radul, A supersymmetric extension of the Kadomtsev-Petviashvili hierarchy, *Comm. Math. Phys.* **98** (1985) 65–77.

[Sa] M. Sato and Y. Sato, Soliton equations as dynamical systems on infinite dimensional Grassmann manifold, *Lect. Notes in Num. Appl. Anal.* **5** (1982) 259–271.

[Scho] F. Schottky, Zur Theorie der Abelschen Functionen von vier Variablen, *J. Reine und Angew. Math.* **102** (1888) 304–352.

[Schu] I. Schur, Über vertauschbare lineare Differentialausdrücke, *Gesammelte Abhandlungen, Band I*, 170–176, Springer-Verlag, 1905.

[Sh] T. Shiota, Characterization of jacobian varieties in terms of soliton equations, *Inv. Math.* **83** (1986) 333–382.

[T] C. H. Taubes, S^1 actions and elliptic general, preprint, 1987.

[UY] K. Ueno and H. Yamada, Supersymmetric extension of the Kadomtsev-Petviashvili hierarchy and the universal super Grassmann manifold, *Advanced Studies in Pure Math., Kinokuniya*, **16** (1987) 373–426.

[vG] B. van Geemen, Siegel modular forms vanishing on the moduli space of curves, *Inv. Math.* **78** (1984) 329–349.

[We] G. E. Welters, A criterion for Jacobi varieties, *Ann. of Math.* **120** (1984) 497–504.

[W1] E. Witten, Supersymmetry and Morse theory, *J. Diff. Geom.* **17** (1982) 661–692.

[W2] E. Witten, The index of Dirac operator in loop space, preprint, 1987.

A Note on the Holonomic System of Invariant Hyperfunctions on a Certain Prehomogeneous Vector Space

Masakazu Muro[1]
Department of Mathematics
Kochi University
Kochi, Japan

Introduction

The theory of holonomic systems and microlocal analysis, which was originated by M. Sato and developed mainly by M. Sato and M. Kashiwara, has had great success in several areas of mathematics. In particular, we cannot do anything in the calculus of invariant hyperfunctions on prehomogeneous vector spaces without microlocal analysis. The microlocal structure of invariant holonomic systems and invariant hyperfunctions has been finely described in some papers in the case of irreducible regular prehomogeneous vector spaces on which reductive groups act, in connection with the computation of functional equations of zeta functions associated

[1] This research is partially supported by Grant-in-Aid for Scientific Research 62740098 of Ministry of Education in Japan.

ISBN 0-12-400466-0

with prehomogeneous vector spaces. However we do not have many examples of calculations of the microlocal structure of relatively invariant hyperfunctions on non-irreducible regular prehomogeneous vector space. Above all, there are remarkable examples among them that raise different phenomena from the case of irreducible regular prehomogeneous vector spaces. They are prehomogeneous vector spaces whose group actions are by parabolic subgroups of reductive groups.

Let $(G_{\mathbf{C}}, \rho, V_{\mathbf{C}})$ be a regular irreducible prehomogeneous vector space, where $G_{\mathbf{C}}$ is a connected reductive algebraic group. Let $B_{\mathbf{C}}$ be a Borel subgroup of $G_{\mathbf{C}}$. We suppose that $(B_{\mathbf{C}}, \rho, V_{\mathbf{C}})$ is also a prehomogeneous vector space. Namely there exists an open dense $B_{\mathbf{C}}$-orbit in $V_{\mathbf{C}}$. Then a theorem by Vinberg [Vi] states that $V_{\mathbf{C}}$ decomposes into a finite number of $B_{\mathbf{C}}$-orbits, which means $(P_{\mathbf{C}}, \rho, V_{\mathbf{C}})$ is a prehomogeneous vector space with finite orbits for any parabolic subgroup $P_{\mathbf{C}}$ in $G_{\mathbf{C}}$. There are not so many examples of this type, but some important examples satisfy this condition. See, for example, [Ru–Sc]. The author thinks that the structures of such prehomogeneous vector spaces should be studied more for the future development of prehomogeneous vector spaces.

In this paper, we shall investigate the microlocal structure of an invariant holonomic system on a typical prehomogeneous vector space satisfying this condition. After introducing the example of the prehomogeneous vector space (Section 1), we illustrate the holonomic system by its holonomic diagram (Section 2), and observe an unusual phenomenon of factors of b-functions (Section 3). In the last section (Section 4) we shall give the explicit formula of the Fourier transforms of the complex powers of relatively invariant polynomials. The author has been informed in a letter from F. Sato that F. Sato carried out this calculation in a different way and obtained the same result.

1. An Example of Prehomogeneous Vector Space

We put

$$\begin{aligned} G_{\mathbf{C}} &:= GO_n(Y, \mathbf{C}) := \{g \in GL_n(\mathbf{C}); \qquad g \cdot Y \cdot {}^t g = \nu(g) \cdot Y\}, \\ V_{\mathbf{C}} &:= \{x = {}^t(x_1, \ldots, x_n) \in \mathbf{C}^n\}, \end{aligned} \tag{1.1}$$

where

$$Y := \begin{bmatrix} & & 1 \\ & U & \\ 1 & & \end{bmatrix}$$

with a non-degenerate complex symmetric $(n-2)\times(n-2)$ matrix U. Here we suppose that $n \geq 3$. The group $G_{\mathbf{C}}$ acts on $V_{\mathbf{C}}$ from the left as a linear representation $\rho : \rho(g);\ x \mapsto g \cdot x$ with $g \in G_{\mathbf{C}}$ and $x \in V_{\mathbf{C}}$ and $\nu(g)$ is a constant depending only on g. Then $(G_{\mathbf{C}}, \rho, V_{\mathbf{C}})$ is a regular prehomogeneous vector space. The irreducible relatively invariant polynomial is given by:

$$P_1(x) := {}^t x \cdot Y \cdot x. \tag{1.2}$$

The corresponding character is $\nu(g)$. We denote it by $\chi_1(g)$ for later convenience. Namely we have $P_1(\rho(g)\cdot x) = \chi_1(g) \cdot P_1(x)$ for all $g \in G_{\mathbf{C}}$. We define an inner product on $V_{\mathbf{C}}$ by:

$$\langle x, y \rangle := {}^t x \cdot Y \cdot y, \qquad \text{for } x, y \in V_{\mathbf{C}}, \tag{1.3}$$

and identify $V_{\mathbf{C}}$ with its dual vector space $V_{\mathbf{C}}^*$ through $\langle\ ,\ \rangle$. The contragredient representation ρ^* makes the prehomogeneous vector space with the same relatively invariant polynomial $P_1(x)$, but the corresponding character is χ_1^{-1}, i.e., $P_1(\rho^*(g)\cdot x) = \chi_1^{-1}(g) \cdot P_1(x)$ for all $g \in G_{\mathbf{C}}$.

Consider the parabolic subgroup $P_{\mathbf{C}}$ in $G_{\mathbf{C}}$:

$$P_{\mathbf{C}} := \left\{ \begin{bmatrix} a & B & d \\ 0 & C & E \\ 0 & 0 & f \end{bmatrix} \in GO_n(Y, \mathbf{C});\ a, d, f \in \mathbf{C},\ B, {}^tE \in M(1, n-2, \mathbf{C}) \text{ and } C \in GL_{n-2}(\mathbf{C}) \right\}. \tag{1.4}$$

Here $M(1, n-2, \mathbf{C})$ means the space of $1\times(n-2)$ complex matrices and $GL_{n-2}(\mathbf{C})$ stands for the space of invertible $(n-2)\times(n-2)$ matrices. Then $(P_{\mathbf{C}}, \rho, V_{\mathbf{C}})$ is still a prehomogeneous vector space, whose irreducible relatively invariant polynomials are:

$$P_1(x) \text{ and } P_2(x) := x_n. \tag{1.5}$$

Any relatively invariant polynomial is obtained as an integer power of $P_1(x)$

and $P_2(x)$. An element g of $P_{\mathbf{C}}$ is written as

$$g = k \times \begin{bmatrix} a & B & d \\ 0 & C & E \\ 0 & 0 & a^{-1} \end{bmatrix},$$

where $k, a \in \mathbf{C}^\times$ and $B, {}^tE \in M(1, n-2, \mathbf{C})$ and $C \in SL_n(\mathbf{C})$. The corresponding character of $P_1(x)$ is $\chi_1(g) := k^2$ and that of $P_2(x)$ is $\chi_2(g) := k \cdot a^{-1}$. Namely we have: $P_1(\rho(g) \cdot x) = \chi_1(g) \cdot P_1(x)$ and $P_2(\rho(g) \cdot x) = \chi_2(g) \cdot P_2(x)$ for all $g \in P_{\mathbf{C}}$. The contragredient representation ρ^* gives a prehomogeneous vector space $(P_{\mathbf{C}}, \rho^*, V_{\mathbf{C}})$, whose irreducible relatively invariant polynomials are $P_1(x)$ and $P_2(x)$. The corresponding characters are χ_1^{-1} and $\chi_1^{-1} \cdot \chi_2$, respectively, i.e., $P_1(\rho^*(g) \cdot y) = \chi_1(g)^{-1} \cdot P_1(y)$ and $P_2(\rho^*(g) \cdot y) = \chi_1(g)^{-1} \chi_2(g) \cdot P_2(y)$ for all $g \in P_{\mathbf{C}}$.

2. Holonomy Diagrams of the Holonomic System of Relatively Invariant Hyperfunctions

Let $\mathscr{G}_{\mathbf{C}}$ and $\mathscr{P}_{\mathbf{C}}$ be the Lie algebras of the complex Lie groups $G_{\mathbf{C}}$ and $P_{\mathbf{C}}$, respectively. We put

$$(1) \quad \mathfrak{M}_s: \left(\left\langle d\rho(A) \cdot x, \frac{\partial}{\partial x} \right\rangle - s\delta\chi_1(A) \right) u(x) = 0 \qquad \text{for all } A \in \mathscr{G}_{\mathbf{C}}.$$

$$(2) \quad \mathfrak{M}_{(s_1, s_2)}: \left(\left\langle d\rho(A) \cdot x, \frac{\partial}{\partial x} \right\rangle - s_1\delta\chi_1(A) - s_2\delta\chi_2(g) \right) u(x) = 0 \qquad \text{for all } A \in \mathscr{P}_{\mathbf{C}}. \tag{2.1}$$

Here s, s_1 and s_2 are complex numbers and $d\rho$, $\delta\chi_1$ and $\delta\chi_2$ are infinitesimal representations of ρ, χ_1 and χ_2, respectively. These are the systems of linear differential equations defining relatively invariant functions corresponding to the characters χ^s and $\chi_1^{s_1} \cdot \chi_2^{s_2}$, respectively.

As we shall state later, $(G_{\mathbf{C}}, \rho, V_{\mathbf{C}})$ and $(P_{\mathbf{C}}, \rho, V_{\mathbf{C}})$ are prehomogeneous vector spaces with a finite number of orbits. As is proved in [Ka 1], then the above systems of differential equations are holonomic systems. Moreover, we can prove that $\mathfrak{M}_s$ and $\mathfrak{M}_{(s_1, s_2)}$ are completely regular holonomic systems. This implies that any relatively invariant hyperfunction on any real forms of $(G_{\mathbf{C}}, \rho, V_{\mathbf{C}})$ or $(P_{\mathbf{C}}, \rho, V_{\mathbf{C}})$ is a tempered distribution.

We will not give the proof of this fact since it is not necessary for the calculations in this paper.

The first purpose of this section is to investigate the orbit structures of $(G_{\mathbf{C}}, \rho, V_{\mathbf{C}})$ and $(P_{\mathbf{C}}, \rho, V_{\mathbf{C}})$ in order to calculate the characteristic varieties of $\mathfrak{M}_s$ and $\mathfrak{M}_{(s_1,s_2)}$. We may suppose that $U := I_{n-2}$ since U is transformed to I_{n-2} by an automorphism in $GL_n(\mathbf{C})$. The orbital decompositions of $V_{\mathbf{C}}$ by $G_{\mathbf{C}}$ and $P_{\mathbf{C}}$ are given in the following proposition.

Proposition 1.1. *The vector space $V_{\mathbf{C}}$ decomposes into the following orbits.*

(1) *By $G_{\mathbf{C}}$ actions:*

(1) $S_{0\mathbf{C}} := \rho(G_{\mathbf{C}}) \cdot {}^t(1, 0, \ldots, 0, 1)$

(2) $S_{1\mathbf{C}} := \rho(G_{\mathbf{C}}) \cdot {}^t(0, \ldots, 1)$

(3) $S_{2\mathbf{C}} := \rho(G_{\mathbf{C}}) \cdot {}^t(0, \ldots, 0)$

(2) *By $P_{\mathbf{C}}$ actions:*

(1) $L_{0\mathbf{C}} := \rho(P_{\mathbf{C}}) \cdot {}^t(1, 0, \ldots, 0, 1)$

(2) $L_{1\mathbf{C}} := \rho(P_{\mathbf{C}}) \cdot {}^t(0, \ldots, 1)$

(3) $L_{2\mathbf{C}} := \rho(P_{\mathbf{C}}) \cdot {}^t(0, 1, 0, \ldots, 0)$

(4) $L_{3\mathbf{C}} := \rho(P_{\mathbf{C}}) \cdot {}^t(0, 1, \sqrt{-1}, 0, \ldots, 0)$

(5) $L_{4\mathbf{C}} := \rho(P_{\mathbf{C}}) \cdot {}^t(1, 0, \ldots, 0)$

(6) $L_{5\mathbf{C}} := \rho(P_{\mathbf{C}}) \cdot {}^t(0, \ldots, 0)$

(3) *The $G_{\mathbf{C}}$-orbits are the unions of the following $P_{\mathbf{C}}$-orbits.*

$S_{0\mathbf{C}} = L_{0\mathbf{C}} \cup L_{2\mathbf{C}}$

$S_{1\mathbf{C}} = L_{1\mathbf{C}} \cup L_{3\mathbf{C}} \cup L_{4\mathbf{C}}$

$S_{3\mathbf{C}} = L_{5\mathbf{C}}$

(4) *We have the same orbital decomposition by the contragredient representation ρ^*.*

The proof of this Proposition is not difficult so we omit it.

The characteristic varieties of $\mathfrak{M}_s$ and $\mathfrak{M}_{(s_1,s_2)}$, which we denote by $\mathrm{ch}(\mathfrak{M}_s)$ and $\mathrm{ch}(\mathfrak{M}_{(s_1,s_2)})$ respectively, are given by the union of the conormal bundles of the orbits:

$$\begin{aligned} &(1)\quad \mathrm{ch}(\mathfrak{M}_s) = \bigcup_{i=0}^{2} T^*_{S_{i\mathbf{C}}} V_{\mathbf{c}}, \\ &(2)\quad \mathrm{ch}(\mathfrak{M}_{(s_1,s_2)}) = \bigcup_{j=0}^{5} T^*_{L_{j\mathbf{C}}} V_{\mathbf{C}}. \end{aligned} \tag{2.2}$$

We denote by $\Xi_{i\mathbf{C}}$ the closure $\overline{T^*_{S_{i\mathbf{C}}} V_{\mathbf{C}}}$ $(i = 0, 1, 2)$ and by $\Lambda_{j\mathbf{C}}$ the closure $\overline{T^*_{L_{j\mathbf{C}}} V_{\mathbf{C}}}$ $(j = 0, 1, 2, 3, 4, 5)$. Each of them is an irreducible Lagrangian

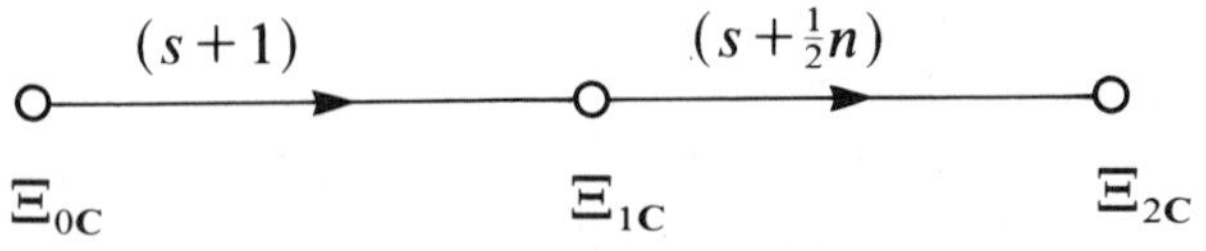

FIGURE 1 The holonomy diagram of $\mathfrak{M}_s$.

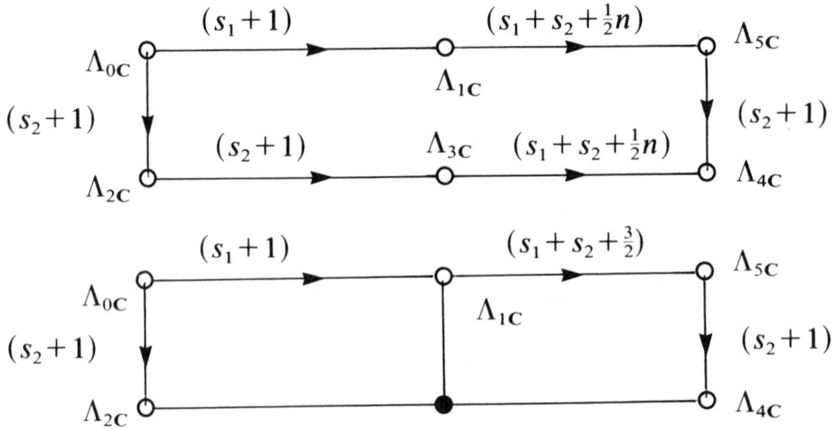

FIGURE 2 (a) The holonomy diagram of $\mathfrak{M}_{(s_1,s_2)}$ $(n \geq 4)$. (b) The holonomy diagram of $\mathfrak{M}_{(s_1,s_2)}$ $(n=3)$.

subvariety in $T^*V_{\mathbf{C}}$. The holonomic systems $\mathfrak{M}_s$ and $\mathfrak{M}_{(s_1,s_2)}$ are simple on each Lagrangian. Then the microlocal structure of the holonomic systems on each Lagrangian is determined by its order. It is routine to compute the order on each Lagrangian by using the formula in [Sm–Ka–Ki–Os]. We give them in Table 1 and Table 2 with the data of intersections of codimension one among the irreducible Lagrangian subvarieties. The complex holonomy diagrams of $\mathfrak{M}_s$ and $\mathfrak{M}_{(s_1,s_2)}$ are given in Figs. 1 and 2. They can be drawn from the data in Tables 1 and 2.

Now we turn to the problem of the computation of b-functions. As stated in [Sm–Sh], the b-function of complex powers of relatively invariant polynomials plays an important role when we construct the meromorphic extensions of the local zeta functions. On the other hand, for the purpose of explicit computations of b-functions, Sato–Kashiwara–Kimura–Oshima

Table 1 The order of $\mathfrak{M}_s$

Lagrangian	Order	Lagrangians with codimension one intersections
Ξ_{0C}	0	Ξ_{1C}
Ξ_{1C}	$-s-\frac{1}{2}$	Ξ_{0C}, Ξ_{2C}
Ξ_{2C}	$-2s-(n/2)$	Ξ_{1C}

Table 2 The order of $\mathfrak{M}_{(s_1,s_2)}$

Lagrangian	Order	Lagrangians with codimension one intersections
Λ_{0C}	0	$\Lambda_{1C}, \Lambda_{2C}$
Λ_{1C}	$-s_1-\frac{1}{2}$	$\Lambda_{0C}, \Lambda_{5C}$
Λ_{2C}	$-s_2-\frac{1}{2}$	$\Lambda_{0C}, \Lambda_{3C}$ **
Λ_{3C} *	$-s_1-s_2-1$	$\Lambda_{2C}, \Lambda_{4C}$
Λ_{4C}	$-2s_1-2s_2-(n+1)/2$	Λ_{3C} ***, Λ_{5C}
Λ_{5C}	$-2s_1-s_2-(n/2)$	$\Lambda_{1C}, \Lambda_{4C}$

* The Lagrangian Λ_{3C} does not appear when $n=3$.
** When $n=3$, Λ_{2C} has an intersection of codimension one jointly with Λ_{1C} and Λ_{4C}.
*** When $n=3$, Λ_{4C} has an intersection of codimension one jointly with Λ_{1C} and Λ_{2C}.

[Sm-Ka-Ki-Os] defined the *local* b-function $b_\Lambda(s)$ on each irreducible Lagrangian subvariety in $\mathrm{ch}(\mathfrak{M})$ where $\mathfrak{M}$ is the holonomic system of the complex powers of relatively invariant polynomials. Then, taking Λ_0 as the conormal bundle of the origin, $b_{\Lambda_0}(s)$ is actually the b-function that we want to compute. Moreover, the local b-function is computed easily from the holonomy diagram. Namely, $b_\Lambda(s)$ is obtained by taking the product of all the factors of b-functions along the route of the holonomy diagram from the Lagrangian of the zero section to Λ. For example, we have

The case of $P_1(x)^s$ of $(G_{\mathrm{C}}, \rho, V_{\mathrm{C}})$.

$$\begin{aligned} b_{\Xi_{0C}}(s) &= 1, \\ b_{\Xi_{1C}}(s) &= (s+1), \\ b_{\Xi_{2C}}(s) &= (s+1)\left(s+\frac{n}{2}\right). \end{aligned} \tag{2.3}$$

The case of $P_1(x)^s1 \cdot P_2(x)^s2$ of $(P_{\mathrm{C}}, \rho, V_{\mathrm{C}})$.

$$\begin{aligned} b_{\Lambda_{0C}}(s_1, s_2) &= 1, \\ b_{\Lambda_{1C}}(s_1, s_2) &= (s_1+1), \\ b_{\Lambda_{2C}}(s_1, s_2) &= (s_2+1), \\ b_{\Lambda_{3C}}(s_1, s_2) &= (s_1+1)(s_2+1), \\ b_{\Lambda_{4C}}(s_1, s_2) &= (s_1+1)(s_2+1)\left(s_1+s_2+\frac{n}{2}\right), \\ b_{\Lambda_{5C}}(s_1, s_2) &= (s_1+1)\left(s_1+s_2+\frac{n}{2}\right). \end{aligned} \tag{2.4}$$

In the case of $P_1(x)^s$ of $(G_{\mathbf{C}}, \rho, V_{\mathbf{C}})$, the conormal bundle of the origin is $\Xi_{2\mathbf{C}}$, hence the b-function of $P_1(x)^s$ is $b_{\Xi_{2\mathbf{C}}}(s) := (s+1)(s+n/2)$. The local b-function $b_{\Xi_{2\mathbf{C}}}(s)$ is the one with the highest degree and is divided by any other local b-functions. In the case of $P_1(x)^{s_1} \cdot P_2(x)^{s_2}$ of $(P_{\mathbf{C}}, \rho, V_{\mathbf{C}})$, the conormal bundle of the origin is $\Lambda_{5\mathbf{C}}$ and the b-function of $P_1(x)^{s_1} \cdot P_2(x)^{s_2}$ is $b_{\Lambda_{5\mathbf{C}}}(s_1, s_2) := (s_1+1)(s_1+s_2+n/2)$. Contrary to the case of $(G_{\mathbf{C}}, \rho, V_{\mathbf{C}})$, $b_{\Lambda_{5\mathbf{C}}}(s_1, s_2)$ is not the local b-function with the highest degree. The highest degree one is $b_{\Lambda_{4\mathbf{c}}}(s_1, s_2)$. Such phenomenon is often observed on non-irreducible regular prehomogeneous vector spaces. It is a remarkable point when we deal with non-irreducible prehomogeneous vector spaces.

3. Fourier Transforms of the Local Zeta Functions

In this last section we shall give the explicit formula of the Fourier transform of the local zeta functions, of a real form of the complex powers of relatively invariant polynomials on the prehomogeneous vector space $(P_{\mathbf{C}}, \rho, V_{\mathbf{C}})$. This is obtained as a by-product when we analyze the microlocal structure of relatively invariant hyperfunctions. We shall not discuss the microlocal structures of the relatively invariant hyperfunctions since they are a little complicated and not necessary to state the final result.

We put

$$\begin{aligned} G_{\mathbf{R}} &:= GO_n(Y, \mathbf{R}) := GO_n(Y, \mathbf{C}) \cap M_n(\mathbf{R}), \\ V_{\mathbf{R}} &:= \mathbf{R}^n, \end{aligned} \tag{3.1}$$

where Y is a *real* $n \times n$ symmetric matrix defined of the form (1.2). Then $P_{\mathbf{R}} := P_{\mathbf{C}} \cap G_{\mathbf{R}}$ is naturally a parabolic subgroup of $G_{\mathbf{R}}$ and $(P_{\mathbf{R}}, \rho, V_{\mathbf{R}})$ is a real form of $(G_{\mathbf{C}}, \rho, V_{\mathbf{C}})$. We may suppose that

$$U = I_{n-2}^{(p)} := \begin{bmatrix} I_p & \\ & -I_q \end{bmatrix},$$

where $q := n-2-p$ since U is transformed to the form $I_{n-2}^{(p)}$ by an automorphism in $Gl_n(V_{\mathbf{R}})$. The restrictions of $P_1(x)$ and $P_2(x)$ to $V_{\mathbf{R}}$ are real valued.

The set $V_{\mathbf{R}} - \{x \in V_{\mathbf{R}};\, P_1(x) \cdot P_2(x) = 0\}$ decomposes into the following four connected components:

$$V_1 := \{x \in V_{\mathbf{R}};\, P_1(x) > 0,\, P_2(x) > 0\}, \qquad V_2 := \{x \in V_{\mathbf{R}};\, P_1(x) < 0,\, P_2(x) > 0\},$$

$$V_3 := \{x \in V_{\mathbf{R}};\, P_1(x) > 0,\, P_2(x) < 0\}, \qquad V_4 := \{x \in V_{\mathbf{R}};\, P_1(x) < 0,\, P_2(x) < 0\}.$$

We put

$$|P(x)|_i^{(s_1, s_2)} := \begin{cases} |P_1(x)|^{s_1} \cdot |P_2(x)|^{s_2} & \text{if } x \in V_i \\ 0 & \text{if } x \notin V_i \ (i = 1, 2, 3, 4). \end{cases}$$

Then $|P(x)|_i^{(s_1, s_2)}$ defines a continuous function on $V_{\mathbf{R}}$ when the real parts $\mathrm{Re}(s_1)$ and $\mathrm{Re}(s_2)$ are sufficiently large. We denote it by $|P(x)|_i^s$ instead of $|P(x)|_i^{(s_1, s_2)}$ for simplicity. As is well known, $|P(x)|_i^s$ is extended to all $(s_1, s_2) \in \mathbf{C}^2$ as a tempered distribution with meromorphic parameters $s := (s_1, s_2)$. More precisely, we have

Theorem 1. *Let* $\Gamma(s)$ *be the gamma-function. Then*

$$\Gamma(s_1+1)^{-1} \cdot \Gamma(s_1+s_2+n/2)^{-1} \cdot \Gamma(s_2+1)^{-1} \cdot |P(x)|_i^s$$

is an entire function with respect to $s := (s_1, s_2) \in \mathbf{C}^2$.

We denote by $|P(x)|_i^{-s-\delta\chi_0}$ the tempered distribution $|P(x)|^{(-s_1-s_2-n/2, s_2)}$. Then we have the following formula of the Fourier transform.

Theorem 2. *We put* $e[\theta] := \exp((\pi/2)\sqrt{-1}\theta)$ *and* $r := p - q$. *Then, we have*

$$\int \sum_{i=1}^{4} a_i \cdot |P(x)|_i^s \exp(-2\pi\sqrt{-1}\langle x, y\rangle)\, dx = \pi^{-2s_1-s_2-n/2} \sum_{j=1}^{4} c_j(s) \cdot |P(y)|_j^{-s-\chi_0},$$

where

$$\begin{bmatrix} c_1(s) \\ c_2(s) \\ c_3(s) \\ c_4(s) \end{bmatrix} = \frac{\Gamma(s_1+1)\Gamma(s_1+s_2+n/2)}{2\pi}$$

$$\times \begin{bmatrix} e\left[-2s_1-s_2+\frac{r-n-2}{2}\right] & e\left[-s_2+\frac{r-n+2}{2}\right] & e\left[2s_1+s_2+\frac{-r+n+2}{2}\right] & e\left[s_2+\frac{-r+n-2}{2}\right] \\ e\left[s_2+\frac{r+n-2}{2}\right] & e\left[2s_1+s_2+\frac{r+n+2}{2}\right] & e\left[-s_2+\frac{-r-n+2}{2}\right] & e\left[-2s_1-s_2+\frac{-r-n-2}{2}\right] \\ e\left[2s_1+s_2+\frac{-r+n+2}{2}\right] & e\left[s_2+\frac{-r+n-2}{2}\right] & e\left[-2s_1-s_2+\frac{r-n-2}{2}\right] & e\left[-s_2+\frac{r-n-2}{2}\right] \\ e\left[-s_2+\frac{-r-n+2}{2}\right] & e\left[-2s_1-s_2+\frac{-r-n-2}{2}\right] & e\left[s_2+\frac{r+n-2}{2}\right] & e\left[2s_1+s_2+\frac{r+n+2}{2}\right] \end{bmatrix},$$

and $dx := |dx_1 \wedge \cdots \wedge dx_n|$.

References

[Ka 1] M. Kashiwara, *Systems of microdifferential equations*, Progress in Math. 34, Birkhäuser, Boston, 1983.

[Ka 2] M. Kashiwara, Microlocal calculus of simple microfunctions, in *Analysis and Algebraic Geometry*, Papers in honor of Professor K. Kodaira. W. L. Baily and T. Shioda (eds.), Iwanami Shoten Publishers, Tokyo, 1977, pp. 369–374.

[Ka-Mi] M. Kashiwara and T. Miwa, Microlocal calculus and Fourier transforms of relative invariants of prehomogeneous vector spaces, (in Japanese), *Surikaiseki Kenkyusho Kokyuroku* **283** (1974) 60–147.

[Ki] T. Kimura, The *b*-functions and holonomy diagrams of irreducible regular prehomogeneous vector spaces, *Nagoya Math. J.* **85** (1982) 1–80.

[Mr 1] M. Muro, Microlocal analysis and calculations on some relatively invariant hyperfunctions related to zeta functions associated with the vector spaces of quadratic forms, *Publ. R.I.M.S., Kyoto Univ.* **22** (1986) 395–463.

[Mr 2] M. Muro, Singular invariant tempered distributions on regular prehomogeneous vector spaces, *J. of Funct. Analy.* **76** (1988) 317–345.

[Mr 3] M. Muro, Relatively invariant tempered distributions on regular prehomogeneous vector spaces, preprint, 1987.

[Mr 4] M. Muro, The dimension of the space of relatively invariant hyperfunctions on regular prehomogeneous vector spaces, *Proc. of Japan Acad.* **63**, Ser. A (1987) 66–68.

[Mu-Ru-Sc] I. Muller, H. Rubenthaler, and G. Schiffmann, Structure des espaces préhomogènes associés à certaines algèbres de Lie graduées, Math. Ann. **274** (1986) 95–123.

[Ru-Sc] H. Rubenthaler and G. Schiffmann, Opérateurs différentials de Shimura et espace préhomogènes, *Invent. Math.* **90** (1987) 409–442.

[Sm-Ka-Ki-Os] M. Sato, M. Kashiwara, T. Kimura, and T. Oshima, Microlocal analysis of prehomogeneous vector spaces, *Invent. Math.* **62** (1980) 117–179.

[Sm-Sh] M. Sato and T. Shintani, On zeta functions associated with prehomogeneous vector spaces, *Ann. of Math.* **100** (1974) 131–170.

[Sh] T. Shintani, On zeta functions associated with the vector space of quadratic forms, *J. Fac. Sci. Univ. Tokyo, SecIA* **22**, No. 1 (1975) 25–65.

[Sf] F. Sato, Zeta functions in several variables associated with prehomogeneous vector spaces I, *Tôhoku Math. J.* **34** (1982) 437–483.

[Vi] E. B. Vinberg, Complexity of actions of reductive groups, *Functional Anal. Appl.* **20**, No. 1 (1986) 1–13.

Applications of Transformation Theory for Nonlinear Integrable Systems to Linear Prediction Problems and Isospectral Deformations

Yoshimasa Nakamura
Department of Mathematics
Gifu University
Gifu, Japan

1. Introduction

In 1967 it came as a surprise that a certain nonlinear evolution equation can be solved in closed form. This discovery lent interest to explicit solutions, and a number of "nonlinear integrable systems" were subsequently found. In 1981 M. Sato introduced the Kadomtsev–Petviashvili (KP) equation hierarchy and completely characterized the solution space of the KP equation being of great interest in terms of an infinite-dimensional Grassmann manifold. It was shown that the transformation group for the KP hierarchy is the automorphism group $GL(\infty)$ of the Grassmann manifold and many other nonlinear integrable systems are derived from the KP

ISBN 0-12-400466-0

hierarchy by reductions. For this reason the theory of M. Sato and Y. Sato [7-9] has received considerable attention and has been the subject of intense study (see [1, 4, 13]).

On the other hand, the method of Riemann–Hilbert (RH) transformations was also developed in the study of symmetries and solutions to nonlinear integrable systems. The RH transformations were successfully applied to the stationary Einstein equation (Hauser–Ernst [3]) and the self-dual Yang–Mills (SDYM) equation (Ueno–Nakamura [12]), which seem to be outside the KP hierarchy. Takasaki [11] introduced an infinite system on an infinite-dimensional Grassmann manifold and formulated an algebraic RH transformation for the SDYM equation, being inspired by the works of M. Sato and Y. Sato. Along with these works, the RH transformation theory has become a powerful tool for nonlinear integrable systems.

The purpose of this paper is to consider applications of the RH transformation theory in mathematical physics to linear prediction problems of stochastic processes and isospectral deformations of linear operators. In Section 2, we make clear an algebraic and geometric structure of the space of linear predictors for stationary and non-stationary processes. The set of linear predictors is identified with a point on a Grassmann manifold. The old predictors and new ones are related by a fractional transformation. In Section 3, a discrete isospectral deformation of matrices is defined as a special member of transformations introduced in Section 2. The deformation is described by nonlinear differential equations of Lax type. We discuss an interesting similarity between our isospectral deformation and the Darboux transformation for Schrödinger operators.

2. Application to Linear Prediction Problems

Let C be a unit circle in the compactified complex plane $\mathbf{C}\cup\{\infty\}$ and $R(z)$ be a rational $(N\times N)$ matrix spectral density function of a complex N-variate stationary stochastic process $\{X_t\,;\,t\in\mathbf{Z}\}$ which is analytic and invertible on C and has the Fourier series expansion $R(z)=\sum_{n\in\mathbf{Z}} R_n z^n$. Let us consider an infinite number of discrete Wiener–Hopf equations $\sum_{j\in\mathbf{N}} \xi_{kj}R_{m-j}=R_{k+m-1}$ for k, $m\in\mathbf{N}$ with $R_n=E[X_lX^*_{n+l}]$, where the asterisk denotes the Hermitian conjugate and E is an expectation operator. For a given past date $\{X_{1-l}\,;\,l\in\mathbf{N}\}$, and consequently for a set of covariance matrices $\{R_n\,;\,n\in\mathbf{Z}\}$, a set of unknowns $\{\xi_{kj}\,;\,j\in\mathbf{N}\}$ is called a least-square

k-step linear optimum predictor in the sense in which $\sum_{j\in\mathbf{N}} \xi_{kj}X_{1-j}$ realizes a minimum of the k-step quadratic prediction error. The spectral factorization method introduced by Wiener [14] has been widely used as a powerful tool in solving Wiener–Hopf equations. Here we note that the spectral factorization is regarded as a special case of RH factorization problems of matrix functions. We then see that solutions to the discrete Wiener–Hopf equations can be obtained by solving an RH problem for $R(z)$. Let C_+ and C_- be the inside and the outside of C in $\mathbf{C}\cup\{\infty\}$, respectively. The RH problem considered here is the problem of factoring $R(z)$ into the form $R(z)=U(z)DL(z)$ on C, where $U(z)$ and $L(z)$ are rational $N\times N$ matrix functions analytic and invertible in $C\cup C_+$ and $C\cup C_-$, respectively, and D is a nonsingular $N\times N$ constant matrix. The set of $U(z)$, $L(z)$ and D is called the solution to the RH problem. Let U_n and $\bar{U}_n$ be Fourier coefficients of $U(z)$ and $U(z)^{-1}$, respectively. It is not hard to see that if there is a solution to the RH problem, then the Wiener–Hopf equations have a solution expressed as $\xi_{kj}=\sum_{l=1}^{j} U_{k+l-1}\bar{U}_{j-l}$, $k, j\in\mathbf{N}$. This formula was originally found by Wiener and Masani [15]. We here introduce four types of block semi-infinite matrices associated with a Fourier series $\sum_{n\in\mathbf{Z}} H_n z^n$ through

$$H_{--}=\begin{bmatrix} \ddots & \ddots & \vdots \\ \ddots & H_0 & H_1 \\ \cdots & H_{-1} & H_0 \end{bmatrix}, \qquad H_{+-}=\begin{bmatrix} \vdots & \vdots & \\ H_2 & H_3 & \cdots \\ H_1 & H_2 & \cdots \end{bmatrix},$$

$$H_{-+}=\begin{bmatrix} \cdots & H_{-2} & H_{-1} \\ \cdots & H_{-3} & H_{-2} \\ & \vdots & \vdots \end{bmatrix}, \qquad H_{++}=\begin{bmatrix} H_0 & H_1 & \cdots \\ H_{-1} & H_0 & \ddots \\ \vdots & \ddots & \ddots \end{bmatrix}. \tag{2.1}$$

In the following discussions, we consider solutions to Wiener–Masani's equations

$$\Xi U_{++}=U_{+-}, \qquad \Xi=\begin{bmatrix} \vdots & \vdots & \\ \xi_{21} & \xi_{22} & \cdots \\ \xi_{11} & \xi_{12} & \cdots \end{bmatrix}, \tag{2.2}$$

in place of the Wiener–Hopf equations and formulate a transformation from the old linear predictors Ξ to new ones denoted by $\tilde{\Xi}$. This correspondence is defined by an RH transformation which relates the old $U(z)$ to a new $\tilde{U}(z)$ and the old spectral density function to a new one. Here the method of RH transformations was used by Hauser and Ernst [3] and Ueno and this author [12] in mathematical physics.

Let $U(z)$ be an $N\times N$ matrix function analytic and invertible in $C\cup C_+$. Let $H(z)$ be an $N\times N$ matrix function analytic that is invertible on C and has the Fourier series expansion $H(z)=\sum_{n\in\mathbf{Z}} H_n z^n$. Consider a RH factorization problem for $U(z)^{-1}H(z)U(z)$ as follows:

$$U(z)^{-1}H(z)U(z)=X(z)Y(z)^{-1} \qquad \text{on } C. \tag{2.3}$$

Here $X(z)$ and $Y(z)$ are analytic and invertible in $C\cup C_+$ and $C\cup C_-$, respectively. We here assume the existence of a solution to (2.3). Define an $N\times N$ matrix $\tilde{U}(z)$ which is analytic and invertible in $C\cup C_+$ by

$$\tilde{U}(z)=U(z)X(z) \quad \text{in } C\cup C_+, \qquad =H(z)U(z)Y(z) \quad \text{in } C\cup C_-. \tag{2.4}$$

Let $\tilde{\Xi}$ be a solution to (2.2) for $\tilde{U}(z)$. We refer to the transformation (2.4) from $U(z)$ to $\tilde{U}(z)$ as the RH transformation induced by $H(z)$. The corresponding action on linear predictors is

$$\tilde{\Xi}=(H_{--}\Xi+H_{+-})(H_{-+}\Xi+H_{++})^{-1}. \tag{2.5}$$

There are several types of RH problems (2.3) actually having solutions. The first is the case where $H(z)$ is analytic and invertible in $C\cup C_+$. The resulting $\tilde{\Xi}$ is related to Ξ as $\tilde{\Xi}=H_{--}\Xi H_{++}^{-1}+H_{+-}H_{++}^{-1}$. We say that $\tilde{\Xi}$ is *gauge equivalent* to Ξ. The second is the infinitesimal RH transformation [12] induced by $H(z)=\exp h(z)$, where $h(z)=\sum_{n\in\mathbf{Z}} h_n z^n$, $h_n\in\mathfrak{sl}(N,\mathbf{C})$, is analytic and invertible on C. This RH transformation acts on linear predictors as follows:

$$\tilde{\Xi}\simeq\Xi+h_{+-}+h_{--}\Xi-\Xi h_{++}-\Xi h_{-+}\Xi, \tag{2.6}$$

where we neglect the terms of higher order in h.

Next we give a geometric interpretation of the linear predictors and the RH transformation. Consider the linear space

$$V=\mathbf{C}^N\otimes\mathbf{C}[z,z^{-1}]=\left\{\sum_{j=-\infty}^{l} v_j z^j;\ v_j\in\mathbf{C}^N,\ l\in\mathbf{Z}\right\}$$

of $\mathbf{C}^N$-valued Laurent series. Decomposing V into $V=V_+\oplus V_-$, where

$$V_+=\left\{\sum_{j=1}^{l} v_j z^j\in V\right\} \quad \text{and} \quad V_-=\left\{\sum_{j=-\infty}^{0} v_j z^j\in V\right\},$$

we have a decomposition

$$\mathrm{End}(V)=\mathrm{End}(V_+)+\mathrm{Hom}(V_+,V_-)+\mathrm{Hom}(V_-,V_+)+\mathrm{End}(V_-).$$

We write the groups of invertible endomorphisms of V, V_+ as $GL(\infty)$, $GL_+(\infty)$, respectively. The Lie algebra $\mathfrak{gl}(\infty)$ can be identified with End(V). Let us consider an infinite-dimensional Grassmann manifold $GM(\infty)$ of subspaces of V modeled on V_+. As in the finite-dimensional case, the manifold can be described by homogeneous coordinates consisting of $\mathbf{Z}\times\mathbf{N}$ rectangular matrices. Each linearly independent column defines a frame. A point of the open subset $GM^{\varnothing}$ in $GM(\infty)$ is identified with an equivalent class $[P]$ under change of frame; $[P]=\{P\circ G;\ P\in \mathrm{Hom}(V_+, V),\ G\in GL_+(\infty)\}$. Here the superscript $\varnothing$ of $GM^{\varnothing}$ indicates the empty Young diagram in the notation of [7–9].

We now turn to an interpretation of linear predictors for stationary processes. Since $\Xi=U_{+-}U_{++}^{-1}$, we can regard linear predictors as affine coordinates of a point of $GM^{\varnothing}$. It should be noted that there is an additional quadratic constraint on Ξ. Let Λ_{--}, Λ_{-+} and Λ_{++} be semi-infinite matrices associated with $1_N z^{-1}$. From $R_{++}\Lambda_{++}=\Lambda_{-+}R_{+-}+\Lambda_{++}R_{++}$, we obtain the identity

$$\Xi\Lambda_{-+}\Xi+\Xi\Lambda_{++}=\Lambda_{--}\Xi. \tag{2.7}$$

This implies that the linear space spanned by columns of Ξ is invariant under the mapping $\zeta: v\in V\to vz\in V$. Thus we have

Theorem 1. *The space of linear predictors for stationary N-variate processes is identified with the set of fixed points $GM_{\zeta}^{\varnothing}\subset GM^{\varnothing}$ under the mapping ζ.*

The set of elements of End(V) commuting with ζ leaves $GM_{\zeta}^{\varnothing}$ invariant and forms a subalgebra of $\mathfrak{gl}(\infty)$. It is identified with the loop algebra $\mathfrak{sl}(N,\mathbf{C})\otimes\mathbf{C}[z,z^{-1}]$ of infinitesimal RH transformations through a mapping $\mu:\mathfrak{sl}(N,\mathbf{C})\otimes\mathbf{C}[z,z^{-1}]\to\mathfrak{gl}(\infty)$. Since $H_{-+}U_{+-}+H_{++}U_{++}$ is invertible provided the RH problem has a solution, we see $\tilde{U}(z)$ determines a point in $GM_{\zeta}^{\varnothing}$. Thus the RH transformation (2.4) induces a $GL(\infty)$ action on $GM_{\zeta}^{\varnothing}$ with the representation (2.5) in the affine coordinates. We have shown that the RH transformation maps a point of $GM_{\zeta}^{\varnothing}$ with the coordinates Ξ to another point of $GM_{\zeta}^{\varnothing}$ denoted by $\tilde{\Xi}$. We next consider a parametric $\mathfrak{sl}(N,\mathbf{C})$ parametrized by $t\in\mathbf{R}$. In the affine coordinates, the infinitesimal action induced by $H(z,t)=\exp h(z,t)$ is given by (2.6). The semi-infinite matrix Riccati equation,

$$\frac{d}{dt}\Xi(t)=h_{+-}+h_{--}\Xi-\Xi h_{++}-\Xi h_{-+}\Xi, \tag{2.8}$$

determines a flow of the corresponding (time-dependent) vector field on $GM_\xi^\varnothing$. For given $h(z,t)$ the corresponding sequence of linear predictors can be interpreted as a dynamical system with the equation of motion (2.8). Let us recall the interesting fact found by M. Sato [7] that soliton equations are regarded as dynamical systems on an infinite-dimensional Grassmann manifold. The linear predictors can be obtained explicitly by integrating the Riccati equation (2.8). Constructions of parametric linear predictors are discussed in [5] by using exponential functions of infinite matrices.

Finally we consider the linear prediction problems for finite-order non-stationary processes being of practical importance. Note that the method of RH transformations discussed above is not applicable to non-stationary processes. Here we propose a method of Bruhat decompositions. Assume that we have observation of a real non-stationary process $\{y_i\} = \{y_1, y_2, \ldots, y_n\}$ with known covariances $r_{ij} = E[y_i y_j]$. We wish to predict the future value of the process. Let us write $y_{n+k|n} = \sum_{j=1}^{n} \xi_{n+k\ n-j+1} y_j$ for $1 \le k \le m$, where the k-step linear predictor $\{\xi_{n+k\,j};\ 1 \le j \le m\}$ is to be chosen to minimize the k-step quadratic prediction error $E[(y_{n+k} - y_{n+k|n})^2]$. As shown in [6], the set of all $(1 \le k \le m)$ linear predictors is given by solving analogues of Wiener–Masani's equations

$$\Xi L_{--} = L_{+-}, \qquad \Xi = \begin{bmatrix} \xi_{n+1\ n} & \cdots & \xi_{n+1\ 1} \\ \vdots & & \vdots \\ \xi_{n+m\ n} & \cdots & \xi_{m+m\ 1} \end{bmatrix}, \tag{2.9}$$

where L_{--} is an $n \times n$ unipotent lower-triangular matrix, and L_{-+} is an $m \times n$ matrix. Here L_{--} and L_{-+} are due to an LD decomposition of the covariance matrix. Set $N = n + m$. Let $M = \left[\begin{smallmatrix} M_{--} & M_{+-} \\ M_{-+} & M_{++} \end{smallmatrix}\right]$ be a $GL(N, \mathbf{R})$ matrix. Let $\mathcal{N}$ and $\bar{\mathcal{N}}$ be the subgroups of $N \times N$ unipotent upper- and lower-triangular matrices, respectively, and $\mathcal{D}$ be the subgroup of $N \times N$ diagonal matrices with non-zero determinant. Let $L \in \bar{\mathcal{N}}$. Consider a Bruhat decomposition of $GL(N, \mathbf{R})$ as follows:

$$L^{-1} M L = X(YD)^{-1}, \qquad X \in \bar{\mathcal{N}},\ D \in \mathcal{D},\ Y \in \mathcal{N}. \tag{2.10}$$

Assuming the existence of factors, let us define an analogue of the RH transformation

$$\tilde{L} = LX = MLYD. \tag{2.11}$$

Since $\tilde{L} \in \bar{\mathcal{N}}$, the $m \times n$ matrix $\tilde{\Xi}$ defined by $\tilde{\Xi} = \tilde{L}_{-+} \tilde{L}_{--}^{-1}$ is regarded as a set of linear predictors for a process with covariance matrix defined by $\tilde{L}$, where $\tilde{L}_{--}$ and $\tilde{L}_{-+}$ are block matrices of $\tilde{L}$. The old predictors and new ones are related by the fractional transformation

$$\tilde{\Xi} = (M_{-+} + M_{++}\Xi)(M_{--} + M_{+-}\Xi)^{-1}. \tag{2.12}$$

By successive uses of the Bruhat decomposition (2.10) and the transformation (2.11) we have a sequence of sets of linear predictors. It is noted that Ξ is regarded as the affine coordinates of a real Grassmann manifold $GM(N, m)$, where columns of Ξ span an m-dimensional subspace in $\mathbf{R}^N$. The fractional transformation (2.12) is a matrix representation of the transitive $GL(N, \mathbf{R})$ action on $GM(N, m)$. Several types of parametric linear predictors are constructed in [6] by making use of abelian and non-abelian actions of $GL(N, \mathbf{R})$.

3. Application to Isospectral Deformations

In this section we discuss an application of successive Bruhat decompositions (2.10) and transformations (2.11) to isospectral deformations of matrices. For simplicity we set $L = I$ in (2.10). Let M_0 admit the Bruhat decomposition $M_0 = X_0(Y_0D_0)^{-1}$, where $X_0 \in \bar{\mathcal{N}}$, $D_0 \in \mathcal{D}$ and $Y_0 \in \mathcal{N}$. Define $L_1 = X_0$ as the transformation from L to L_1. Let M_1 defined by $M_1 = L_1^{-1}M_0L_1$ admit the Bruhat decomposition $M_1 = X_1(Y_1D_1)^{-1}$, where $X_1 \in \bar{\mathcal{N}}$, $D_1 \in \mathcal{D}$ and $Y_1 \in \mathcal{N}$. Define $L_2 = L_1X_1$ as the transformation from L_1 to L_2. We then consider the Bruhat decomposition of M_2 defined by $M_2 = L_2^{-1}M_0L_2$; $M_2 = X_2(Y_2D_2)^{-1}$, where $X_2 \in \bar{\mathcal{N}}$, $D_2 \in \mathcal{D}$ and $Y_2 \in \mathcal{N}$. For general $k \in \mathbf{N}$, we formulate the Bruhat decomposition $M_k = X_k(Y_kD_k)^{-1}$, $X_k \in \bar{\mathcal{N}}$, $D_k \in \mathcal{D}$ and $Y_k \in \mathcal{N}$, and the transformation from L_k to L_{k+1} by $L_{k+1} = L_kX_k$. Here the next M_{k+1} is defined by $M_{k+1} = L_{k+1}^{-1}M_0L_{k+1}$. Observe that $M_{k+1} = (Y_kD_k)^{-1}X_k$. We see that this iteration generates a sequence of matrices $\{M_k\}$, where each M_k is similar to M_0:

$$M_k = X_{k-1}^{-1}X_{k-2}^{-1} \cdots X_0^{-1}M_0X_0 \cdots X_{k-1}. \tag{3.1}$$

The main purpose of this section is to prove

Theorem 2. Spec M_k = Spec M_0 *for* $k \in \mathbf{N}$ *in terms of nonlinear integrable systems.*

It is easy to see $\det(X_k(Y_kD_k)^{-1} - \lambda) = \det((Y_kD_k)^{-1}X_k - \lambda)$. Since invertible matrices are dense, this identity holds in general. Thus M_k and M_{k+1}

have the same characteristic polynomial and hence have the same eigenvalue. In the subsequent discussions we give an alternative proof of this. To study the discrete case (3.1), we want to construct a one-parameter family of $\bar{\mathcal{N}}$-valued matrices $X(t)$ with $X(0)=I$ so that

$$M(t)=X(t)^{-1}M_0X(t) \tag{3.2}$$

would satisfy a differential equation which describes an isospectral deformation of M_0. Here $X(t)$ is defined by the Bruhat decomposition

$$\exp(tF(M_0))=X(t)(Y(t)D(t))^{-1}, \tag{3.3}$$

with $X(t)\in\bar{\mathcal{N}}$, $D(t)\in\mathcal{D}$ and $Y(t)\in\mathcal{N}$, where $F(M_0)$ is an analytic function whose domain contains the eigenvalues of M_0. Since $[M_0, \exp(tF(M_0))]=0$, we see $M(t)=(Y(t)D(t))^{-1}M_0Y(t)D(t)$. Taking derivatives on both sides of (3.3), we have $F(M)=X^{-1}\,d/dt\,X-(YD)^{-1}\,d/dt(YD)$, where $M=M(t)$ and so on. Let $F_L(M)$ be the strictly lower-triangular part of $F(M)$ and $F_U(M)=F(M)-F_L(M)$. Assuming (3.3), we have $d/dt\,X=XF_L(M)$, $X(0)=I$ and $d/dt\,(YD)=YDF_U(M)$, $Y(0)D(0)=I$, and consequently from (3.2) we obtain a Lax equation

$$\frac{d}{dt}M=[M, F_L(M)], \qquad M(0)=M_0. \tag{3.4}$$

This is analogous to the celebrated (one-dimensional finite non-periodic) Toda equation $d/dt\,A(t)=[A, A_L-A_L^T]$, where A is a Jacobi (symmetric tridiagonal) matrix and A_L^T is the transposed matrix of A_L. We remark that Toda equations describe isospectral deformations of Jacobi matrices. Following the theory of Toda equations, let us set $M(t)\Phi(t)=\lambda(t)\Phi(t)$. From $d/dt\,\Phi(t)=-F_L(M)\Phi(t)$, we have $\lambda(t)=\lambda(0)$. This implies that $\operatorname{Spec} M(t)=\operatorname{Spec} M_0$ for $t\in[0,\varepsilon]$. In [10], Symes discussed the QR algorithm for calculating eigenvalues of Jacobi matrices and reproved a global existence of Toda flows. The result presented is a generalization of that in [10]. Solutions to (3.4), however, may have singularities in t. The isospectral deformation equation (3.4) is exemplified by $d/dt\,M=[M, \log_L M]$ or $d/dt\,M=[M, \exp_L M]$.

We now give a proof of Theorem 2. For a given M_0, we assume the existence and the uniqueness of solutions to (3.4) for $t\in[0,1]$. If $M_0=X_0(Y_0D_0)^{-1}$ with $X_0\in\bar{\mathcal{N}}$, $D_0\in\mathcal{D}$ and $Y_0\in\mathcal{N}$, then the solution $M(t)=X(t)^{-1}X_0(Y_0D_0)^{-1}X(t)$ gives $M(1)=(Y_0D_0)^{-1}X_0$. Thus we see $M(1)=M_1$ and $\operatorname{Spec} M_1=\operatorname{Spec} M_0$. Let M_1 be the next initial value. Along the same line of thought as above we have $\operatorname{Spec} M_2=\operatorname{Spec} M_1$. Generally we can see

Spec $M_{k+1}=\text{Spec } M_k$ for every $k\in\mathbf{N}$. We conclude that the successive Bruhat decompositions (2.10) and the transformations (2.11) induced by the same M_0 define an isospectral sequence of matrices similar to M_0.

Recently, Duistermaat and Grünbaum [2] solved the following problem, which has its origin in mathematical approaches to tomography: Find potentials $V(x)$ such that for some eigenfunctions $\Psi(x, \lambda)$ of the Schrödinger equation

$$\left(-\frac{\partial^2}{\partial x^2}+V(x)\right)\Psi(x,\lambda)=\lambda^2\Psi(x,\lambda) \tag{3.5}$$

we have an equation of the form

$$\sum_{r=0}^{m} A_r(\lambda)\frac{\partial^r}{\partial\lambda^r}\Psi(x,\lambda)=\Theta(x)\Psi(x,\lambda). \tag{3.6}$$

They show that all the potentials in question are rational and can be obtained from $V(x)=0$ or $-(4x^2)^{-1}$ by successive uses of Darboux transformations. It is also proved that rational solutions $V=V(x,t)$ to the Korteweg–de Vries (K–dV) equation,

$$\frac{\partial}{\partial t}V=\frac{\partial^3}{\partial x^3}V-6V\frac{\partial}{\partial x}V, \tag{3.7}$$

with initial value $V(x)$ have the property in question and give a half of all the potentials with the property. The other half is related to an integro-differential equation. The proof of the K–dV case, $V_k(x,t)=-2\,\partial^2/\partial x^2\log\theta_k$, is carried out with the help of the time variables of the K–dV hierarchy found by M. Sato [7], where θ_k are given by character polynomials of the irreducible representations of general linear groups.

The Darboux transformation from a given potential $V(x)$ to a new one denoted by $\tilde{V}(x)$ plays an essential role in [2]. The potentials are related as follows. Set $\mathcal{M}=-\partial^2/\partial x^2+V(x)$. We can always factorize $\mathcal{M}$ into two first-order operators; $\mathcal{M}=\mathcal{P}\circ\mathcal{Q}$ with $P=-\partial/\partial x-\partial/\partial x\log\phi(x)$ and $\mathcal{Q}=\partial/\partial x-\partial/\partial x\log\phi(x)$, where $\phi(x)$ is a non-zero eigenfunction for the zero eigenvalue of $\mathcal{M}$. Permuting the factors lets us define another Schrödinger operator $\tilde{\mathcal{M}}=\mathcal{Q}\circ\mathcal{P}=-\partial^2/\partial x^2+\tilde{V}(x)$, where $\tilde{V}(x)=V(x)-2\,\partial^2/\partial x^2\log\phi(x)$. For each λ ($\neq 0$) in the resolvent set of $\mathcal{P}\circ\mathcal{Q}$, the operator $\mathcal{Q}\circ(\mathcal{P}\circ\mathcal{Q}-\lambda)^{-1}\circ\mathcal{P}-I$ is a two-sided inverse of $\mathcal{Q}\circ\mathcal{P}-\lambda$, so that λ also belongs to the resolvent set of $\mathcal{Q}\circ\mathcal{P}$. The opposite inclusion holds by symmetry. Thus $\mathcal{M}$ and $\tilde{\mathcal{M}}$ have the same resolvent set away from zero and hence the same eigenvalue away from zero provided $\mathcal{M}$ and $\tilde{\mathcal{M}}$ are bounded. We have an isospectral sequence of linear operators $\mathcal{M}_k = \mathcal{P}_k \circ \mathcal{Q}_k$, $k \in \mathbf{N} \cup \{0\}$, or formally,

$$\mathcal{M}_k = \mathcal{P}_{k-1}^{-1} \circ \mathcal{P}_{k-2}^{-1} \circ \cdots \circ \mathcal{P}_0^{-1} \circ \mathcal{M}_0 \circ \mathcal{P}_0 \circ \cdots \circ \mathcal{P}_{k-1}. \tag{3.8}$$

The Schrödinger operators with the K-dV potentials $V_k(x, t)$ are obtained from $\mathcal{M}_0 = -\partial^2/\partial x^2$ by finitely many Darboux transformations. We can see an interesting similarity between the discrete isospectral deformation of matrices defined by the Bruhat decompositions and a transformation for Schrödinger operators defined by the Darboux transformations. Although what we get from the proof of Theorem 2 is only a characterization of isospectral deformations for given matrices in terms of the Lax equation (3.4), we should remember that the K-dV equation (3.7) is an isospectral deformation equation for the Schrödinger equation (3.5). Thus it is reasonable to expect that transformations defined by factoring and and permuting factors make clear a mysterious relationship between isospectral deformations of linear operators and nonlinear integrable systems. The result presented in this section may have implications in these more general situations.

References

[1] E. Date, M. Jimbo, M. Kashiwara, and T. Miwa, Transformation groups for soliton equations, in *Non-linear Integrable Systems—Classical Theory and Quantum Theory* (M. Jimbo and T. Miwa, eds.), World Scientific, 1983, pp. 39–119.

[2] J. J. Duistermaat and F. A. Grünbaum, Differential equation in the spectral parameter, *Commun. Math. Phys.* **103** (1986) 177–240.

[3] I. Hauser and F. J. Ernst, A homogeneous Hilbert problem for the Kinnersley-Chitre transformations, *J. Math. Phys.* **21** (1980) 1126–1140.

[4] M. Mulase, Complete integrability of the Kadomtsev-Petviashivili equation, *Adv. Math.* **54** (1984) 57–66.

[5] Y. Nakamura, Riemann-Hilbert transformations for a Toeplitz matrix equation: Some ideas and applications to linear prediction problems, preprint, Gifu Univ., 1987.

[6] Y. Nakamura, Group actions on linear predictors for nonstationary processes, *IMA J. Math. Control Inform.* **5** (1988) 69–75.

[7] M. Sato, Soliton equations as dynamical systems on an infinite dimensional Grassmann manifold, *RIMS Kôkyûroku, Kyoto Univ.* **439** (1981) 30–46.

[8] M. Sato, Soliton equations and the Universal Grassmann Manifold (In Japanese), Notes by M. Noumi, *Sophia Univ. Kôkyûroku in Math.* **18** (1984).

[9] M. Sato and Y. Sato, Soliton equations as dynamical systems on infinite dimensional Grassmann manifold, in *Nonlinear Partial Differential Equations in Applied Science; Proceeding of the U.S.-Japan Seminar, Tokyo, 1982* (H. Fujita, P. D. Lax and G. Strang, eds.), North-Holland, 1983, pp. 259–271.

[10] W. W. Symes, The QR algorithm and scattering for the finite nonperiodic Toda lattice, *Physica* **4D** (1982) 275–280.

New Local Supersymmetry in the Framework of Einstein Gravity

Noboru Nakanishi
Research Institute for Mathematical Sciences
Kyoto University
Kyoto, Japan

1. Quantum Field Theory

Quantum field theory is the fundamental theory describing particle physics. Its basic object is a *field*, which is a finite set of operator-valued generalized functions (i.e., distributions or hyperfunctions) of spacetime x^μ. According to the statistics of particles, fields are classified into two kinds, *Bosonic* (commutation type) and *Fermionic* (anticommutation type).

The operand of fields is a *state*. The totality of states forms an infinite-dimensional complex linear space equipped with an indefinite inner product. In order to have the probabilistic interpretability of quantum theory, however, it is necessary to introduce a *physical subspace* in which the inner product is positive semi-definite. The physical subspace is usually defined by the totality of the states that are annihilated by a particular operator; this condition is called a *subsidiary condition*.

Although there are several approaches to quantum field theory, I here adopt the *Lagrangian formalism*, which is the most traditional and standard

Algebraic Analysis, Volume II

ISBN 0-12-400466-0

one. The Lagrangian formalism is not mathematically rigorous, but its formulation is quite systematic and suitable for discussing symmetry principles. The Lagrangian formalism starts with an *action*, which is a spacetime integral of a *Lagrangian density*, a local function of fields. Given an action, one derives *field equations* and *equal-time commutation* (for Bosonic fields) and *anticommutation relations* (for Fermionic fields), which define fields at the operator level. The operator algebra of fields is represented in the space of states mentioned above. One can then, at least in principle, calculate a (pseudo-)unitary operator, called the *S-matrix*. The *physical S-matrix*, which is of direct physical significance, is defined by restricting the S-matrix to the physical subspace. The conservation of total probability requires the physical S-matrix to be unitary.

2. Symmetry Principles

The fundamental problem of the Lagrangian formalism is how to determine the expression for the action. The usual requirements are symmetry principles.

In general, *symmetry* is the invariance of the action under a group of transformations of fields. If symmetry is accompanied with coordinate transformations of spacetime x^μ, it is called *spacetime symmetry*; otherwise it is *internal symmetry*. If transformations of symmetry explicitly involve arbitrary functions of x^μ, it is called *local symmetry*; otherwise it is called *global symmetry*. Local internal symmetry is called *gauge invariance*.

It is well established that particle physics is invariant under the *Poincaré group*, which consists of translations and Lorentz transformations. Poincaré invariance is a prototype of global spacetime symmetry. As for internal symmetry, the electromagnetic $U(1)$ symmetry accounts for the electric-charge conservation law, and its local version requires the introduction of the electromagnetic field.

Unfortunately, the real world, which should be described by the physical S-matrix, scarcely exhibits invariance other than the ones above. It is important to note, however, that symmetry is the invariance property of the action but not of the physical S-matrix. Symmetry may be broken at the level of representation (*spontaneous breakdown*) or may become trivial by the restriction to the physical subspace (*ghost symmetry*). For example, the electroweak theory, which is experimentally successful, has local $SU(2)_{\text{Left}} \times$

$U(1)_{\text{Right}}$ symmetry, which is spontaneously broken up to the electromagnetic $U(1)$ symmetry.

If there is global symmetry of continuous transformations, the *Noether theorem* implies the existence of a set of *conserved currents*, i.e., divergence-less Lorentz-vectors consisting of local functions of fields. The spatial integrals of their time components define *symmetry generators*, which are shown to induce the infinitesimal version of the original symmetry if the commutator between each of them and each of the fields is calculated by means of the equal-time commutation and anticommutation relations. The totality of symmetry generators forms a Lie algebra, which corresponds to the Lie group of the original symmetry. For example, the *Poincaré algebra* consists of *translation generators* P_μ and *Lorentz generators* $M_{\mu\nu}$ $(=-M_{\nu\mu})$, and its structure is symbolically[1] written as $[P, P]=0$, $[M, P]=P$, and $[M, M]=M$.

One of the most interesting recent discoveries is the possible existence of *Fermionic symmetry generators*. Bosonic and Fermionic symmetry generators form a graded Lie algebra, which is called a *superalgebra* in physics, and the corresponding symmetry is called *supersymmetry* in the broad sense.

It is known that local symmetry in the strict sense can exist only at the classical (i.e., pre-quantized) level. That is, quantum field theory does not admit local symmetry, which should be replaced by a global Fermionic symmetry, called *BRS invariance.*[2] Any locally symmetric part of the action is manifestly BRS invariant, but BRS invariance admits the addition of a *gauge-fixing* plus *FP-ghost*[3] part, which explicitly breaks the local symmetry. The choice of gauge fixing is rather arbitrary,[4] but it is preferable to preserve the global version of the original local symmetry. Once gauge fixing is set up, the FP-ghost part is uniquely determined by the requirement of BRS invariance. The BRS invariance of the action implies the existence of a nilpotent *BRS generator*, which consistently defines a subsidiary condition [7, 11]. The physical S-matrix defined in the BRS-invariant subspace is shown to be independent of the choice of gauge fixing, provided that what particles are observable is prescribed. The BRS invariance itself is ghost symmetry.

Although there is no well established principle for prescribing what internal symmetry is really relevant, the most appealing way for determining

[1] Here "symbolically" means to omit explicit expressions for the structure constants.
[2] "BRS" is the abbreviation of Becchi-Rouet-Stora.
[3] "FP ghost" is the abbreviation of (Feynman-DeWitt-) Faddeev-Popov ghost.
[4] But I expect that it should be unique in the ultimate theory.

it seems to be to extend the Poincaré algebra so as to include internal symmetry. Unfortunately, however, there is a no-go theorem [4]: If the physical S-matrix has a physically reasonable structure, there is no indecomposable Lie algebra which properly includes the Poincaré algebra; the only exception is *supersymmetry* in the narrow sense [12]. In this superalgebra, there is a set of Fermionic spinor generators Q_A^α and $\bar{Q}_{\dot{B}}^\beta$, where A and $\dot{B}$ are spinor indices and $\alpha, \beta = 1, 2, \ldots, N$. The commutation and anticommutation relations are symbolically as follows: $[P, Q] = 0$, $[P, \bar{Q}] = 0$, $[M, Q] = Q$, $[M, \bar{Q}] = \bar{Q}$, $\{Q, \bar{Q}\} = P$, $\{Q, Q\} = 0$, and $\{\bar{Q}, \bar{Q}\} = 0$.

The supersymmetry predicts the existence of *supermultiplets* of particles; as a consequence of $\{Q, \bar{Q}\} = P$, each supermultiplet contains an equal number of integral-spin Bosons and half-odd-spin Fermions having exactly the same mass, in sharp contradiction to the elementary particles in the real world. Hence the supersymmetry must be broken. But its spontaneous breakdown implies the existence of a massless spin-$\frac{1}{2}$ Fermion (different from neutrinos), which is also non-existent in the real world. This trouble can be avoided if the supersymmetry is promoted to local symmetry, owing to the *Higgs mechanism*. Since the supersymmetry is spacetime symmetry, one then encounters local spacetime symmetry, which necessarily involves the invariance under general coordinate transformations. Thus one must take account of the theory of gravity, which is the subject of the next section.

Finally, I make a comment on the widespread misunderstanding concerning the no-go theorem. It is not a strong result as is usually believed. Since it is based on the invariance of the physical S-matrix, it can say nothing about spontaneously broken symmetry nor the ghost one. Indeed, it is possible to extend the Poincaré algebra nontrivially in various ways if such symmetries are admitted [9, 11].

3. Einstein Gravity and Supergravity

General relativity, which is called *Einstein gravity* in order to discriminate it from its various modified versions, is an extremely successful theory; it is undoubtedly the correct theory describing gravity at the classical level.

Since gravity interacts with elementary particles, Einstein gravity must be unified into the framework of quantum field theory. The invariance under general coordinate transformations is nothing but local spacetime symmetry. To quantize Einstein gravity, therefore, one should introduce a gauge-fixing plus FP-ghost part into the action so as to replace general coordinate invariance by BRS invariance. Here if one requires general linear invariance (this requirement is fulfilled only when one introduces an auxiliary field, called the *B-field*), one obtains an outstandingly beautiful theory, called the *manifestly covariant canonical formalism of quantum gravity* [9, 11].

Since the spinor representation of the Lorentz group cannot be linearly extended to such a larger group as the group of general coordinate transformations, the Dirac field, which is a spinor under Lorentz group, is regarded as a scalar in the framework of gravity. The metric field $g_{\mu\nu}(x)$ is decomposed into a product of two *vierbein fields*. The vierbein field, denoted by $h_{\mu a}(x)$ $(a = 0, 1, 2, 3)$, forms a Minkowskian orthonormal basis of the tangent space at x^μ of the spacetime pseudo-Riemannian manifold at the classical level. Lorentz transformations in the tangent space are called *local Lorentz transformations*, and $g_{\mu\nu}$ is, of course, invariant under them. The theory must be local-Lorentz invariant at the classical level because otherwise the metric field cannot be uniquely characterized in the vierbein formalism. Since local Lorentz invariance is local internal symmetry, quantization is carried out by introducing a BRS-invariant guage-fixing plus FP-ghost part into the action [9, 11]. Here the global internal Lorentz symmetry is kept unbroken.

As stated above, the Dirac field is a *world* (i.e., spacetime) scalar; it is a spinor only under internal Lorentz transformations. Then, at the classical level, it is impossible to explain why the Dirac field must become a world spinor in particle physics. In the framework of the manifestly covariant canonical formalism of quantum gravity, however, this fundamental problem is beautifully resolved: Both the world general linear invariance and the global internal Lorentz invariance are spontaneously broken up to the global spacetime Lorentz invariance, under which the Dirac field behaves as a world spinor [8]. Thus it is very important to recognize that the world Poincaré invariance is not the basic symmetry of the fundamental theory including gravity; it becomes relevant only at the level of representation, just like the electromagnetic $U(1)$ symmetry in the electroweak theory.

Now, I return to the problem of extending the supersymmetry. As discussed in Section 2, the supersymmetry theory should be promoted to its local version. Such a theory can indeed be constructed; it is called *supergravity* [12]. But one should note that supergravity is not a natural extension of Einstein gravity. The point is this: The group of general coordinate transformations is, of course, the local version of the translation group, but the former already includes the Lorentz group. If the local Lorentz group is regarded as the local version of the Lorentz group, one does not have the local version of the Poincaré group because one has $[M, P] = 0$ rather than $[M, P] = P$. In order to preserve the structure of the Poincaré group, one must introduce a new notion of *local internal translations*, whose global version cannot exist. Thus supergravity is a gauge theory based on the supersymmetric extension of the local *internal* Poincaré group. It becomes an extension of Einstein gravity by introducing another requirement that the vierbein field plays the role of "converter" between spacetime and internal.

Although supergravity has been intensively investigated by many physicists as a possible candidate for the unified theory, I believe that it cannot be qualified to be a fundamental theory because, in my opinion, it has the following *intrinsic* defects:

(*i*) Since supergravity is a gauge theory, all indices in $\{Q, \bar{Q}\} = P$ must be internal ones. Its global version is non-existent[5] because there are no global internal translations; that is, the genuine translation generators are not encountered in the superalgebra.

(*ii*) As emphasized above, it is unreasonable to regard the local Poincaré invariance as the fundamental symmetry.

(*iii*) Perhaps it is impossible to find a globally supersymmetric gauge fixing in supergravity. Indeed, if one did, then by quantization one could construct generators of non-existent global internal translations.

(*iv*) It is impossible to extend the manifestly covariant canonical formalism of quantum gravity to supergravity in such a way that the beautiful properties of the former are not destroyed.

Finally, I briefly make a comment on higher-dimensional theories such as the *Kaluza-Klein theory*, *Kaluza-Klein supergravity*, *superstring theory*, etc., since they are quite fashionable recently. In the theory formulated in D (>4) dimensions, one separates D-4 spatial dimensions from the four-dimensional spacetime by hand. More precisely, one assumes, without any reasonable argument concerning its plausibility, that the D-4 dimensions are *spontaneously compactified* into a nonsensically extremely small size in all directions in such a way that the compactified space is orthogonal to the four-dimensional spacetime everywhere and has uniformly the same structure everywhere. Since such an assumption is much too artificial, I

[5] The spacetime version of $\{Q, \bar{Q}\} = P$ may be recovered in the sense of the expectation value in the physical subspace [6], but such a result is not very significant because what is essential is the algebraic nature of generators.

cannot believe that higher-dimensional theories may have any physical reality, unless an entirely new idea is invented to resolve their artificiality.

The superstring theory seems to be free of divergence difficulty in each order of loop expansion in spite of its high dimensionality. The usual ultraviolet divergences inherent to quantum field theory are of exponential type in string theories, and are removed by analytic continuation without ambiguity. Although the unitarity is not spoiled, this procedure destroys the measure property (i.e., complete additivity) which probability density should have. Accordingly, string theories are unlikely to be probabilistically interpretable.

The superstring theory is quite a wonderful theory as long as it is not regarded as the physics of the real world.

4. New Local Supersymmetry

I have recently proposed a new supersymmetric extension of the vierbein formalism of Einstein gravity so as to be free of the difficulties stated in Section 3 for supergravity [10]. I do not intend to claim that it may, in its present form, become a candidate for the unifield theory. My aim is to show that even in the conservative framework there still exists a natural extension of the well established theory, without using such an artificial assumption as spontaneous compactification.

My basic observation is the fact that general coordinate transformations *commute* with local Lorentz transformations. I believe, therefore, that it is more natural to consider a supersymmetric extension of the local Lorentz invariance *alone*, rather than to extend the local Poincaré invariance. In the case of Poincaré symmetry, one has $\{Q, \bar{Q}\} = P$, but one cannot set $\{Q, Q\} = M$ because $[P, Q] = 0$ but $[P, M] = P \neq 0$ [4]. If P is not considered, however, it is possible to set $\{Q, Q\} = M$. This is the characteristic anticommutation relation of the new local supersymmetry.

In the spinor notation, Lorentz generators M_{ab} $(= -M_{ba})$ can be decomposed into M_{AB} $(= M_{BA})$ and $\bar{M}_{\dot{A}\dot{B}}$ $(= \bar{M}_{\dot{B}\dot{A}})$, where spinor indices take the values 1 and 2. Since dotted quantities are simply conjugates of undotted ones, I write undotted quantities only hereafter.

The commutation relation for Lorentz generators is known to be

$$[M_{AB}, M_{CD}] = -i(\varepsilon_{AC}M_{BD} + \varepsilon_{BC}M_{AD} + \varepsilon_{AD}M_{BC} + \varepsilon_{BD}M_{AC}), \qquad (1)$$

where ε_{AB} $(= -\varepsilon_{BA},\ \varepsilon_{12} = \varepsilon^{12} = 1)$ is the spinor-space metric.

I now introduce supersymmetry generators Q^l_A $(l=1,2,\ldots,N)$ and accompanied-symmetry ones X^{lm} $(=-X^{ml})$, and propose the following superalgebra:

$$[M_{AB}, Q^l_C]=-i(\varepsilon_{AC}Q^l_B+\varepsilon_{BC}Q^l_A), \tag{2}$$

$$[M_{AB}, X^{lm}]=0, \tag{3}$$

$$\{Q^l_A, Q^m_B\}=\delta^{lm}M_{AB}+\varepsilon_{AB}X^{lm}, \tag{4}$$

$$[Q^l_A, X^{mn}]=-i(\delta^{lm}Q^n_A-\delta^{ln}Q^m_A), \tag{5}$$

$$[X^{lm}, X^{np}]=-i(\delta^{mn}X^{lp}-\delta^{ln}X^{mp}+\delta^{lp}X^{mn}-\delta^{mp}X^{ln}). \tag{6}$$

From (6), one sees that the accompanied-symmetry algebra is $\mathfrak{so}(N)$;[6] other Lie algebras[7] are incompatible except for $N=4$,[8] provided that all components of X^{lm} $(=-X^{ml})$ are linearly independent.[9]

The vierbein field $h_{\mu a}$ is rewritten as $h_{\mu\dot{B}C}$ in the spinor notation. It is found that the supertransformation of $h_{\mu\dot{B}C}$ is nonlinear and is given by

$$[Q^l_A, h_{\mu\dot{B}C}]=i(\xi^l_C h_{\mu\dot{B}A}-\tfrac{1}{2}\xi^l_A h_{\mu\dot{B}C}), \tag{7}$$

where ξ^l_A is a new Fermionic spinor field. The Jacobi identities are satisfied if and only if

$$\{Q^l_A, \xi^m_B\}=-\delta^{lm}\varepsilon_{AB}-i\xi^m_A\xi^l_B. \tag{8}$$

The metric field

$$g_{\mu\nu}=\tfrac{1}{2}h_\mu^{\ \dot{B}C}h_{\nu\dot{B}C} \tag{9}$$

is then *super-invariant*, i.e., $g_{\mu\nu}$ commutes not only with M_{AB} and X^{lm} but also with Q^l_A.

A supermultiplet of fields $\{F_{BC}\ (=F_{CB}), \Psi^m_B, K^{mn}\ (=-K^{nm})\}$, constitutes an *adjoint representation* if its members transform like $\{M_{BC}, Q^m_B, X^{mn}\}$, that is, if

$$[Q^l_A, F_{BC}]=-i(\varepsilon_{AB}\Psi^l_C+\varepsilon_{AC}\Psi^l_B), \tag{10}$$

$$\{Q^l_A, \Psi^m_B\}=\delta^{lm}F_{AB}+\varepsilon_{AB}K^{lm}, \tag{11}$$

$$[Q^l_A, K^{mn}]=-i(\delta^{lm}\Psi^n_A-\delta^{ln}\Psi^m_A). \tag{12}$$

[6] Precisely speaking, it should be written as $\mathfrak{so}(N, \mathbf{C})$ because one must take account of $\bar{X}^{\dot{l}\dot{m}}$. The superalgebra defined by (1)-(6) is called $\mathfrak{osp}(N, 2)$.

[7] $\mathfrak{so}(n, N-n)$ can be reduced to $\mathfrak{so}(N)$ by transforming generators.

[8] For $N=4$, $\mathfrak{so}(3)\times[\mathfrak{so}(2)]^3$ is possible [5].

[9] If one admits that X^{lm}'s are linearly dependent, it is possible to have $\mathfrak{gl}(N/2)$ for N even, for instance. Mathematically, all simple superalgebras are known.

For two supermultiplets $\{F_{BC}, \Psi_B^m, K^{mn}\}$ and $\{F'_{BC}, \Psi'^m_B, K'^{mn}\}$, of adjoint representations, their super-invariant inner product is given by

$$F'^{BC}F_{BC} + 2i\Psi'^{Bm}\Psi_B^m - K'^{mn}K^{mn}. \tag{13}$$

There exist the following three special types of adjoint representations:

$$\{f_{BC}\ (=f_{CB}),\ \xi^{Cm}f_{BC},\ -i\xi^{Bm}\xi^{Cn}f_{BC}\}, \tag{14}$$

$$\{i(\xi_B^m\varphi_C^m + \xi_C^m\varphi_B^m),\ \varphi_B^m - i\xi_B^n\xi^{Cm}\varphi_C^n,\ -i(\xi^{Bm}\varphi_B^n - \xi^{Bn}\varphi_B^m)\}, \tag{15}$$

$$\{i\xi_B^m\xi_C^n k^{mn},\ \xi_B^n k^{mn},\ k^{mn}\ (=-k^{nm})\}, \tag{16}$$

where

$$[Q_A^l, f_{BC}] = -i(\varepsilon_{AB}\xi^{Dl}f_{CD} + \varepsilon_{AC}\xi^{Dl}f_{BD}), \tag{17}$$

$$\{Q_A^l, \varphi_B^m\} = -i\varepsilon_{AB}\xi^{Cl}\varphi_C^m + i\delta^{lm}\xi_A^n\varphi_B^n, \tag{18}$$

$$[Q_A^l, k^{mn}] = -i(\delta^{lm}\xi_A^p k^{np} - \delta^{ln}\xi_A^p k^{mp}). \tag{19}$$

Abe and I [2] have shown that (14)-(16) are mutually orthogonal with respect to the inner product (13) and that any adjoint representation satisfying (10)-(12) is decomposed into a sum of them.

In conclusion, I briefly mention the construction of the super-invariant action (done in collaboration with Abe).

First, since $g_{\mu\nu}$ is super-invariant, the Einstein–Hilbert action of Einstein gravity is locally super-invariant without modification, in contrast to supergravity in which the affine connection acquires torsion.

Second, the Dirac-field action can by itself be extended to a locally super-invariant one [1]. The equality between Boson and Fermion degrees of freedom is not required because of the absence of $\{Q, \bar{Q}\} = P$.

Third, the gauge-fixing plus FP-ghost part of the action can be satisfactorily constructed in a globally super-invariant and BRS-invariant manner [2, 3]. Corresponding to the three types (14)-(16), the gauge-fixing part consists of three terms, each of which is formally the super-invariant inner product of a supermultiplet of connections and that of the covariant derivatives of B-fields. If the field ξ_B^m is set equal to zero, the gauge-fixing part reduces to that of the quantum theory of the vierbein formalism.

Thus it is now possible to extend the manifestly covariant canonical formalism of quantum gravity to a supersymmetric theory in quite a natural way.

References

[1] M. Abe and N. Nakanishi, New local supersymmetry of the vierbein formalism and the Dirac theory, *Prog. Theor. Phys.* **78** (1987) 704–718.

[2] M. Abe and N. Nakanishi, Complete gauge fixing in the new local supersymmetry of the vierbein formalism of Einstein gravity, *Prog. Theor. Phys.* **79** (1988) 227–239.

[3] M. Abe and N. Nakanishi, BRS-invariant Lagrangian density in the new local supersymmetry of the vierbein formalism of Einstein gravity, *Prog. Theor. Phys.* **79** (1988) 240–249.

[4] R. Haag, J. T. Łopouszański, and M. Sohnius, All possible generators of supersymmetries of the S-matrix, *Nucl. Phys.* **B88** (1975) 257–274.

[5] Y. Kaminaga, On the internal symmetry in the new supersymmetry proposed by Nakanishi, unpublished.

[6] T. Kugo, T. Kuramoto, and S. Uehara, How does anticommutator $\{Q_\alpha, \bar{Q}_\beta\}$ realize energy-momentum P_μ in quantum supergravity?, *Z. Phys.* **C19** (1983) 241–250.

[7] T. Kugo and I. Ojima, Local covariant operator formalism of non-Abelian gauge theories and quark confinement problem, *Prog. Theor. Phys.* Suppl. **66** (1979) 1–130.

[8] N. Nakanishi, Indefinite-metric quantum field theory of general relativity. XIV—Sixteen-dimensional Noether supercurrents and general linear invariance, *Prog. Theor. Phys.* **66** (1981) 1843–1857.

[9] N. Nakanishi, Manifestly covariant canonical formalism of quantum gravity—Systematic presentation of the theory, *Publications RIMS* (*Kyoto Univ.*) **19** (1983) 1095–1137.

[10] N. Nakanishi, Local supersymmetry different from supergravity, *Prog. Theor. Phys.* **77** (1987) 1533–1541.

[11] N. Nakanishi and I. Ojima, *Covariant Operator Formalism of Gauge Theories and Quantum Gravity*, World Scientific, Singapore, to appear.

[12] J. Wess and J. Bagger, *Supersymmetry and Supergravity*, Princeton University Press, Princeton, New Jersey, 1983.

$K3$ Surfaces Related to Root Systems in E_8

Isao Naruki
Research Institute for Mathematical Sciences
Kyoto University
Kyoto, Japan

Introduction

In this note we construct a universal family of rational elliptic surfaces from which we also obtain various families of $K3$ surfaces by restricting it to special sections of the base space and forming the double coverings of surfaces branched over suitable singular fibers.

For each of these families of $K3$ surfaces, the dimension of the parameter space is equal to 20 minus the Picard number of the generic member, which is exactly the dimension of the target domain of the period mapping; the monodromy, represented on the module of transcendental cycles, is of great significance since it almost determines the mapping. Moreover the families are so explicit that one can in principle compute the monodromy by the standard method of Zariski and Van Kampen. The parameter space of the universal family is the maximal torus of the exceptional simple group E_8 which can also be described as the partial compactification of the moduli space of marked Del Pezzo surfaces of degree one with a distinguished

Algebraic Analysis, Volume II

ISBN 0-12-400466-0

nodal anticanonical divisor. The idea of such moduli spaces of Del Pezzo surfaces is due to Looijenga [L] and Demazure–Pinkham [DP]. We only needed to restrict ourselves to this special type of anticanonical divisor for the sake of the rationality of the parameter space which makes the explicit computation possible. Moreover dealing with rational elliptic surfaces is advantageous in the sense that one can then apply the theory of elliptic surfaces due to Kodaira [K] to them. At the present state of the theory, it is still too difficult to answer the monodromy question for the whole family, so we might have to restrict ourselves to lower-dimensional sections of the parameter space. But, to show how fruitful this point of view is for the algebro-geometric construction of quotients of domains of Type IV, we will mention an example in which some interesting Hilbert modular groups appear.

1. Del Pezzo Surfaces

A smooth complex surface S is called a Del Pezzo surface if the anticanonical bundle $-K$ is ample. K^2 is then between 1 and 9 and it is called the degree of the surface. The orthogonal complement $K^\perp = \{C \in H_2(S, \mathbf{Z}) \cong H^2(S, \mathbf{Z});\ C \cdot K = 0\}$ is a negative-definite even integral lattice of rank $9 - K^2$ with determinant K^2; $K^\perp$ is isometric to the root lattice, denoted by Γ, of E_8, E_7, E_6, D_5, A_4, $A_2 + A_1$ according as $K^2 = 1, 2, 3, 4, 5, 6$ [DP]. A marking α of S is an isometry of Γ onto $K^\perp$, and the pair (S, α) is then called a marked Del Pezzo surface. Now suppose that D is an anticanonical divisor of S, and for the moment that D is nonsingular. By the restriction to D, the marking α gives rise to a homomorphism of Γ to the connected component $\mathrm{Pic}^0(D)$ of the Picard group ($\cong D$) of D, thus assigning an element of $\mathrm{Hom}(\Gamma, E)$ ($E := \mathrm{Pic}^0(D)$) to the triplets (S, α, D). By this assignment, the (compactified) moduli space of isomorphism classes of (S, α, D) such that $\mathrm{Pic}^0(D)$ is isomorphic to a fixed one-dimensional abelian variety E is described to be $\mathrm{Hom}(\Gamma, E)$ (Demazure–Pinkham [DP], Looijenga [L]).

But the theory is naturally generalized without any essential change to the case where E is replaced by the algebraic group $\mathbf{C}$ or $\mathbf{C}^* := \mathbf{C} - \{0\}$, in which case the anticanonical divisor D should be a cuspidal or nodal rational curve. In the next section we will confine ourselves to the case where $K^2 = 1$, $\Gamma = \Gamma(E_8)$, mainly to the explicit construction of the total deformation space of Del Pezzo surfaces over the torus $\mathrm{Hom}(\Gamma(E_8), \mathbf{C}^*)$.

(The corresponding construction over the algebra $\mathrm{Hom}(\Gamma(E_k), \mathbf{C})$ ($k = 8, 7, 6$) is already given by the author in the appendix to Saito [Sa].)

2. Construction of Universal Family

We begin with the remark that, for a Del Pezzo surface of degree one, the anticanonical linear system $|-K|$ is one-dimensional and has exactly one base point, and that the blowing up of the point, which is nothing other than the fibration $|-K| \to P_1(\mathbf{C})$, is a rational elliptic surface. Therefore $\mathrm{Hom}(\Gamma(E_8), \mathbf{C}^*)$ is interpreted to be a reasonable moduli space of rational elliptic surfaces, and what we actually construct here is a deformation of rational elliptic surfaces over it. This viewpoint is naturally of advantage since, by it, we are able to use the theory of elliptic surfaces (Kodaira [K_1], [K_2]).

As is proved in Naruki [N_2], any rational elliptic surface can be blown down to the complex projective plane $P_2(\mathbf{C})$. Any fiber is then mapped onto a cubic, and in particular any nodal anticanonical divisor to a nodal cubic. (Note that the nodal anticanonical divisors are exactly the singular fibers of Type I_1.) So, to construct the desired family, we first fix a nodal cubic:

$$C_\infty\colon f_\infty(x, y, z) = 0$$

$$f_\infty(x, y, z) := xyz + x^3 + y^3.$$

We also distinguish an inflexion point on C_∞:

$$p_0 := (1, -1, 0).$$

As an elliptic curve, the regular locus $C^*_\infty := C_\infty - \{\text{the node: } (0, 0, 1)\}$ should be given the natural group structure isomorphic to $\mathbf{C}^*$ in such a way that p_0 is to be the identity and that, for the three intersection points of C^*_∞ and a generic line, their product is equal to the identity. Such an isomorphism is unique up to the projective automorphisms of C_∞. It is for example given to be the mapping

$$\mathbf{C}^* \ni t \to (t^2, -t, t^3 - 1) \in C^*_\infty. \tag{2.1}$$

For later use we denote this mapping of $\mathbf{C}$: t to $\mathbf{C}^3$: (x, y, z) by φ:

$$\varphi(t) := (t^2, -t, t^3 - 1). \tag{2.2}$$

Now the following proposition is fundamental for our construction:

Proposition 2.1. *Let $V(k)$ be the vector space of homogeneous polynomials in the variables x, y, z of degree k and $W(l)$ the space of polynomials in the single variable t of degree at most l. We further denote by $\check{W}(3m)$ the subspace consisting of such elements $a_0t^{3m}+a_1t^{3m-1}+\cdots+a_{3m}$ that we have*

$$a_0+(-1)^{m+1}a_{3m}=0. \tag{2.3}$$

Then, for the mappings φ^* induced by φ, we have the following exact sequences:

$$0 \longrightarrow V(1) \xrightarrow{\varphi^*} \check{W}(3) \longrightarrow 0$$

$$0 \longrightarrow V(2) \xrightarrow{\varphi^*} \check{W}(6) \longrightarrow 0$$

$$0 \longrightarrow \mathbf{C}f_\infty \longrightarrow V(3) \xrightarrow{\varphi^*} \check{W}(9) \longrightarrow 0.$$

Now, by the mapping (2.1), we identify C_∞^* with the group $\mathbf{C}^*$, so that in particular a point $(t)=(t_1, t_2, \ldots, t_9)$ of the product group $(\mathbf{C}^*)^9$ is regarded as an ordered set of nine points on the curve C_∞^*. The following corollary answers the question: when are the nine points $t_1, t_2, \ldots, t_9$ the nine intersection points of the curve C_∞ and another cubic in $P_2(\mathbf{C})$?

Corollary 2.2. *With the multiplicity respected, the nine points are the intersection of C_∞ and a cubic if and only if the following condition is fulfilled:*

$$t_1t_2\cdots t_9=1.$$

As a sextuplet (resp. *triplet of*) *points on C_∞^*, $t_1, t_2, \ldots, t_6$* (resp. t_1, t_2, t_3) $\in$ $\mathbf{C}^*$ *are on a conic* (resp. *line*) *if and only if*

$$t_1t_2\cdots t_6=1 \qquad (\text{resp. } t_1t_2t_3=1).$$

By Proposition 2.1 we also see that the conic or the line is unique for the last two cases in the corollary. For the first case, however, the cubic is not unique, but its ambiguity stays exactly within the pencil generated by C_∞ and it. Our idea is to assign this pencil to the parameter $\tau=(t_1, t_2, \ldots, t_9)$ for which the above condition is satisfied, the parameter space is now the algebraic torus

$$T:=\left\{(t_1, t_2, \ldots, t_9)\in(\mathbf{C}^*)^9;\ \prod_{i=1}^{9} t_i=1\right\}.$$

For $\tau=(t_1, t_2, \ldots, t_9)$ we set

$$g_\tau(t):=\prod_{i=1}^{9}(t-t_i),$$

and we thus obtain the mapping

$$T \ni \tau \to g_\tau \in \check{W}(9),$$

whose image is the subset consisting of monic elements of $\check{W}(9)$ for which the condition (2.3) is satisfied. Note that this vector space is spanned by t^9-1, t^i $(i=1,2,\ldots,8)$. We also introduce the linear section σ of $\varphi^*\colon V(3) \to \check{W}(9)$ by setting

$$\sigma(t^9-1) := z(z^2-3xy)$$

$$\sigma(t) := -yz^2 - x^2z + xy^2 \qquad \sigma(t^8) := xz^2 + y^2z - x^2y$$

$$\sigma(t^2) := -y(x^2+yz) \qquad \sigma(t^7) := x(y^2+xz)$$

$$\sigma(t^3) := -y^3 \qquad \sigma(t^6) := x^3$$

$$\sigma(t^4) := xy^2 \qquad \sigma(t^5) := -x^2y.$$

Combining the two mappings, we obtain

$$T \ni \tau \to \sigma(g_\tau) \in V(3).$$

To express that $\sigma(g_\tau)$ is a polynomial in x, y, z we rather write f_τ for it:

$$f_\tau(x, y, z) = \sigma(g_\tau)(x, y, z).$$

The curve $f_\tau = 0$ is a cubic which satisfies the condition of Corollary 2.2. Now the pencil generated by C_∞ and this is the one-parameter family of elliptic curves

$$C_{\tau,s}\colon sf_\infty + f_\tau = 0 \qquad (s \in P_1(\mathbf{C}) - \{\infty\})$$

$$C_{\tau,\infty} = C_\infty\colon f_\infty = 0.$$

The disjoint union of $C_{\tau,s}$ $(s \in P_1(\mathbf{C}))$ is naturally imbedded into $P_1(\mathbf{C}) \times P_2(\mathbf{C})\colon (s; x, y, z)$ as a hypersurface (= surface in this case) and we denote this algebraic surface by P_τ:

$$P_\tau := \coprod_{s \in P_1(\mathbf{C})} C_{\tau,s}.$$

We call P_τ the pencil *associated with* (or simply *over*) the parameter $\tau \in T$. Obviously we obtain the two natural projections $P_\tau \to P_1(\mathbf{C})$ and $P_\tau \to P_2(\mathbf{C})$. The first one gives the structure of elliptic surface, when P_τ is minimally desingularized. The second one allows us to regard P_τ as a blowing up of $P_2(\mathbf{C})$. In fact, outside the nine points $t_1, t_2, \ldots, t_9$ $(\tau = (t_1, t_2, \ldots, t_9))$ every point of $P_2(\mathbf{C})$ is on exactly one member of the pencil, and over each t_i

there lies the locus $\{(s, t_i)\}_{s \in P_1(\mathbf{C})}$ as a global section. More precisely, P_τ is the blowing up of $P_2(\mathbf{C})$ whose center is the intersection ideal of the curves C_∞ and $C_{\tau,0}$. By this we also see that, over a k-fold intersection point of the curves, there lies exactly one rational double point of type A_{k-1}, and that all singular points of P_τ are of this kind. (See for this Naruki [N_2].) In particular the pencil is nonsingular if and only if $t_1, t_2, \ldots, t_9$ are distinct. Now the important question arises: Can the family P_τ be simultaneously resolved for $\tau \in T$? In other words this is to ask whether one can make the disjoint union $\coprod_{\tau \in T} \hat{P}_\tau$ an algebraic variety in a natural way. We have here denoted the minimal desingularization of P_τ by $\hat{P}_\tau$. The answer will be given in the next section.

3. Simultaneous Resolution and Period Mapping

We restrict the family P_τ of rational surface to the following open and dense subset of the parameter space:

$$T^0 := \{(t_1, t_2, \ldots, t_9) \in T;\ t_1, t_2, \ldots, t_9 \text{ are distinct}\}.$$

As is remarked in the previous section, we have $\hat{P}_\tau = P_\tau$ for $\tau \in T^0$, and the mapping $\coprod_{\tau \in T^0} P_\tau \to T^0$ is a smooth fiber bundle. Now we want to show that the homological monodromy associated with this fibration is trivial. But this is almost evident: Note first that we have the uniform blowing down $P_\tau \to P_2(\mathbf{C})$ for $\tau \in T^0$. This implies that the following classes are sections of the locally constant sheaf $H_2(P_\tau, \mathbf{Z})$ $\tau \in T^0$: L denotes the total transform of the class of lines and $[t_i]$ denotes the homology class of the global section over the point t_i (recall that $t_1, t_2, \ldots, t_9$ are regarded as points on $P_2(\mathbf{C})$ by $\mathbf{C}^* \simeq C_\infty^* \hookrightarrow P_2(\mathbf{C})$). It is also obvious that these classes form a base for the second homology group:

$$H_2(P_\tau, \mathbf{Z}) = \mathbf{Z}L \oplus \sum_{i=1}^{9} \mathbf{Z}[t_i].$$

This triviality of the monodromy implies the simultaneous resolution of the family $\coprod_{\tau \in T} P_t$ (Brieskorn [B]). Thus the disjoint union $\coprod_{\tau \in T} \hat{P}_\tau$ has the natural structure of an algebraic manifold and it is a smooth fibration over the whole parameter space T. We have also proved that the locally constant sheaf $H_2(\hat{P}_\tau, \mathbf{Z})$ $\tau \in T$ is actually constant and that the sections L, $[t_i]$ $i = 1, 2, \ldots, 9$ defined only over T^0 now extend uniquely to T so that they form a base for each stalk of the sheaf.

To obtain the right family of surfaces over the maximal torus $T(E_8)=\mathrm{Hom}(\Gamma(E_8), \mathbf{C}^*)$, we still have to explain what is the period mapping of T to $T(E_8)$. As is proved in [N₂], for $\hat{P}_\tau$ $(\tau \in T)$, the anticanonical class $-K$ is equal to the class F of fibers; we have

$$F=-K=3L-\sum_{i=1}^{9}[t_i]$$

$$F^\perp = K^\perp \qquad K^2=F^2=0.$$

In [N_2], it is also proved that the intersection form restricts itself onto $F^\perp/\mathbf{Z}F$ and that this lattice is isometric to $\Gamma(E_8)$. We set

$$\tilde{r}(i,j) := [t_i]-[t_j]$$

$$\tilde{r}(i,j,k) := L-[t_i]-[t_j]-[t_k],$$

and denote their classes in $F^\perp/\mathbf{Z}F$ by $r(i,j)$ and $r(i,j,k)$, deleting the tilde. Since these are all of self-intersection number -2, we see by enumeration that we have, up to sign, obtained all the roots of E_8. In particular we can choose an extended Dynkin diagram from them:

$$\begin{array}{c} r(1,2)\text{—}r(2,3)\text{—}r(3,4)\text{—}r(4,5)\text{—}\cdots\text{—}r(8,9) \\ \qquad\qquad\quad | \\ \qquad\qquad\quad r(1,2,3). \end{array}$$

Now, assume again that $\tau=(t_1, t_2, \ldots, t_9) \in T^0$, and observe the six oriented intersection points of the two divisors C_∞ and $L-[t_i]-[t_j]-[t_k]$ on the surface P_τ, where L is regarded as the transform of a generic line by $P_\tau \to P_2(\mathbf{C})$ and $[t_i]$, $[t_j]$, $[t_k]$ the global sections corresponding to t_i, t_j, t_k. With the minus sign, the points t_i, t_j, t_k are by definition among them; the other positive intersection points are exactly the proper transform of the three intersection points of the plane curve C_∞ and the generic line, which we can regard as three points, denoted by s_1, s_2 and s_3, on the group $\mathbf{C}^*$ via $\mathbf{C}^* \simeq C_\infty^*$. Thus the homology class $r(i,j,k)$ of P_τ cuts on the group curve $C_\infty^* \cong \mathbf{C}^*$ the divisor $(s_1)+(s_2)+(s_3)-(t_i)-(t_j)-(t_k)$ whose image in $\mathrm{Pic}^0(C_\infty^* \cong \mathbf{C}^*$ is exactly $s_1s_2s_3/t_it_jt_k$. But we know by Corollary 2.2 that $s_1s_2s_3=1$. We have thus assigned $(t_it_jt_k)^{-1} \in \mathbf{C}$ to the surface P_τ $\tau=(t_1, t_2, \ldots, t_9)$ as the period associated with the root $r(i,j,k)$ of E_8. In a similar way we also see that the ratio t_i/t_j is assigned to P_τ as the period associated with $r(i,j)$. To sum up, we have defined, by the additivity, an element of $\mathrm{Hom}(\Gamma(E_8), \mathbf{C}^*) = T(E_8)$ for the parameter τ defining the surface P_τ. We denote this element of the maximal torus by $\int(\tau)$. It is now obvious that the mapping $\tau \mapsto \int(\tau)$ is a surjective homomorphism of the groups, and that the kernel consists of diagonal elements $(\omega^\nu, \omega^\nu, \ldots, \omega^\nu)$ $\nu = 0, 1, 2$, $\omega = \exp(2\pi i/3)$, which we identify with the group $\{1, \omega, \omega^2\}$. We have thus the exact sequence

$$1 \to \{1, \omega, \omega^2\} \to T \xrightarrow{\int} T(E_8) \to 1.$$

To obtain the family of surfaces over $T(E_8)$ we need only divide the family $\coprod_{\tau \in T} \hat{P}_\tau \to T$ by the action of the cyclic group $\{1, \omega, \omega^2\}$. But we have already chosen the section $\sigma: \check{W}(9) \to V(3)$ so carefully that we have

$$f_{\omega \cdot \tau}(\omega \cdot (x, y, z)) = f_\tau(x, y, z),$$

where we let $\{1, \omega, \omega^2\}$ act on T and $\mathbf{C}^3: (x, y, z)$ by the law: $\omega \cdot (t_1, t_2, \ldots, t_9) = (\omega t_1, \omega t_2, \ldots, \omega t_9)$ $\omega \cdot (x, y, z) = (\omega^2 z, \omega y, z)$. Note that the last action leaves the polynomial $f_\infty(x, y, z) = xyz + x^3 + y^3$ and the curve $C_\infty: f_\infty = 0$ invariant. Thus the group $\{1, \omega, \omega^2\}$ acts (freely) on the family $\coprod_{\tau \in T} \hat{P}_\tau \to T$ and the quotient of the action is the desired fibration over the maximal torus $T(E_8)$.

4. Section-Module, Fiber-Module and Main Property of the Family

We devote this section to a little general discussion about elliptic surfaces. Let S be a smooth elliptic surface which has at least one global section. We fix a global section of S and denote it by s_0. Then the set of global sections, denoted here by $\mathcal{S}$, receives the unique abelian group structure for which s_0 is the zero and which is compatible with the group structure of fibers. We call $\mathcal{S}$ the *section-module* of S (with the reference section s_0). Now we also want to introduce another important module which is in a sense complementary to $\mathcal{S}$. Let F_i $i = 1, 2, \ldots, m$ be the singular fibers of S and F_{ij} $j = 1, 2, \ldots, n_i$ be the irreducible components of F_i. By $[F_{ij}]$ we denote the class of the homology cycle F_{ij} in the quotient module $F^\perp/\mathbf{Z}F$ where F is the class of fibers as before and $F^\perp$ is the orthogonal complement of $\mathbf{Z}F$ in the Neron-Severi group $\mathrm{NS}(S)$ of S (the group of algebraic equivalence classes of divisors). We set

$$\mathcal{F}_i := \sum_{j=1}^{n_i} \mathbf{Z}[F_{ij}]$$

$$\mathcal{F} := \sum_{i=1}^{m} \mathcal{F}_i.$$

By the classification of Kodaira [K_1] each lattice $\mathscr{F}_i$ is isometric to the root lattice of one of the simple groups A_k $(k \geq 0)$, D_k $(k \geq 4)$, E_k $(k = 6, 7, 8)$. (For A_0 the corresponding lattice should be interpreted to be zero.) In particular $\mathscr{F}$ is a negative-definite lattice as the direct sum of mutually orthogonal $\mathscr{F}_i$ $(i = 1, 2, \ldots, m)$. We call $\mathscr{F}$ the *fiber-module* of S since it almost determines the types of singular fibers; it does not distinguish only between I and II (which correspond to A_0), between I_2 and III (which correspond to A_1), between I_3 and IV (which correspond to A_2). We note now that, for any s_1, s_2 of $\mathscr{S}$, the homology class $[s_2] - [s_1]$ is orthogonal to F, and that we can consider its class $[[s_2] - [s_1]]$ in any quotient module of $F^\perp/\mathbf{Z}F$. The following is fundamental in the study of elliptic surfaces:

Proposition 4.1 (Shioda [Sh]). *With the notation as above, the mapping*

$$\mathscr{S} \ni s \mapsto [[s] - [s_0]] \in (F^\perp/\mathbf{Z}F)/\mathscr{F}$$

is a bijective homomorphism of abelian groups. The Picard number $\rho(S)$, the rank of $NS(S)$, is given by

$$\rho(S) = \operatorname{rank}_{\mathbf{Z}} \mathscr{S} + \operatorname{rank}_{\mathbf{Z}} \mathscr{F} + 2.$$

The second assertion follows from the first by the fact that the lattice $F^\perp/\mathbf{Z}F$ is isometric to the orthogonal complement in NS(S) of the unimodular lattice generated by F and $[s_0]$ $(F^2 = 0,\ F \cdot [s_0] = 1)$.

Remark. By Proposition 4.1 we see that the orthogonal complement in $F^\perp/\mathbf{Z}F$ of the fiber-module $\mathscr{F}$ is naturally imbedded into the section-module $\mathscr{S}$ and that the index of the image is finite. By using the classification of singular fibers we can further prove that the image is exactly the submodule

$$\mathscr{S}' := \{s \in \mathscr{S};\ s \text{ hits the same irreducible component at each } F_i \text{ as } s_0\}.$$

We can moreover see that the addition in $\mathscr{S}'$ coincides with the addition in homology classes, i.e., we have

$$[[s_1 + s_2] - [s_0]] = [[s_1] - [s_0]] + [[s_2] - [s_0]] \qquad s_1, s_2 \in \mathscr{S}'. \tag{4.1}$$

We call $\mathscr{S}'$ the *module of special sections.*

Now we are able to formulate the fundamental property of the family $\coprod_{\tau \in T} \hat{P}_\tau$. We denote the fiber-module of $\hat{P}_\tau$ by $\mathscr{F}_\tau$. $\mathscr{F}_\tau$ controls the types of singular fibers. The rank of $\mathscr{F}_\tau$ varies much with the parameter τ; it is anyway a submodule of the constant object $K^\perp/\mathbf{Z}K = F^\perp/\mathbf{Z}K$. The roots

$r(i,j)$ $(1 \le i \ne j \le g)$, $\pm r(i,j,k)$ $(\{i,j,k\} \subseteq \{1,2,\dots,g\})$ are the sections of the constant sheaf $K^\perp/\mathbf{Z}K^\perp$; as the periods, they are also characters on the abelian Lie groups T and $T(E_8)$. We denote by R the set of roots. Now the main property of our family is stated as follows:

Proposition 4.2. *A root $r \in R$ lies in $\mathscr{F}_\tau$ if and only if the value $r(\tau)$ at τ of r is equal to* 1.

The proof is omitted.

5. Monodromy Module

In this section we are concerned with the introduction of a few modules which play important roles in the computation of the monodromy. Under the notation and the assumption of the previous section we observe the larger orthogonal complement

$$F^{\perp(H)} := \{C \in H_2(S,\mathbf{Z});\ C \cdot F = 0\}$$

and the quotient module $F^{\perp(H)}/\mathbf{Z}F$. (For any submodule M of $\mathrm{NS}(S)$, we denote its orthogonal complement in $H_2(S,\mathbf{Z})$ by $M^{\perp(H)}$.) We have the inclusions

$$\mathscr{F} \hookrightarrow \mathscr{F} \oplus \mathscr{S}' \hookrightarrow F^\perp/\mathbf{Z}F \hookrightarrow F^{\perp(H)}/\mathbf{Z}F.$$

By the *monodromy-module* of S we mean the orthogonal complement of $\mathscr{F}$ in the lattice $F^{\perp(H)}/\mathbf{Z}F$:

$$\mathscr{M}(S) := \{C \in F^{\perp(H)}/\mathbf{Z}F;\ C \cdot \mathscr{F} = 0\}.$$

Since $\mathscr{S}'$ is the orthogonal complement of $\mathscr{F}$ in the smaller lattice $F^\perp/\mathbf{Z}F$, we obtain the inclusion

$$\mathscr{S}' \hookrightarrow \mathscr{M}(S).$$

The orthogonal complement of $\mathscr{S}'$ in $\mathscr{M}(S)$ is exactly the orthogonal complement $\mathrm{NS}(S)^{\perp(H)}$ which we call the *module of transcendental cycles.* We denote it by $\mathscr{T}(S)$. Thus, to describe $\mathscr{T}(S)$, it suffices to describe $\mathscr{S}'$, $\mathscr{M}(S)$ and the above inclusion. The reason why $\mathscr{M}(S)$ is called the monodromy-module of S is that it is determined by the monodromy representation $\pi_1(\Delta^*, p_0) \to SL(2,\mathbf{Z})$, where Δ is the base curve of S, Δ^* is the largest subset of Δ over which $S \to \Delta$ is a smooth fibration and $p_0 \in \Delta^*$ is the reference point. We close this section by giving a precise formula for $\mathcal{M}(S)$. Let $\{p_1, p_2, \ldots, p_m\} = \Delta - \Delta^*$ and suppose that $F_i = \pi^{-1}(p_i)$, where π denotes the projection $S \to \Delta$. By a suitable choice of paths we can assume that the group $\pi_1(\Delta^*, p_0)$ is generated by α_j, β_j ($j = 1, 2, \ldots, g$, $g = \text{genus}(\Delta)$), γ_i ($i = 1, 2, \ldots, m$) with the single relation

$$\prod_{j=1}^{g} (\alpha_j \beta_j \alpha_j^{-1} \beta_j^{-1}) \cdot (\gamma_1 \gamma_2 \ldots \gamma_m) = 1,$$

where γ_i is a simple path surrounding once the point p_i. We set $\Gamma := \pi_1(\Delta^*, p_0)$ for brevity and consider the group cohomology $H^1_\rho(\Gamma, \mathbf{Z}^2)$ associated with the monodromy representation $\Gamma \to SL(2, \mathbf{Z}) \cong \text{Aut}(H^1(\pi^{-1}(p_0), \mathbf{Z})$, the intersection form). By $\bar{H}^1_\rho(\Gamma, \mathbf{Z}^2)$ we denote the subgroup of $H^1_\rho(\Gamma, \mathbf{Z}^2)$ which consists of the images of those cocycles c for which $c(\gamma_i) \in (\rho(\gamma_i) - 1) \cdot \mathbf{Z}^2$.

Proposition 5.1 (Naruki [N_1]). *With the notation above we have the natural isomorphism*

$$\mathcal{M}(S) \cong \bar{H}^1_\rho(\Gamma, \mathbf{Z}^2).$$

In this note we treat only the case where $g = 0$, i.e., Δ is $P_1(\mathbf{C})$. We remark here that one can find a geometric version of Proposition 5.1 in [K_2].

6. Double Coverings Branched over Singular Fibers

In this section we study a special class of double coverings by base change for elliptic surfaces over $P_1(\mathbf{C})$, i.e., those which ramify over two singular fibers of Class I. Let S be an elliptic surface over $P_1(\mathbf{C})$ with the projection π and $F_i = \pi^{-1}(p_i)$, $i = 1, 2, \ldots, m$ the singular fibers of S. As in Section 4 we assume that S has at least one global section, so that S is in particular free from multiple singular fibers. Now assume that $m \geq 2$ and let $\rho : P_1(\mathbf{C}) \to P_1(\mathbf{C})$ be the double covering of the base curve branched over the points p_1, p_m. The minimal desingularization of the fiber product $P_1(\mathbf{C}) \underset{P_1(\mathbf{C})}{\times} S$, which has in general singular points of type A_1, is almost an elliptic surface by the projection to the first factor. By blowing down successively the (-1)-curves appearing in the ramified fibers if they exist, we obtain an elliptic surface in the usual sense, which we denote, together with its projection, by $\tilde{\pi}:\tilde{S}\to P_1(\mathbf{C})$. We have the inverse images $\{\tilde{p}_1\}=\rho^{-1}(p_1)$, $\{\tilde{p}_m\}=\rho^{-1}(p_m)$, $\{\tilde{p}'_i,\tilde{p}''_i\}=\rho^{-1}(p_i)$, $i=2,\ldots,m-1$ and the corresponding singular fibers of $\tilde{S}$: $\tilde{F}_1=\tilde{\pi}^{-1}(\tilde{p}_1)$, $\tilde{F}_m=\tilde{\pi}^{-1}(\tilde{p}_m)$, $\tilde{F}'_i=\tilde{\pi}^{-1}(\tilde{p}'_i)$, $\tilde{F}''_i=\tilde{\pi}^{-1}(\tilde{p}''_i)$, $i=2,\ldots,m-1$. For $2\le i\le m-1$, $\tilde{F}'_i$, $\tilde{F}''_i$ are of the same type as F_i. Now, for simplicity, we assume that F_1, F_m are of Types I_a, I_b. This implies that $\tilde{F}_1$, $\tilde{F}_m$ are Types I_{2a}, I_{2b}. (To see this one need only remark that the monodromy matrix of I_{2c} is the square of that for I_c.) Thus we have proved

$$\operatorname{rank}\tilde{\mathscr{F}}_1=2\operatorname{rank}\mathscr{F}_1+1$$

$$\operatorname{rank}\tilde{\mathscr{F}}_m=2\operatorname{rank}\mathscr{F}_m+1$$

$$\operatorname{rank}\tilde{\mathscr{F}}'_i=\operatorname{rank}\tilde{\mathscr{F}}''_i=\operatorname{rank}\mathscr{F}_i \qquad i=2,\ldots,m-1,$$

where $\mathscr{F}_i$; $\tilde{\mathscr{F}}_l$, $\tilde{\mathscr{F}}_m$, $\tilde{\mathscr{F}}'_i$, $\tilde{\mathscr{F}}''_i$ are the direct summands of the fiber modules of S, $\tilde{S}$ corresponding to the fibers F_i, $\tilde{F}_1$, $\tilde{F}_m$, $\tilde{F}'_i$, $\tilde{F}''_i$. From these follows the identity between the ranks of the fiber modules $\mathscr{F}=\sum_{i=1}^m\mathscr{F}_i$, $\tilde{\mathscr{F}}=\tilde{\mathscr{F}}_i\oplus\tilde{\mathscr{F}}_m\oplus\sum_{i=2}^{m-1}(\tilde{\mathscr{F}}'_i\oplus\tilde{\mathscr{F}}''_i)$:

$$\text{rank of }\tilde{\mathscr{F}}=2\operatorname{rank}\mathscr{F}+2. \tag{6.1}$$

Now we want to study the relationship between the section modules of S and $\tilde{S}$. Let us fix a global section of s_0 of $\pi:S\to P_1(\mathbf{C})$ to give an abelian structure to the set $\mathscr{S}$ of global sections of S. By this we also make the set $\tilde{\mathscr{S}}$ of global sections of $\tilde{S}$ an abelian group; the zero of $\tilde{\mathscr{S}}$ is the lifting $\tilde{s}_0$ of s_0. Note that, by the above assumption, we have the unique morphism $\tilde{\rho}:\tilde{S}\to S$ such that $\rho\circ\tilde{\pi}=\pi\circ\tilde{\rho}$ and that $\tilde{s}_0=\tilde{\rho}^{-1}(s_0)$. We also have the covering automorphism on $\tilde{S}$, i.e., the involution σ such that $\tilde{\rho}\circ\sigma=\tilde{\rho}$ and the mapping $\tilde{S}/\sigma\to S$, induced by $\tilde{\rho}$, is a birational morphism. For $s\in\mathscr{S}$ we denote the lifting $\tilde{\rho}^{-1}(s)$ by $\tilde{s}$, and by identifying s with $\tilde{s}$, we regard $\mathscr{S}$ as imbedded into $\tilde{\mathscr{S}}$. We have the obvious characterization of $\mathscr{S}$:

$$\mathscr{S}=\{t\in\tilde{\mathscr{S}};\ \sigma(t)=t\}.$$

We will now prove that $\mathscr{S}$ is of finite index in $\tilde{\mathscr{S}}$, under the assumption that the self-intersection number $s_0\cdot s_0$ is negative. We observe the following three subgroups of $\mathscr{S}$:

$$\tilde{\mathscr{S}}':=\{t\in\tilde{\mathscr{S}};\ t\text{ hits the same irreducible component as }\tilde{s}_0\text{ at each fiber}\}$$

$$\tilde{\mathscr{S}}'(\pm)=\{t\in\tilde{\mathscr{S}}';\ \sigma(t)=\pm t\}.$$

Since we have the inclusions $\tilde{\mathscr{S}}'(+)\subseteq\mathscr{S}$, $2\tilde{\mathscr{S}}'\subseteq\tilde{\mathscr{S}}'(+)+\tilde{\mathscr{S}}'(-)$, and since $\tilde{\mathscr{S}}'$ is of finite index in $\tilde{\mathscr{S}}$, it suffices to show that $\tilde{\mathscr{S}}'(-)=\{0\}$, under the assumption. By Kodaira [K_2], there exists an integer μ such that $K=\mu F$ for S. Then we have $\tilde{K}=2(\mu+1)\tilde{F}$, where $\tilde{K}$, $\tilde{F}$ are the canonical class and the class of fibers of $\tilde{S}$, and from which it follows by the adjunction that $s\cdot s=-(\mu+2)$ for $s\in\mathscr{S}$, and that $t\cdot t=-2(\mu+2)$ for $t\in\tilde{\mathscr{S}}$. On the other hand we have by (4.1) the identity in the homology:

$$[\sigma(t)]+[t]=2[\tilde{s}_0] \qquad t\in\tilde{\mathscr{S}}'(-),$$

since $\sigma(t)+t=0=2\tilde{s}_0$ in the abelian group $\tilde{\mathscr{S}}$. This implies that $2(\sigma(t)\cdot t)-4(\mu+2)=-8(\mu+2)$, i.e., $\sigma(t)\cdot t=2(s_0\cdot s_0)$. Thus we see that, if $s_0\cdot s_0<0$, then $\sigma(t)$ and t could not be different, which implies that $t\in\tilde{\mathscr{S}}'(+)\cap\tilde{\mathscr{S}}'(-)=\{0\}$, which was to be proved.

Proposition 6.1. *Assume that the self-intersection number is negative for the global sections of S. Then we have the identity*

$$\operatorname{rank}\mathscr{S}=\operatorname{rank}\tilde{\mathscr{S}}.$$

In particular, this, together with (6.1), *implies the identity between the Picard numbers of S and* $\tilde{S}$:

$$\rho(\tilde{S})=2\rho(S)-\operatorname{rank}\mathscr{S}.$$

7. Computation of Monodromy

In this section we want to explain what kind of subfamilies of $\coprod_\tau \hat{P}_\tau$ should be studied and how the monodromy can be computed for them and their double covering formations. We first choose a root subsystem $R\subseteq R(E_8)$ ($R(E_8)$: root system of E_8). Then this naturally determines the subgroup

$$G:=\{\tau\in T(E_8);\ r(\tau)=1 \quad (\forall r\in R)\}.$$

(Recall that every root is a character of the group $T(E_8)$ or T.) We also obtain the lifting of G to T by the natural homomorphism $\varphi: T\to T(E_8)$ ($T(E_8)=T/\{1,\omega,\omega^2\}$):

$$G':=\varphi^{-1}(G)=\{\tau\in T;\ r(\tau)=1 \quad (\forall r\in R)\}.$$

The dimension of G and G' is equal to the corank of R in E_8 and we again obtain a root subsystem by setting

$$\bar{R}:=\{r\in R(E_8);\ r(\tau)=1 \quad (\forall\tau\in G)\}.$$

$\bar{R}$ contains R, and is of the same rank; but there often occurs the case where $\bar{R}$ is strictly larger than R, which we however exclude by putting the assumption that $\bar{R} = R$. Then there exists a connected component H of the Lie group G such that the set

$$H(*) := \{\tau \in H;\ R_\tau = R\}$$

is open and dense in H, where we have set $R_\tau := \{r \in R(E_8);\ r(\tau) = 1\}$. We also set $H' := \varphi^{-1}(H)$, $H'(*) := \varphi^{-1}(H(*))$. Now our object of study is the subfamily

$$P_{H'} := \coprod_{\tau \in H'} \hat{P}_\tau \to H'$$

or its quotient by the automorphism group in Section 3,

$$P_H := P_{H'}/\{1, \omega, \omega^2\} \to H,$$

both of which are associated essentially with that root subsystem fixed above. Recall that every member of the universal family is an elliptic surface over $P_1(\mathbf{C})$ whose coordinate s does not depend on the parameter of the family. Thus we obtain the commutative diagram

$$\begin{array}{ccc} P_H & & \\ & \searrow^{\eta} & \\ \pi \downarrow & & P_1(\mathbf{C}) \times H, \\ & \swarrow_{\delta} & \\ H & & \end{array}$$

where η is the elliptic fibration and δ the trivial projection. By Σ we denote the discriminant set for η:

$$\Sigma = \{p \in P_1(\mathbf{C}) \times H;\ \eta^{-1}(p) \text{ is not a smooth elliptic curve}\}.$$

We have to remark here that the restriction of δ to $\Sigma(*) := \delta^{-1}(H(*)) \cap \Sigma$ is in general not an unramified covering of $H(*)$. (In fact there are some types of confluence of singular fibers in deformations of elliptic surfaces which do not affect the rank of the fiber module. See Naruki [N_3] for this.) We can anyway consider the maximal open subset of H over which the map $\Sigma \to H$ is unramified. We denote this subset by $H((*))$ and we set

$$\Sigma((*)) := \delta^{-1}(H((*))) \cap \Sigma.$$

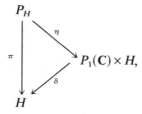

It is obvious that the complement of $H((*))$ is a divisor in H, to which $H - H(*)$ contributes several components in general. Now recall that, for all $\hat{P}_\tau$, there is fixed the singular fiber C_∞ of Type I_1 at the infinity $s=\infty$. Thus we have the commutative diagram

$$\begin{array}{ccc} \Sigma((*))-\{\infty\}\times H((*)) & \hookrightarrow & \mathbf{C}\times H((*)) \\ & {\scriptstyle \delta_\Sigma}\searrow \quad \swarrow{\scriptstyle \delta} & \\ & H((*)), & \end{array}$$

where the left sloped arrow is an unramified covering. If we choose $t_0 \in H((*))$ as the point of reference, then we obtain the braid representation associated with the diagram:

$$h: \pi_1(H((*)), t_0) \to B_k,$$

where k is the sheet number of the covering and B_k stands for the braid group of k strings. Choose $p_0 \in \delta^{-1}(t_0)-\Sigma$ and let $\Gamma_0 = \pi_1(\delta^{-1}(t_0)-\Sigma_{t_0}, p_0)$ and assume that p_0 can be extended to a section of δ which maps into the complement $P_1(\mathbf{C})\times H((*))-\Sigma$. Let further ρ_0 denote the monodromy homomorphism of Γ_0 into $SL(2, \mathbf{Z})$ associated with the elliptic fibration of the surface $\hat{P}_{t_0}$. Since B_k operates on Γ_0 as automorphisms and the action of every image element of h is compatible with ρ_0, we obtain the induced homomorphism

$$h^*: \pi_1(H((*)), t_0) \to \mathrm{Aut}(\bar{H}^1_{\rho_0}(\hat{P}_{t_0}, \mathbf{Z}^2)),$$

which is nothing other than the monodromy representation on the monodromy module. Since $\hat{P}_{t_0}$ is rational, this is exactly the representation over the section module. This itself is however not of great interest since the image group is finite.

To obtain infinite monodromy groups, we will do the double covering formation as is described in Section 6, and for this we have to assume that, besides $\{\infty\}\times H((*))$, we have an extra section of the unramified covering $\delta\,|\,\Sigma((*)) \to H((*))$:

$$\sigma_0: H((*)) \to \Sigma((*)) \qquad \delta\circ\sigma_0 = \mathrm{id}.$$

We can then form the double covering $\tilde{P}_t$ of $\hat{P}_t$ for any $t\in H((*))$ which is branched exactly over the singular fibers $\eta^{-1}(\sigma_0(t))$, $C_\infty = \eta^{-1}((\infty, t))$. If $\eta^{-1}(\sigma_0(t))$ is of Type I_a for fixed $a>0$, then we obtain a family of $K3$ surfaces over $H((*))$:

$$\tilde{\pi}: \tilde{P}_{H,\sigma_0} := \coprod_{t\in H((*))} \tilde{P}_t \to H((*)).$$

It is clear that, for this family of elliptic surfaces, we can also define the homomorphisms

$$\tilde{h}: \pi_1(H((*)), t_0) \to B_{2k-1},$$

$$\tilde{h}^*: \pi_1(H((*)), t_0) \to \mathrm{Aut}(\bar{H}^1_{\tilde{\rho}_0}(\tilde{P}_{t_0}, \mathbf{Z}^2)),$$

where $\tilde{\rho}_0$ is the representation of $\pi_1(\tilde{P}_{t_0}, \tilde{p}_0)$ ($\tilde{p}_0 \in \tilde{P}_{t_0}$) into $SL(2, \mathbf{Z})$ associated with the $K3$ elliptic surface $\tilde{P}_{t_0}$. We have already seen that the section modules $\mathscr{S}'$, $\tilde{\mathscr{S}}'$ for $\hat{P}_{t_0}$, $\tilde{P}_{t_0}$ are submodules of $\bar{H}^1_{\rho_0}(\hat{P}_{t_0}, \mathbf{Z}^2)$, $\bar{H}^1_{\tilde{\rho}_0}(\tilde{P}_{t_0}, \mathbf{Z}^2)$ and that, by the lifting $\bar{H}^1_{\rho_0}(\hat{P}_{t_0}, \mathbf{Z}^2) \to \bar{H}^1_{\tilde{\rho}_0}(\tilde{P}_{t_0}, \mathbf{Z}^2)$, $\mathscr{S}$ is a submodule of $\tilde{\mathscr{S}}$ of finite index. Since $\mathscr{S}' = \bar{H}^1_{\rho_0}(\hat{P}_{t_0}, \mathbf{Z}^2)$ as is remarked above, the desired representation over the module of transcendental cycles coincides, modulo torsion, with that on $H^1_{\tilde{\rho}_0}(\tilde{P}_{t_0}, \mathbf{Z}^2)/H^1_{\rho_0}(\hat{P}_{t_0}, \mathbf{Z}^2)$ which can be calculated from the homomorphisms h^* and $\tilde{h}^*$. Furthermore $\tilde{h}$ is in principle obtained in a functorial way from h and $\tilde{\rho}_0$ from ρ_0. For the determination of h one has to calculate $\pi_1(H((*)), t_0)$ first by observing closely the discriminant set $D = H - H((*))$ in H. Then the homomorphism h is computed by choosing a suitable generic curve C in H and observing how the curve $\delta^{-1}(C) \cap \Sigma$ in the surface $\delta^{-1}(C)$ behaves with the projection $\delta^{-1}(C) \to C$. By this kind of observation and by using [N_3], one can easily determine ρ_0 also.

8. Example

For the test case we choose in this last section the root subsystem $A_3 + 3A_1$ of E_8 generated by the following roots:

$$r_1 := r(2,3) \qquad r_2 := r(3,4) \qquad r_3 := r(1,2,3)$$

$$r_4 := r(5,6) \qquad r_5 := r(7,8) \qquad r_6 := r(1,5,6),$$

where r_1, r_2, r_3 form a Dynkin diagram of A_3 and r_4, r_5, r_6 are mutually orthogonal and orthogonal to r_1, r_2, r_3. By the definition the subgroup G' of the torus $T = \{\tau = (t_1, t_2, \ldots, t_9); \prod_i t_i = 1\}$ associated with this root system is given by the equations $t_2 = t_3 = t_4$, $t_5 = t_6$, $t_7 = t_8$, $t_1 t_2 t_3 = 1$, $t_1 t_5 t_6 = 1$, $\prod_i t_i = 1$. By setting $s := t_2 = t_3 = t_4$, $t := t_5 = t_6$, $u := t_7 = t_8$, we obtain the relations $s^2 = t^2$, $t_9 = (s^3 u^2)^{-2}$. The first equation shows that G' has exactly two connected components G_1', G_2' corresponding to the equalities $s = t$, $s = -t$. If we choose $s = t$, then we have the roots $r(2,5)$, $r(2,6)$, $r(3,5)$, etc., which assume the constant value 1 on G_1' and which are not contained in the root system $A_3 + 3A_1$. Thus we have to choose G_2' as the connected component H' for which the restricted family $P_{H'} = \coprod_{\tau \in H'} \hat{P}_\tau$ is to be investigated in connection with the original root system $A_3 + 3A_1$. As was remarked before, we have to divide H', $P_{H'}$ by the automorphisms $\{1, \omega, \omega^2\}$ to obtain the corresponding subset H in the maximal torus $T(E_8)$ and the family of surfaces P_H over it. It is now evident that we can for example take the characters s^3, s/u as the coordinates of the set H. The discriminant set $H - H((*))$ decomposes into several irreducible algebraic curves, among which there appear some not coming from any roots. Unfortunately, for these curves, not all singular points or intersection points receive real coordinates yet, which might make the computation of π_1 complicated. To remedy this difficulty, we shall further divide the family $P_H \to H$ by suitable automorphisms. Without mentioning any detailed computation we state that one can in fact transfer $P_H \to H$ to a family defined over the affine plane: (x, y) where x and y are introduced by the identities

$$x := \frac{(s^8 u^4 + 1)}{2s^4 u^2},$$

$$y := \frac{(s^2 + u^2)}{2su}.$$

We denote this affine plane by A and the family of elliptic surfaces over A by P_A. The parameter space of elliptic curves for this family is the product $P_1(\mathbf{C}) \times A$. With ρ being a suitable affine coordinate of $P_1(\mathbf{C})$ we now have the coordinates $(\rho; x, y)$ by which we can write down the discriminant set Σ for the elliptic fibration. It has five irreducible components:

$$\Sigma_\infty\colon \rho = \infty$$

$$\Sigma_1\colon \rho = -\frac{(y-1)^2}{2}$$

$$\Sigma_2\colon \rho = -\frac{(x-1)(y-1)}{2}$$

$$\Sigma_3\colon \rho^2 - 2(x + 2y + 1)\rho + (x-1)(x+1-2y^2) = 0$$

$$\Sigma_4\colon \rho = 0.$$

Σ_∞ corresponds to the singular fiber C_∞ which we have fixed at infinity and Σ_1 to the other singular fiber of Type I_1. Among the three singular fibers of Type I_2 $(= A_1)$, there is the special one which is distinguished for some reason; Σ_2 corresponds to it and Σ_3 to the remaining two. Σ_4 corresponds

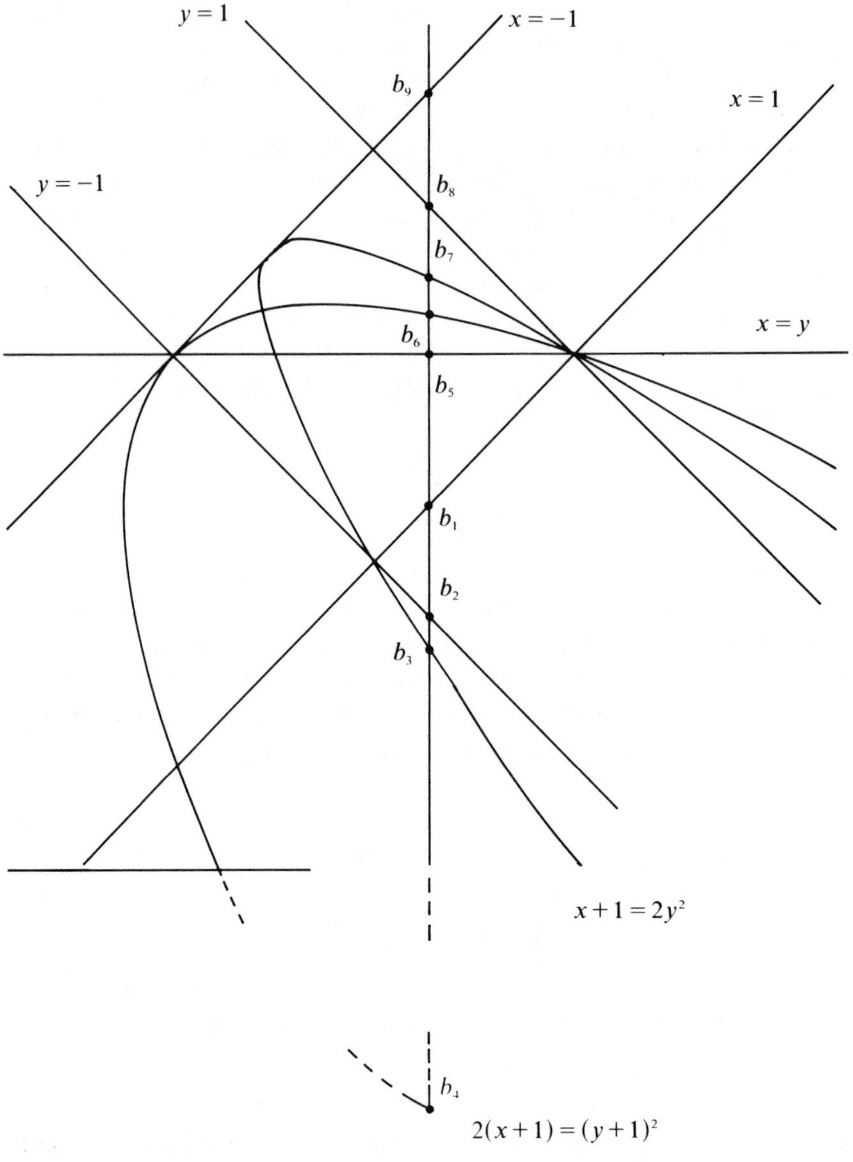

FIGURE 1

to the singular fiber of Type I_4 ($=A_3$). The maximal open subset $A^* \subseteq A$ over which the projection $\Sigma \to A$ is an unramified covering is now the complement of the arrangement of the following curves:

$$x=\pm 1 \qquad y=y=\pm 1$$
$$x+1=2y^2$$
$$2(x+1)=(y+1)^2.$$

We cut this arrangement transversally by the line $x+y=\frac{1}{2}$ and give names $b_1, b_2, \ldots, b_9$ to the intersection points as is indicated in Fig. 1.

In the complex feature of the line $x+y=\frac{1}{2}$ we further draw closed paths $\beta_1, \beta_2, \ldots, \beta_9$ all issuing from the reference point to near $+i_\infty$ and surrounding once counterclockwise the corresponding points $b_1, b_2, \ldots, b_9$. When regarded as paths in A^*, they generate $\pi_1(A^*, t_0)$ and the relations among them are given by

$$\beta_7\beta_9\beta_7\beta_9=\beta_9\beta_7\beta_9\beta_7$$
$$\beta_3=\beta_9^{-1}\beta_7\beta_9 \qquad [\beta_3, \beta_6]=1$$
$$[\beta_3, \beta_5]=1 \qquad [\beta_8, \beta_9]=1$$
$$\beta_1\beta_3\beta_2=\beta_3\beta_2\beta_1=\beta_3\beta_1(\beta_3\beta_2\beta_3^{-1})$$
$$[(\beta_3\beta_2\beta_3^{-1})\beta_5, \beta_6\beta_9]=1$$
$$[\beta_5, \beta_6\beta_9(\beta_3\beta_2\beta_3^{-1})]=1$$
$$[\beta_9, (\beta_3\beta_2\beta_3^{-1})\beta_6\beta_9\beta_6]=1$$
$$\beta_4=\beta_9^{-1}\beta_6\beta_9 \qquad [\beta_1, \beta_4]=1$$
$$\beta_1\beta_5\beta_6\beta_7\beta_8=\beta_5\beta_6\beta_7\beta_8\beta_1=\beta_6\beta_7\beta_8\beta_1\beta_5$$
$$=\beta_7\beta_8\beta_1\beta_5\beta_6=\beta_8\beta_1\beta_5\beta_6\beta_7.$$

Now, by forming the double coverings branched over the singular fibers corresponding to Σ_∞, Σ_4, we obtain the family of elliptic $K3$ surfaces over A^* which we denote by $\tilde{P}_{A^*,\Sigma_4} \to A^*$ as before. Since the Picard number is equal to 18 for the $K3$ surfaces, the target domain of the period mapping for the family is considered to be the product of the upper half domain $H \times H$: (s, t). Through the monodromy representation, the paths β_i induce the automorphisms β_i^* on $H \times H$. For the homogeneous coordinates $(s_1, s_2; t_1, t_2)$ introduced by $s=s_1/s_2$, $t=t_1/t_2$, β_i^* $(i=1, 2, \ldots, 9)$ have the following expressions:

$$\beta_1^*\begin{pmatrix} s_1 \\ s_2 \end{pmatrix} \to \begin{pmatrix} 1 & 2+\sqrt{2} \\ & 1 \end{pmatrix}\begin{pmatrix} s_1 \\ s_2 \end{pmatrix}$$

$$\beta_2^*\begin{pmatrix} s_1 \\ s_2 \end{pmatrix} \to \begin{pmatrix} 1 & \\ & 1 \end{pmatrix}\begin{pmatrix} t_1 \\ t_2 \end{pmatrix}$$

$$\beta_3^*\begin{pmatrix} s_1 \\ s_2 \end{pmatrix} \to \begin{pmatrix} 1 & \sqrt{2} \\ & 1 \end{pmatrix}\begin{pmatrix} t_1 \\ t_2 \end{pmatrix}$$

$$\beta_4^* = \beta_5^* = \beta_6^* = \text{identity}$$

$$\beta_7^*\begin{pmatrix} s_1 \\ s_2 \end{pmatrix} \to \begin{pmatrix} 1 & \\ \sqrt{2} & 1 \end{pmatrix}\begin{pmatrix} t_1 \\ t_2 \end{pmatrix}$$

$$\beta_8^*\begin{pmatrix} s_1 \\ s_2 \end{pmatrix} \to \begin{pmatrix} 1-\sqrt{2} & 2+\sqrt{2} \\ -2+\sqrt{2} & 1+\sqrt{2} \end{pmatrix}\begin{pmatrix} s_1 \\ s_2 \end{pmatrix}$$

$$\beta_9^*\begin{pmatrix} s_1 \\ s_2 \end{pmatrix} \to \begin{pmatrix} & 1 \\ -1 & \end{pmatrix}\begin{pmatrix} t_1 \\ t_2 \end{pmatrix}.$$

Here we have given only the half of the expressions since every β_i^* is either of the form

$$\begin{pmatrix} s_1 \\ s_2 \end{pmatrix} \to A\begin{pmatrix} s_1 \\ s_2 \end{pmatrix} \quad \begin{pmatrix} t_1 \\ t_2 \end{pmatrix} \to A^c\begin{pmatrix} t_1 \\ t_2 \end{pmatrix},$$

or of the form

$$\begin{pmatrix} s_1 \\ s_2 \end{pmatrix} \to A\begin{pmatrix} t_1 \\ t_2 \end{pmatrix} \quad \begin{pmatrix} t_1 \\ t_2 \end{pmatrix} \to A^c\begin{pmatrix} s_1 \\ s_2 \end{pmatrix},$$

where A^c denotes the conjugate of A by the Galois group of $\mathbf{Q}(\sqrt{2})/\mathbf{Q}$.

By replacing Σ_4 by Σ_1 we can also consider the family $\tilde{P}_{A^*,\Sigma_1} \to A^*$ and we obtain the representation β_i^* $i = 1, 2, \ldots, 9$ associated with this. The result is summarized as follows:

$$\beta_1^*\begin{pmatrix} s_1 \\ s_2 \end{pmatrix} \to \begin{pmatrix} & -1 \\ 1 & \end{pmatrix}\begin{pmatrix} t_1 \\ t_2 \end{pmatrix}$$

$$\beta_2^*\begin{pmatrix} s_1 \\ s_2 \end{pmatrix} \to \begin{pmatrix} t_1 \\ t_2 \end{pmatrix}$$

$$\beta_3^* = \text{identity}$$

$$\beta_4^*\begin{pmatrix} s_1 \\ s_2 \end{pmatrix} \to \begin{pmatrix} \sqrt{2}-1 & \\ & \sqrt{2}+1 \end{pmatrix}\begin{pmatrix} t_1 \\ t_2 \end{pmatrix}$$

$$\beta_5^*\begin{pmatrix} s_1 \\ s_2 \end{pmatrix} \to \begin{pmatrix} 1 & -2\sqrt{2} \\ & 1 \end{pmatrix}\begin{pmatrix} t_1 \\ t_2 \end{pmatrix}$$

$$\beta_6\begin{pmatrix} s_1 \\ s_2 \end{pmatrix} \to \begin{pmatrix} 1+\sqrt{2} & -4 \\ & -1+\sqrt{2} \end{pmatrix}\begin{pmatrix} t_1 \\ t_2 \end{pmatrix}$$

$$\beta_7^* = \text{identity}$$

$$\beta_8^*\begin{pmatrix} s_1 \\ s_2 \end{pmatrix} \to \begin{pmatrix} 3-\sqrt{2} & -2+\sqrt{2} \\ 2-\sqrt{2} & -1+\sqrt{2} \end{pmatrix}\begin{pmatrix} s_1 \\ s_2 \end{pmatrix}$$

$$\beta_9^*\begin{pmatrix} s_1 \\ s_2 \end{pmatrix} \to \begin{pmatrix} 1+\sqrt{2} & -2 \\ & -1+\sqrt{2} \end{pmatrix}\begin{pmatrix} t_1 \\ t_2 \end{pmatrix}.$$

For both of these cases, the monodromy group is thus commensurate with the Hilbert modular group associated with the real quadratic field $\mathbf{Q}(\sqrt{2})$. To close this note we remark that, if one chooses the section Σ_2 instead of Σ_4 or Σ_1, then the monodromy group is commensurate with the product $SL(2, \mathbf{Z}) \times SL(2, \mathbf{Z})$. I have to admit that I needed to use a computer for performing a part of calculation.

Acknowledgements

The author is grateful to Professor Mikio Sato for the interest and discussion in this topic. He also sincerely thanks Kyoji Saito for the introduction to the method of Zariski and Van Kampen.

References

[B] E. Brieskorn, Die Auflösung der rationalen Singularitäten holomorpher Abbildungen, *Math. Ann.* **178** (1968) 255–270.

[DP] M. Demazure, H. Pinkham, and others. *Seminaire sur les Singularités des Surface*, Springer L.N., 777 (1980), Berlin-Heidelberg-New York.

[K_1] K. Kodaira, On compact analytic surfaces I–III, *Ann. of Math.* **71** (1960) 111–152; **77** (1963) 563–626; **78** (1963) 1–40.

[K_2] K. Kodaira, *Theory of Complex Analytic Surfaces*, Sem. Note 32, Tokyo Univ., 1974 (in Japanese).

[L] E. Looijenga, Rational surfaces with an anticanonical cycle, *Ann. of Math.* (2), **114** (1981) 267–322.

[N_1] I. Naruki, E_8 und die binäre Ikosaedergruppe, *Invent. Math.* **42** (1977) 273–283.
[N_2] I. Naruki, Configurations related to maximal rational elliptic surfaces, Complex Analytic Singularities, *Advanced Study in Pure Math.* **8** (1986) 314–340, Kinokuniya.
[N_3] I. Naruki, On confluence of singular fibers in elliptic fibrations, *Publ. RIMS, Kyoto Univ.* **23** (1987) 409–431.
[Sa] K. Saito, Algebraic surfaces for regular systems of weights, RIMS Preprint 563, Kyoto Univ., 1987.
[Sh] T. Shioda, On elliptic modular surfaces, *J. Math. Soc. Japan* **25** (1972) 20–59.

Wronskian Determinants and the Gröbner Representation of a Linear Differential Equation: An Approach to Nonlinear Integrable Systems

Masatoshi Noumi
Department of Mathematics
Sophia University
Tokyo, Japan

Introduction

In his lectures at Kyoto University from 1984 on, Professor Mikio Sato presented a new idea to develop the theory of integrable systems of nonlinear differential equations ([7]). In his prospect, the correspondence between linear differential equations and their solutions gives rise to various types of isomorphisms between "Grassmann manifolds," and a theory of nonlinear integrable systems can be constructed on the bridge where Grassmann manifolds of different origins are linked. Following his idea, some people in Kyoto and Tokyo are studying to find a "universal framework" for the theory of higher-dimensional integrable systems which should play a role as the KP hierarchy in the one-dimensional case.

ISBN 0-12-400466-0

The purpose of this article is to give a foundation for this "Grassmannian formalism," confining ourselves to *holonomic* cases. In Section 1, we investigate basic properties of Wronskian determinants in several variables, in the category of $\mathcal{D}$-modules over a differential field $\mathcal{K}$. Those are used in Section 2 to establish a prototype of such an isomorphism between Grassmann manifolds as above (Theorem 2.5). In order to interpret the Grassmannian formalism of our case in explicit terms, we study in Section 3 the canonical forms of cyclic $\mathcal{D}$-modules with respect to a well-ordering of the lattice $\mathbf{N}^r$; this gives a generalization of the theory of Gröbner bases ([1]) for polynomial ideals to a noncommutative case. Combining the arguments of preceding sections, we present in Section 4 a framework for the generalization of a *restricted version* of the KP hierarchy to higher-dimensional cases (Theorem 4.5).

The content of this article was suggested by the lectures of Professor Sato; the author intended to give an introduction with proofs to his ideas, far from his vision though it may be. Most of it was also studied by a group of people in Kyoto including Dr. K. Takasaki, Mr. N. Suzuki and Mr. T. Ohyama. In understanding Professor Sato's ideas, the author was helped by Professor K. Ueno, Mr. H. Harada and those people in Kyoto. It is his great pleasure to thank them for fruitful discussions. Lastly, financial support from Research Aid of Inoue Foundation for Science is gratefully acknowledged.

1. Wronskian Determinants

Let $\mathcal{K}$ be a *differential field* (of any characteristic) endowed with a family $\partial=(\partial_0,\ldots,\partial_{r-1})$ of r mutually commutative derivations $\partial_k:\mathcal{K}\to\mathcal{K}$ $(0\le k<r)$. We denote by $\mathcal{C}=\{c\in\mathcal{K};\,\partial_0(c)=\cdots=\partial_{r-1}(c)=0\}$ the *field of constants* of $\mathcal{K}$ and by $\mathcal{D}$ the *ring of differential operators* in ∂ with coefficients in $\mathcal{K}$. By definition, $\mathcal{D}$ is a free left $\mathcal{K}$-module with basis $\{\partial_0^{\alpha^{(0)}}\cdots\partial_{r-1}^{\alpha^{(r-1)}};\ \alpha^{(0)},\ldots,\alpha^{(r-1)}\in\mathbf{N}\}$ and its multiplicative structure is determined by the commutation rules

$$[\partial_k,\partial_l]=0 \quad\text{and}\quad [\partial_k,f]=\partial_k(f) \quad \text{for all } f\in\mathcal{K}. \tag{1.1}$$

Note that $\mathcal{K}$ has a natural structure of a left $\mathcal{D}$-module and that one has $\mathrm{Hom}_{\mathcal{D}}(\mathcal{K},\mathcal{K})=\mathcal{C}$ since $\mathcal{D}/\sum_{0\le k<r}\mathcal{D}\partial_k\rightrightarrows\mathcal{K}$. Throughout this article, we denote by L the set $\mathbf{N}^r$ of multi-indices and freely use the notation of multi-indices: $\partial^\alpha=\partial_0^{\alpha^{(0)}}\cdots\partial_{r-1}^{\alpha^{(r-1)}}$ for each $\alpha=(\alpha^{(0)},\ldots,\alpha^{(r-1)})\in L=\mathbf{N}^r$.

Let $\underline{f}=(f_0,\ldots,f_{m-1})\in\mathscr{K}$ be an m-tuple of "functions" (i.e., elements of $\mathscr{K}$). Then for each m-tuple $\underline{\alpha}=(\alpha_0,\ldots,\alpha_{m-1})\in L^m$ of multi-indices, we define the *Wronskian determinant* $\mathrm{Wr}_{\underline{\alpha}}(\underline{f})=\mathrm{Wr}_{\alpha_0,\ldots,\alpha_{m-1}}(f_0,\ldots,f_{m-1})$ of $\underline{f}$ with indices $\underline{\alpha}$ by

$$\mathrm{Wr}_{\alpha_0,\ldots,\alpha_{m-1}}(f_0,\ldots,f_{m-1})=\det(\partial^{\alpha_i}(f_j);\ 0\le i,j<m)\in\mathscr{K}. \tag{1.2}$$

It is well known that in the case where $r=1$, $f_0,\ldots,f_{m-1}$ are linearly independent over $\mathscr{C}$ if and only if the "principal" Wronskian determinant $\mathrm{Wr}_{0,\ldots,m-1}(f_0,\ldots,f_{m-1})$ does not vanish. In the case where $r>1$, however, the situation becomes a little delicate.

We say that a subset S of the partially ordered set $L=\mathbf{N}^r$ is an *order ideal* if it satisfies the condition

$$\alpha\in L,\qquad \beta\in S\ \text{ and }\ \alpha\le\beta\ \Rightarrow\ \alpha\in S. \tag{1.3}$$

where the symbol $\le$ stands for the natural partial order of L. This condition is equivalent to saying that the complement $E=L\backslash S$ is a *monoideal* of the monoid (i.e., the additive semigroup with 0) $L=\mathbf{N}^r$: $\alpha\in E$ and $\beta\in L\Rightarrow\alpha+\beta\in L$.

Theorem 1.1. *Let $\underline{f}=(f_0,\ldots,f_{m-1})\in\mathscr{K}^m$ be an m-tuple of elements of $\mathscr{K}$. Then $f_0,\ldots,f_{m-1}$ are linearly independent over the field of constants $\mathscr{C}$ if and only if there exists an order ideal $S=\{\alpha_0,\ldots,\alpha_{m-1}\}$ of $L=\mathbf{N}^r$ such that the Wronskian determinant $\mathrm{Wr}_{\underline{\alpha}}(\underline{f})$ does not vanish, where $\underline{\alpha}=(\alpha_0,\ldots,\alpha_{m-1})\in L^m$.*

This theorem was conjectured by Professor K. Ueno, and proved independently by Mr. N. Suzuki and by the author. This section will be devoted to preliminary investigations on Wronskian determinants, and our proof of Theorem 1.1 will be completed in Section 4, Proposition 4.2.

A motive to introduce the notion of Wronskian determinants is to describe, in an explicit manner, the system of linear differential equations to be satisfied by a given family of functions.

Let V be an m-dimensional $\mathscr{C}$-vector subspace of $\mathscr{K}$ and choose a $\mathscr{C}$-basis $\underline{f}=(f_0,\ldots,f_{m-1})$ for V. Then we define a left ideal J_V of $\mathscr{D}$ by

$$J_V=\{P\in\mathscr{D};\ Pf=0\text{ for all }f\in V\}=\{P\in\mathscr{D};\ Pf_j=0\text{ for }0\le j<m\}, \tag{1.4}$$

and set $M_V=\mathscr{D}/J_V$. Note here that one has an exact sequence of left $\mathscr{D}$-modules

$$0\to J_V\to\mathscr{D}\xrightarrow{w}\mathrm{Hom}_{\mathscr{C}}(V,\mathscr{K}), \tag{1.5}$$

where $w(P)(f)=Pf$ for all $P\in\mathscr{D}$ and $f\in V$. (For any $\mathscr{C}$-vector space V, $\mathrm{Hom}_{\mathscr{C}}(V,\mathscr{K})$ has a natural structure of a left $\mathscr{D}$-module induced by that of $\mathscr{K}$.) Moreover, under the identification $\mathrm{Hom}_{\mathscr{C}}(V,\mathscr{K})\rightleftarrows\mathscr{K}^m$ by the $\mathscr{C}$-basis $\underline{f}$ for V, the $\mathscr{K}$-homomorphism w is represented by the *Wronskian matrix*

$$\mathrm{WR}(\underline{f})=\mathrm{WR}(f_0,\ldots,f_{m-1})=(\partial^\alpha(f_j))_{\alpha\in L,0\le j<m}\in\mathrm{Mat}(L,m;\mathscr{K}),\qquad(1.6)$$

with respect to the left $\mathscr{K}$-basis $(\partial^\alpha)_{\alpha\in L}$ for $\mathscr{D}$.

Lemma 1.2. *The $\mathscr{D}$-homomorphism w of* (1.5) *is surjective. Equivalently, for any m-tuple $\underline{f}=(f_0,\ldots,f_{m-1})\in\mathscr{K}^m$, there exists an m-tuple $\underline{\alpha}=(\alpha_0,\ldots,\alpha_{m-1})\in L^m$ such that $\mathrm{Wr}_{\underline{\alpha}}(\underline{f})\neq 0$, iff $f_0,\ldots,f_{m-1}$ are linearly independent over $\mathscr{C}$.*

Proof. We prove the lemma by induction on $m=\dim_{\mathscr{C}} V$. Suppose that w is *not* surjective. Then there exists a non-zero $\mathscr{K}$-homomorphism $g:\mathrm{Hom}_{\mathscr{C}}(V,\mathscr{K})\to\mathscr{K}$ which vanishes on the image of w. Namely, one can find a non-zero vector $\underline{g}=(g_0,\ldots,g_{m-1})\in\mathscr{K}^m$ such that

$$\sum_{0\le j<m}\partial^\alpha(f_j)g_j=0\qquad\text{for all }\alpha\in L.\qquad(1.7)$$

Without losing generality, one may assume that $g_0=-1$. Then (1.7) reads

$$\partial^\alpha(f_0)=\sum_{1\le j<m}\partial^\alpha(f_j)g_j\qquad\text{for all }\alpha\in L.\qquad(1.8)$$

Differentiating these by ∂_k, one obtains

$$\sum_{1\le j<m}\partial^\alpha(f_j)\partial_k(g_j)=0\qquad\text{for all }\alpha\in L\text{ and }0\le k<r.\qquad(1.9)$$

By the induction hypothesis, there exists an $(m-1)$-tuple $(\alpha_1,\ldots,\alpha_{m-1})\in L^{m-1}$ such that $\mathrm{Wr}_{\alpha_1,\ldots,\alpha_{m-1}}(f_1,\ldots,f_{m-1})\neq 0$; hence one has $\partial_k(g_j)=0$ for all $0\le j<m$ and $0\le k<r$. This means that the g_j's are all constants, which is a contradiction since (1.8) contains the relation $f_0=\sum_{1\le j<m}f_jg_j$.

Q.E.D.

Proposition 1.3. *For any finite-dimensional $\mathscr{C}$-subspace V of $\mathscr{K}$, one has a natural isomorphism of $\mathscr{D}$-modules*

$$w:M_V=\mathscr{D}/J_V\rightleftarrows\mathrm{Hom}_{\mathscr{C}}(V,\mathscr{K});\qquad P\bmod J_V\mapsto(f\mapsto Pf).\qquad(1.10)$$

Especially one has $\dim_{\mathscr{K}} M_V=\dim_{\mathscr{C}} V$. Moreover, one has

$$V=\{f\in\mathscr{K};\ Pf=0\text{ for all }P\in J_V\}.\qquad(1.11)$$

The isomorphism (1.10) is a direct consequence of Lemma 1.2. Applying the "solution functor" $\mathrm{Hom}_{\mathscr{D}}(\cdot, \mathscr{K})$ to (1.10), one has

$$\mathrm{Hom}_{\mathscr{D}}(M_V, \mathscr{K}) \rightrightarrows \mathrm{Hom}_{\mathscr{D}}(\mathrm{Hom}_{\mathscr{C}}(V, \mathscr{K}), \mathscr{K}) = V. \tag{1.12}$$

(Since $\mathrm{Hom}_{\mathscr{D}}(\mathscr{K}, \mathscr{K}) = \mathscr{C}$, $V \rightrightarrows \mathrm{Hom}_{\mathscr{D}}(\mathrm{Hom}_{\mathscr{C}}(V, \mathscr{K}), \mathscr{K})$ for any finite-dimensional $\mathscr{C}$-vector space V.) The isomorphism (1.12) is equivalent to the equality (1.11).

Proposition 1.2 forms a part of the correspondence between linear differential equations and their solution spaces. This point of view will be enlarged in the next section.

For each subset S of $L = \mathbf{N}^r$, we define a left $\mathscr{K}$-submodule $\mathscr{K}\langle S\rangle$ of $\mathscr{D}$ by

$$\mathscr{K}\langle S\rangle = \bigoplus_{\sigma \in S} \mathscr{K}\partial^\sigma \subset \mathscr{D}. \tag{1.13}$$

which is a bilateral $\mathscr{K}$-submodule if S is an order ideal. Keeping the notations $V, \underline{f}$ and $M_V = \mathscr{D}/J_V$ as before, let $S = \{\sigma_0, \ldots, \sigma_{m-1}\}$ be a subset of L with $\# S = m$. Then, in view of the $\mathscr{K}$-isomorphism $M_V \rightrightarrows \mathscr{K}^m$, one sees that the following three conditions are equivalent.

$$\begin{array}{ll} \text{(a)} & (\partial^\sigma u)_{\sigma \in S} \text{ forms a } \mathscr{K}\text{-basis for } M_V. \\ \text{(b)} & \mathscr{D} = J_V \oplus \mathscr{K}\langle S\rangle \text{ as a left } \mathscr{K}\text{-module.} \\ \text{(c)} & \mathrm{Wr}_{\underline{\sigma}}(\underline{f}) \neq 0 \text{ for } \underline{\sigma} = (\sigma_0, \ldots, \sigma_{m-1}). \end{array} \tag{1.14}$$

Here $u = 1 \bmod J_V$ is the canonical generator for $M_V = \mathscr{D}/J_V$. The condition (a) means that M_V can be represented as a system of linear differential equations of first order for the column vector $\mathbf{u} = (\partial^\sigma u)_{\sigma \in S}$ of unknown functions. We fix a subset S of L satisfying the equivalent conditions (1.14). Then, for each $\alpha \in L$, one can find a unique differential operator $W_{S,\alpha}$ in J_V (namely $W_{S,\alpha} u = 0$) of the form

$$W_{S,\alpha} = \partial^\alpha - \sum_{\sigma \in S} w_{S,\alpha/\sigma} \partial^\sigma \in \partial^\alpha + \mathscr{K}\langle S\rangle. \tag{1.15}$$

(Note that $W_{S,\sigma} = 0$ for $\sigma \in S$.) Moreover, the family $(W_{S,\alpha})_{\alpha \in L \setminus S}$ forms a left $\mathscr{K}$-basis for the ideal J_V, as can be easily seen. It should be remarked that the operator $W_{S,\alpha}$ is determined by the equation

$$\mathrm{Wr}_{\sigma_0, \ldots, \sigma_{m-1}, \alpha}(f_0, \ldots, f_{m-1}, u) = \mathrm{Wr}_{\underline{\sigma}}(\underline{f})\, W_{S,\alpha} u \tag{1.16}$$

in M_V. Expanding the Wronskian determinant on the left-hand side along the column of u, one obtains the expression of the coefficients $w_{S,\alpha/\sigma} (\sigma \in S)$

of $W_{S,\alpha}$ in terms of Wronskian determinants: For each i,

$$w_{S,\alpha/\sigma_i} = \mathrm{Wr}_{\sigma_0,\ldots,\alpha,\ldots,\sigma_{m-1}}(\underline{f})/\mathrm{Wr}_{\underline{\sigma}}(\underline{f}), \tag{1.17}$$

where the index α is put in place of σ_i. We refer to the numerator of the right-hand side as $\mathrm{Wr}_{\underline{\sigma};\sigma\to\alpha}(\underline{f})$, for short, in the case where $\sigma = \sigma_i \in S$.

It is indeed true that some finite members of $W_{S,\sigma}$ $(\alpha \in L \backslash S)$ generate the ideal J_V, but we cannot foresee, at this stage, which members do. To specify a generator system for J_V, we need to restrict the choice of S; this point will be discussed later, in the framework of Gröbner representations. (See Proposition 4.2 and Example 4.3.)

2. Grassmannian Formalism

In this section, we will generalize Proposition 1.3 to a categorical equivalence between certain systems of linear differential equations and their solution spaces. From this correspondence, we derive an isomorphism between two Grassmann manifolds, one consisting of $\mathscr{D}$-modules and the other of $\mathscr{C}$-vector spaces.

Keeping the notations in Section 1, let $\mathscr{K}$ be a differential field and $\mathscr{D}$ the ring of differential operators with coefficients in $\mathscr{K}$. For a left $\mathscr{D}$-module M, we denote by $\mathrm{Sol}(M) = \mathrm{Hom}_{\mathscr{D}}(M, \mathscr{K})$ the $\mathscr{C}$-vector space of solutions to the system M in $\mathscr{K}$. Then the contravariant functor

$$\mathrm{Sol}: (\text{left } \mathscr{D}\text{-modules}) \to (\mathscr{C}\text{-vector spaces}) \tag{2.1}$$

has an adjoint functor $\mathrm{Hom}_{\mathscr{C}}(\cdot, \mathscr{K})$; namely, one has a functorial isomorphism

$$\mathrm{Hom}_{\mathscr{D}}(M, \mathrm{Hom}_{\mathscr{C}}(V, \mathscr{K})) = \mathrm{Hom}_{\mathscr{C}}(V, \mathrm{Hom}_{\mathscr{D}}(M, \mathscr{K})) \tag{2.2}$$

for any left $\mathscr{D}$-module M and $\mathscr{C}$-vector space V. Equivalently, one has a functorial $\mathscr{D}$-homomorphism

$$\omega(M): M \to \mathrm{Hom}_{\mathscr{C}}(\mathrm{Hom}_{\mathscr{D}}(M, \mathscr{K}), \mathscr{K}); \qquad v \mapsto (\varphi \mapsto \varphi(v)) \tag{2.3}$$

for each left $\mathscr{D}$-module M and a functorial $\mathscr{C}$-homomorphism

$$\gamma(V): V \to \mathrm{Hom}_{\mathscr{D}}(\mathrm{Hom}_{\mathscr{C}}(V, \mathscr{K}), \mathscr{K}); \qquad f \mapsto (\psi \mapsto \psi(f)) \tag{2.4}$$

for each $\mathscr{C}$-vector space V. Moreover, these two functorial morphisms are compatible in the following sense: with an *ad hoc* notation $\Phi = \mathrm{Hom}_{\mathscr{D}}(\cdot, \mathscr{K})$

and $\Psi = \mathrm{Hom}_{\mathscr{C}}(\cdot, \mathscr{K})$, one has

$$\Phi(\omega(M)) \circ \gamma(\Phi(M)) = \mathrm{id}_{\Phi(M)} \quad \text{and} \quad \Psi(\gamma(V)) \circ \omega(\Psi(V)) = \mathrm{id}_{\Psi(V)}. \tag{2.5}$$

Note that, as for $\gamma(V)$, it is a $\mathscr{C}$-isomorphism if V is finite-dimensional since $\mathrm{Hom}_{\mathscr{D}}(\mathscr{K}, \mathscr{K}) = \mathscr{C}$.

We say that a left $\mathscr{D}$-module M is *holonomic* if it is finite-dimensional as a $\mathscr{K}$-vector space. (Here we adopt this condition as the definition of holonomicity.)

Proposition 2.1. *If M is a holonomic $\mathscr{D}$-module, then the $\mathscr{D}$-homomorphism $\omega(M)$ of (2.3) is surjective. Especially if one has $\dim_{\mathscr{C}} \mathrm{Sol}(M) \le \dim_{\mathscr{K}} M$.*

Proof. First we consider the case where $M = \mathscr{D}/J$ with a left ideal J of $\mathscr{D}$. In this case, one can naturally identify $\mathrm{Sol}(M) = \mathrm{Hom}_{\mathscr{D}}(M, \mathscr{K})$ with the $\mathscr{C}$-subspace V of $\mathscr{K}$ defined by $V = \{f \in \mathscr{K};\ Pf = 0 \text{ for all } P \in J\}$. If U is a finite-dimensional $\mathscr{C}$-subspace of V, then one has a surjection $M = \mathscr{D}/J \to M_U = \mathscr{D}/J_U$ since $J \subset J_U$. Combining this with Proposition 1.3, one sees that $\dim_{\mathscr{C}} U = \dim_{\mathscr{K}} M_U \le \dim_{\mathscr{K}} M$. Since $\dim_{\mathscr{C}} U$ is uniformly bounded for any finite-dimensional $\mathscr{C}$-subspace U, V itself must be finite-dimensional. Applying now Proposition 1.3 to $V = \mathrm{Sol}(M)$, one knows that $\omega(M)$ is surjective since it equals the composition of the surjection $M = \mathscr{D}/J \to M_V = \mathscr{D}/J_V$ and the isomorphism (1.10). From this cyclic case, one can proceed to the general case by induction on the number of generators for M. In fact, consider an exact sequence $0 \to M' \to M \to M'' \to 0$ of holonomic $\mathscr{D}$-modules. Then one has a commutative diagram

$$\begin{array}{ccccccccc}
0 & \longrightarrow & M' & \longrightarrow & M & \longrightarrow & M'' & \longrightarrow & 0 \quad \text{(exact)} \\
& & \downarrow{\scriptstyle \omega(M')} & & \downarrow{\scriptstyle \omega(M)} & & \downarrow{\scriptstyle \omega(M'')} & & \\
& & \mathrm{Hom}_{\mathscr{C}}(V', \mathscr{K}) & \longrightarrow & \mathrm{Hom}_{\mathscr{C}}(V, \mathscr{K}) & \longrightarrow & \mathrm{Hom}_{\mathscr{C}}(V'', \mathscr{K}) & \longrightarrow & 0 \quad \text{(exact)}
\end{array} \tag{2.6}$$

where V', V and V'' stand for $\mathrm{Sol}(M')$, $\mathrm{Sol}(M)$ and $\mathrm{Sol}(M'')$ respectively. The second row of (2.6) is exact since the functor $\mathrm{Sol} = \mathrm{Hom}_{\mathscr{D}}(\cdot, \mathscr{K})$ is left exact and $\mathrm{Hom}_{\mathscr{C}}(\cdot, \mathscr{K})$ is exact. By the Five Lemma, one sees that $\omega(M)$ is surjective if so are $\omega(M')$ and $\omega(M'')$. Q.E.D.

By Proposition 2.1, we know that, for a holonomic $\mathscr{D}$-module M, the following three conditions are equivalent:

$$\begin{aligned}
&\text{(a)}\quad \dim_{\mathscr{C}} \mathrm{Sol}(M) = \dim_{\mathscr{K}} M, \\
&\text{(b)}\quad \omega(M)\colon M \simeq \mathrm{Hom}_{\mathscr{C}}(\mathrm{Hom}_{\mathscr{D}}(M, \mathscr{K}), \mathscr{K}), \\
&\text{(c)}\quad \text{As a } \mathscr{D}\text{-module, } M \simeq \mathrm{Hom}_{\mathscr{C}}(V, \mathscr{K}) \text{ for some finite-dimensional } \mathscr{C}\text{-vector space } V.
\end{aligned} \tag{2.7}$$

We say that a holonomic $\mathscr{D}$-module M is *$\mathscr{K}$-solvable* if it satisfies the equivalent conditions of (2.7). Since $\gamma(V)$ is a $\mathscr{C}$-isomorphism for any finite-dimensional $\mathscr{C}$-vector space V, we have

Proposition 2.2. *The functor* Sol *gives an anti-equivalence between the category of $\mathscr{K}$-solvable holonomic left $\mathscr{D}$-modules and the category of finite-dimensional $\mathscr{C}$-vector spaces.*

This proposition itself is more or less tautological indeed, unless one should know when a holonomic $\mathscr{D}$-module is $\mathscr{K}$-solvable. In this connection, the following lemma is useful.

Lemma 2.3. *Let $0 \to M' \to M \to M'' \to 0$ be an exact sequence of holonomic $\mathscr{D}$-modules. If M is $\mathscr{K}$-solvable, so are M' and M''.*

Proof. We use the same notation as in the latter half of the proof of Proposition 2.1. Denote by m', m and m'' the dimensions of M', M and M'' over $\mathscr{K}$, respectively. Then one has $\dim_{\mathscr{C}} V' \leq m'$, $\dim_{\mathscr{C}} V'' \leq m''$ and $\dim_{\mathscr{C}} V = m = m' + m''$. On the other hand, one has $\dim_{\mathscr{C}} V \leq \dim_{\mathscr{C}} V' + \dim_{\mathscr{C}} V''$ since the second row of (2.6) is exact. Hence one has $\dim_{\mathscr{C}} V' + \dim_{\mathscr{C}} V'' = m' + m''$; this can hold only if $\dim_{\mathscr{C}} V' = m'$ and $\dim_{\mathscr{C}} V'' = m''$. Q.E.D.

Remark 2.4. In the setting of Lemma 2.3, M is not necessarily $\mathscr{K}$-solvable even if so are both M' and M''. For example, consider the case where $\mathscr{K} = \mathbf{C}(x)$ is the field of rational functions in one variable x and $\partial = d/dx$ $(r = 1)$. Set $P = \partial^2 + (1/x)\partial$; then one has an exact sequence

$$0 \to \mathscr{D}/\mathscr{D}\left(\partial + \frac{1}{x}\right) \xrightarrow{\cdot \partial} \mathscr{D}/\mathscr{D}P \to \mathscr{D}/\mathscr{D}\partial \to 0, \tag{2.8}$$

according to the factorization $P = (\partial + 1/x)\partial$. The equation $Pu = 0$ is *not* solvable in $\mathbf{C}(x)$, while the other two are solvable in $\mathbf{C}(x)$.

By virtue of the categorical equivalence of Proposition 2.2, we can prove an ismorphism between Grassmann manifolds.

For any left $\mathscr{D}$-module M, we denote by $\mathrm{Quot}_{\mathscr{D}}(m; M)$ (resp. $\mathrm{Quot}_{\mathscr{D}}(m; M)^{\mathscr{K}\text{-solv}}$) the set of all ($\mathscr{K}$-solvable) holonomic quotients M' of M with $\dim_{\mathscr{K}} M' = m$. On the other hand, for any $\mathscr{C}$-vector space V, we denote by $\mathrm{Sub}_{\mathscr{C}}(m; V)$ the set of all m-dimensional $\mathscr{C}$-subspaces of V. Then we have

Theorem 2.5. (1) *Let $\mathscr{L}$ be an arbitrary left $\mathscr{D}$-module and set $\mathscr{F} = \mathrm{Sol}(\mathscr{L})$. Then, for each $m \in \mathbf{N}$, one has a natural bijection*

$$\mathrm{Sol}\colon \mathrm{Quot}_{\mathscr{D}}(m; \mathscr{L})^{\mathscr{K}\text{-solv}} \xrightarrow{\sim} \mathrm{Sub}_{\mathscr{C}}(m; \mathscr{F}). \tag{2.9}$$

(2) *Let M be a $\mathscr{K}$-solvable holonomic $\mathscr{D}$-module and set $V = \mathrm{Sol}(M)$. Then, for each $m \in \mathbf{N}$, one has a natural bijection*

$$\mathrm{Sol}\colon \mathrm{Quot}_{\mathscr{D}}(m; M) \xrightarrow{\sim} \mathrm{Sub}_{\mathscr{C}}(m; V). \tag{2.10}$$

The statement (2) of Theorem 2.5 follows from (1) since any quotient of M is $\mathscr{K}$-solvable by Lemma 2.3. We will show that Sol induces a bijection between the set of all $\mathscr{K}$-solvable holonomic quotients of $\mathscr{L}$ and the set of all finite-dimensional $\mathscr{C}$-subspaces of $\mathscr{F}$. To make the argument clear, we use the notation $\Phi = \mathrm{Hom}_{\mathscr{D}}(\cdot, \mathscr{K})$ and $\Psi = \mathrm{Hom}_{\mathscr{C}}(\cdot, \mathscr{K})$ as in (2.5). Note that the contravariant functor Φ is left exact and Ψ is exact.

If M is a $\mathscr{K}$-solvable holonomic quotient of $\mathscr{L}$ with canonical surjection $u\colon \mathscr{L} \to M$, then $\Phi(M) = \mathrm{Hom}_{\mathscr{D}}(M, \mathscr{K})$ is a $\mathscr{C}$-subspace of $\mathscr{F}$ with canonical injection $\Phi(u)\colon \Phi(M) \to \Phi(\mathscr{L}) = \mathscr{F}$. Conversely, let V be a finite-dimensional $\mathscr{C}$-subspace of $\mathscr{F}$ with canonical injection $\iota\colon V \to \mathscr{F}$ and set $M = \Psi(V) = \mathrm{Hom}_{\mathscr{C}}(V, \mathscr{K})$. We will show that $\Psi(\iota)\colon \Psi(\mathscr{F}) \to M = \Psi(V)$, combined with $\omega(\mathscr{L})\colon \mathscr{L} \to \Psi\Phi(\mathscr{L}) = \Psi(F)$ induces a *surjection* $u = \Psi(\iota) \circ \omega(\mathscr{L})\colon \mathscr{L} \to M = \Psi(V)$ and that the correspondence $\iota \mapsto u = \Psi(\iota) \circ \omega(\mathscr{L})$ gives the inverse to $u \mapsto \Phi(u)$. Set $M' = \mathrm{Coker}(u)$; this is also a $\mathscr{K}$-solvable holonomic $\mathscr{D}$-module by Lemma 2.3. Apply the left exact functor Φ to the exact sequence $\mathscr{L} \to \Psi(V) \to M' \to 0$; then one has a diagram

$$\begin{array}{ccccccl} & & 0 \longrightarrow & V & \xrightarrow{\ \iota\ } & \mathscr{F} & \quad \text{(exact)} \\ & & & \downarrow{\scriptstyle \gamma(V)} & & \|{\scriptstyle \mathrm{id}_{\mathscr{F}}} & \\ 0 \longrightarrow & \Phi(M') & \longrightarrow & \Phi\Psi(V) & \underset{\Phi(u)}{\longrightarrow} & \Phi(\mathscr{L}) & \quad \text{(exact).} \end{array} \tag{2.11}$$

As to $u = \Psi(\iota) \circ \omega(\mathscr{L})\colon \mathscr{L} \to M = \Psi(V)$, one has $\Phi(u) = \Phi(\omega(\mathscr{L})) \circ \Phi\Psi(\iota)$, hence

$$\Phi(u) \circ \gamma(V) = \Phi(\omega(\mathscr{L})) \circ \Phi\Psi(\iota) \circ \gamma(V) = \Phi(\omega(\mathscr{L})) \circ \gamma(\Phi(\mathscr{L})) \circ \iota = \iota$$

by the functoriality of γ and the compatibility of (2.5). This means that the diagram (2.11) is commutative. Since $\gamma(V)$ is a $\mathscr{C}$-isomorphism, one has $\Phi(M')=0$. As we remarked above, M' is $\mathscr{K}$-solvable, hence $M' \rightrightarrows \Psi\Phi(M')=0$ as required. At the same time, we have shown that $\Phi(u):\Phi\Psi(V)\rightarrow\mathscr{F}$ coincides with $\iota: V\rightarrow\mathscr{F}$ as a subobject of $\mathscr{F}$. The other part of our claim is easier. Let M be a $\mathscr{K}$-solvable holonomic quotient of $\mathscr{L}$ with canonical surjection $u:\mathscr{L}\rightarrow M$; then one has $\omega(M)\circ u=\Psi\Phi(u)\circ\omega(\mathscr{L})$ by the functoriality of ω. Since $\omega(M)$ is a $\mathscr{D}$-isomorphism, $u:\mathscr{L}\rightarrow M$ and $\Psi\Phi(u)\circ\omega(\mathscr{L}):\mathscr{L}\rightarrow\Psi\Phi(M)$ coincide as quotient objects of $\mathscr{L}$; this completes the proof of Theorem 2.5.

Theorem 2.5 implies, in particular, a natural bijection

$$\mathrm{Sol}:\mathrm{Quot}_{\mathscr{D}}(m;\mathscr{D}^n)^{\mathscr{K}\text{-solv}} \rightrightarrows \mathrm{Sub}_{\mathscr{C}}(m;\mathscr{K}^n), \tag{2.12}$$

for each $m, n\in\mathbf{N}$.

As Professor M. Sato pointed out, an isomorphism of this type can be regarded as a parametrization, by the Grassmann manifold on the right-hand side, of the totality of solutions to a certain system of nonlinear equations. Moreover, one can develop a theory of nonlinear integrable systems on this isomorphism. This point of view, which we call the Grassmannian formalism, will be interpreted in Section 4, through the discussion of canonical forms of linear differential equations.

3. Gröbner Representations

In this section, we discuss the canonical form of a cyclic $\mathscr{D}$-module (i.e., a $\mathscr{D}$-module generated by a single element) with respect to a total order on $\mathbf{N}^r$. Arguments below give a generalization to a noncommutative case, of Gröbner bases for ideals of polynomials over a field. The theory of Gröbner bases was initiated by B. Buchberger [1] and has been developed by computer algebraists. The "Gröbner representation" (or the reduced Gröbner basis) of a system of linear differential equations is a generator system, uniquely associated to the corresponding ideal of differential operators. The Gröbner basis in the case of differential operators is also discussed by N. Takayama [8]; we refer the reader to his paper for the algorithm to construct Gröbner bases for ideals of differential operators.

Before the discussion of $\mathscr{D}$-modules, we review two lemmas on the "filtration-graduation" argument. Let L be a monoid satisfying the condition $\alpha+\gamma=\beta+\gamma\Rightarrow\alpha=\beta$. Moreover, suppose that L is endowed with a well-ordering $\preccurlyeq$ compatible with the additive structure of L in the following sense:

$$0 \preccurlyeq \alpha \text{ for all } \alpha \in L; \qquad \alpha \preccurlyeq \beta,\ \gamma \in L \Rightarrow \alpha + \gamma \preccurlyeq \beta + \gamma. \tag{3.1}$$

This implies: $\alpha \prec \beta \Leftrightarrow \alpha + \gamma \prec \beta + \gamma$ for any $\alpha, \beta, \gamma \in L$. In this setting, by "$L$-filtered ring," we mean a ring A endowed with an increasing filtration $(F_\alpha A)_{\alpha \in L}$ by submodules, exhaustive (i.e., $\bigcup_{\alpha \in L} F_\alpha A = A$) and satisfying: $1 \in F_0 A$ and $F_\alpha A \cdot F_\beta = F_{\alpha+\beta} A$ $(\alpha, \beta \in L)$. (Hereafter, all filtrations will be assumed to be increasing and exhaustive.) Then, by "filtered A-module," we mean a left A-module M endowed with a filtration $(F_\alpha M)_{\alpha \in L}$ by submodules such that $F_\alpha A \cdot F_\beta M \subset F_{\alpha+\beta} M$. A morphism of filtered A-modules is an A-homomorphism $\varphi : M \to N$ such that $\varphi(F_\alpha M) \subset F_\alpha N$ for all $\alpha \in L$; the morphism φ is said to be *strict* if $\varphi(F_\alpha M) = \varphi(M) \cap F_\alpha N$ for all $\alpha \in L$. For a filtered A-module M, its associated graded module $\mathrm{gr}(M) = \bigoplus_{\alpha \in L} \mathrm{gr}_\alpha(M)$ is defined by $\mathrm{gr}_\alpha(M) = F_\alpha M / F_{\prec\alpha} M$ where $F_{\prec\alpha} M = \bigcup_{\beta \prec \alpha} F_\beta M$ for $\alpha \succ 0$ and $F_{\prec 0} M = 0$. Then $\mathrm{gr}(A)$ is an L-graded ring and $\mathrm{gr}(M)$ is a graded $\mathrm{gr}(A)$-module in the natural sense. A morphism of filtered A-modules $\varphi : M \to N$ induces a morphism of graded $\mathrm{gr}(A)$-modules $\mathrm{gr}(\varphi) : \mathrm{gr}(M) \to \mathrm{gr}(N)$, functorially. By using the induction with respect to the well-ordering $\preccurlyeq$, one can prove

Lemma A. *Let $M' \to^{\varphi} M \to^{\psi} M''$ be a sequence of filtered A-modules with $\psi \circ \varphi = 0$. Then the graduation $\mathrm{gr}(M') \to^{\mathrm{gr}(\varphi)} \mathrm{gr}(M) \to^{\mathrm{gr}(\psi)} \mathrm{gr}(M'')$ is exact if and only if $M' \to M \to M''$ is exact and both φ and ψ are strict.*

A filtered A-module M is said to be *quasi-free* if it has a free A-basis $(e_i)_{i \in I}$ and its filtration is given by $F_\alpha M = \bigoplus_{i \in I, \beta \in L, \alpha_i + \beta \preccurlyeq \alpha} (F_\beta A) e_i$, for some $\alpha_i \in L$ $(i \in I)$, giving the L-degree α_i to e_i; then its graduation $\mathrm{gr}(M)$ is quasi-free in the following sense: $\mathrm{gr}_\alpha(M) = \bigoplus_{i \in I, \beta \in L, \alpha_i + \beta = \alpha} \mathrm{gr}_\beta(\mathscr{D}) e_i$ for all $\alpha \in L$.

Lemma B. *Let $\psi : M \to M''$ be a* strict *morphism of filtered A-modules. Let M' be a quasi-free filtered A-module and $f : \mathrm{gr}(M') \to \mathrm{gr}(M)$ a morphism of graded $\mathrm{gr}(A)$-modules. If $\mathrm{gr}(\psi) \circ f = 0$, there exists a morphism $\varphi : M' \to M$ of filtered A-modules such that $\psi \circ \varphi = 0$ and $\mathrm{gr}(\varphi) = f$.*

Now, we take the monoid $\mathbf{N}^r$ for L and fix a well-ordering $\preccurlyeq$ of $L = \mathbf{N}^r$ satisfying (3.1). Note that such a well-ordering is compatible with the natural partial order $\leq$ of L: $\alpha \leq \beta (\alpha, \beta \in L) \Rightarrow \alpha \preccurlyeq \beta$. Then the ring $\mathscr{D}$ becomes an L-filtered ring, endowed with the filtration defined by $F_\alpha \mathscr{D} = \bigoplus_{\beta \preccurlyeq \alpha} \mathscr{K} \partial^\beta$ for each $\alpha \in L$. Note that this filtration of $\mathscr{D}$ satisfies an additional condition on the commutators

$$[F_\alpha \mathscr{D}, F_\beta \mathscr{D}] \subset F_{<\alpha+\beta}\mathscr{D} \qquad \text{for all } \alpha, \beta \in L, \tag{3.2}$$

so that the graded ring $\mathrm{gr}(\mathscr{D})$ is identified with the polynomial ring $K[\xi]$, $\xi = (\xi_0, \ldots, \xi_{r-1})$, where $\xi_k = \partial_k \bmod F_{<\varepsilon_k}$ for $0 \leq k < r$ ($\varepsilon_k = (\delta_{k,l})_{0 \leq l < r}$ is the kth unit vector of $L = \mathbf{N}^r$). Moreover, each $\mathrm{gr}_\alpha(\mathscr{D}) = \mathscr{K}\xi^\alpha$ is a one-dimensional $\mathscr{K}$-vector space; this fact makes the argument below the easier. The field $\mathscr{K}$ will also be regarded as an L-filtered ring by the trivial filtration $F_\alpha \mathscr{K} = \mathscr{K} (\alpha \in L)$.

For each non-zero differential operator $P \in \mathscr{D}$, we define its *$\preccurlyeq$-order*, denoted by $\mathrm{ord}_\preccurlyeq(P)$, to be the $\alpha \in L$ such that $P \in F_\alpha \mathscr{D}$ and $P \notin F_{<\alpha}\mathscr{D}$; by convention, we set $\mathrm{ord}_\preccurlyeq(0) = -\infty$ and understand that $-\infty < 0$ and $-\infty + \alpha = -\infty$ for all $\alpha \in L$. An operator $P \in \mathscr{D}$ is said to be *monic* of $\preccurlyeq$-order α if $P \equiv \partial^\alpha \bmod F_{<\alpha}\mathscr{D}$.

Remark 3.1. Recall that any ascending chain of monoideals in $\mathbf{N}^r$ is stationary. Using this, one can show that any *total order* $\preccurlyeq$ satisfying (3.1) is necessarily a well-ordering. A typical example of such a well-ordering is the lexicographic order of the direct product $L = \mathbf{N}^r$ and another is the one composed of the length of multi-indices and the lexicographic order. Moreover, let $\lambda: \mathbf{N}^r \to \mathbf{Q}_{\geq 0}^N$ an injective additive mapping, where $\mathbf{Q}_{\geq 0}$ is the monoid of non-negative rational numbers. Then the pull-back by λ of the lexicographic order of $\mathbf{Q}_{\geq 0}^N$ gives a well-ordering satisfying (3.1).

First we propose a division lemma "along the well-ordering $\preccurlyeq$."

Lemma 3.2 (Division Lemma). *Let A be a finite subset of $L = \mathbf{N}^r$ and define an order ideal $S \subset L$ by $S = L \backslash \bigcup_{\alpha \in A}(\alpha + L)$. Let $(P_\alpha)_{\alpha \in A}$ be a family of differential operators $P_\alpha \in \mathscr{D}$, monic of $\preccurlyeq$-order α ($\alpha \in A$). Then, for each $P \in \mathscr{D}$, there exist operators Q_α ($\alpha \in A$) and R in $\mathscr{D}$ such that*

$$P = \sum_{\alpha \in A} Q_\alpha P_\alpha + R \qquad \text{and } R \in \mathscr{K}\langle S \rangle. \tag{3.3}$$

If $\mathrm{ord}_\preccurlyeq(P) = \beta$, *$Q_\alpha$ and R can be chosen so that* $\mathrm{ord}_\preccurlyeq(Q_\alpha) + \alpha \preccurlyeq \beta$ ($\alpha \in A$) *and* $\mathrm{ord}_\preccurlyeq(R) \preccurlyeq \beta$.

For each subset A of L, we denote by E_A the $\mathscr{K}$-vector space with basis $(e_\alpha)_{\alpha \in A}$ indexed by A. Consider e_α as an element of $\preccurlyeq$-order (or L-degree) α; then $\mathscr{D} \otimes_\mathscr{K} E_A$ has a natural structure of a quasi-free filtered $\mathscr{D}$-module.

Given a family $(P_\alpha)_{\alpha\in A}$ of operators as in Lemma 3.2, we define a morphism of filtered $\mathscr{D}$-modules $\delta_0: \mathscr{D}\otimes_{\mathscr{K}} E_A \to \mathscr{D}$ by $\delta_0(e_\alpha)=P_\alpha$ for each $\alpha\in A$. Apply Lemma A to the morphism of filtered $\mathscr{K}$-modules $\delta_0\oplus\iota: \mathscr{D}\otimes_{\mathscr{K}} E_A\oplus\mathscr{K}\langle S\rangle\to\mathscr{D}$, where $\iota:\mathscr{K}\langle S\rangle\to\mathscr{D}$ is inclusion and $\mathscr{K}\langle S\rangle$ is filtered by $F_\alpha\mathscr{K}\langle S\rangle=\mathscr{K}\langle S\rangle\cap F_\alpha\mathscr{D}(\alpha\in L)$. Then $\delta_0\oplus\iota$ is surjective and strict, since its graduation is surjective; this proves Lemma 3.2.

In proving Lemma 3.2, one does not need the fact that $\mathscr{K}$ is a *field*; the Division Lemma above is valid for differential operators with coefficients in an arbitrary differential algebra. An algorithm for this division is given similar to the case of polynomials (M-reduction).

In general, one cannot hope for the uniqueness of $R\in\mathscr{K}\langle S\rangle$ in (3.3). The theory of Gröbner bases characterizes such families $(P_\alpha)_{\alpha\in A}$ that assure the uniqueness of residues in this type of Division Lemma.

Let J be an arbitrary left ideal of $\mathscr{D}$ and let M be the cyclic $\mathscr{D}$-module $\mathscr{D}/J$ with generator $u=1 \bmod J$. To M, we give the filtration induced by the surjection $\cdot u:\mathscr{D}\to M=\mathscr{D}/J: F_\alpha M=(F_\alpha\mathscr{D})u$ for all $\alpha\in L$. Then, one has $\dim_{\mathscr{K}}\mathrm{gr}_\alpha(M)\le 1$ for each $\alpha\in L$, since $\mathrm{gr}(\mathscr{D})\to\mathrm{gr}(M)$ is surjective. Noting this, we define a subset $\mathrm{S}(M)$ of $L=\mathbf{N}^r$ by

$$\mathrm{S}(M)=\{\alpha\in L;\ \mathrm{gr}_\alpha(M)\neq 0\}\subset L. \tag{3.4}$$

Then the complement $L\backslash\mathrm{S}(M)$ is the set of $\preccurlyeq$-orders $\{\mathrm{ord}_{\preccurlyeq}(P);\ P\in J$ and $P\neq 0\}$. Since $\xi^\mu:\mathrm{gr}_\alpha(M)\to\mathrm{gr}_{\alpha+\mu}(M)$ is surjective for any $\alpha,\mu\in L$, the set $\mathrm{S}(M)$ is an order ideal of $L=\mathbf{N}^r$; we call $\mathrm{S}(M)$ the *canonical order ideal* of the cyclic $\mathscr{D}$-module M with respect to the well-ordering $\preccurlyeq$. Since the injection $\cdot u$: $\mathscr{K}\langle\mathrm{S}(M)\rangle\to M$ induces an isomorphism on graduation, one has $\cdot u$: $F_\alpha\mathscr{K}\langle\mathrm{S}(M)\rangle\xrightarrow{\sim} F_\alpha M$ $(\alpha\in L)$ and $\cdot u$: $\mathscr{K}\langle\mathrm{S}(M)\rangle\xrightarrow{\sim} M$ by Lemma A.

Here we give a characterization of the canonical order ideal. We introduce a total order, denoted by $\preccurlyeq$, on the power set of L as follows. Let S and T be two subsets of L. Assuming that $S\neq T$, let α be the minimum of the symmetric difference $(S\backslash T)\cup(T\backslash S)$ under the well-ordering $\preccurlyeq$. Then we define to be $S<T$ or $S>T$ according as $\alpha\in S$ or $a\in T$. (Under this total order, L is the minimum and the empty set is the maximum.) By using the fact $\cdot u$: $F_\alpha\mathscr{K}\langle\mathrm{S}(M)\rangle\xrightarrow{\sim} F_\alpha M(\alpha\in L)$, one can easily show

Proposition 3.3. *The canonical order ideal* $\mathrm{S}(M)$ *of a cyclic* $\mathscr{D}$*-module* M *is the minimum, under the total order* $\preccurlyeq$ *of the power set of* L *defined as above, among all subsets* $T\subset L$ *such that the* $\mathscr{K}$*-homomorphism* $\cdot u:\mathscr{K}\langle T\rangle\to M$ *is injective.*

For an order ideal S of L, we denote by $\mathrm{A}(S) \subset L$ the set of all minimal elements of the complement $L \backslash S$ under the partial order $\leq$. Then $\mathrm{A}(S)$ is a finite set in which any two distinct elements are incomparable under $\leq$ and one has $S = L \backslash \bigcup_{\alpha \in \mathrm{A}(S)} (\alpha + L)$. Let $S = \mathrm{S}(M)$ be the canonical order ideal of a cyclic $\mathscr{D}$-module $M = \mathscr{D}/J$. For each $\alpha \in L \backslash S$, consider the operator W_α in J uniquely determined by the condition $W_\alpha \in \partial^\alpha + \mathscr{K}\langle S \rangle$; then it must be monic of $\preccurlyeq$*-order* α, since $F_\alpha M = F_{<\alpha} M = (F_{<\alpha} \mathscr{K}\langle S \rangle) u$. Namely, W_α has the form

$$W_\alpha = \partial^\alpha - \sum_{\sigma \in S, \beta < \alpha} w_{\alpha/\sigma} \partial^\sigma \tag{3.5}$$

for $\alpha \in L \backslash S$. Let us show that, in this case, the finite family $(W_\alpha)_{\alpha \in \mathrm{A}(S)}$ generates the ideal J. Setting $A = \mathrm{A}(S)$, consider the sequence of filtered $\mathscr{D}$-modules

$$\mathscr{D} \bigoplus_{\mathscr{K}} \mathrm{E_A} \xrightarrow{\delta_0} \mathscr{D} \xrightarrow{\cdot u} M \longrightarrow 0, \qquad \text{where } \delta_0(e_\alpha) = W_\alpha \quad (\alpha \in A). \tag{3.6}$$

Then (3.6) induces an exact sequence on its graduation; (3.6) is a strict exact sequence by Lemma A. Hence, one knows that the family $(W_\alpha)_{\alpha \in \mathrm{A}(S)}$ generates J. We call this family $(W_\alpha)_{\alpha \in \mathrm{A}(S)}$ the *canonical generator system* for J, or the *Gröbner representation* of M with respect to the well-ordering $\preccurlyeq$. (This generalizes the notion of "reduced Gröbner bases" for polynomial ideals.) Note that, as for the canonical generator system $(W_a)_{\alpha \in \mathrm{A}(S)}$, uniqueness of residues $R \in \mathscr{K}\langle S \rangle$ as in (3.3) is automatic, since $\mathscr{K}\langle S \rangle \xrightarrow{\sim} M = \mathscr{D}/J$.

Theorem 3.4. *Let A be a finite subset of $L = \mathbf{N}^r$ and define an order ideal S of L by $S = L \backslash \bigcup_{\alpha \in A} (\alpha + L)$. Let $(W_\alpha)_{\alpha \in A}$ be a family of differential operators $W_\alpha \in \mathscr{D}$, monic of $\preccurlyeq$-order α $(\alpha \in A)$. Denote by J the left ideal of $\mathscr{D}$ generated by W_α $(\alpha \in A)$ and set $M = \mathscr{D}/J$ and $u = 1 \bmod J$. Then S contains the canonical order ideal $\mathrm{S}(M)$. Moreover, the following four conditions are equivalent.*

(a) *The set S coincides with $\mathrm{S}(M)$.*

(b) *Uniqueness of residues*: $\cdot u : \mathscr{K}\langle S \rangle \xrightarrow{\sim} M$ *(or $\mathscr{D} = J \oplus \mathscr{K}\langle S \rangle$).*

(c) *For any $P \in J$, there exist operators Q_α $(\alpha \in A)$ in $\mathscr{D}$ such that $P = \sum_{\alpha \in A} Q_\alpha W_\alpha$ and $\mathrm{ord}_\preccurlyeq(Q_\alpha) + \alpha \preccurlyeq \mathrm{ord}_\preccurlyeq(P)$ $(\alpha \in A)$.*

(d) *Compatibility condition*: *For any couple $(\alpha, \beta) \in A^2$ there exist operators $C_{\alpha,\beta;\gamma}(\gamma \in A)$ in $\mathscr{D}$ such that $\mathrm{ord}_\preccurlyeq(C_{\alpha,\beta;\gamma}) + \gamma < \sup\{\alpha, \beta\}$ for all $\alpha, \beta, \gamma \in A$ and that*

$$\partial^\mu W_\beta - \partial^\nu W_\alpha = \sum_{\gamma \in A} C_{\alpha,\beta;\gamma} W_\gamma, \tag{3.7}$$

where $\mu = \sup\{\alpha, \beta\} - \beta$, $\nu = \sup\{\alpha, \beta\} - \alpha$ under the partial order of L, for each couple $(\alpha, \beta) \in A^2$.

This type of theorem for differential operators was also obtained by Mr. Y. Ohyama in the case where $\preccurlyeq$ is the lexicographic order of $L=\mathbf{N}^r$. Condition (d) is known, in polynomial cases, as the condition that any critical pair (or S-polynomial) in $(W_\alpha)_{\alpha\in A}$ should be M-reducible by $(W_\alpha)_{\alpha\in A}$.

The statement $S\supset \mathrm{S}(M)$ is immediate. The equivalence (a)$\Leftrightarrow$(b) follows from the fact that the composition of the inclusion $\mathscr{K}\langle \mathrm{S}(M)\rangle\to\mathscr{K}\langle S\rangle$ and the surjection $\mathscr{K}\langle S\rangle\to M$ (by the Division Lemma) is the isomorphism $\mathscr{K}\langle \mathrm{S}(M)\rangle \rightleftarrows M$. In the setting of the theorem, the sequence (3.6) is exact. Note that condition (a) means the exactness of the graduation of (3.6), while (c) means the strictness of δ_0. Hence one has (a)$\Leftrightarrow$(c) by Lemma A. Consider the following sequence of quasi-free graded gr($\mathscr{D}$)-modules:

$$\mathscr{K}[\xi]\otimes_{\mathscr{K}}\textstyle\bigwedge^2 E_A \xrightarrow{d_1} \mathscr{K}[\xi]\otimes_{\mathscr{K}} E_A \xrightarrow{d_0} \mathscr{K}[\xi]. \tag{3.8}$$

Here we identify gr($\mathscr{D}$) with $\mathscr{K}[\xi]$. We define the morphisms d_0 and d_1 as follows: $d_0(e_\alpha)=\xi^\alpha$ for $\alpha\in A$ and

$$d_1(e_\alpha\wedge e_\beta)=\xi^{\sup\{\alpha,\beta\}-\beta}\,e_\beta-\xi^{\sup\{\alpha,\beta\}-\alpha}\,e_\alpha \tag{3.9}$$

for each couple $(\alpha,\beta)\in A^2$ considering $e_\alpha\wedge e_\beta$ $(\alpha\neq\beta)$ as an element of L-degree $\sup\{\alpha,\beta\}$. Then one can easily show that (3.8) is exact. Note that $d_0=\mathrm{gr}(\delta_0)$ and that condition (d) above means the existence of a morphism of quasi-free filtered $\mathscr{D}$-modules $\delta_1:\mathscr{D}\otimes_{\mathscr{K}}\bigwedge^2 E_A\to\mathscr{D}\otimes_{\mathscr{K}} E_A$ such that $\delta_0\circ\delta_1=0$ and $\mathrm{gr}(\delta_1)=d_1$. Since the graduation is exact, Lemma A combined with Lemma B implies the equivalence (c)$\Leftrightarrow$(d). This completes the proof of Theorem 3.6. The above argument also implies that, if (d) is satisfied, then the lifting of (3.8),

$$\mathscr{D}\otimes_{\mathscr{K}}\textstyle\bigwedge^2 E_A \xrightarrow{\delta_1} \mathscr{D}\otimes_{\mathscr{K}} E_A \xrightarrow{\delta_0} \mathscr{D}, \tag{3.10}$$

is necessarily exact (and strict). Namely, the left linear relations among W_α $(\alpha\in A)$ over $\mathscr{D}$ are generated by those in (3.7).

Corollary. *There is a one-to-one correspondence, by the Gröbner representation, between the following two sets:*

(a) *The set of all quotient left $\mathscr{D}$-modules of $\mathscr{D}$. (Or the set of all left ideals of $\mathscr{D}$.)*

(b) *The set of all pairs $(S,(W_\alpha)_{\alpha\in \mathrm{A}(S)})$, where S is an order ideal of $L=\mathbf{N}^r$ and $(W_\alpha)_{\alpha\in\mathrm{A}(S)}$ is a family of operators W_α in $\mathscr{D}$ of the form*

$$W_\alpha=\partial^\alpha-\sum_{\sigma\in S,\sigma<\alpha} w_{\alpha/\sigma}\partial^\sigma \text{ (finite sum)},\qquad w_{\alpha/\sigma}\in\mathscr{K}, \tag{3.11}$$

satisfying condition (d) of Theorem 3.4.

We remark that Theorem 3.4 and the corollary can be generalized to "multi-component" cases, namely, to the case where the left $\mathscr{D}$-submodules of a free $\mathscr{D}$-module $\mathscr{D}^n$ are concerned, by introducing a suitable well-ordering on the disjoint union $L_0 \cup \cdots \cup L_{n-1}$, where $L_i = L = \mathbf{N}^r$ for $0 \leq i < n$.

4. Applications to Nonlinear Integrable Systems

In holonomic cases, the corollary to Theorem 3.4 gives a decomposition of the Grassmann manifold $\mathrm{Quot}_{\mathscr{D}}(m; \mathscr{D})$ of all holonomic quotients M of $\mathscr{D}$ with $\dim_{\mathscr{K}} M = m$ (cf. Section 2), into the disjoint union

$$\mathrm{Quot}_{\mathscr{D}}(m; \mathscr{D}) = \bigcup_{S \in \Omega_m} \mathrm{Quot}_{\mathscr{D}}(S; \mathscr{D}), \tag{4.1}$$

where Ω_m is the set of order ideals of L consisting of m elements and $\mathrm{Quot}_{\mathscr{D}}(S; \mathscr{D})$ stands for the set of all holonomic quotients M of $\mathscr{D}$ with $\mathrm{S}(M) = S$. Moreover, each "stratum" can be described by a system of nonlinear differential equations. In fact, a point of $\mathrm{Quot}_{\mathscr{D}}(S; \mathscr{D})$ corresponds to a family $(W_\alpha)_{\alpha \in A}$, where $A = \mathrm{A}(S)$, of differential operators

$$W_\alpha = \partial^\alpha - \sum_{\alpha \in S, \sigma < \alpha} w_{\alpha/\sigma} \partial^\sigma \tag{4.2}$$

in $\mathscr{D}$, satisfying the compatibility condition: For any couple $(\alpha, \beta) \in A^2$, there exist operators $C_{\alpha,\beta;\gamma} (\gamma \in A)$ in $\mathscr{D}$ such that

$$\begin{aligned} &(i) \quad \mathrm{ord}_{\leqslant}(C_{\alpha,\beta;\gamma}) + \gamma < \sup\{\alpha, \beta\} \qquad \text{for all } \gamma \in A, \\ &(ii) \quad \partial^\mu W_\beta - \partial^\nu W_\alpha = \sum_{\gamma \in A} C_{\alpha,\beta;\gamma} W_\gamma, \\ &\qquad \text{where } \mu = \sup\{\alpha, \beta\} - \beta \text{ and } \nu = \sup\{\alpha, \beta\} - \alpha. \end{aligned} \tag{4.3}$$

Considering the coefficients $w_{\alpha/\sigma}$ of W_α as differential indeterminates, apply the Division Lemma 3.2 to the left-hand side of (4.3.*ii*). Then one has an expression

$$\partial^\mu W_\beta - \partial^\nu W_\alpha = \sum_{\gamma \in A} C_{\alpha,\beta;\gamma} W_\gamma + R_{\alpha,\beta} \qquad \text{for each } (\alpha, \beta) \in A^2, \tag{4.4}$$

with $C_{\alpha,\beta;\gamma}$ satisfying the condition (4.3.*i*), so that the residue $R_{\alpha,\beta}$ has the form

$$R_{\alpha,\beta} = \sum_{\substack{\sigma \in S \\ \sigma < \sup\{\alpha,\beta\}}} r_{\alpha,\beta;\sigma} \partial^\sigma. \tag{4.5}$$

Note that the coefficients $r_{\alpha,\beta;\sigma}$ ($\alpha, \beta \in A$, $\sigma \in S$) are expressed as differential polynomials in $w_{\alpha/\sigma}$ ($\alpha \in A$, $\sigma \in S$) (with coefficients in $\mathbf{Z}$), since the Division Lemma 3.2 is valid for differential operators with coefficients in an arbitrary differential algebra. Then the system of nonlinear differential equations $R_{\alpha,\beta}=0$ ($\alpha, \beta \in A$), or $r_{\alpha,\beta;\sigma}=0$ ($\alpha, \beta \in A$, $\sigma \in S$) for $w_{\alpha/\sigma}(\alpha \in A, \sigma \in S)$ is satisfied if and only if the family $(W_\alpha)_{\alpha \in A}$ of (4.2) satisfies the condition (4.3). (Though $r_{\alpha,\beta;\sigma}$ may not be uniquely determined as differential polynomials, the system of nonlinear equations $R_{\alpha,\beta}=0$, for all $\alpha, \beta \in A$, makes sense for any differential field $\mathscr{K}$.) We remark that the number of equations, as well as the number of unknown variables, reduce to be finite in this case.

Remark 4.1. Suppose that a family $(W_\alpha)_{\alpha \in A}$ of differential operators is given in the form (4.2). Then, by using the Division Lemma 3.2, one can derive, from the equations $W_\alpha u=0$ for $\alpha \in A$, a system of linear differential equations of first order for the column vector $\mathbf{u}=(\partial^\sigma u)_{\sigma \in S}$ in the form

$$\partial_k \mathbf{u} = A_k \mathbf{u}, \qquad \text{where } A_k \in \mathrm{Mat}(S, S; \mathscr{K}) \text{ for } 0 \le k < r. \tag{4.6}$$

Then the compatibility condition (4.3) for $(W_\alpha)_{\alpha \in A}$ is equivalent to the integrability condition of Frobenius,

$$[\partial_k - A_k, \partial_l - A_l] = 0 \qquad \text{for } 0 \le k, l < r. \tag{4.7}$$

In fact, it is easy to show that (4.7) is equivalent to condition (b) of Theorem 3.4.

Next, we consider the $\mathscr{K}$-solvable cases. For each m-dimensional $\mathscr{C}$-subspace V of $\mathscr{K}$, we have to determine which is the canonical order ideal of the cyclic $\mathscr{D}$-module $M_V = \mathscr{D}/J_V$. Note that $\mathrm{S}(M_V)$ consists of m elements. The set of all subset S of L with $\# S = m$ can be identified with the set of all m-tuples $\underline{\alpha} = (\alpha_0, \ldots, \alpha_{m-1}) \in L^m$ such that $\alpha_0 < \alpha_1 < \cdots < \alpha_{m-1}$. In this set, the total order among subsets of L, defined before Proposition 3.3, coincides with the restriction of the lexicographic order of L^m induced from the well-ordering $\preccurlyeq$ of L, under the above identification. Hence, by the equivalence (1.14) and Proposition 3.3, we have

Proposition 4.2. *Let V be an m-dimensional $\mathscr{C}$-subspace of $\mathscr{K}$ and $\underline{f} = (f_0, \ldots, f_{m-1})$ a $\mathscr{C}$-basis for V. Let $\underline{\sigma} = (\sigma_0, \ldots, \sigma_{m-1}) \in L^m$ be the m-tuple defined by*

$$\underline{\sigma} = \min\{\underline{\alpha} = (\alpha_0, \ldots, \alpha_{m-1}) \in L^m;\ \mathrm{Wr}_{\underline{\alpha}}(\underline{f}) \neq 0\}, \tag{4.8}$$

under the lexicographic order of L^m induced from the well-ordering $\leqslant$ of L; one has $\sigma_0 < \cdots < \sigma_{m-1}$. Then the set $S = \{\sigma_0, \ldots, \sigma_{m-1}\}$ coincides with the canonical order ideal of $\mathrm{S}(M_V)$. Moreover, the Gröbner representation $(W_\alpha)_{\alpha \in \mathrm{A}(S)}$ of M_V is determined as follows. In the expression (4.2), the coefficients $w_{\alpha/\sigma}$ are given as quotients of Wronskian determinants

$$w_{\alpha/\sigma} = \mathrm{Wr}_{\underline{\sigma};\sigma\to\alpha}(\underline{f})/\mathrm{Wr}_{\underline{\sigma}}(\underline{f}) \qquad \text{for all } \sigma \in \mathrm{A}(S),\ \sigma \in S, \tag{4.9}$$

where $\mathrm{Wr}_{\underline{\sigma};\sigma\to\alpha}(\underline{f})$ stands for the Wronskian determinant $\mathrm{Wr}_{\underline{\sigma}}(\underline{f})$ with $\sigma_i = \sigma$ replaced by α.

Note that $w_{\alpha/\sigma} = 0$ for $\sigma \in S$ with $\sigma > \alpha$ because of the choice of $\underline{\sigma}$. Since the set $S = \{\sigma_0, \ldots, \sigma_{m-1}\}$ is in fact an order ideal, the proposition implies Theorem 1.1.

Example 4.3. Let $\mathscr{K} = \mathbf{C}(x, y)$ be the differential field of rational functions in two variables x and y, endowed with the derivations $\partial_0 = \partial_x = \partial/\partial x$ and $\partial_1 = \partial_y = \partial/\partial y$ $(r = 2)$. Let V be the four-dimensional $\mathbf{C}$-subspace of $\mathbf{C}(x, y)$ with $\mathbf{C}$-basis $\underline{f} = (f_0, f_1, f_2, f_3)$ defined by

$$f_0 = 1, \quad f_1 = y, \quad f_2 = x \quad \text{and} \quad f_3 = x^p y^q, \tag{4.10}$$

where we assume that $p, q \in \mathbf{Z}$, $p + q = 1$ and $(p, q) \neq (0, 1), (1, 0)$. The order ideals of $\mathbf{N}^2$ consisting of four elements are five (= the partition number of 4) in all. Taking the lexicographic order of $\mathbf{N}^2$ for the well-ordering $\leqslant$, let S_i $(i = 0, 1, 2, 3, 4)$ be the five order ideals, so that $S_0 < S_1 < \cdots < S_4$. Then a direct calculation shows that the order ideal $S_1 = \{(0, 0), (0, 1), (0, 2), (1, 0)\}$ is the canonical order ideal for this case, with respect to the lexicographic order. Then, the Gröbner representation $(W_{S_1,\alpha})_{\alpha \in \mathrm{A}(S_1)}$ of $M_V = \mathscr{D}/J_V$ consists of three operators since $\mathrm{A}(S_1) = \{(0, 3), (1, 1), (2, 0)\}$. Note that the Wronskian determinant indexed by S_2 (the square) does not vanish; hence $u, \partial_y u, \partial_x u$ and $\partial_x \partial_y u$ form a $\mathscr{K}$-basis for M_V. However, the family $(W_{S_2,\alpha})_{\alpha \in \mathrm{A}(S_2)}$ consisting of two operators *cannot* generate the ideal J_V. In fact, the differential equation $W_{S_2,\alpha} u = 0$ for $\alpha \in \mathrm{A}(S_2)$ has infinitely many linearly independent solutions in $\mathbf{C}(x, y)$. (See Proposition 1.3.)

Recall that one has a natural bijection

$$\mathrm{Sol}\colon \mathrm{Quot}_{\mathscr{D}}(m; \mathscr{D})^{\mathscr{K}\text{-solv}} \xrightarrow{\sim} \mathrm{Sub}_{\mathscr{C}}(m; \mathscr{K}), \tag{4.11}$$

by Theorem 2.5.(1) for $\mathscr{D}$ and $\mathscr{K} = \mathrm{Sol}(\mathscr{D})$. Similarly to (4.1), the Grassmann manifold $\mathrm{Sub}_{\mathscr{C}}(m; \mathscr{K})$ is decomposed as follows:

$$\mathrm{Sub}_{\mathscr{C}}(m; \mathscr{K}) = \bigcup_{S \in \Omega_m} \mathrm{Sub}_{\mathscr{C}}(S; \mathscr{K}) \text{ (disjoint union).} \tag{4.12}$$

Here $\mathrm{Sub}_{\mathscr{C}}(S; \mathscr{K})$ denotes the totality of m-dimensional $\mathscr{C}$-subspaces V of $\mathscr{K}$ such that $\mathrm{Wr}_{\underline{\alpha}}(\underline{f}) = 0$ for all $\underline{\alpha} \in L^m$ with $\underline{\alpha} < \underline{\sigma}$, and $\mathrm{Wr}_{\underline{\sigma}}(\underline{f}) \neq 0$, where $\underline{f}$ is a $\mathscr{C}$-basis for V and $\underline{\sigma} \in L^m$ is the m-tuple corresponding to S. We understand that the codimension of $\mathrm{Sub}_{\mathscr{C}}(S; \mathscr{K})$ in $\mathrm{Sub}_{\mathscr{C}}(m; \mathscr{K})$ becomes higher as S becomes greater with respect to the total order of the power set of L. Set $\mathrm{Quot}_{\mathscr{D}}(S; \mathscr{D})^{\mathscr{K}\text{-solv}} = \mathrm{Quot}_{\mathscr{D}}(S; \mathscr{D}) \cap \mathrm{Quot}_{\mathscr{D}}(m; \mathscr{D})^{\mathscr{K}\text{-solv}}$; then (4.11) induces the bijection

$$\mathrm{Sol} : \mathrm{Quot}_{\mathscr{D}}(S; \mathscr{D})^{\mathscr{K}\text{-solv}} \rightrightarrows \mathrm{Sub}_{\mathscr{C}}(S; \mathscr{K}). \tag{4.13}$$

The inverse of this mapping is given in Proposition 4.2.

Finally, we apply this type of Grassmannian formalism to the following special case. Let $\mathscr{O}$ be the ring $\mathscr{C}[[x]]$ of formal power series in r variables $x = (x_0, \ldots, x_{r-1})$ over a field of characteristic zero. Let $\mathscr{K}$ be the field of fractions of $\mathscr{O}$; $\mathscr{K}$ is regarded as a differential field with the derivations $\partial_x = (\partial_0, \ldots, \partial_{r-1})$ where $\partial_k = \partial_{x_k} = \partial/\partial x_k$ for $0 \le k < r$.

We say that a holonomic $\mathscr{D}$-module $M = \mathscr{D}/J$ is *$\mathscr{O}$-solvable* if M is $\mathscr{K}$-solvable and the solution space $\mathrm{Sol}(M)$ is contained in $\mathscr{O}$, regarded as a subspace of $\mathscr{K}$. Then the bijection (4.11) induces a bijection between the $\mathscr{O}$-solvable holonomic quotients and the finite-dimensional $\mathscr{C}$-subspaces of $\mathscr{O}$:

$$\mathrm{Sol} : \mathrm{Quot}_{\mathscr{D}}(m; \mathscr{D})^{\mathscr{O}\text{-solv}} \rightrightarrows \mathrm{Sub}_{\mathscr{C}}(m; \mathscr{O}). \tag{4.14}$$

We pay attention to the "big cell" of the infinite-dimensional Grassmann manifold $\mathrm{Sub}_{\mathscr{C}}(m; \mathscr{O})$. Let $S_0 = \{\sigma_0, \ldots, \sigma_{m-1}\}$, $0 = \sigma_0 < \cdots < \sigma_{m-1}$, be the *minimum* of all subsets of L, consisting of m elements, and set $\underline{\sigma} = (\sigma_0, \ldots, \sigma_{m-1})$; this set S_0 is necessarily an order ideal. We denote by $\mathrm{Sub}_{\mathscr{C}}(m; \mathscr{O})^{\varnothing}$ the totality of m-dimensional $\mathscr{C}$-subspaces V such that the *value* of the Wronskian determinant $\mathrm{Wr}_{\underline{\sigma}}(\underline{f})|_{x=0}$ does not vanish for a $\mathscr{C}$-basis $\underline{f}$ for V. Note that $\mathrm{Sub}_{\mathscr{C}}(m; \mathscr{O})^{\varnothing}$ is strictly contained in the biggest "stratum," say $\mathrm{Sub}_{\mathscr{C}}(S_0; \mathscr{O})$, consisting of all V with $\mathrm{Wr}_{\underline{\sigma}}(\underline{f}) \neq 0$ $(\mathrm{Sub}_{\mathscr{C}}(S_0; \mathscr{O}) = \mathrm{Sub}_{\mathscr{C}}(S_0; \mathscr{K}) \cap \mathrm{Sub}_{\mathscr{C}}(m; \mathscr{O}))$.

Proposition 4.4. *With the notations as above, the inverse image of the big cell* $\mathrm{Sub}_{\mathscr{C}}(m; \mathscr{O})^{\varnothing}$ *of* $\mathrm{Sub}_{\mathscr{C}}(m; \mathscr{O})$, *by the bijection* (4.14) *coincides with the set of all points in* $\mathrm{Quot}_{\mathscr{D}}(S_0; \mathscr{D})$ *whose Gröbner representation* $(W_\alpha)_{\alpha \in \mathrm{A}(S_0)}$

is in the form

$$W_\alpha = \partial^\alpha - \sum_{\sigma \in S_0} w_{\alpha/\sigma} \partial^\sigma, \tag{4.15}$$

where $w_{\alpha/\sigma} \in \mathcal{O}$ *for all* $\alpha \in \mathrm{A}(S_0)$ *and* $\sigma \in S_0$.

Note that, if $\alpha \in \mathrm{A}(S_0)$, then $\alpha > \sigma$; hence all the $w_{\alpha/\sigma}$ $(\sigma \in S_0)$ make sense. One inclusion of the proposition is clear by the expression (4.9) of $w_{\alpha/\sigma}$ by Wronskian determinants. The other follows from the Theorem of Frobenius. In fact, the system $W_\alpha u = 0$ $(\alpha \in \mathrm{A}(S_0))$ can be transformed into a system (4.6) of first order with integrability condition by the Division Lemma 3.2 (Remark 4.1). If $(W_\alpha)_{\alpha \in \mathrm{A}(S_0)}$ is regular in the sense of (4.15), the matrices A_k $(0 \le k < r)$ are necessarily in $\mathrm{Mat}(S_0, S_0; \mathcal{O})$ as we remarked after Lemma 3.2. Moreover, the fundamental system Φ of solutions for (4.6) coincides with the Wronskian matrix, indexed by S_0, of the row $\underline{f}$ corresponding to u. Hence, one has $\mathrm{Wr}_{\underline{\sigma}}(\underline{f})|_{x=0} = \Phi|_{x=0} \neq 0$; this means that the solution space of $W_\alpha u = 0$ $(\alpha \in \mathrm{A}(S_0))$ belongs to $\mathrm{Sub}_{\mathscr{C}}(m; \mathcal{O})^{\varnothing}$.

Introducing a time variable t, we consider the case where $\mathscr{C}$ is the field of fractions of $\mathbf{C}[[t]]$. Setting $\mathcal{O}_0 = \mathbf{C}[[x]]$, let P be a differential operator with coefficients in $\mathcal{O}_0$. Then the $\mathbf{C}$-homomorphism $P\cdot : \mathcal{O}_0 \to \mathcal{O}_0$ defines a global vector field on the Grassmann manifold $\mathrm{Sub}_{\mathbf{C}}(m; \mathcal{O}_0)$; the tangent vector at $V_0 \in \mathrm{Sub}_{\mathbf{C}}(m; \mathcal{O})$ is the induced $\mathbf{C}$-homomorphism $V_0 \to \mathcal{O}_0 / V_0$. Exponentiating this, we define the formal flow $e^{tP} V_0$ with initial point $V_0 \in \mathrm{Sub}_{\mathbf{C}}(m; \mathcal{O}_0)$ to be the $\mathscr{C}$-subspace of $\mathcal{O} = \mathscr{C}[[x]]$ spanned by the set $\{e^{tP} f_0; f_0 \in V_0\}$. One can easily show that, for a "$\mathscr{C}$-valued point" $V \in \mathrm{Sub}_{\mathscr{C}}(m; \mathcal{O})$, the equation $(d/dt - P)V \subset V$ holds if and only if the ideal $J_V \subset \mathscr{D}$ of V satisfies the *Sato equation*

$$\frac{dW}{dt} + WP \in J_V \qquad \text{for all } W \in J_V, \tag{4.16}$$

by the correspondence (4.14). Let S_0 and $\underline{\sigma}$ be as in Proposition 4.4. Then, one sees immediately that, for a point V of $\mathrm{Sub}_{\mathscr{C}}(m; \mathcal{O})$, the following three conditions are equivalent:

(a) $V = e^{tP} V_0$ for some $V_0 \in \mathrm{Sub}_{\mathbf{C}}(m; \mathcal{O}_0)^{\varnothing}$.

(b) $(d/dt - P)V \subset V$ and V has a $\mathscr{C}$-basis $\underline{f}$ in $\mathbf{C}[[x, t]]^m$ such that $\mathrm{Wr}_{\underline{\sigma}}(\underline{f})|_{x=t=0} \neq 0$.

(c) M_V belongs to $\mathrm{Quot}_{\mathscr{C}}(S_0; \mathscr{D})$ and its Gröbner representation $(W_\alpha)_{\alpha \in A_0}$, $A_0 = \mathrm{A}(S_0)$, satisfies the following.

  (*i*) The coefficients of W_α are in $\mathbf{C}[[x, t]]$ for $\alpha \in A_0$.

(*ii*) There exist operators $B_{\alpha,\beta}(\alpha, \beta \in A_0)$ in $\mathscr{D}$ such that

$$\frac{dW_\alpha}{dt} + W_\alpha P = \sum_{\beta \in A_0} B_{\alpha,\beta} W_\beta \qquad \text{for all } \alpha \in A_0. \tag{4.17}$$

By the property of the Gröbner representation, one can find a unique operator $P_\alpha \in \mathscr{D}$ such that $WP \equiv P_\alpha \bmod J_V$ and $P \in \mathscr{K}\langle S_0 \rangle$. Then the condition (4.17) means that $dW_\alpha/dt + P_\alpha = 0$ for all $\alpha \in A$: this gives a system of nonlinear equations $dW_{\alpha/\sigma}/dt + p_{\alpha,\sigma} = 0$ $(\alpha \in A_0, \sigma \in S_0)$ in a similar sense as we explained the compatibility condition (4.3). Here the $p_{\alpha,\sigma}(\alpha \in A_0, \sigma \in S_0)$ are expressed as polynomials in $w_{\alpha/\sigma}(\alpha \in A_0, \sigma \in S_0)$ and their derivatives with respect to $\partial_x = (\partial_{x_0}, \ldots, \partial_{x_{r-1}})$. Combining this remark with Proposition 4.4 for $\mathscr{C} = \mathbf{C}$, we have

Theorem 4.5. *With the notations as above, let* $(\mathring{W}_\alpha)_{\alpha \in A_0}$ *be a family of operators of the form* (4.15) *with coefficients in* $\mathscr{O}_0 = \mathbf{C}[[x]]$. *If* $(\mathring{W}_\alpha)_{\alpha \in A_0}$ *satisfies the compatibility condition* (4.3), *then there exists a unique family* $(W_\alpha)_{\alpha \in A_0}$ *of the form* (4.15) *with coefficients in* $\mathbf{C}[[x, t]]$, *with initial condition* $W_\alpha|_{t=0} = \mathring{W}_\alpha$ *for* $\alpha \in A_0$, *satisfying the Sato equation* (4.17) *as well as the compatibility condition* (4.3). *In this sense, the totality of formal solutions to the Sato equation* (4.17), *with compatibility condition* (4.3), *is parametrized by the big cell* $\mathrm{Sub}_{\mathbf{C}}(m; \mathscr{O}_0)^{\varnothing}$ *of the infinite-dimensional Grassmann manifold* $\mathrm{Sub}_{\mathbf{C}}(m; \mathscr{O}_0)$.

Theorem 4.5 generalizes the framework of a restricted version of the KP hierarchy, studied by Mr. H. Harada [2], to higher-dimensional cases.

References

[1] B. Buchberger, An algorithmic criterion for the solvability of algebraic systems of equations (German). *Aequationes mathematicae* **4**/3 (1970) 374–383.

[2] H. Harada, New subhierarchies of the KP hierarchy in the Sato theory II, Truncation of the KP hierarchy, *J. Phys. Soc. Japan* **56** (1987) 3847–3852.

[3] M. Sato, Soliton equations as dynamical systems on an infinite dimensional Grassmann manifold, *RIMS Kokyuroku* **439**, RIMS Kyoto University (1981) 30–46.

[4] M. Sato and Y. Sato, Soliton equations as dynamical systems on an infinite dimensional Grassmann manifold, *Proc. U.S.-Japan Seminar "Nonlinear Partial Differential Equations in Applied Sciences," Tokyo 1982*, Kinokuniya, North-Holland, 1982, pp. 259–271.

[5] M. Sato, Soliton equations and the universal Grassmann manifold, *Sophia Kokyuroku in Math.* **18**, Sophia University (1984). (Notes by M. Noumi, in Japanese.)

[6] M. Sato, Lectures at Kyoto University, 1984.

[7] K. Takasaki, Differential equations and Grassmann manifolds—from Prof. Sato's lectures, *RIMS Kokyuroku* **597**, RIMS Kyoto University (1986) 109–126.

[8] N. Takayama, Gröbner basis and the problem of contiguous relation, to appear in *Japan. J. Appl. Math.*

Higher-Codimensional Boundary Value Problem and *F*-Mild Hyperfunctions

Toshinori Ôaku
Department of Mathematics
Yokohama City University
Yokohama, Japan

Introduction

Let M be a real analytic manifold and N be its closed real analytic submanifold of codimension ≥ 2. We take complexifications X, Y of M, N respectively so that Y is a closed complex submanifold of X. Let $\mathcal{M}$ be a coherent $\mathcal{D}_X$-module (i.e., a system of linear partial differential equations with analytic coefficients) on X. Assume Y is non-characteristic with respect to $\mathcal{M}$. Our purpose is to define the boundary value to N of a hyperfunction solution of $\mathcal{M}$ defined on a "wedge-like" domain with edge N. In fact, it is defined as a hyperfunction solution on N of the tangential system of $\mathcal{M}$ to Y. For this purpose we introduce the notion of F-mild hyperfunctions for higher-codimensional boundary, which extends that in the one codimensional case studied by Ôaku [6]. F-mild hyperfunctions give an

Algebraic Analysis, Volume II

ISBN 0-12-400466-0

explicit meaning to the boundary value of a hyperfunction solution of $\mathcal{M}$ through defining functions.

In Section 1 we develop the theory of F-mild hyperfunctions for higher-codimensional boundary. We adopt the viewpoint that Sato's hyperfunctions are defined concretely as "boundary values" of holomorphic functions (defining functions) (see, e.g., Kaneko [1]). An F-mild hyperfunction is, roughly speaking, a hyperfunction defined on a wedge-like domain (in M) with edge on N whose defining functions extend across the boundary Y in a complex domain. We define the sheaf of F-mild hyperfunctions on the normal sphere bundle S_NM, which represents the directions toward N. From such a concrete viewpoint as mentioned above, one of the most useful facts is the so-called edge-of-the-wedge theorem, which serves as a strong criterion for defining functions to define a zero hyperfunction. Hence our main aim in Section 1 is to prove the edge-of-the-wedge theorem for F-mild hyperfunctions. This enables us to define the boundary value (as a hyperfunction on N) of an F-mild hyperfunction. We also prove that an F-mild hyperfunction defined on S_NM is extended to a hyperfunction on a neighborhood of N with real analytic parameters normal to N.

In Section 2 we formulate the higher-codimensional boundary value problem for systems of linear partial differential equations by using F-mild hyperfunctions. We prove that the boundary value to N of a hyperfunction solution of $\mathcal{M}$ is defined independently of the direction toward N. A local uniqueness theorem for the boundary value problem is also proved at the same time. We remark that an abstract and more general formulation has been given by Schapira [10] except for the local uniqueness theorem. As an application we give a direct proof of a celebrated theorem of Kawai [5] on the extension of hyperfunction solutions.

1. F-Mild Hyperfunctions for Higher-Codimensional Boundary

Let M be an n-dimensional real analytic manifold and N be its d-codimensional closed real analytic submanifold (for the sake of simplicity we assume $d \geq 2$). We put $m = n - d$. We denote by

$$\mathrm{Mon}_N(M) = (M - N) \cup S_NM$$

the real monoidal transform of M with center N and by

$$\tau : \mathrm{Mon}_N(M) \to M, \qquad \tau : S_NM \to N$$

the natural projections, where $S_N M$ denotes the normal sphere bundle of N in M. Then $\mathrm{Mon}_N(M)$ becomes a real analytic manifold with real analytic boundary $S_N M$ (see Sato–Kawai–Kashiwara [9] for the real monoidal transform). Roughly speaking, $S_N M$ represents the directions along which the boundary value is taken. We denote by $\mathscr{B}_M$ the sheaf of hyperfunctions on M and by $\iota: M-N \to \mathrm{Mon}_N(M)$ the natural inclusion map. Let X and Y be complexifications of M and N respectively such that Y is a closed submanifold of X. We assume that there exists a real analytic submanifold $\tilde{M}$ of X containing both Y and M such that the triplet $(N, M, \tilde{M})$ is locally isomorphic to the triplet $(\{0\}\times \mathbf{R}^m, \mathbf{R}^n, \mathbf{R}^d \times \mathbf{C}^m)$ by a local coordinate system $z=(z_1, \ldots, z_n)$ of X around each point of N. We call such a local coordinate system z *admissible.* We use the notation $z=(z', z'')$ with $z'=(z_1, \ldots, z_d)$, $z''=(z_{d+1}, \ldots, z_n)=x''+\sqrt{-1}y''$ $(x'', y'' \in \mathbf{R}^m)$, $|z|=\max_{1\le j\le n}|z_j|$ and so on for an admissible local coordinate system z.

Definition 1.1. $\mathscr{B}_{N|M}=(\iota_*(\mathscr{B}_M|_{M-N}))|_{S_N M}$.

Hence a section of $\mathscr{B}_{N|M}$ is a hyperfunction defined on a wedge-like domain with edge in N. Now we define the sheaf of F-mild hyperfunctions as a subsheaf of $\mathscr{B}_{N|M}$.

Definition 1.2. Let x^* be a point of $S_N M$ and z be an admissible local coordinate system of X around $\mathring{x}=\tau(x^*)$ such that $z(\mathring{x})=0$. Then a germ $u(x)$ of $\mathscr{B}_{N|M}$ at x^* is called *F-mild* (with respect to $\tilde{M}$) at x^* if there exist open convex cones (with vertices at the origin) $\Gamma_1, \ldots, \Gamma_J$ in $\mathbf{R}^m$ with an integer J, and holomorphic functions $F_j(z)$ defined on a neighborhood (in X) of $D(\varepsilon, \tau(U), \Gamma_j)$ with an $\varepsilon>0$ and an open neighborhood U of x^* in $\mathrm{Mon}_N(M)$ such that

$$u(x)=\sum_{j=1}^{J} F_j(x', x''+\sqrt{-1}\Gamma_j 0) \tag{1}$$

as a hyperfunction on $\tau(U)\cap\{x\in M-N;\ |x|<\varepsilon\}$. Here we put

$$D(\varepsilon, V, \Gamma)=\{(x', z'')\in \mathbf{R}^d\times\mathbf{C}^m;\ |z|<\varepsilon, (x', x'')\in V, \operatorname{Im} z''\in\Gamma\}$$

for $\varepsilon>0$, a subset V of M and a cone Γ of $\mathbf{R}^m$. We denote by $\mathscr{B}^F_{N|M}$ the sheaf of sections of $\mathscr{B}_{N|M}$ which are F-mild at each point of their defining domains. Sections of $\mathscr{B}^F_{N|M}$ are called *F-mild hyperfunctions.*

On the other hand, since $S_N M$ is the 1-codimensional boundary of $\mathrm{Mon}_N(M)$, there exists the sheaf $\mathscr{B}^F_{S_N M|\mathrm{Mon}_N(M)}$ of F-mild hyperfunctions for the 1-codimensional boundary defined by Ôaku [6]. We denote $\mathscr{B}^F_{S_N M|\mathrm{Mon}_N(M)}$ by $\mathscr{B}^F_1$ for short. Then $\mathscr{B}^F_1$ is a subsheaf of $\mathscr{B}_{N|M}$ by definition.

Proposition 1.1. *Let x^* be a point of $S_N M$. Then F-mildness (with respect to $\tilde{M}$) at x^* of a section of $\mathscr{B}_{N|M}$ does not depend on the admissible local coordinate system z of X around $\mathring{x} = \tau(x^*)$ taken in Definition 1.2.*

Proof. Let $z = (z', z'')$ and $\tilde{z} = (\tilde{z}', \tilde{z}'')$ be two admissible local coordinate systems of X around $\mathring{x}$ such that

$$x^* = \left(0, \frac{\partial}{\partial x_1} 0\right) = \left(0, \frac{\partial}{\partial \tilde{x}_1} 0\right).$$

Since $\operatorname{Im} z' = \operatorname{Im} \tilde{z}' = 0$ on $\tilde{M}$, z' depends only on $\tilde{z}'$ as a function of $\tilde{z}$. This implies the conclusion.

Our main purpose in this section is to prove the following theorem. (We use the notation $\Gamma_1 + \Gamma_2 = \{y'' + \tilde{y}''; y'' \in \Gamma_1, \tilde{y}'' \in \Gamma_2\}$ for open cones Γ_1, Γ_2 of $\mathbf{R}^m$.)

Theorem 1.1 (Edge-of-the-wedge theorem for F-mild hyperfunctions). *Let x^* be a point of $S_N M$. We take an admissible local coordinate system z around $\mathring{x} = \tau(x^*)$ such that $z(\mathring{x}) = 0$. Let U be an open neighborhood of x^* in $\mathrm{Mon}_N(M)$, Γ_j $(j = 1, \dots, J)$ be open convex cones of $\mathbf{R}^m$, and $\varepsilon > 0$. Let $F_j(z)$ be holomorphic functions defined on a neighborhood of $D(\varepsilon, \tau(U), \Gamma_j)$ such that*

$$\sum_{j=1}^{J} F_j(x', x'' + \sqrt{-1}\Gamma_j 0) = 0$$

holds as a hyperfunction on $\{x \in M - N; |x| < \varepsilon\} \cap \tau(U)$. Then for any open convex cones Γ'_j such that $\Gamma'_j \Subset \Gamma_j$ (i.e., the closure of $S^{m-1} \cap \Gamma'_j$ is contained in Γ_j), there exist an open neighborhood V of x^* in $\mathrm{Mon}_N(M)$, a $\delta > 0$, and holomorphic functions $F_{jk}(z)$ defined on a neighborhood of $D(\delta, \tau(V), \Gamma'_j + \Gamma'_k)$ such that

$$F_j(z) = \sum_{k=1}^{J} F_{jk}(z) \qquad (j = 1, \dots, J),$$
$$F_{jk}(z) + F_{kj}(z) = 0 \qquad (j, k = 1, \dots, J).$$

In order to prove this theorem we first show the following.

Lemma 1.1. *$\mathscr{B}^F_{N|M}$ is a subsheaf of $\mathscr{B}^F_1$.*

Proof. Let x^* be a point of $S_N M$ and z be an admissible local coordinate system around $\mathring{x} = \tau(x^*)$ such that $z(\mathring{x}) = 0$. We may assume

$$x^* = \left(0, \frac{\partial}{\partial x_1} 0\right).$$

Let $F(z)$ be a holomorphic function defined on a neighborhood of $D(\varepsilon, \tau(U), \Gamma)$ with an $\varepsilon > 0$, a neighborhood U of x^* in $\mathrm{Mon}_N(M)$, and an open cone Γ of $\mathbf{R}^m$. We set

$$G(z_1, \zeta_2, \ldots, \zeta_d, z'') = F(z_1, z_2\zeta_2, \ldots, z_1\zeta_d, z'').$$

Then G is holomorphic on a neighborhood in $\mathbf{C}^n$ of

$$\{(x_1, \zeta_2, \ldots, \zeta_d, z'') \in \mathbf{R}^d \times \mathbf{C}^m;\, 0 \le x_1 < \delta, |\zeta_j| < \delta\ (j = 2, \ldots, d),$$
$$|z''| < \varepsilon, \operatorname{Im} z'' \in \Gamma\}$$

with a $\delta > 0$. By applying the local version of Bochner's tube theorem to $G(z_1^2, \zeta_2, \ldots, \zeta_d, z'')$, we know that $G(z_1, \zeta_2, \ldots, \zeta_d, z'')$ is holomorphic on a neighborhood of

$$D_1(\delta', \tilde{\Gamma}) = \{(x_1, \zeta_2, \ldots, \zeta_d, z'') \in \mathbf{R} \times \mathbf{C}^{n-1};\, 0 \le x_1 < \delta', |\zeta_j| < \delta'$$
$$(j = 2, \ldots, d), |z''| < \delta', (\operatorname{Im} \zeta_2, \ldots, \operatorname{Im} \zeta_d, \operatorname{Im} z'') \in \tilde{\Gamma}\}$$

for a $\delta' > 0$ and an open cone $\tilde{\Gamma}$ of $\mathbf{R}^{n-1}$ containing $\{0\} \times \Gamma$ with $0 \in \mathbf{R}^{d-1}$. Thus G defines a germ of $\mathscr{B}_1^F$ at x^*.

Proof of Theorem 1.1 for the case $J = 2$. We may assume

$$x^* = \left(0, \frac{\partial}{\partial x_1} 0\right).$$

Put

$$G_j(z_1, \zeta_2, \ldots, \zeta_d, z'') = F_j(z_1, z_1\zeta_2, \ldots, z_1\zeta_d, z'')$$

for $j = 1, 2$. Then G_j are holomorphic on a neighborhood of $D_1(\varepsilon_1, \tilde{\Gamma}_j)$ for an $\varepsilon_1 > 0$ and open convex cones $\tilde{\Gamma}_j$ of $\mathbf{R}^{n-1}$ containing $\{0\} \times \Gamma_j$ $(j = 1, 2)$. In view of the edge-of-the-wedge theorem for $\mathscr{B}_1^F$ (Theorem 1 of Ôaku [6]), for any open cones $\Gamma_j' \Subset \Gamma_j$ $(j = 1, 2)$, there exists a holomorphic function $G(z_1, \zeta_2, \ldots, \zeta_d, z'')$ defined on a neighborhood of $D_1(\varepsilon_2, \tilde{\Gamma})$, with an $\varepsilon_2 > 0$ and an open convex cone $\tilde{\Gamma}$ of $\mathbf{R}^{n-1}$ containing $\{0\} \times (\Gamma_1' + \Gamma_2')$, such that $G = G_1$ on $D_1(\varepsilon_2, \tilde{\Gamma}_1)$ and $G = -G_2$ on $D_1(\varepsilon_2, \tilde{\Gamma}_2)$. Put

$$F(z) = G\left(z_1, \frac{z_2}{z_1}, \ldots, \frac{z_d}{z_1}, z''\right).$$

Then F is holomorphic on a neighborhood of

$$\{(x', z'') \in \mathbf{R}^d \times \mathbf{C}^m; 0 < x_1 < \varepsilon_2, |x_j| < \delta|x_1| \ (j = 2, \ldots, d),$$
$$|z''| < \varepsilon_2, \operatorname{Im} z'' \in \Gamma_1' + \Gamma_2'\}$$
$$\cup \left\{ z \in \mathbf{C}^n; |z_1| < \varphi(z''), \left|\frac{z_j}{z_1}\right| < \varphi(z'') \ (j = 2, \ldots, d), \operatorname{Im} z'' \in \Gamma_1' + \Gamma_2' \right\},$$

where φ is a positive-valued lower semicontinuous function on $\{z'' \in \mathbf{C}^m; |z''| < \varepsilon_2, \operatorname{Im} z'' \in \Gamma_1' + \Gamma_2'\}$. In particular, F can be developed into a Laurent series with respect to z' in the form

$$F(z) = \sum_{\alpha \in \mathbf{Z}^d} a_\alpha(z'') z'^\alpha,$$

where $\alpha = (\alpha_1, \ldots, \alpha_d)$ are multi-indices such that $\alpha_j \geq 0$ for $j = 2, \ldots, d$. $F(z)$ is continued as a holomorphic function to a neighborhood of $\{0\} \times \{z'' \in \mathbf{C}^m; |z''| < \varepsilon_2, \operatorname{Im} z'' \in \Gamma_1' \cup \Gamma_2'\}$ since $F = F_1$ or $F = -F_2$ there. Hence $a_\alpha(z'') = 0$ if $\alpha_1 < 0$, $|z''| < \varepsilon_2$, and $\operatorname{Im} z'' \in \Gamma_1' \cup \Gamma_2'$. By the unique continuation property of holomorphic functions, we get $a_\alpha = 0$ if $\alpha_1 < 0$. Thus $F(z)$ is holomorphic on a neighborhood of $D(\varepsilon_2, \tau(V), \Gamma_1' + \Gamma_2')$ for a neighborhood V of x^*. This completes the proof of Theorem 1.1 when $J = 2$.

Before proving Theorem 1.1 for $J \geq 3$, we shall prove a weak version of Theorem 1.1,

Lemma 1.2. *Under the same assumptions as Theorem* 1.1, *for any open convex cones* Γ_j' *such that* $\Gamma_j' \Subset \Gamma_j$, *there exist a* $\delta > 0$ *and holomorphic functions* $F_{jk}(z)$ *defined on a neighborhood of*

$$D_0(\delta, \Gamma_j' + \Gamma_k') = \{(0, z'') \in \{0\} \times \mathbf{C}^m; |z''| < \delta, \operatorname{Im} z'' \in \Gamma_j' + \Gamma_k'\}$$

such that

$$F_j(z) = \sum_{k=1}^{J} F_{jk}(z) \qquad (j = 1, \ldots, J),$$

$$F_{jk}(z) + F_{kj}(z) = 0 \qquad (j, k = 1, \ldots, J).$$

Proof. We may assume

$$x^* = \left(0, \frac{\partial}{\partial x_1} 0\right).$$

Put

$$G_j(z_1, \zeta_2, \ldots, \zeta_d, z'') = F_j(z_1, z_1\zeta_2, \ldots, z_1\zeta_d, z'').$$

Then by the proof of Lemma 1.1, G_j are holomorphic on a neighborhood of $D_1(\varepsilon_1, \tilde{\Gamma}_j)$ for an $\varepsilon_1 > 0$ and open convex cones $\tilde{\Gamma}_j$ containing $\{0\} \times \Gamma_j$. Hence by the edge-of-the-wedge theorem for $\mathscr{B}_1^F$, we can find, for any open convex cones $\Gamma'_j \Subset \Gamma_j$, holomorphic functions $G_j(z_1, \zeta_2, \ldots, \zeta_d, z'')$ defined on a neighborhood of $D_1(\varepsilon_2, \tilde{\Gamma}_{jk})$ with an $\varepsilon_2 > 0$ and open convex cones $\tilde{\Gamma}_{jk}$ containing $\{0\} \times (\Gamma'_j + \Gamma'_k)$ such that

$$G_j = \sum_{k=1}^{J} G_{jk} \qquad (k = 1, \ldots, J),$$

$$G_{jk} + G_{kj} = 0 \qquad (j, k = 1, \ldots, J).$$

In particular, G_{jk} are holomorphic on

$$\{(z_1, \zeta_2, \ldots, \zeta_d, z'') \in \mathbf{C}^n;\ |z_1| < \varphi(z''),\ |\zeta_\nu| < \varphi(z'')\ (\nu = 2, \ldots, d),$$
$$|z''| < \varepsilon_2,\ \mathrm{Im}\ z'' \in \Gamma'_j + \Gamma'_k\},$$

where φ is a positive-valued lower semicontinuous function on

$$D''(\varepsilon_2, \Gamma'_j + \Gamma'_k) = \{z'' \in \mathbf{C}^m;\ |z''| < \varepsilon_2,\ \mathrm{Im}\ z'' \in \Gamma'_j + \Gamma'_k\}.$$

Hence

$$H_{jk}(z) = G_{jk}\left(z_1, \frac{z_2}{z_1}, \ldots, \frac{z_d}{z_1}, z''\right)$$

are developed into the Laurent series of the form

$$H_{jk}(z) = \sum_{\alpha \in \mathbf{Z}^d} a_{jk,\alpha}(z'') z'^{\alpha},$$

with holomorphic functions $a_{jk,\alpha}$ on $D''(\varepsilon_2, \Gamma'_j + \Gamma'_k)$, where $\alpha = (\alpha_1, \ldots, \alpha_d)$ satisfies $\alpha_2, \ldots, \alpha_d \geq 0$. Set

$$F_{jk}(z) = \sum_{\alpha \in \mathbf{N}^d} a_{jk,\alpha}(z'') z'^{\alpha},$$

with $\mathbf{N} = \{0, 1, 2, \ldots\}$. Then F_{jk} are holomorphic on

$$\{z \in \mathbf{C}^n;\ |z| < \delta,\ |z_1| < \varphi(z''),\ |z_\nu| < \varphi(z'')^2\ (\nu = 2, \ldots, d),$$
$$|z''| < \varepsilon_2,\ \mathrm{Im}\ z'' \in \Gamma'_j + \Gamma'_k\}.$$

Since

$$F_j = \sum_{k=1}^{J} H_{jk}, \qquad H_{jk} + H_{kj} = 0,$$

it follows that

$$F_j = \sum_{k=1}^{J} F_{jk}, \qquad F_{jk} + F_{kj} = 0.$$

This completes the proof.

Now we need the following proposition concerning a kind of micro-analyticity of F-mild hyperfunctions in order to prove Theorem 1.1 for $J \geq 3$. For this purpose we put

$$W(z'', \xi) = \frac{(m-1)!}{(-2\pi\sqrt{-1})^m} \times \frac{(1-\sqrt{-1}\langle z'', \xi\rangle)^{m-2}\{1-\sqrt{-1}\langle z'', \xi\rangle - (z''^2 - \langle z'', \xi\rangle^2)\}}{\{\langle z'', \xi\rangle + \sqrt{-1}(z''^2 - \langle z'', \xi\rangle^2)\}^m}$$

for $z'' = (z_{d+1}, \ldots, z_n) \in \mathbf{C}^m$ and $\xi \in S^{m-1} = \{\xi = (\xi_1, \ldots, \xi_m) \in \mathbf{R}^m;\ \xi_1^2 + \cdots + \xi_m^2 = 1\}$, where $\langle z'', \xi\rangle = z_{d+1}\xi_1 + \cdots + z_n\xi_m$ and $z''^2 = z_{d+1}^2 + \cdots + z_n^2$.

Proposition 1.2. *Let x^* be a point of $S_N M$ and z be an admissible local coordinate system around $\mathring{x} = \tau(x^*)$ such that $z(\mathring{x}) = 0$. Let $F_j(z)$ $(j = 1, \ldots, J)$ be holomorphic functions defined on a neighborhood of $D(\varepsilon, \tau(U), \Gamma_j)$ with an $\varepsilon > 0$, open convex cones Γ_j of $\mathbf{R}^m$, and an open neighborhood U of x^* in $\mathrm{Mon}_N(M)$ such that $\tau(U) \supset \{(0, x'') \in N; |x''| < \varepsilon\}$. Put*

$$F(z, \xi) = \sum_{j=1}^{J} \int_{C(\varepsilon/2, \mathring{y}_j'')} F_j(z', w'') W(z'' - w'', \xi)\, dw'',$$

with $C(\varepsilon/2, \mathring{y}_j'') = \{w'' \in \mathbf{C}^m; |\mathrm{Re}\, w''| \leq \varepsilon/2, \mathrm{Im}\, w'' = \mathring{y}_j''\}$ for $\mathring{y}_j'' \in \Gamma_j$. Then the following two conditions are equivalent for any $\mathring{\xi} \in S^{m-1}$:

(*i*) *There exists a $\delta > 0$ such that $F(z, \xi)$ is analytic on a neighborhood of $(0, \mathring{\xi}) \in \mathbf{C}^n \times S^{m-1}$ if $\mathring{y}_j'' \in \Gamma_j$ and $|\mathring{y}_j''| < \delta$ for $j = 1, \ldots, J$.*

(*ii*) *There exist holomorphic functions $G_k(z)$ $(k = 1, \ldots, K)$ defined on a neighborhood of $D(\delta, \tau(V), \Xi_k)$ for a $\delta > 0$, an open neighborhood V of x^* in $\mathrm{Mon}_N(M)$, and open convex cones Ξ_k of $\mathbf{R}^m$ such that*

$$\mathring{\xi} \notin \Xi_k^0 = \{\xi \in S^{m-1}; \langle y'', \xi\rangle \geq 0 \text{ for any } y'' \in \Xi_k\}$$

for any $k = 1, \ldots, K$ and such that

$$\sum_{k=1}^{K} G_k(x', x'' + \sqrt{-1}\Xi_k 0) = \sum_{j=1}^{J} F_j(x', x'' + \sqrt{-1}\Gamma_j 0)$$

on $\tau(V) \cap (M - N)$.

Proof. Assume (*i*). Let $\Delta_0 \subset S^{m-1}$ be a closed neighborhood of $\mathring{\xi}$ such that $F(z, \xi)$ is analytic on a neighborhood of $\{0\} \times \Delta_0$. Let $\Delta_1, \ldots, \Delta_{m+1}$ be closed convex sets of S^{m-1} such that $S^{m-1} = \Delta_0 \cup \Delta_1 \cup \cdots \cup \Delta_{m+1}$, $|\Delta_j \cap \Delta_k| =$

0 if $j \neq k$. Choosing $\mathring{y}_j'' \in \Gamma_j$ so that $|\mathring{y}_j''| < \delta$, put

$$G_k(z) = \int_{\Delta_k} F(z, \xi)\, d\sigma(\xi),$$

where $d\sigma(\xi)$ is the volume element of S^{m-1}. Then G_k are holomorphic on a neighborhood of $D(\varepsilon_1, \tau(V), \operatorname{int}\Delta_k^0)$ for an $\varepsilon_1 > 0$. Moreover G_0 is holomorphic on a neighborhood of $0 \in \mathbf{C}^m$. By virtue of the inverse formula of Radon transformations (cf. Kataoka [4] and Kaneko [1]),

$$\sum_{j=1}^{J} F_j(x', x'' + \sqrt{-1}\Gamma_j 0) = \sum_{k=0}^{K} G_k(x', x'' + \sqrt{-1} \operatorname{int}\Delta_k^0 0)$$

holds on $\tau(V) \cap (M - N)$ for an open neighborhood V of x^*. Since $\mathring{\xi}$ is not contained in Δ_k for $k = 1, \ldots, K$, we get (ii).

Conversely assume (ii). Put $F_{J+k} = -G_k$ and $\Gamma_{J+k} = \Xi_k$ for $k = 1, \ldots, K$. Lemma 1.2 assures that for any open convex cones $\Gamma_j' \Subset \Gamma_j$, there exist holomorphic functions $F_{jk}(z)$ defined on a neighborhood of $D_0(\varepsilon_2, \Gamma_j' + \Gamma_k')$ with an $\varepsilon_2 > 0$ such that

$$F_j(z) = \sum_{k=1}^{J+K} F_{jk}(z) \qquad (j = 1, \ldots, J+K),$$

$$F_{jk}(z) + F_{kj}(z) = 0 \qquad (j, k = 1, \ldots, J+K).$$

We may assume $\tau(V) \supset \{(0, x'') \in N;\ |x''| < \delta\}$, $\delta < \varepsilon$, $\delta < \varepsilon_2$. Put

$$F(z, \xi) = \sum_{j=1}^{J} \int_{C(\delta/2, \mathring{y}_j'')} F_j(z', w'')\, W(z'' - w'', \xi)\, dw'',$$

$$G(z, \xi) = \sum_{k=1}^{K} \int_{C(\delta/2, \mathring{y}_{J+k}'')} G_k(z', w'')\, W(z'' - w'', \xi)\, dw''$$

for $\mathring{y}_j'' \in \Gamma_j$ $(j = 1, \ldots, J+K)$. Then we have

$$\begin{aligned} F(z, \xi) - G(z, \xi) &= \sum_{j=1}^{J+K} \sum_{k=1}^{J+K} \int_{C(\delta/2, \mathring{y}_j'')} F_{jk}(z', w'')\, W(z'' - w'', \xi)\, dw'' \\ &= \sum_{j<k} \int_{C(\delta/2, \mathring{y}_j'') - C(\delta/2, \mathring{y}_k'')} F_{jk}(z', w'')\, W(z'' - w'', \xi)\, dw'' \\ &= \sum_{j<k} \int_{C_{jk}} F_{jk}(z', w'')\, W(z'' - w'', \xi)\, dw'', \end{aligned}$$

where $C_{jk}=\{w''\in\mathbf{C}^m;\ |\mathrm{Re}\, w''|=\delta/2,\ \mathrm{Im}\, w''=(1-t)\mathring{y}''_j+t\mathring{y}''_k$ for some $0\le t\le 1\}$ with an appropriate orientation. It is easy to see that the integrals of the last line are holomorphic on a neighborhood of $0\in\mathbf{C}^n$ if $|\mathring{y}''_j|$ are small enough. Since $G(z,\xi)$ is analytic on a neighborhood of $(0,\mathring{\xi})$, so is $F(z,\xi)$. This completes the proof.

Proof of Theorem 1.1 for $J\ge 3$. We prove Theorem 1.1 by induction on J. We may assume $\tau(U)\supset\{(0,x'');\ |x''|<\varepsilon\}$. Assume Theorem 1.2 for $J-1$ and put

$$F(z,\xi)=\int_{C(\varepsilon/2,\mathring{y}''_J)} F_J(z',w'')\,W(z''-w'',\xi)\,dw''$$

for $\mathring{y}''_J\in\Gamma_J$. Then since

$$-F_J(x',x''+\sqrt{-1}\Gamma_J0)=\sum_{j=1}^{J-1}F_j(x',x''+\sqrt{-1}\Gamma_j0),$$

$F(z,\xi)$ is analytic on a neighborhood of

$$\{0\}\times(S^{m-1}-(\Gamma_J^0\cap(\Gamma_1^0\cup\cdots\cup\Gamma_{J-1}^0)))$$

if $|\mathring{y}''_J|$ is sufficiently small, by virtue of Proposition 1.2. Let Γ''_j be open convex cones such that $\Gamma'_j\Subset\Gamma''_j\Subset\Gamma_j$ and let Δ_j $(j=1,\dots,J-1)$ be closed sets of S^{m-1} such that

$$\Delta_j\Subset(\Gamma''_j+\Gamma''_J)^0,\qquad \bigcup_{j=1}^{J-1}\Delta_j\Supset(\Gamma_1^0\cup\cdots\cup\Gamma_{J-1}^0)\cap\Gamma_J^0,$$

and such that the measure of $\Delta_j\cap\Delta_k$ is zero if $j\ne k$. Put

$$G_j(z)=\int_{\Delta_j}F(z,\xi)\,d\sigma(\xi)$$

for $j=0,1,\dots,J-1$, where $\Delta_0=S^{m-1}-(\Delta_1\cup\cdots\cup\Delta_{J-1})$. Then

$$F_J(z)=\sum_{j=0}^{J-1}G_j(z)$$

holds and $G_0(z)$ is holomorphic on a neighborhood of $0\in\mathbf{C}^n$. Put

$$H_1(z)=F_1(z)+G_0(z)+G_1(z),$$
$$H_j(z)=F_j(z)+G_j(z)\qquad(j=2,\dots,J-1).$$

Then we have

$$\sum_{j=1}^{J-1}H_j(x',x''+\sqrt{-1}\Gamma''_j0)=0$$

on $\tau(U')\cap(M-N)$ for a neighborhood U' of x^*. By the induction hypothesis, we can find holomorphic functions H_{jk} $(j, k=1,\ldots,J-1)$ defined on a neighborhood of $D(\delta, \tau(V), \Gamma'_j+\Gamma'_k)$ with a $\delta>0$ and an open neighborhood V of x^* such that

$$H_j=\sum_{k=1}^{J-1} H_{jk}, \qquad H_{jk}+H_{kj}=0 \qquad (j, k=1,\ldots,J-1).$$

Put $F_{jk}=H_{jk}$ $(j, k=1,\ldots,J-1)$, $F_{1,J}=-F_{J,1}=-G_0-G_1$, and $F_{j,J}=-F_{J,j}=-G_j$ $(j=2,\ldots,J-1)$. Then F_{jk} satisfy the properties we need. This completes the proof of Theorem 1.1.

Proposition 1.3. *There exists a sheaf homomorphism*

$$\gamma^F:\mathscr{B}^F_{N|M}\to\tau^{-1}\mathscr{B}_N$$

defined by

$$\gamma^F(u)=\sum_{j=1}^{J} F_j(0, x''+\sqrt{-1}\Gamma_j 0)$$

if a section u of $\mathscr{B}^F_{N|M}$ is expressed by (1) *in Definition 1.2.*

Proof. $\gamma^F(u)$ does not depend on the choice of defining functions of u in view of Theorem 1.1.

This proposition means that the boundary value $\gamma^F(u)$ does not depend on the direction along which the boundary value is taken. More precisely, let $u(x)$ be a section of $\mathscr{B}^F_{N|M}$ on an open set U of S_NM with connected fibers. Then there exists a section $v(x')$ of $\mathscr{B}_N$ on $\tau(U)$ such that $\gamma^F(u|_V)=\tau^{-1}(v)|_V$ for any open subset V of U.

As a special case of F-mild hyperfunctions, let us consider F-mild hyperfunctions on a whole fiber of τ.

Proposition 1.4. *Suppose $d\geq 2$. Let $u(x)$ be a hyperfunction on $M-N$ which is F-mild at each point of S_NM. Then there exists a unique hyperfunction $v(x)$ on M such that $v(x)=u(x)$ on $M-N$ and such that $\mathrm{SS}(v)\cap S^*_NM=\varnothing$, where $\mathrm{SS}(v)$ denotes the singular spectrum of v, and S^*_NM denotes the conormal sphere bundle of N in M.*

Proof. The uniqueness of v follows immediately from Holmgren's uniqueness theorem for hyperfunctions. Hence it suffices to prove the existence of v on a neighborhood of each point of N. Let $\mathring{x}$ be a point of N and let z be an admissible local coordinate system such that $z(\mathring{x})=0$. Since $u(x)$ is F-mild at each point of $\tau^{-1}(\mathring{x})$, there exist open cones U^ν $(\nu=1,\ldots,\nu_0)$ of $\mathbf{R}^d$ such that $U^1\cup\cdots\cup U^{\nu_0}=\mathbf{R}^d-\{0\}$, an $\varepsilon>0$, and holomorphic functions $F_j^\nu(z)$ $(j=1,\ldots,J_\nu)$ defined on a neighborhood of

$$\{(x',z'')\in\mathbf{R}^d\times\mathbf{C}^m;\ |x'|<\varepsilon,\ |z''|<\varepsilon,\ x'\in U^\nu\cup\{0\},\ \operatorname{Im} z''\in\Gamma_j^\nu\}$$

with open convex cones Γ_j^ν of $\mathbf{R}^m$ such that

$$u(x)=\sum_{j=1}^{J_\nu} F_j^\nu(x',x''+\sqrt{-1}\Gamma_j^\nu 0)$$

on $\{x\in\mathbf{R}^n;\ |x|<\varepsilon,\ x'\in U^\nu\}$. Choosing sufficiently small $\mathring{y}_j''^\nu\in\Gamma_j^\nu$ we put

$$F^\nu(z,\xi)=\sum_{j=1}^{J_\nu}\int_{C(\varepsilon/2,\mathring{y}_j''^\nu)} F_j^\nu(z',w'')\,W(z''-w'',\xi)\,dw''.$$

Then $F^\nu(z,\xi)$ is analytic on a neighborhood of

$$\{(x',z'',\xi)\in\mathbf{R}^d\times\mathbf{C}^m\times S^{m-1};\ |x'|<\varepsilon,\ x'\in U^\nu\cup\{0\},$$
$$|z''|<\varepsilon/4,\ \langle y'',\xi\rangle>y''^2-\langle y'',\xi\rangle^2\}$$

if ε is small enough. Proposition 1.2 implies that there exists a $\delta>0$ $(\delta<\varepsilon/4)$ so that $F^\nu(z,\xi)-F^\mu(z,\xi)$ is analytic on $B_\delta\times S^{m-1}=\{(z,\xi)\in\mathbf{C}^n\times\mathbf{S}^{m-1};\ |z|<\delta\}$ if $U^\nu\cap U^\mu\neq\varnothing$. It is easy to see that $F^\nu-F^\mu$ is analytic on $B_\delta\times S^{m-1}$ for any μ and ν by taking a chain of U^i's connecting U^ν and U^μ. Hence there exist analytic functions $G^\nu(z,\xi)$ on $B_\delta\times S^{m-1}$ such that $G^\nu-G^\mu=F^\nu-F^\mu$. Then $F^\nu-G^\nu$ define an analytic function $H(z,\xi)$ on

$$\{(x',z'',\xi)\in\mathbf{R}^d\times\mathbf{C}^m\times\mathbf{S}^{m-1};\ |x'|<\delta,\ |z''|<\delta,\ \langle y'',\xi\rangle>y''^2-\langle y'',\xi\rangle^2\}.$$

Let Δ_j $(j=1,\ldots,J)$ be proper convex closed sets of S^{m-1} such that $\Delta_1\cup\cdots\cup\Delta_J=S^{m-1}$ and that the measure of $\Delta_j\cap\Delta_k$ is zero if $j\neq k$. Set

$$H_j(z)=\int_{\Delta_j} H(z,\xi)\,d\sigma(\xi).$$

Then

$$w(x)=\sum_{j=1}^{J} H_j(x',x''+\sqrt{-1}\,\operatorname{int}(\Delta_j^0)0)$$

defines a hyperfunction on $\{x\in\mathbf{R}^n;\ |x|<\delta\}$ with real analytic parameters x'. In view of the inverse formula for Radon transforms, we get

$$u(x)-w(x)=\int_{S^{m-1}} G^\nu(x,\xi)\,d\sigma(\xi)$$

on $\{x; |x| < \delta/2, x' \in U''\}$. Since the integral on the right side is analytic on $\{x; |x| < \delta/2\}$, $u(x) - w(x)$ is continued to an analytic function on $\{x: |x| < \delta/2\}$. This completes the proof.

2. Higher-Codimensional Boundary Value Problem

In order to formulate the boundary value problem, we introduce some sheaves attached to the boundary. We use the same notation as in Section 1. Let $\tilde{\iota}: \tilde{M} - Y \to \mathrm{Mon}_Y(\tilde{M})$ be the inclusion map and let $\mathcal{BO} = \mathcal{H}^d_{\tilde{M}}(\mathcal{O}_X) \otimes \omega_{\tilde{M}}$ be the sheaf on $\tilde{M}$ of hyperfunctions with holomorphic parameters, where $\mathcal{O}_X$ denotes the sheaf on X of holomorphic functions, and $\omega_{\tilde{M}}$, the sheaf of orientation of $\tilde{M}$.

Definition 2.1.

$$\mathcal{BO}_{Y|\tilde{M}} = (\tilde{\iota}_*(\mathcal{BO}|_{\tilde{M}-Y}))|_{S_Y\tilde{M}}, \qquad \tilde{\mathcal{B}}_{N|M} = \mathcal{H}^m_{S_NM}(\mathcal{BO}_{Y|\tilde{M}}) \otimes \omega_N.$$

Definition 2.2. $\tilde{\mathcal{B}}^A = \mathcal{H}^m_N(\mathcal{O}_X|_Y) \otimes \omega_N$.

It is easy to see that there exist natural homomorphisms

$$\mathcal{B}_{N|M} \to \tilde{\mathcal{B}}_{N|M}, \qquad \tau^{-1}\tilde{\mathcal{B}}^A \to \tilde{\mathcal{B}}_{N|M}.$$

In the same way as in the proof of Lemma 2.1 of Ôaku [7], we get

Lemma 2.1. *The homomorphism $\tau^{-1}\tilde{\mathcal{B}}^A \to \tilde{\mathcal{B}}_{N|M}$ is injective.*

Lemma 2.2. *There exists an injective sheaf homomorphism*

$$\mathcal{B}^F_{N|M} \to \tau^{-1}\tilde{\mathcal{B}}^A.$$

Proof. This homomorphism is well-defined by virtue of Theorem 1.1. The injectivity follows from the edge-of-the-wedge theorem for $\tilde{\mathcal{B}}^A$ (cf. [7]) and the proof of Proposition 1.2.

The method of the proof of Proposition 2.3 of [7] gives

Lemma 2.3. *The homomorphism $\mathcal{B}_{N|M}/\mathcal{B}^F_{N|M} \to \tilde{\mathcal{B}}_{N|M}/\tau^{-1}\tilde{\mathcal{B}}^A$ is injective.*

Now let $\mathcal{M}$ be a coherent $\mathcal{D}_X$-module on X such that Y is non-characteristic with respect to $\mathcal{M}$, i.e., the intersection of the characteristic variety of $\mathcal{M}$ and of the conormal bundle $T^*_Y X$ is contained in the zero section. Here $\mathcal{D}_X$ denotes the sheaf on X of rings of linear partial differential operators (of finite order) with holomorphic coefficients. Then the tangential system $\mathcal{M}_Y$ of $\mathcal{M}$ to Y is defined as a coherent $\mathcal{D}_Y$-module (cf. Kashiwara [2]).

Lemma 2.4. *Under the above assumptions we have*

$$\mathcal{H}om_{\mathcal{D}_X}(\mathcal{M}, \tilde{\mathcal{B}}_{N|M}/\tau^{-1}\tilde{\mathcal{B}}^A)=0.$$

Proof. Let $\tilde{\tau}: S_Y\tilde{M} \to Y$ be the canonical projection. First we shall show an isomorphism

$$\mathbf{R}\mathcal{H}om_{\mathcal{D}_X}(\mathcal{M}, \tilde{\tau}^{-1}(\mathcal{B}\mathcal{O}|_Y)) \xrightarrow{\sim} \mathbf{R}\mathcal{H}om_{\mathcal{B}_X}(\mathcal{M}, \mathcal{B}\mathcal{O}_{Y|\tilde{M}}). \tag{2}$$

For this purpose we define a sheaf $\mathcal{F}$ on $S_Y\tilde{M}$ as the kernel sheaf of the natural homomorphism of $\tilde{\tau}^{-1}(\mathcal{B}\mathcal{O}|_Y)$ to $\mathcal{B}\mathcal{O}_{Y|\tilde{M}}$. Let x^* be a point of $S_Y\tilde{M}$ and let z be an admissible local coordinate system such that

$$x^* = \left(0, \frac{\partial}{\partial x_1} 0\right).$$

Then by definition we have

$$\mathcal{F}_{x^*} = \varinjlim_{\varepsilon} H^0_{U_\varepsilon - V_\varepsilon}(U_\varepsilon; \mathcal{B}\mathcal{O}),$$

with $U_\varepsilon = \{(x', z'') \in \tilde{M};\ |x'| < \varepsilon,\ |z''| < \varepsilon\}$ and $V_\varepsilon = \{(x', z'') \in U_\varepsilon;\ \varepsilon x_1 > |x_2| + \cdots + |x_d|\}$. Hence we get

$$\mathbf{R}\mathcal{H}om_{\mathcal{D}_X}(\mathcal{M}, \mathcal{F})_{x^*} = \varinjlim_{\varepsilon} \mathbf{R}\mathcal{H}om_{\mathcal{D}_X}(\mathcal{M}, \Gamma_{U_\varepsilon - V_\varepsilon}(U_\varepsilon; \mathcal{B}\mathcal{O})) = 0,$$

applying Corollary 2.2.2 of Kashiwara–Schapira [3] to $\mathcal{M}$ combined with the partial Cauchy–Riemann system defining $\mathcal{B}\mathcal{O}$. This proves (2). Since Y is non-characteristic with respect to $\mathcal{M}$, we get

$$\mathbf{R}\mathcal{H}om_{\mathcal{D}_X}(\mathcal{M}, \mathcal{B}\mathcal{O}_{Y|\tilde{M}}) = \tilde{\tau}^{-1}\mathbf{R}\mathcal{H}om_{\mathcal{D}_X}(\mathcal{M}, \mathcal{B}\mathcal{O}|_Y) = \tilde{\tau}^{-1}\mathbf{R}\mathcal{H}om_{\mathcal{D}_X}(\mathcal{M}, \mathcal{O}_X|_Y).$$

Applying the derived functor $\mathbf{R}\Gamma_{S_N M}$ we get the conclusion.

It is easy to see that there is a natural sheaf homomorphism

$$\gamma^F : \mathcal{H}om_{\mathcal{D}_X}(\mathcal{M}, \mathcal{B}^F_{N|M}) \to \tau^{-1}\mathcal{H}om_{\mathcal{D}_Y}(\mathcal{M}_Y, \mathcal{B}_N)$$

induced by the boundary value homomorphism $\gamma^F : \mathcal{B}^F_{N|M} \to \tau^{-1}\mathcal{B}_N$. Lemmas 2.3 and 2.4 imply

Proposition 2.1. $\mathcal{H}om_{\mathcal{D}_X}(\mathcal{M}, \mathcal{B}_{N|M}) = \mathcal{H}om_{\mathcal{D}_X}(\mathcal{M}, \mathcal{B}^F_{N|M})$.

The Cauchy–Kowalevsky theorem by Kashiwara [2] gives

Proposition 2.2. *There exists an isomorphism*

$$\tilde{\gamma} : \mathcal{H}om_{\mathcal{D}_X}(\mathcal{M}, \tilde{\mathcal{B}}^A) \rightleftarrows \tau^{-1}\mathcal{H}om_{\mathcal{D}_Y}(\mathcal{M}_Y, \mathcal{B}_N)$$

compatible with γ^F.

Combined together these facts give the boundary value homomorphism of the following

Theorem 2.1. *There exists an injective sheaf homomorphism*

$$\gamma : \mathcal{H}om_{\mathcal{D}_X}(\mathcal{M}, \mathcal{B}_{N|M}) \to \tau^{-1}\mathcal{H}om_{\mathcal{D}_Y}(\mathcal{M}_Y, \mathcal{B}_N),$$

which coincides with the γ^F *above.*

Remark. (*i*) It is easy to see that γ is independent of $\tilde{M}$ since so is $\mathcal{B}^F_1$.
(*ii*) Schapira [10] has defined this boundary value homomorphism in a more general situation by using the theory of microlocalization of sheaves developed by Kashiwara and Schapira.

Finally let us give a new proof to an extension theorem of Kawai as an application of our theory.

Theorem 2.2 (Theorem 1 of Kawai [5]). *Let N be a real analytic submanifold of codimension* ≥ 2 *of a real analytic manifold M and let X, Y be complexifications of M, N respectively. Suppose that* $\mathcal{M}$ *is a coherent* $\mathcal{D}_X$*-module on X for which Y is non-characteristic and suppose that* $u(x)$ *is a hyperfunction solution of* $\mathcal{M}$ *(i.e., a section of* $\mathcal{H}om_{\mathcal{D}_X}(\mathcal{M}, \mathcal{B}_M)$*) on* $M - N$*. Then there exists a unique hyperfunction solution* $\tilde{u}(x)$ *of* $\mathcal{M}$ *on M which coincides with* $u(x)$ *on* $M - N$*.*

Proof. Since $\tilde{u}(x)$ is unique, we may assume the existence of $\tilde{M}$. By Proposition 2.1, $u(x)$ defines an F-mild hyperfunction on S_NM. Hence Proposition 1.4 implies the existence of a hyperfunction $\tilde{u}(x)$ such that $\mathrm{SS}(\tilde{u}(x)) \cap S^*_NM = \varnothing$ and such that $\tilde{u}(x)$ coincides with $u(x)$ on $M - N$.

It follows from Holmgren's theorem for hyperfunctions that $\tilde{u}(x)$ is a hyperfunction solution of $\mathcal{M}$ on M. This completes the proof.

Remark. This theorem can also be proved by arguments in [8, Section 1] applied to a hypersurface containing N.

References

[1] A. Kaneko, *Introduction to Hyperfunctions, I, II*, University of Tokyo Press, Tokyo, 1980, 1982.

[2] M. Kashiwara, *Systems of Microdifferential Equations*, Birkhäuser, Boston-Basel-Stuttgart, 1983.

[3] M. Kashiwara and P. Schapira, Micro-hyperbolic systems, *Acta Math.* **142** (1979) 1-55.

[4] K. Kataoka, On the theory of Radon transformations of hyperfunctions, *J. Fac. Sci. Univ. Tokyo* **28** (1981) 331-413.

[5] T. Kawai, Extension of solutions of systems of linear differential equations, *Publ. RIMS Kyoto Univ.* **12** (1976) 215-227.

[6] T. Ôaku, *F*-mild hyperfunctions and Fuchsian partial differential equations, *Advanced Studies in Pure Math.* **4** (1984) 223-242.

[7] T. Ôaku, Boundary value problems for systems of linear partial differential equations and propagation of micro-analyticity, *J. Fac. Sci. Univ. Tokyo* **33** (1986) 175-232.

[8] T. Ôaku, Removable singularities of solutions of linear partial differential equations—systems and Fuchsian equations, *J. Fac. Sci. Univ. Tokyo* **33** (1986) 403-428.

[9] M. Sato, T. Kawai and M. Kashiwara, Microfunctions and pseudodifferential equations, *Lecture Notes in Math.* **287**, Springer, 1973, 265-529.

[10] P. Schapira, Front d'onde analytique au bord II, *Séminaire équations aux dérivées partielles 1985-1986*, Centre de math., Ecole polytechnique.

Hodge Numbers of a Kummer Covering of $\mathbf{P}^2$ Ramified along a Line Configuration

Takayuki Oda
Department of Mathematics
Faculty of Science
Niigata University
Niigata, Japan

0. Introduction

Let us consider a line configuration $\bigcup_{i=1}^{k} L_i$ in the two-dimensional projective space $\mathbf{P}^2$ over the complex number field $\mathbf{C}$, and form a Kummer covering X' of $\mathbf{P}^2$ ramified along every line L_i with exponent n. The minimal desingularization X of X' has an action of $A = \mathrm{Gal}(X'/\mathbf{P}^2)$. Hence we can decompose the cohomology groups $H^i(X)$ of X into eigenspaces $H^i(\alpha)$ with respect to characters α of A. We define a notion of generic characters α. It is proved that $H^1(\alpha) = H^3(\alpha) = \{0\}$ for such characters. We give an explicit formula of Hodge numbers $h^{i,j}(\alpha) = \dim_{\mathbf{C}} H^{i,j}(\alpha)$ $(i+j=2)$ in terms of certain invariants of α for generic characters α.

Let us explain the outline of this paper. In Section 1, we give basic notations. In Section 2, Theorem 1, which gives $\mathrm{e}(\alpha) =$

ISBN 0-12-400466-0

$\sum_{i=0}^{4}(-1)^i \dim_{\mathbf{C}} H^i(\alpha)$, is formulated, and proved in Section 3. In Section 4, recalling the results of Kohno [3] and Ishida [2], we prove a formula for $h^{i,2-i}(\alpha)$ for generic α.

Except for some explicit computation, the main ingredients of this paper come from the results of Saito [4], Kohno [3], and Ishida [2]. We thank them and Prof. K. Aomoto for valuable information.

1. Basic Notation

For this section, we refer to Hirzebruch [1] for basic results. Let us consider the 2-dimensional projective space $\mathbf{P}^2$ defined over $\mathbf{C}$. Let $L_1, \ldots, L_k$ be k distinct lines in $\mathbf{P}^2$. Each line L_i is defined by a linear equation $l_i = 0$ with linear form l_i. Let t_r $(r \geq 2)$ be the number of r-tuple points on this line configuration. Then

$$\tfrac{1}{2}k(k-1) = \sum_{r=2}^{\infty} t_r \frac{r(r-1)}{2}.$$

Fix a natural number $n \geq 2$. Let us consider the maximal abelian covering X'_n of $\mathbf{P}^2$ ramified only along each L_i with ramification index n. Then the function field of X'_n is given by

$$K = \mathbf{C}(x, y, \sqrt[n]{l_i/l_j} \quad (1 \leq i, j \leq k)).$$

X'_n is the normalization of $\mathbf{P}^2$ in the function field K. X'_n has singularities which lie over some r-tuple $(r \geq 3)$ point on $\bigcup_{i=1}^k L_i$.

Let S be the set of singular points of $\bigcup L_i$, and let $S' = S$-{double points}. Let F be the surface obtained from $\mathbf{P}^2$ by blowing up at all the points in S'. Then the Euler number $\mathrm{e}(F)$ of F is given by

$$\mathrm{e}(F) = \mathrm{e}(\mathbf{P}^2) + \sum_{r \geq 3} t_r.$$

Let X_n be the minimal resolution of X'_n. Then the natural map

$$\pi : X_n \to F$$

is a finite flat morphism. Put

$$A = \mathrm{Gal}(X_n/F) = \mathrm{Gal}(K/\mathbf{C}(x, y)).$$

Let μ_n be the group of n-th root of unity. Then $A \simeq \mu_n^{\oplus k}/\mu_n$ via the correspondence

$$\sigma \in A \mapsto ((\zeta_1, \zeta_2, \ldots, \zeta_k) \bmod \mu_n) \in \mu_n^{\oplus k}/\mu_n,$$

where $\sigma(\sqrt[n]{l_i/l_j}) = (\zeta_i/\zeta_j)\sqrt[n]{l_i/l_j}$ for each i, j. Let $\alpha \in A^*$ be a character of A. Then we put

$$H^i(\alpha) = \{\omega \in H^i(X_n, \mathbf{C}) \mid g^*(\omega) = \alpha(g)\omega, \text{ for all } g \in A\}.$$

We have

$$H^i(X_n, \mathbf{C}) = \bigoplus_{\alpha \in A^*} H^i(\alpha) \qquad \text{(eigenspace decomposition)}$$

Our first result is to give an explicit formula for

$$\mathrm{e}(\alpha) = \sum_{i=0}^{4} (-1)^i \dim_{\mathbf{C}} H^i(\alpha).$$

2. Formulation of Theorem 1

Let S_r be the set of r-tuple points in $L_1 \cup L_2 \cup \cdots \cup L_k$. Assume that $r \geq 3$. Let $p \in S_r$, and put $E_p = \rho^{-1}(p)$, where $\rho: F \to \mathbf{P}^2$ is the blowing up. The line E_p on F intersects with the proper transform L_i^* of L_i at one point if and only if $p \in L_i$. Set

$$S(E_p) = \bigcup_{i=1}^{k} (L_i^* \cap E_p).$$

Then

$$|S(E_p)| = r.$$

For each L_i^*, if $r \geq 3$, we put

$$S_r(L_i^*) = \bigcup_{p \in S_r} (L_i^* \cap E_p),$$

and if $r = 2$, we put

$$S_2(L_i^*) = \bigcup_{\substack{j=1 \\ j \neq i}}^{k} (L_i^* \cap L_j^*) \quad (\simeq S_2(L_i)).$$

Set

$$D = \left(\bigcup_{i=1}^{k} L_i^*\right) \cup \left(\bigcup_{r \geq 3} \left(\bigcup_{p \in S_r} E_p\right)\right).$$

Then D is a divisor of normal crossing. Let D^{sing} be the set of singular points on D.

Put

$$L_i^0 = L_i^* - L_i^* \cap D^{\text{sing}}, \qquad E_p^0 = E_p - E_p \cap D^{\text{sing}}.$$

Let Λ_p be the set of lines which pass through $p \in S_r$. Then for each $p \in S_r$, $|\Lambda_p| = r$. Moreover set $\Lambda_p^* = \{L_i^* \mid L_i \in \Lambda_p\}$. We write

$$\Lambda = \{L_1, \ldots, L_k\}.$$

Then $A \simeq \operatorname{Coker}(\Delta : \mu_n \to \mu_n^\Lambda)$, where Δ is the diagonal homomorphism, and μ_n^Λ is the set of maps from Λ to μ_n, which is a direct sum of k copies of μ_n.

The inertia groups of A at divisors L_i^* and E_p are given as follows, respectively:

For $L_i^* = L^*$, $H_{L_i} = H_L = \mu_n^L \Delta(\mu_n)/\Delta(\mu_n)$ $(1 \leq i \leq k)$.

For E_p with $p \in S_r$ $(r \geq 3)$, define the diagonal homomorphism $\Delta_p : \mu_n \to \bigoplus_{L \in \Lambda_p} \mu_n^L$, then $K_{r,p} = \Delta_p(\mu_n)\Delta(\mu_n)/\Delta(\mu_n)$.

Definition. Let $\alpha \in A^*$, and let B be a subgroup of A. Then we put

$$\delta(\alpha \mid B) = \begin{cases} 1, & \text{if } \alpha_{|B} = 1; \\ 0, & \text{if } \alpha_{|B} \neq 1. \end{cases}$$

The following theorem gives $\mathrm{e}(\alpha)$ in terms of $\delta(* \mid *)$.

Theorem 1 (Lefschetz fixed point theorem).

$$\mathrm{e}(\alpha) = 3 - 2k + \sum_{r \geq 2} (r-1)t_r + \sum_{L \in \Lambda} \mathrm{e}(L^0)\delta(\alpha \mid H_L) + \sum_{r \geq 3} (2-r) \sum_{p \in S_r} \delta(\alpha \mid K_{r,p})$$

$$+ \sum_{p \in S_2} \delta\left(\alpha \mid \prod_{L \in \Lambda_p} H_L\right) + \sum_{r \geq 3} \sum_{p \in S_r} \sum_{L \in \Lambda_p} \delta(\alpha \mid K_{r,p} \times H_L).$$

Definition. A character $\alpha \in A^*$ of A is called *generic* if $\delta(\alpha \mid H_L) = 0$ and $\delta(\alpha \mid K_{r,p}) = 0$ for any L and any $p \in S_r$.

Corollary. *If α is generic, then* $\mathrm{e}(\alpha) = 3 - 2k + \sum_{r \geq 2} (r-1)t_r$.

3. Proof of Theorem 1

For each $g \in A$, set

$$\operatorname{Lef}(g) = \sum_{i=0}^{4} (-1)^i \operatorname{trace}(g|_{H^i(X_n, \mathbf{C})}).$$

Then

$$\mathrm{Lef}(g)=\sum_{\alpha\in A^*}\mathrm{e}(\alpha)\alpha(g).$$

Hence

$$\mathrm{e}(\alpha)=\frac{1}{|A|}\sum_{g\in A}\bar{\alpha}(g)\,\mathrm{Lef}(g).$$

If $g\in A$ is not contained in

$$\{1\}\cup\Big(\bigcup_{L\in\Lambda}H_L-\{1\}\Big)\cup\Big(\bigcup_{r\geq 3}\bigcup_{p\in S_r}K_{r,p}-\{1\}\Big)\cup\Big(\bigcup_{p\in S_2}\Big(\prod_{L\in\Lambda_p}H_L-\bigcup_{L\in\Lambda_p}H_L\Big)\Big)$$
$$\cup\Big(\bigcup_{r\geq 3}\bigcup_{p\in S_r}\bigcup_{L\in\Lambda_p}(K_{r,p}\times H_L-K_{r,p}\cup H_L)\Big),$$

then

$$\mathrm{Lef}(g)=0.$$

Hence, $\mathrm{e}(\alpha)$ is given as a sum:

$$\begin{aligned}\mathrm{e}(\alpha)&=\frac{1}{|A|}\bar{\alpha}(1)\,\mathrm{Lef}(1) &&\text{(Main term)}\\ &\quad+\sum_{L\in\Lambda}\frac{1}{|A|}\sum_{g\in H_L-\{1\}}\bar{\alpha}(g)\,\mathrm{Lef}(g) &&(H\text{-terms})\\ &\quad+\sum_{r\geq 3}\sum_{p\in S_r}\frac{1}{|A|}\sum_{g\in K_{r,p}-\{1\}}\bar{\alpha}(g)\,\mathrm{Lef}(g) &&(K\text{-terms})\\ &\quad+\sum_{p\in S_2}\frac{1}{|A|}\sum_{g\in\prod_{L\in\Lambda_p}H_L-\bigcup_{L\in\Lambda_p}H_L}\bar{\alpha}(g)\,\mathrm{Lef}(g) &&(H\times H\text{-terms})\\ &\quad+\sum_{r\geq 3}\sum_{p\in S_r}\sum_{L\in\Lambda_p}\frac{1}{|A|}\sum_{g\in H_L\times K_{r,p}-H_L\cup K_{r,p}}\bar{\alpha}(g)\,\mathrm{Lef}(g). &&(H\times K\text{-terms})\end{aligned}$$

We compute the contribution from the unity of A.

Lemma A.

$$\text{Main term}=\frac{1}{|A|}\bar{\alpha}(1)\,\mathrm{Lef}(1)=\frac{1}{n^{k-1}}\mathrm{e}(X_n)$$
$$=n^{-k+1}\Big[n^{k-1}\Big(3-2k+\sum_{r\geq 2}(r-1)t_r\Big)+n^{k-2}\Big(2k-\sum_{r\geq 2}rt_r\Big)+n^{k-3}t_2$$
$$+n^{k-2}\sum_{r\geq 3}(2-r)t_r+n^{k-3}\sum_{r\geq 3}rt_r\Big].$$

3.1. Computation of H-Terms

Consider the stratification

$$\bigcup_{i=1}^{k} L_i^* = \left(\bigcup_{i=1}^{k} L_i^0\right) \cup \left(\bigcup_{i=1}^{k} (D^{\text{sing}} \cap L_i^*)\right).$$

Then

$$\mathrm{Lef}(g) = \mathrm{Lef}(g)\big|_{\pi^{-1}(\bigcup_{i=1}^{k} L_i^0)} + \mathrm{Lef}(g)\big|_{\pi^{-1}(\bigcup_{i=1}^{k}(D^{\text{sing}} \cap L_i^*))}.$$

Let us call the first sum $(H-1)$. Then

$$(H-1) = n^{k-2} \sum_{i=1}^{k} \mathrm{e}(L_i^0) n\delta(\alpha \mid H_i) - n^{k-2} \sum_{i=1}^{k} \mathrm{e}(L_i^0).$$

Note here that

$$\sum_{i=1}^{k} \mathrm{e}(L_i^0) = \sum_{L \in \Lambda} \mathrm{e}(L^0) = 2k - \sum_{r \geq 2} rt_r.$$

On the other hand, $(H-0)$ is given by

$$(H-0) = n^{k-3} \sum_{i=1}^{k} t_2(L_i) n\delta(\alpha \mid H_i) - n^{k-3} \sum_{i=1}^{k} t_2(L_i) \qquad \text{(double points)}$$

$$+ n^{k-3} \sum_{r \geq 3} \left\{ \sum_{i=1}^{k} t_r(L_i) n\delta(\alpha \mid H_i) - \sum_{i=1}^{k} t_r(L_i) \right\}. \qquad (r\text{-tuple points})$$

Lemma B.

$$H\text{-term} = \left[n^{k-1} \sum_{L \in \Lambda} \mathrm{e}(L^0) - n^{k-2}\left(2k - \sum_{r \geq 2} rt_r\right) + n^{k-2}\left\{ \sum_{r \geq 2} \sum_{L \in \Lambda} t_r(L)\delta(\alpha \mid H_L) \right\} - n^{k-3} \sum_{r \geq 2} rt_r \right] n^{-k+1}.$$

3.2. Computation of K-Terms

Let $p \in S_r$ $(r \geq 3)$. Consider $\pi: X_n \to F$. Let C_r be an irreducible component of $\pi^{-1}(E_p)_{\text{red}}$. Then π induces a finite flat morphism $C_r \to E_p$ of degree n^2. Set $E_p^{\text{sing}} = E_p \cap D^{\text{sing}} = \{p_1, \ldots, p_r\}$. Put $C_r^0 = C_r - C_r \cap \pi^{-1}(\{p_1, \ldots, p_r\})$. Then the Euler number $\mathrm{e}(C_r^0)$ of C_r^0 equals $n^{r-1}(2-r)$. According to the stratification

$$\pi^{-1}(E_p) = \pi^{-1}(E_p - E_p^{\text{sing}}) \cup \pi^{-1}(E_p^{\text{sing}}),$$

we compute the number Lef(g). We denote by $(K-1)$ the term from 1-dimensional strata, by $(K-0)$ the term from 0-dimensional strata.

The results are as follows:

$$(K-1) = n^{k-r-1}\,\mathrm{e}(C_r^0) \sum_{p\in S_r} n\delta(\alpha\,|\,K_{r,p}) - n^{k-2}(2-r)t_r$$

$$= n^{k-1}(2-r) \sum_{p\in S_r} \delta(\alpha\,|\,K_{r,p}) - n^{k-2}(2-r)t_r.$$

Note here that $|S_r| = t_r$.

$$(K-0) = n^{k-r-1} n^{r-2} r \sum_{p\in S_r} n\delta(\alpha\,|\,K_{r,p}) - n^{k-3} r t_r.$$

Lemma C.

$$K\text{-terms} = \Bigg[n^{k-1} \sum_{r\geq 3} (2-r) \sum_{p\in S_r} \delta(\alpha\,|\,K_{r,p}) - n^{k-2} \sum_{r\geq 3} (2-r)t_r$$

$$+ n^{k-2} \sum_{r\geq 3} r \sum_{p\in S_r} \delta(\alpha\,|\,K_{r,p}) - n^{k-3} \sum_{r\geq 3} r t_r \Bigg] n^{-k+1}.$$

3.3. Computation of $H\times H$-Terms

The $H\times H$-terms consist of contributions from isolated fixed points.

$$H\times H\text{-terms} = n^{k-3}\left(n^2 \sum_{p\in S_2} \delta\left(\alpha \,\middle|\, \prod_{L\in\Lambda_P} H_L\right) - n \sum_{p\in S_2} \sum_{L\in\Lambda_p} \delta(\alpha\,|\,H_L) + t_2 \right).$$

Lemma D.

$$H\times H\text{-terms} = \Bigg[n^{k-1} \sum_{p\in S_2} \sum_{L\in\Lambda_p} \delta\left(\alpha \,\middle|\, \prod_{L\in\Lambda_p} H_L\right)$$

$$- n^{k-2} \sum_{L\in\Lambda_p} t_2(L)\delta(\alpha\,|\,H_L) + n^{k-3} t_2 \Bigg] n^{-k+1}.$$

Here $t_2(L)$ is the number of double points on L in $L_1\cup L_2\cup\cdots\cup L_k$.

3.4. Computation of $H\times K$-Terms

Assume that $r\geq 3$. The $H\times K$-terms also consist of contributions from isolated fixed points. The result is as follows.

$$H\times K\text{-terms}=\Big[n^{k-3}\Big\{n^2\sum_{p\in S_r}\sum_{L\in\Lambda_p}\delta(\alpha\,|\,K_{r,p}\times H_L)$$

$$-n\sum_{p\in S_r}\sum_{L\in\Lambda_p}(\delta(\alpha\,|\,K_{r,p})+\delta(\alpha\,|\,H_L))+rt_r\Big\}\Big]n^{-k+1}$$

$$=\Big[n^{k-1}\sum_{p\in S_r}\sum_{L\in\Lambda_p}\delta(\alpha\,|\,K_{r,p}\times H_L)-n^{k-2}\sum_{p\in S_r}r\delta(\alpha\,|\,K_{r,p})$$

$$-n^{k-2}\sum_{p\in S_r}\sum_{L\in\Lambda_p}\delta(\alpha\,|\,H_L)+n^{k-3}rt_r\Big]n^{-k+1}.$$

Note here that the third sum in the above formula is transformed as follows:

$$\sum_{p\in S_r}\sum_{L\in\Lambda_p}\delta(\alpha\,|\,H_L)=\sum_{L\in\Lambda}t_r(L)\delta(\alpha\,|\,H_L).$$

Lemma E.

$$H\times K\text{-terms}=\Big[n^{k-1}\sum_{p\in S_r}\sum_{L\in\Lambda_p}\delta(\alpha\,|\,K_{r,p}\times H_L)-n^{k-2}\sum_{L\in\Lambda}t_r(L)\delta(\alpha\,|\,H_L)$$

$$-n^{k-2}r\sum_{p\in S_r}\delta(\alpha\,|\,K_{r,p})+n^{k-3}rt_r\Big]n^{-k+1}.$$

The sum of the formulae in Lemmas A, B, C, D, and E gives Theorem 1. (Q.E.D.)

Remark. Let us check our computation by putting $\alpha=$ the identity character. e(1) should be equal to the Euler number $\mathrm{e}(F)$ of F.

$$\mathrm{e}(1)=3-2k+\sum_{r\geq 2}(r-1)t_r+\sum_{L\in\Lambda}\mathrm{e}(L^0)+\sum_{r\geq 3}(2-r)t_r+t_2+\sum_{r\geq 3}rt_r$$

$$=3-2k+\sum_{r\geq 2}(r-1)t_r+\Big(2k-\sum_{r\geq 2}rt_r\Big)+2\sum_{r\geq 3}t_r-\sum_{r\geq 3}rt_r+t_2+\sum_{r\geq 3}rt_r$$

$$=3-\sum_{r\geq 2}t_r+2\sum_{r\geq 3}t_r+t_2$$

$$=3+\sum_{r\geq 3}t_r=\mathrm{e}(F)!$$

4. Hodge Type of $H^2(\alpha)$ for Generic α

In the first place, let us recall the following vanishing theorem substantially due to Kohno [3].

Theorem 2. *If $\alpha \in A^*$ is generic, then*

$$H^1(\alpha) = H^3(\alpha) = \{0\}.$$

Proof. The Poincaré duality

$$H^1(X_n, \mathbf{C}) \times H^3(X_n, \mathbf{C}) \to \mathbf{C}$$

implies the perfect pairing

$$H^1(\alpha) \times H^3(\alpha^{-1}) \to \mathbf{C}.$$

Since α^{-1} is generic if and only if α is generic, it suffices to show that $H^3(\alpha) = \{0\}$ for generic α. Set $U = \mathbf{P}^2 - \bigcup_{I=1}^{k} L_i$, and $U_n = p^{-1}(U)$ for the covering map $p: X_n \to F$. Put $D = F - U$ and $D_n = X_n - U_n$. Let $j_n: U_n \to X_n$ and $i_n: D_n \to X_n$ be the open and closed immersions respectively. Then

$$0 \to (j_n)_!(\mathbf{C}_{U_n}) \to \mathbf{C}_{X_n} \to (i_n)_*(\mathbf{C}_{D_n}) \to 0 \qquad \text{(exact)},$$

where $\mathbf{C}_*$ is the constant sheaf on $*$. Hence

$$H_c^3(U_n, \mathbf{C}) = H^3(X_n, (j_n)_!(\mathbf{C}_{U_n})) = H^3(X_n, \mathbf{C}),$$

because the real dimension of D_n is 2. Since $p: X_n \to F$ is finite,

$$H_c^3(U_n, \mathbf{C}) = H_c^3(U, p_*\mathbf{C}_{|U}) = H^3(F, j_!(p_*\mathbf{C}_{|U}))$$

and

$$H^3(X_n, \mathbf{C}) = H^3(F, p_*\mathbf{C}).$$

Thus

$$H^3(X_n, \mathbf{C}) = H^3(F, j_!(p_*\mathbf{C}_{|U})). \qquad (*)$$

Recall the natural surjection

$$\pi_1(U, *) \to A = \mathrm{Gal}(X_n/F).$$

The character α of A induces a character of the fundamental group $\pi_1(U, *)$ of U, which in turn gives a complex local system M_α of rank 1 over U. Let $q: U_n \to U$ be the restriction of p to U. Then

$$p_*\mathbf{C}_{|U} = q_*\mathbf{C}_{U_n} = \bigoplus_{\alpha \in A^*} M_\alpha.$$

Thus the compatible action of A on both sides of the isomorphism $(*)$ implies

$$H^3(\alpha) = H^3(F, j_!M_\alpha) = H_c^3(U, M_\alpha).$$

By Poincaré duality on U, we have a perfect pairing

$$H_c^3(U, M_\alpha)\times H^1(U, M_{\alpha^{-1}})\to \mathbf{C}.$$

Hence it suffices to show that $H^1(U, M_\alpha)=\{0\}$ for generic α.

Investigating the local monodromy along irreducible components of D, we have (cf. Kohno [3], Proof of Theorem 1, from now on)

$$j_! M_\alpha = j_* M_\alpha \qquad \text{for generic } \alpha.$$

Hence

$$H_c^1(U, M_\alpha)=H^1(F, j_! M_\alpha)=H^1(F, j_* M_\alpha)=H^1(U, M_\alpha) \qquad \text{for generic } \alpha.$$

In order to conclude the proof, we want to show that $H_c^1(U, M_\alpha)=\{0\}$, which follows from Poincaré duality,

$$H^3(U, M_{\alpha^{-1}})\times H_c^1(U, M_\alpha)\to \mathbf{C},$$

and the Lefschetz theorem $H^3(U, M_\alpha)=\{0\}$ for the affine variety U of dimension 2.

Corollary 1. *If α is generic, then* $\dim_{\mathbf{C}} H^2(\alpha)=3-2k+\sum_{r\geq 2}(r-1)t_r$.

This corollary follows immediately from Theorem 2 and the Corollary to Theorem 1.

Let $H^{p,q}$ be the (p, q)-type component of the Hodge structure $H^{p+q}(X_n, \mathbf{C})$. Put

$$H^{p,q}(\alpha)=H^{p+q}(\alpha)\cap H^{p,q},$$

and set

$$h^{p,q}(\alpha)=\dim_{\mathbf{C}} H^{p,q}(\alpha).$$

Then by the above theorem, we have

$$h^{1,0}(\alpha)=h^{0,1}(\alpha)=0 \qquad \text{for generic } \alpha\in A^*.$$

From now on we refer to Ishida [2]. Let $\pi: X_n\to F$ be the covering map. Then $\pi_* O_{X_n}$ is an A-linearized sheaf, and has an eigenspace decomposition

$$\pi_* O_X = \bigoplus_{\alpha\in A^*} L_\alpha.$$

Each L_α is an invertible sheaf on F. Then we have a canonical isomorphism

$$H^{0,1}(\alpha) \simeq H^1(F, L_\alpha).$$

Put

$$q(\alpha) = \dim_{\mathbf{C}} H^1(F, L_\alpha),$$

then the above theorem implies that

Corollary 2. $q(\alpha) = h^{0,1}(\alpha) = 0$ *for generic* $\alpha \in A^*$.

In order to formulate Theorem 3, we need some notation. We parametrize the character group A^* as follows. Fix a primitive nth root of unity ζ in $\mathbf{C}$. Then any element σ of $A = \mu_n^\Lambda/\Delta(\mu_n)$ is represented by

$$\sigma = ((\zeta^{s_1}, \zeta^{s_2}, \ldots, \zeta^{s_k}) \bmod \Delta(\mu_n)) \qquad \text{with each } s_1 \in \mathbf{Z}.$$

Put $\Lambda = \{L_i, 1 \leq i \leq k\}$, and consider the subset B^* of $\mathbf{Z}^\Lambda$ defined by

$$B^* = \left\{(a_L)_{L\in\Lambda} \,\middle|\, a_L \in \mathbf{Z}, 0 \leq a_L < n \text{ for each } L \in \Lambda, \text{ and } \sum_{L\in\Lambda} a_L \equiv 0 (\operatorname{mod} n)\right\}$$

$$= \left\{(a_1, \ldots, a_k) \in \mathbf{Z}^k \,\middle|\, 0 \leq a_i < n (1 \leq i \leq k), \sum_{i=1}^{k} a_i \equiv 0 (\operatorname{mod} n)\right\}.$$

We identify B^* with the character group A^* in the following manner.

An element $(a_L)_{L\in\Lambda} = (a_1, \ldots, a_k)$ of B^* corresponds to $\alpha \in A^*$ by

$$\alpha(\sigma) = \zeta^{(\sum_{i=1}^{k} a_i s_i)} \in \mathbf{C} \qquad \text{for any } \sigma = ((\zeta^{s_1}, \ldots, \zeta^{s_k}) \bmod \Delta(\mu_n)).$$

For $\alpha = (a_L) = (a_1, \ldots, a_k) \in A^*$, we put

$$\nu(\alpha) = \frac{1}{n}\left(\sum_{L\in\Lambda} a_L\right) = \frac{1}{n}\left(\sum_{i=1}^{k} a_i\right).$$

Moreover, for each point $p \in S_r \subset S$, we set

$$\beta_p(\alpha) = \left[\frac{1}{n}\left(\sum_{L\in\Lambda_p} a_L\right)\right].$$

Finally, we write

$$s(\alpha) = \tfrac{1}{2} \sum_{p\in S} \beta_p(\alpha)(\beta_p(\alpha) - 1).$$

Theorem 3. *Assume that $\alpha \in A^*$ is generic. Then the Hodge numbers $h^{2,0}(\alpha)$, $h^{1,1}(\alpha)$, $h^{0,2}(\alpha)$ are given by*

$$h^{2,0}(\alpha)=\tfrac{1}{2}\nu(\alpha^{-1})(\nu(\alpha^{-1})-3)-s(\alpha^{-1})+1,$$

$$h^{1,1}(\alpha)=1-\tfrac{3}{2}k+\sum_{r\geq 2}(r-1)t_r-\tfrac{1}{2}\{\nu(\alpha)^2+\nu(\alpha^{-1})^2\}-s(\alpha)-s(\alpha^{-1}),$$

$$h^{0,2}(\alpha)=\tfrac{1}{2}\nu(\alpha)(\nu(\alpha)-3)-s(\alpha)+1.$$

Proof. We have

$$h^{2,0}(\alpha)+h^{1,1}(\alpha)+h^{0,2}(\alpha)=h^2(\alpha)=3-2k+\sum_{r\geq 2}(r-1)t_r.$$

Hence we know $h^{1,1}(\alpha)$, if we know $h^{2,0}(\alpha)$ and $h^{0,2}(\alpha)$. By Hodge symmetry,

$$h^{2,0}(\alpha)=h^{0,2}(\bar{\alpha})=h^{0,2}(\alpha^{-1}).$$

Hence, it suffices to show the formula for $h^{0,2}(\alpha)$.

Lemma (Riemann–Roch Theorem). *Put*

$$\chi(\alpha)=h^0(\alpha)-h^{0,1}(\alpha)+h^{0,2}(\alpha).$$

Then

$$\chi(\alpha)=\chi(L_\alpha)=\tfrac{1}{2}\nu(\alpha)(\nu(\alpha)-3)-s(\alpha)+1.$$

Proof. Let $\rho: F\to \mathbf{P}^2$ be the blowing up along multiple points S on $\cup L_i$. Then by Ishida [2], we have

$$L_\alpha \simeq \rho^* O_{\mathbf{P}^2}(-\nu(\alpha)) \otimes_{O_F} O_F\left(\sum_{p\in S}\beta_p(\alpha)E_p\right),$$

where E_p is the exceptional curve at $p\in S$. On the other hand, the canonical sheaf $\Omega_F^2=O_F(K)$ of F is given by

$$\Omega_F^2 \simeq \rho^* O_{\mathbf{P}^2}(-3) \otimes_{O_F} O_F\left(\sum_{p\in S}E_p\right).$$

Therefore the intersection number is given by

$$\tfrac{1}{2}L_\alpha(K+L_\alpha)=\tfrac{1}{2}\nu(\alpha)(\nu(\alpha)-3)-s(\alpha).$$

Since F is a rational surface, $\chi(O_F)=1$. Hence our lemma follows immediately from the Riemann–Roch Theorem.

Let us conclude the proof of Theorem 3. By Corollary 2 of Theorem 2, $h^{0,1}(\alpha)=0$ for generic α. If $\alpha \neq 1$, $h^0(\alpha)=0$. Thus $\chi(\alpha)=h^{0,2}(\alpha)$ for generic α, which settles the proof of Theorem 3.

Remark. Note that $\nu(\alpha)+\nu(\alpha^{-1})=k$ for generic characters α.

References

[1] F. Hirzebruch, Arrangements of lines and algebraic surfaces. *Progress in Math.* **36**, Arithmetic and Geometry.

[2] M. Ishida, The irregularities of Hirzebruch's examples of surfaces of general type with $c_1^2=3c_2$. *Math. Annalen* **262** (1983) 407–420.

[3] T. Kohno, Homology of a local system on the complement of hyperplanes. *Proc. Japan Acad.* **62**, Ser. A (1986) 144–147.

[4] Shuji Saito, General fixed point formula for algebraic surface and the theory of Swan representations for two-dimensional local rings. To appear in *Amer. J. Math.*

On a Stochastic Difference Equation for the Multi-Dimensional Weakly Stationary Process with Discrete Time

Yasunori Okabe*
Department of Mathematics
Hokkaido University
Sapporo, Japan

1. Introduction

Research in the problem of linear prediction of a one-dimensional weakly stationary process X can be classified according to whether the length of the prediction interval is infinitely long (Kolmogorov [4], Wiener [28] and Dym–McKean [3]) or finitely limited (Krein [5] and Dym–McKean [2]). The method is to reformulate the above problem in terms of a real Hilbert space $L^2(\mathbf{R}, \Delta(d\lambda))$ obtained by closing the linear hull of $\{e^{it\cdot}; t \in \mathbf{R}\}$, where $\Delta(d\lambda)$ is the spectral measure of X. In particular, besides a traditional spectral theory, Krein has developed the so-called Krein's theory for the

* Partially supported by the Grant-in-Aid for Scientific Research of the Ministry of Education, Science and Culture of Japan.

ISBN 0-12-400466-0

inverse spectral problem, and Dym and McKean, Jr. have reconstructed the theory of Hilbert spaces of entire functions developed by de Branges ([1]).

On the other hand, the author has studied the problem of linear prediction for X from the viewpoint of the theory of stochastic differential equations ([10], [12]-[27]). The bud of its spirit can be found in [11], where Sato's hyperfunctions play an important role in the study of the infinitely multiple Markovian property. It is important, not only from the point of view of statistical physics, but also from a probabilistic point of view, to derive a stochastic equation of motion describing the time evolution of X. In particular, the process X with reflection positivity has been investigated in detail with the object of clarifying a mathematical structure of the so-called fluctuation-dissipation theorem in statistical physics (Kubo [6] and Mori [9]). In the course of these investigations, it has been found that the time evolution of X can be described by two kinds of Langevin equations with a notable difference in character of random forces ([18], [19]): One is the first KMO-Langevin equation having a white noise as a random force, and the other is the second KMO-Langevin equation where a colored noise, called the *Kubo noise*, is taken to be a random force. It is a key to obtain the structure theorem of the outer function of X. Since the Fourier transform of the outer function gives the canonical representation kernel for X, the author's studies are related to the case of long-time prediction interval, except [16] treated by an innovation method.

Following the same spirit, Miyoshi ([7] and [8]) has derived and then characterized a stochastic differential equation for the multi-dimensional weakly stationary process, by referring to de Branges's theory for a multi-dimensional case and using Krein's method, which is said to be the $(\alpha, \beta, \gamma, \delta)$-Langevin equation and treats the case of finite-time prediction interval.

Recently, the author ([24], [25] and [26]) derived and then characterized two kinds of stochastic difference equations for one-dimensional weakly stationary time series with reflection positivity, by using the result obtained in the continuous-time case, which are called the first and second KMO-Langevin equation, related to the case of long-time prediction interval.

The purpose of the present paper is to derive and then characterize a stochastic difference equation for multi-dimensional weakly stationary time series of a general type, by using the innovation method. We call it the *KM_2O-Langevin equation* and treat the case of finite-time prediction interval. The notable point is that a random force in the KM_2O-Langevin equation is not always a white noise, different from the one in the $(\alpha, \beta, \gamma, \delta)$-Langevin equation.

We will state the content of this paper. Let $X = (X(n); n \in \mathbf{Z})$ be a d-dimensional weakly stationary process with mean vector zero and covariance matrix R. We define for each $n \in \mathbf{N}$ a block Toeplitz matrix S_n $(\in M(nd; \mathbf{R}))$ by

$$S_n = \begin{pmatrix} R(0) & R(1) & \cdots & & R(n-1) \\ {}^tR(1) & R(0) & R(1) & & \\ \vdots & & \ddots & \ddots & \vdots \\ & & {}^tR(1) & R(0) & R(1) \\ {}^tR(n-1) & & \cdots & {}^tR(1) & R(0) \end{pmatrix}. \tag{1.1}$$

We suppose that S_n $(n \in \mathbf{N})$ are invertible in what follows. In Section 2, we will introduce a *forward* (resp. *backward*) *innovation process* $I_+ = (I_+(n); n \in \mathbf{N})$ (resp. $I_- = (I_-(n); n \in \mathbf{N})$) associated with X. It is noted that I_+ (resp. I_-) is an orthogonal process. And then we derive two kinds of stochastic difference equations describing the time evolution of X:

$$X(n) = -\sum_{k=1}^{n-1} \gamma_+(n, k)X(k) - \delta_+(n)X(0) + I_+(n) \qquad (n \in \mathbf{N}), \tag{1.2}$$

$$X(-n) = -\sum_{k=1}^{n-1} \gamma_-(n, k)X(-k) - \delta_-(n)X(0) + I_-(n) \qquad (n \in \mathbf{N}), \tag{1.3}$$

where $\gamma_+(\cdot, *)$, $\gamma_-(\cdot, *)$, $\delta_+(\cdot)$ and $\delta_-(\cdot)$ belong to $M(d; \mathbf{R})$.

We call (1.2) (resp. (1.3)) a *forward* (resp. *backward*) *KM_2O-Langevin equation* for X. It is noted that a class of nonlinear Langevin equations for strongly stationary time series is derived from our approach (Remark 2.2).

We will in Section 3 obtain fundamental recursive relations among $\gamma_+(\cdot, *)$, $\gamma_-(\cdot, *)$, $\delta_+(\cdot)$ and $\delta_-(\cdot)$ (Theorem 3.1): for any $n, k \in \mathbf{N}$, $n > k$,

$$\gamma_+(n, k) = \gamma_+(n-1, k-1) + \delta_+(n)\gamma_-(n-1, n-1-k), \tag{1.4}$$

$$\gamma_-(n, k) = \gamma_-(n-1, k-1) + \delta_-(n)\gamma_+(n-1, n-1-k), \tag{1.5}$$

where $\gamma_+(n, 0) = \delta_+(n)$ and $\gamma_-(n, 0) = \delta_-(n)$.

We denote by $V_+(n)$ (resp. $V_-(n)$) the covariance matrix of $I_+(n)$ (resp. $I_-(n)$) $(n \in \mathbf{N})$. By using (1.4) and (1.5), we will in Section 4 obtain fundamental recursive relations among $V_+(\cdot)$, $V_-(\cdot)$, $R(0)$, $\delta_+(\cdot)$ and $\delta_-(\cdot)$ (Theorem 4.1): for any $n \in \mathbf{N}^*$,

$$V_+(n+1) = (I - \delta_+(n+1)\delta_-(n+1))\,V_+(n), \tag{1.6}$$

$$V_-(n+1) = (I - \delta_-(n+1)\delta_+(n+1))\,V_-(n), \tag{1.7}$$

$$V_+(n)\,{}^t\delta_-(n+1) = \delta_+(n+1)\,V_-(n), \tag{1.8}$$

where $V_+(0) = V_-(0) = R(0)$.

By taking advantage of the innovation method, we will in Section 5 give a *forward* (resp. *backward*) *prediction formula* for X (Theorems 5.1 and 5.2). It will be found that prediction matrices and prediction error matrices are determined by $R(0)$, $\delta_+(\cdot)$ and $\delta_-(\cdot)$.

As a converse setting of Sections 2–5, we will in Section 6 show a reconstruction theorem (Theorem 6.1), which states that for a given system $\{V, \delta_+(n);\ n \in \mathbf{N}\}$ in $M(d;\mathbf{R})$ and a d-dimensional orthogonal process $I_+ = (I_+(n);\ n \in \mathbf{N}^*)$, the recursive relations (1.4)–(1.8) characterize the weakly stationary property of the unique solution $X_+ = (X(n);\ n \in \mathbf{N}^*)$ of the forward KM_2O-Langevin equation (1.2), where $I_+(0) = X(0)$, $V = R(0)$ and I_+ becomes a forward innovation process associated with X_+.

In a forthcoming paper ([27]), we will construct an outer matrix function and then derive a KMO–Langevin equation for a multi-dimensional weakly stationary time series. And together with some relations between the KM_2O-Langevin equation and the KMO–Langevin equation, generalized fluctuation-dissipation theorems will be proved based on both Langevin equations.

2. KM_2O–Langevin Equations

Let $X = (X(n);\ n \in \mathbf{Z})$ be an $\mathbf{R}^d$-valued weakly stationary time series on a probability space $(\Omega, \mathcal{B}, P)$ with mean vector zero and covariance matrix R:

$$R(n) = E(X(n)\,{}^tX(0)) \qquad (n \in \mathbf{Z}). \tag{2.1}$$

Note that

$${}^tR(n) = R(-n) \qquad (n \in \mathbf{Z}). \tag{2.2}$$

Let M be the closed subspace of $L^2(\Omega, \mathcal{B}, P)$ defined by

$$M = \text{the closed linear hull of } \{X_j(n);\ 1 \le j \le d,\ n \in \mathbf{Z}\}. \tag{2.3}$$

We then have the unitary group $(U(m);\ m \in \mathbf{Z})$ acting on M such that

$$U(m)(X(n)) = X(n+m) \qquad (m, n \in \mathbf{Z}). \tag{2.4}$$

For each $n \in \mathbf{N}$ we define a block Toeplitz matrix $S_n \in M(nd;\mathbf{R})$ by

$$S_n = \begin{pmatrix} R(0) & R(1) & \cdots & & R(n-1) \\ {}^tR(1) & R(0) & R(1) & & \\ \vdots & \ddots & \ddots & \ddots & \vdots \\ & & {}^tR(1) & R(0) & R(1) \\ {}^tR(n-1) & & \cdots & {}^tR(1) & R(0) \end{pmatrix}. \tag{2.5}$$

Since

$$S_n = E(Y_n {}^tY_n), \tag{2.6}$$

where $Y_n = {}^t(X_1(n-1), \ldots, X_d(n-1), \ldots, X_1(0), \ldots, X_d(0))$, we can see that either of the following (2.7) and (2.8) holds:

$$S_n \in GL(nd; \mathbf{R}) \qquad \text{for any } n \in \mathbf{N} \tag{2.7}$$

$$\text{there exists } n_0 \in \mathbf{N} \text{ such that } S_n \in GL(nd; \mathbf{R}) \text{ for any } n \in \{1, \ldots, n_0\}$$
$$\text{and } S_n \notin GL(nd; \mathbf{R}) \text{ for any } n \in \{n_0+1, n_0+2, \ldots\}. \tag{2.8}$$

We treat the case where condition (2.7) holds in what follows.

Remark 2.1. If R has a spectral density matrix $\Delta = \Delta(\theta)$ ($\theta \in [-\pi, \pi)$) such that $\Delta(\theta) \in GL(d; \mathbf{C})$ for almost all $\theta \in [-\pi, \pi)$, it can be seen that condition (2.7) holds.

For each $n \in \mathbf{N}$ we define the following block Toeplitz matrix $T_n \in M(nd; \mathbf{R})$ by

$$T_n = \begin{pmatrix} R(0) & {}^tR(1) & & \cdots & & {}^tR(n-1) \\ R(1) & R(0) & {}^tR(1) & & & \\ \vdots & & \ddots & \ddots & \ddots & \vdots \\ & & & R(1) & R(0) & {}^tR(1) \\ R(n-1) & & \cdots & & R(1) & R(0) \end{pmatrix}. \tag{2.9}$$

Since

$$T_n = E(Z_n {}^tZ_n), \tag{2.10}$$

where $Z_n = {}^t(X_1(0), \ldots, X_d(0), \ldots, X_1(n-1), \ldots, X_d(n-1))$, it holds from (2.6) and (2.7) that

$$T_n \in GL(nd; \mathbf{R}) \qquad \text{for any } n \in \mathbf{N}. \tag{2.11}$$

For each $n \in \mathbf{N}^* = \{0, 1, 2, \ldots\}$ we define two closed linear subspaces $M_0^+(n)$ and $M_0^-(n)$ of M by

$$M_0^+(n) = \text{the linear hull of } \{X_j(m);\ 1 \le j \le d, 0 \le m \le n\}, \tag{2.12}$$

$$M_0^-(n) = \text{the linear hull of } \{X_j(-m);\ 1 \le j \le d, 0 \le m \le n\}. \tag{2.13}$$

Then we introduce two $\mathbf{R}^d$-valued stochastic processes $I_+ = (I_+(n); n \in \mathbf{N})$ and $I_- = (I_-(n); n \in \mathbf{N})$ on $(\Omega, \mathcal{B}, P)$ by

$$I_+(n) = X(n) - P_{M_0^+(n-1)} X(n), \tag{2.14}$$

$$I_-(n) = X(-n) - P_{M_0^-(n-1)} X(-n), \tag{2.15}$$

where $P_{M_0^+(n-1)}$ (resp. $P_{M_0^-(n-1)}$) stands for the orthogonal projection on $M_0^+(n-1)$ (resp. $M_0^-(n-1)$).

It follows from condition (2.7) that there uniquely exist two systems $\{\gamma_+(n,k), \delta_+(m); k, m, n \in \mathbf{N}, n > k\}$ and $\{\gamma_-(n,k), \delta_-(m); k, m, n \in \mathbf{N}, n > k\}$ of members in $M(d; \mathbf{R})$ such that for any $n \in \mathbf{N}$,

$$X(n) = -\sum_{k=1}^{n-1} \gamma_+(n,k)X(k) - \delta_+(n)X(0) + I_+(n), \tag{2.16}$$

$$X(-n) = -\sum_{k=1}^{n-1} \gamma_-(n,k)X(-k) - \delta_-(n)X(0) + I_-(n). \tag{2.17}$$

In particular, it holds that for any $n \in \mathbf{N}$,

$$\{X(0), I_+(m); m \in \mathbf{N}\} \quad \text{is orthogonal in } M^d, \tag{2.18}$$

$$\{X(0), I_-(m); m \in \mathbf{N}\} \quad \text{is orthogonal in } M^d, \tag{2.19}$$

$$M_0^+(n) = \text{the linear hull of } \{X_j(0), I_{+j}(l); 1 \le j \le d, 1 \le l \le n\}, \tag{2.20}$$

$$M_0^-(n) = \text{the linear hull of } \{X_j(0), I_{-j}(l); 1 \le j \le d, 1 \le l \le n\}, \tag{2.21}$$

where $I_+(l) = {}^t(I_{+1}(l), \ldots, I_{+d}(l))$ and $I_-(l) = {}^t(I_{-1}(l), \ldots, I_{-d}(l))$.

We call I_+ (resp. I_-) a *forward* (resp. *backward*) *innovation process* associated with X. Furthermore, by regarding (2.16) (resp. (2.17)) as a stochastic difference equation describing the time evolution of X, we call (2.16) (resp. (2.17)) a *forward* (resp. *backward*) *KM_2O-Langevin equation* for X.

Remark 2.2. Let $Y = (Y(n); n \in \mathbf{Z})$ be a one-dimensional strongly stationary time series on $(\Omega, \mathcal{B}, P)$ such that

$$Y(0) \in L^4(\Omega, \mathcal{B}, P), \tag{2.22}$$

$$E(Y(0)) = 0. \tag{2.23}$$

We define three weakly stationary time series $X^{(j)} = (X^{(j)}(n); n \in \mathbf{Z})$ $(j = 1, 2, 3)$ by

$$X^{(1)}(n) = \begin{pmatrix} Y(n) \\ Y(n)^2 - E(Y(0)^2) \end{pmatrix}, \tag{2.24}$$

$$X^{(2)}(n) = \begin{pmatrix} Y(n) \\ Y(n)Y(n-1) - E(Y(1)Y(0)) \end{pmatrix}, \tag{2.25}$$

$$X^{(3)}(n) = \begin{pmatrix} Y(n) \\ Y(n)^2 - E(Y(0)^2) \\ Y(n)Y(n-1) - E(Y(1)Y(0)) \end{pmatrix}. \tag{2.26}$$

If we suppose condition (2.7) for $X^{(j)}$ $(j=1,2,3)$, we can derive three nonlinear Langevin equations for Y through KM_2O-Langevin equations for $X^{(j)}$ $(j=1,2,3)$. The problem of nonlinear prediction for Y will be investigated based on these nonlinear Langevin equations in the near future.

3. Relations Among $\gamma_+(\cdot, *)$, $\gamma_-(\cdot, *)$, $\delta_+(\cdot)$ and $\delta_-(\cdot)$

For convenience, we set for each $n \in \mathbf{N}$

$$\gamma_+(n, 0) = \delta_+(n), \tag{3.1}$$

$$\gamma_-(n, 0) = \delta_-(n). \tag{3.2}$$

Theorem 3.1. *For any $n, k \in \mathbf{N}$, $n > k$,*

(*i*) $\gamma_+(n, k) = \gamma_+(n-1, k-1) + \delta_+(n)\gamma_-(n-1, n-1-k)$,
(*ii*) $\gamma_-(n, k) = \gamma_-(n-1, k-1) + \delta_-(n)\gamma_+(n-1, n-1-k)$.

Proof. Fix any $n \in \{2, 3, \dots\}$. By multiplying both sides of equation (2.16) by ${}^tX(m)$ $(m = 0, 1, \dots, n-1)$ and then taking an expectation with respect to the probability P, we see from (2.18) and (2.20) that for any $m \in \{0, 1, \dots, n-1\}$,

$$R(n-m) = -\sum_{k=1}^{n-1} \gamma_+(n, k)R(k-m) - \delta_+(n)\,{}^tR(m). \qquad (3.3)_m$$

By replacing n in $(3.3)_m$ by $n-1$, we have, for any $m \in \{0, 1, \dots, n-2\}$,

$$R(n-1-m) = -\sum_{k=0}^{n-2} \gamma_+(n-1, k)R(k-m). \qquad (3.4)_m$$

On the other hand, by multiplying both sides of equation (2.17) with n replaced by $n-1$ by ${}^tX(-m)$ $(m = 0, 1, \dots, n-2)$ and then taking an expectation with respect to P, we see from (2.19) and (2.21) that for any $m \in \{0, 1, \dots, n-2\}$,

$${}^tR(n-1-m) = -\sum_{k=0}^{n-2} \gamma_-(n-1, k)\,{}^tR(k-m). \qquad (3.5)_m$$

Now, let any $m \in \{1, 2, \dots, n-1\}$ be fixed. By combining $R(n-m)$ (resp. ${}^tR(m)$) in $(3.3)_m$ with $R(n-m)$ in $(3.4)_{m-1}$ (resp. ${}^tR(m)$ in $(3.5)_{n-1-m}$), we have

$$-\sum_{k=0}^{n-2} \gamma_+(n-1, k)R(k+1-m)$$

$$= -\sum_{k=1}^{n-1} \gamma_+(n, k)R(k-m) + \delta_+(n) \sum_{k=0}^{n-2} \gamma_-(n-1, k)^t R(k-n+1+m)$$

$$= -\sum_{k=0}^{n-2} (\gamma_+(n, k+1) - \delta_+(n)\gamma_-(n-1, n-2-k))R(k+1-m),$$

and so

$$\sum_{k=0}^{n-2} (\gamma_+(n, k+1) - \gamma_+(n-1, k) - \delta_+(n)\gamma_-(n-1, n-2-k))R(k+1-m) = 0.$$

Therefore, it follows from condition (2.7) that (i) holds. Similarly, we can prove (ii). Q.E.D.

For future use in Section 6, for any $m \in \mathbf{N}^*$ and $n \in \mathbf{N}$, we set

$$\eta_+(m, n) = \gamma_+(m+n, m), \tag{3.6}$$

$$\eta_-(m, n) = \gamma_-(m+n, m). \tag{3.7}$$

Immediately from Theorem 3.1, we have

Theorem 3.1′. *For any $m \in \mathbf{N}^*$ and $n \in \mathbf{N}$,*

(i) $\eta_+(m+1, n) = \eta_+(m, n) + \delta_+(m+n+1)\eta_-(n-1, m+1),$
(ii) $\eta_-(m+1, n) = \eta_-(m, n) + \delta_-(m+n+1)\eta_+(n-1, m+1).$

4. Relations Among $V_+(\cdot)$, $V_-(\cdot)$, $\delta_+(\cdot)$ and $\delta_-(\cdot)$

For convenience, we put

$$I_+(0) = I_-(0) = X(0), \tag{4.1}$$

and denote by $V_+(n)$ (resp. $V_-(n)$) the covariance matrix of $I_+(n)$ (resp. $I_-(n)$) ($n \in \mathbf{N}^*$):

$$V_+(n) = E(I_+(n)^t I_+(n)), \tag{4.2}$$

$$V_-(n) = E(I_-(n)^t I_-(n)). \tag{4.3}$$

Lemma 4.1.

$$(i)\quad \det S_n = \prod_{k=0}^{n-1} \det V_+(k) \qquad \text{for any } n \in \mathbf{N},$$

$$(ii)\quad \det T_n = \prod_{k=0}^{n-1} \det V_-(k) \qquad \text{for any } n \in \mathbf{N},$$

$$(iii)\quad V_+(n),\ V_-(n) \in GL(nd; \mathbf{R}) \qquad \text{for any } n \in \mathbf{N}.$$

Proof. (i) for $n=1$ is trivial. Let any $n \in \{2, 3, \ldots\}$ be fixed. By multiplying both sides of equation (2.16) by ${}^tX(n)$ and then taking an expectation with respect to P, we see from (2.18) and (2.20) that

$$R(0) = -\sum_{k=1}^{n-1} \gamma_+(n,k)\,{}^tR(n-k) - \delta_+(n)\,{}^tR(n) + V_+(n). \tag{4.4}$$

By making a matrix representation of (4.4) and $(3.3)_m$ $(m = n-1, n-2, \ldots, 1, 0)$, we have

$$\begin{pmatrix} I\ \gamma_+(n,n-1) & \cdots & \gamma_+(n,1)\delta_+(n) \\ I & & 0 \\ & \ddots & \\ 0 & & I \end{pmatrix} S_{n+1} = \begin{pmatrix} V_+(n) & 0\cdots 0 \\ {}^tR(1) & \\ \vdots & S_n \\ {}^tR(n) & \end{pmatrix},$$

which yields (i). (ii) is also proved similarly. (iii) follows from (2.7), (2.11), (i) and (ii). Q.E.D.

Lemma 4.2. *For any* $n \in \mathbf{N}^*$,

$$(i)\quad R(n+1) = -\delta_+(n+1)\,V_-(n) - \sum_{k=0}^{n-1} \gamma_+(n,k)R(k+1),$$

$$(ii)\quad {}^tR(n+1) = -\delta_-(n+1)\,V_+(n) - \sum_{k=0}^{n-1} \gamma_-(n,k)\,{}^tR(k+1).$$

Proof. Since $V_+(n)$ is a symmetric matrix, it follows from (4.4) that for any $n \in \mathbf{N}^*$,

$$V_+(n) = R(0) + \sum_{k=0}^{n-1} R(n-k)\,{}^t\gamma_+(n,k). \tag{4.5}$$

Similarly,

$$V_-(n) = {}^tR(0) + \sum_{k=0}^{n-1} {}^tR(n-k)\,{}^t\gamma_-(n,k). \tag{4.6}$$

By replacing n by $n+1$ in $(3.3)_0$ and then using Theorem 3.1(i),

$$R(n+1)=-\delta_+(n+1)\left(\sum_{k=0}^{n-1}\gamma_-(n,k)R(n-k)+R(0)\right)-\sum_{k=0}^{n-1}\gamma_+(n,k)R(k+1),$$

which, together with (4.6), implies (i). (ii) is also proved similarly.

Q.E.D.

Now, we will show the following fundamental

Lemma 4.3. *For any $n\in\mathbf{N}$,*

$$\sum_{k=0}^{n-1}\eta_+(k,n-k)R(k+1)=\sum_{k=0}^{n-1}R(k+1)^t\eta_-(k,n-k).$$

The proof is divided into 13 steps. We denote by A_n the right-hand side minus the left-hand side in Lemma 4.3:

$$A_n=\sum_{k=0}^{n-1}R(k+1)^t\eta_-(k,n-k)-\sum_{k=0}^{n-1}\eta_+(k,n-k)R(k+1). \tag{4.7}$$

Step 1.

(i) $V_+(n+1)=V_+(n)-\delta_+(n+1)V_-(n)^t\delta_+(n+1)+A_n$ $(n\in\mathbf{N})$,

(ii) $V_-(n+1)=V_-(n)-\delta_-(n+1)V_+(n)^t\delta_-(n+1)-{}^tA_n$ $(n\in\mathbf{N})$,

(iii) $V_+(n)^t\delta_-(n+1)-\delta_+(n+1)V_-(n)=A_n$ $(n\in\mathbf{N})$.

Proof. By applying Theorem 3.1 to (4.5),

$$V_+(n+1)=V_+(n)+\left(\sum_{k=0}^{n-1}R(k+1)^t\eta_-(k,n-k)+R(n+1)\right)^t\delta_+(n+1),$$

which, together with Lemma 4.2(i), implies (i). (ii) is proved similarly. By using Lemma 4.2 again, we get (iii).

Q.E.D.

Step 2.

(i) $R(1)=-\delta_+(1)R(0)$,

(ii) ${}^tR(1)=-\delta_-(1)R(0)$,

(iii) $R(0)^t\delta_-(1)=\delta_+(1)R(0)$,

(iv) $V_+(1)=(I-\delta_+(1)\delta_-(1))R(0)$,

(v) $V_-(1)=(I-\delta_-(1)\delta_+(1))R(0)$.

Proof. By multiplying both sides of (2.16) and (2.17) for $n=1$ by ${}^tX(0)$ and then taking an expectation with respect to P, we have (*i*) and (*ii*). (*iii*) follows from (*i*) and (*ii*). Next, by multiplying (2.16) for $n=1$ by ${}^tX(1)$ and then taking an expectation, we see from (2.18), (2.20), (*i*) and (*ii*) in Step 2 that (*iv*) holds. (*v*) is proved similarly. Q.E.D.

Step 3. $A_1=0$.

This follows from (*iii*) in Step 2.

Step 4. *For any* $n\in\{2,3,\ldots\}$,

$$A_n=\delta_+(n)\mathrm{I}_n^{(1)t}\delta_-(n)+\delta_+(n)\mathrm{II}_n^{(1)}+\mathrm{III}_n^{(1)t}\delta_-(n)+\mathrm{IV}_n^{(1)},$$

where

$$\mathrm{I}_n^{(1)}=V_-(n-1)^t\delta_+(n-1)-\delta_-(n-1)V_+(n-1),$$

$$\mathrm{II}_n^{(1)}=-\left(R(1)+\sum_{j=1}^{n-2}\eta_-(n-1-j,j)R(j+1)\right)$$
$$+\delta_-(n-1)\sum_{j=0}^{n-2}R(j+1)^t\eta_-(j,n-1-j)-V_-(n-1)^t\eta_-(n-2,1),$$

$$\mathrm{III}_n^{(1)}=R(1)+\sum_{j=1}^{n-2}R(j+1)^t\eta_+(n-1-j,j)$$
$$-\left(\sum_{j=0}^{n-2}\eta_+(j,n-1-j)R(j+1)\right)^t\delta_+(n-1)$$
$$+\eta_+(n-2,1)V_+(n-1),$$

$$\mathrm{IV}_n^{(1)}=\sum_{j=0}^{n-3}\eta_+(j,n-1-j)R(j+2)$$
$$+\left(\sum_{j=0}^{n-3}\eta_+(j,n-1-j)R(j+1)\right)^t\eta_-(n-2,1)$$
$$-\sum_{j=0}^{n-3}R(j+2)^t\eta_-(j,n-1-j)$$
$$-\eta_+(n-2,1)\sum_{j=0}^{n-3}R(j+1)^t\eta_-(j,n-1-j).$$

This decomposition follows from Theorem 3.1′ and Lemma 4.2.

Step 5. *For any* $n \in \{2, 3, \ldots\}$ *and any* $k \in \{1, 2, \ldots, n-1\}$,

$$\mathrm{IV}_n^{(k)} = \delta_+(n-k)\mathrm{I}_n^{(k+1)t}\delta_-(n-k) + \delta_+(n-k)\mathrm{II}_n^{(k+1)} + \mathrm{III}_n^{(k+1)t}\delta_-(n-k) + \mathrm{IV}_n^{(k+1)},$$

where

$$\begin{aligned}
\mathrm{I}_n^{(k+1)} = &- \sum_{j=0}^{k-1} \eta_-(j, n-k-1-j)\left(R(k-j) + \sum_{i=1}^{n-2k-1} R(k-j+i)^t \eta_+(n-k-1-i, i)\right) \\
&+ \sum_{j=0}^{k-1} \left(R(k-j) + \sum_{i=1}^{n-2k-1} \eta_-(n-k-1-i, i) \times R(k-j+1)\right) \eta_+(j, n-k-j) \\
&- \eta_-(k, n-2k-1)\, V_+(n-k-1) + V_-(n-k-1)^t \eta_+(k, n-2k-1),
\end{aligned}$$

$$\begin{aligned}
\mathrm{II}_n^{(k+1)} = &\, R(k+1) + \sum_{j=1}^{n-2k-2} \eta_-(n-k-1-j, j) R(k+1+j) \\
&- \sum_{j=0}^{k} \eta_-(j, n-k-1-j)\left(\sum_{i=0}^{n-2k-2} R(k-j+1+i)^t \eta_-(i, n-k-1-i)\right) \\
&+ \sum_{j=1}^{k} \left(R(k+1-j) + \sum_{i=1}^{n-2k-2} \eta_-(n-k-1-i, i) \times R(k+1-j+i)\right)^t \eta_-(n-k-1-j, j) \\
&+ V_-(n-k-1)^t \eta_-(n-2k-2, k+1),
\end{aligned}$$

$$\begin{aligned}
\mathrm{III}_n^{(k+1)} = &- \left(R(k+1) + \sum_{j=1}^{n-2k-2} R(k+1+j)^t \eta_+(n-k-1-j, j)\right) \\
&+ \sum_{j=0}^{k} \left(\sum_{i=0}^{n-2k-2} {}^t\eta_+(i, n-k-1-i) \times R(k-j+1+i)\right) {}^t\eta_+(j, n-k-1-j) \\
&- \sum_{j=1}^{k} \eta_+(n-k-1-j, j)\left(R(k+1-j) \right.
\end{aligned}$$

$$+\sum_{i=1}^{n-2k-2} R(k+1-j+i)\eta_+(n-k-1-i,i)\Big)$$

$$-\eta_+(n-2k-2,k+1)V_+(n-k-1),$$

$$\mathrm{IV}_n^{(k+1)}=\sum_{j=0}^{n-2k-3}\eta_+(j,n-k-1-j)R(k+2+j)$$

$$-\sum_{j=0}^{n-2k-3}R(k+2+j)^t\eta_-(j,n-k-1-j)$$

$$-\sum_{j=1}^{k+1}\eta_+(n-k-1-j,j)$$

$$\times\left(\sum_{i=0}^{n-2k-3}R(k+1+i)^t\eta_-(i,n-k-1-i)\right)$$

$$+\sum_{j=1}^{k+1}\left(\sum_{i=0}^{n-2k-3}\eta_+(i,n-k-1-i)R(k+1+i)\right)^t\eta_-(n-k-1-j,j).$$

Similarly to Step 4, this follows from Theorem 3.1′ and Lemma 4.2.

Step 6.

(i) For any even $n=2N$ ($N\in\mathbf{N}$), $\mathrm{IV}_n^{(N)}=0$.
(ii) For any odd $n=2N+1$ ($N\in\mathbf{N}$),

$$\mathrm{IV}_n^{(N)}=\delta_+(N+1)\Big\{-V_+(N)-\sum_{j=0}^{N-1}V_+(N-1-j)^t\delta_-(N-j)^t\delta_+(N-j)$$

$$+V_-(N)+\sum_{j=0}^{N-1}\delta_-(N-j)\delta_+(N-j)V_-(N-1-j)\Big\}^t\delta_-(N+1).$$

Proof. Immediately from Step 5, we have (i). (ii) can be proved as follows: By Step 5,

$$\mathrm{IV}_n^{(N)}=\delta_+(N+1)\left(R(N+1)+\sum_{j=1}^{N}R(N+1-j)^t\eta_-(N+1-j,j)\right)$$

$$-\left(R(N+1)+\sum_{j=1}^{N}\eta_+(N+1-j,j)R(N+1-j)\right)^t\delta_-(N+1).$$

By Theorem 3.1′ and Lemma 4.2,

$$R(N+1)+\sum_{j=1}^{N} R(N+1-j)^t\eta_-(N+1-j,j)$$

$$=\left(-V_+(N)+\sum_{j=0}^{N-1} R(N-j)^t\eta_+(j,N-j)\right)^t\delta_-(N+1),$$

$$R(N+1)+\sum_{j=1}^{N} \eta_+(N+1-j,j)R(N+1-j)$$

$$=\delta_+(N+1)\left(-V_-(N)+\sum_{j=0}^{N-1} \eta_-(j,N-j)R(N-j)\right).$$

And so

$$\mathrm{IV}_n^{(N)}=\delta_+(N+1)\left\{-V_+(N)+\sum_{j=0}^{N-1} R(N-j)^t\eta_+(j,N-j)\right.$$

$$\left.-V_-(N)+\sum_{j=0}^{N-1} \eta_-(j,N-j)R(N-j)\right\}^t\delta_-(N+1).$$

Furthermore, by using Theorem 3.1′ and Lemma 4.2 again,

$$\sum_{j=0}^{N-1} R(N-j)^t\eta_+(j,N-j)$$

$$=-V_+(N-1)^t\delta_-(N)^t\delta_+(N)+\sum_{j=0}^{N-2} R(N-1-j)^t\eta_+(j,N-1-j),$$

$$\sum_{j=0}^{N-1} \eta_-(j,N-j)R(N-j)$$

$$=-\delta_-(N)\delta_+(N)V_-(N-1)+\sum_{j=0}^{N-2} \eta_-(j,N-1-j)R(N-1-j),$$

which yields

$$\mathrm{IV}_n^{(N)}=\delta_+(N+1)\left\{-V_+(N)-V_+(N-1)^t\delta_-(N)^t\delta_+(N)\right.$$

$$+\sum_{j=0}^{N-2} R(N-1-j)^t\eta_+(j,N-1-j)$$

$$+V_-(N)+\delta_-(N)\delta_+(N)V_-(N-1)$$

$$\left.-\sum_{j=0}^{N-2} \eta_-(j,N-1-j)R(N-1-j)\right\}^t\delta_-(N+1).$$

By repeating the same procedure, we see from Step 2 that (i) holds. (ii) follows immediately from Step 5. Q.E.D.

Step 7. *For any fixed* $n_0 \in \mathbf{N}^*$, if $V_+(n_0)^t\delta_-(n_0+1) = \delta_+(n_0+1)V_-(n_0)$, *then* $V_-(n_0+1)^t\delta_+(n_0+1) = \delta_-(n_0+1)V_+(n_0+1)$.

Proof. By Step 1, (ii), (iv) and (v) in Step 2,

$$\begin{aligned} V_-(n_0+1)^t\delta_+(n_0+1) &= (I-\delta_-(n_0+1)\delta_+(n_0+1))V_-(n_0)^t\delta_+(n_0+1) \\ &= \delta_-(n_0+1)(I-\delta_+(n_0+1)\delta_-(n_0+1))V_+(n_0) \\ &= \delta_-(n_0+1)V_+(n_0+1). \end{aligned}$$

Q.E.D.

Step 8. $\mathrm{I}_n^{(1)} = \mathrm{II}_n^{(1)} = \mathrm{III}_n^{(1)} = \mathrm{IV}_n^{(1)} = 0$ *for any* $n \in \{2, 3\}$.

Proof. By (iii) in Step 2 (resp. Step 3) with Step 7, we have $\mathrm{I}_2^{(1)} = 0$ (resp. $\mathrm{I}_3^{(1)} = 0$). $\mathrm{II}_2^{(1)}$ and $\mathrm{III}_2^{(1)}$ follow from Step 2. A simple calculation, together with Lemma 4.2 and Step 2, yields $\mathrm{II}_3^{(1)}$ and $\mathrm{III}_3^{(1)}$. $\mathrm{IV}_n^{(1)}$ $(n=2,3)$ follow from Steps 2 and 6. Q.E.D.

Now, by a mathematical induction, we shall show the following statement $(*)_n$ $(n=2,3,\ldots)$: for any $m \in \{2, 3, \ldots, n\}$,

$$(*)_n \begin{cases} (i) & \mathrm{I}_m^{(k)} = \mathrm{II}_m^{(k)} = \mathrm{III}_m^{(k)} = 0 \qquad (1 \le k \le M) \qquad \text{if } m = 2M+1 \\ & \mathrm{IV}_m^{(M)} = 0 \\ (ii) & \mathrm{I}_m^{(k)} = \mathrm{II}_m^{(k)} = \mathrm{III}_m^{(k)} = 0 \qquad (1 \le k \le M) \qquad \text{if } m = 2M. \end{cases}$$

Step 8 implies that $(*)_n$ $(n=2,3)$ hold. For any fixed $n_0 \in \{2, 3, \ldots\}$, let us suppose that $(*)_{n_0}$ holds. Moreover, we consider the case where n_0 is even, $n_0 = 2N_0$. The case where n_0 is odd is proved similarly. What we have to show is that

$$(*)'_{n_0,k} \qquad \mathrm{I}_{n_0+1}^{(k)} = \mathrm{II}_{n_0+1}^{(k)} = \mathrm{III}_{n_0+1}^{(k)} = 0 \qquad (1 \le k \le N_0),$$

$$(*)''_{n_0,N_0} \qquad \mathrm{IV}_{n_0+1}^{(N_0)} = 0.$$

Since $(*)_{n_0}$ implies that A_m $(2 \le m \le n_0)$ hold, it follows from Steps 1 and 7 that for any $m \in \{2, \ldots, n_0\}$,

$$V_+(m+1) = (I-\delta_+(m+1)\delta_-(m+1))V_+(m), \tag{4.8}$$

$$V_-(m+1) = (I-\delta_-(m+1)\delta_+(m+1))V_-(m), \tag{4.9}$$

$$V_+(m)^t\delta_-(m+1) = \delta_+(m+1)V_-(m), \tag{4.10}$$

$$V_+(m+1)^t\delta_-(m+1) = \delta_+(m+1)V_-(m+1). \tag{4.11}$$

Step 9. $I^{(1)}_{n_0+1}=0.$

This follows from Step 4 and (4.11).

Step 10. (i) $\mathrm{II}^{(1)}_{n_0+1}=0$; (ii) $\mathrm{III}^{(1)}_{n_0+1}=0.$

Proof. We prove only (i), since (ii) is shown similarly. By Theorem 3.1′, Lemma 4.2(ii), and the assumption $\mathrm{II}^{(1)}_{n_0}=0$,

$$\mathrm{II}^{(1)}_{n_0+1}=-V_-(n_0-1)^t\eta_-(n_0-2,1)-\delta_-(n_0-1)V_+(n_0-1)^t\delta_-(n_0)\\+V_-(n_0)^t\eta_-(n_0-1,1)+\delta_+(n_0)\left(\sum_{j=0}^{n_0-2}\eta_+(j,n_0-1-j)R(2+j)\right.\\\left.-\sum_{j=0}^{n_0-1}R(1+j)^t\eta_-(j,n_0-j)\right).$$

Since

$$V_-(n_0-1)^t\eta_-(n_0-2,1)+\delta_-(n_0-1)V_+(n_0-1)^t\delta_-(n_0)\\=V_-(n_0-1)^t\eta_-(n_0-1,1),$$

we apply Theorem 3.1′ again to find that

$$\mathrm{II}^{(1)}_{n_0+1}=(V_-(n_0)-V_-(n_0-1))^t\eta_-(n_0-1,1)\\+\delta_-(n_0)\left\{\sum_{j=0}^{n_0-3}\eta_+(j,n_0-1-j)R(2+j)\right.\\-\sum_{j=0}^{n_0-3}R(2+j)^t\eta_-(j,n_0-1-j)\\+\eta_+(n_0-2,1)R(n_0)-R(n_0)^t\eta_-(n_0-2,1)\\\left.-\left(R(1)+\sum_{j=0}^{n_0-2}R(n_0-j)^t\eta_+(j,n_0-1-j)\right)^t\delta_-(n_0)\right\}.$$

By using $\mathrm{IV}^{(1)}_{n_0}=0$, coming from $(*)_{n_0}$ and Step 5, we see that

$$\mathrm{II}^{(1)}_{n_0+1}=(V_-(n_0)-V_-(n_0-1))^t\eta_-(n_0-1,1)\\+\delta_-(n_0)\left\{\eta_+(n_0-2,1)\left(\sum_{j=0}^{n_0-3}R(1+j)^t\eta_-(j,n_0-1-j)+R(n_0)\right)\right.\\-\left(\sum_{j=0}^{n_0-3}\eta_+(j,n_0-1-j)R(1+j)+R(n_0)\right)\\\left.-\left(R(1)+\sum_{j=0}^{n_0-2}R(n_0-j)^t\eta_+(j,n_0-1-j)\right)^t\delta_-(n_0)\right\}.$$

By using Lemma 4.2(i), (ii) and $\mathrm{III}_{n_0}^{(1)} = 0$,

$$\begin{aligned}\mathrm{II}_{n_0+1}^{(1)} &= (V_-(n_0) - V_-(n_0-1))^t \eta_-(n_0-1, 1) \\ &\quad + \delta_-(n_0)\delta_+(n_0) V_-(n_0-1)^t \eta_-(n_0-2, 1) \\ &\quad - \delta_-(n_0)\Bigg(\eta_+(n_0-2, 1) V_+(n_0-1) \\ &\quad + R(1) + \sum_{j=0}^{n_0-2} R(n_0-j)^t \eta_+(j, n_0-1-j)\Bigg)^t \delta_-(n_0) \\ &= (V_-(n_0) - V_-(n_0-1))^t \eta_-(n_0-1, 1) \\ &\quad + \delta_-(n_0)\delta_+(n_0) V_-(n_0-1)^t \eta_-(n_0-2, 1) \\ &\quad - \delta_-(n_0)\Bigg(\sum_{j=0}^{n_0-2} \eta_+(j, n_0-1-j) R(1+j) \\ &\quad + R(n_0)\Bigg)^t \delta_+(n_0-1)^t \delta_-(n_0).\end{aligned}$$

Therefore, it follows from Theorem 3.1′, Lemma 4.2(i) and (4.10) that

$$\begin{aligned}\mathrm{II}_{n_0+1}^{(1)} &= -V_-(n_0)^t(\eta_-(n_0-1, 1) - \eta_-(n_0-2, 1)) \\ &\quad - V_-(n_0-1)^t \delta_+(n_0-1)^t \delta_-(n_0) \\ &\quad + \delta_-(n_0)\delta_+(n_0) V_-(n_0-1)^t \delta_+(n_0-1)^t \delta_-(n_0) \\ &= (V_-(n_0) - V_-(n_0-1))^t \delta_+(n_0-1)^t \delta_-(n_0) \\ &\quad + \delta_-(n_0)\delta_+(n_0) V_-(n_0-1)^t \delta_+(n_0-1)^t \delta_-(n_0) \\ &= 0.\end{aligned}$$

Q.E.D.

By Steps 9 and 10, we have proved that $(*)'_{n_0,1}$ holds. Moreover, for any fixed k, $1 \le k < k+1 \le N_0$, by assuming that $(*)'_{n_0,k}$ holds, we will show that $(*)'_{n_0,k+1}$ holds.

Step 11. $\mathrm{I}_{n_0+1}^{(k+1)} = 0$.

Proof. By applying $\mathrm{II}_{n_0}^{(k)} = 0$ (resp. $\mathrm{III}_{n_0}^{(k)} = 0$) to the coefficient of $\delta_+(n_0-k)$ (resp. $\delta_-(n_0-k)$), we see from Theorem 3.1′, (4.8) and (4.9) that

$$\mathrm{I}_{n_0+1}^{(k+1)} = -\delta_-(n_0-k)\mathrm{I}' - \mathrm{I}''^t \delta_+(n_0-k) + \mathrm{I}''', \tag{4.12}$$

where

$$
\begin{aligned}
\mathrm{I}' = & \sum_{j=0}^{k-1}\left(\sum_{i=0}^{n_0-2k} \eta_+(i, n_0-k-i)R(k-j+i)\right){}^t\eta_+(j, n_0-k-j) \\
& -\sum_{j=1}^{k-1}(\eta_+(n_0-k-j, j)-\eta_+(n_0-k-1-j, j)) \\
& \times\left(R(k-j)+\sum_{i=1}^{n_0-2k} R(k-j+i)^t\eta_+(n_0-k-i, i)\right) \\
& -(\eta_+(n_0-2k, k)-\eta_+(n_0-2k-1, k))V_+(n_0-k) \\
& +\delta_+(n_0-k)V_-(n_0-k-1)^t\eta_+(k-1, n_0-2k), \\
\mathrm{I}'' = & -\sum_{j=0}^{k-1}\eta_-(j, n_0-k-j)\left(\sum_{i=0}^{n_0-2k} R(k-j+i)^t\eta_-(i, n_0-k-i)\right) \\
& +\sum_{j=1}^{k-1}\left(R(k-j)+\sum_{i=1}^{n_0-2k}\eta_-(n_0-k-i, i)R(k-j+i)\right) \\
& \times{}^t(\eta_-(n_0-k-j, j)-\eta_-(n_0-k-1-j, j)) \\
& +V_-(n_0-k)^t(\eta_-(n_0-2k, k)-\eta_-(n_0-2k-1, k)) \\
& -\eta_-(k-1, n_0-2k)V_+(n_0-k-1)^t\delta_-(n_0-k), \\
\mathrm{I}''' = & -\sum_{j=0}^{k-2}\eta_-(j, n_0-k-1-j)\left(R(k-1-j)\right. \\
& \left.+\sum_{i=1}^{n_0-2k} R(k-1-j+i)^t\eta_+(n_0-k-i, i)\right) \\
& +\sum_{j=0}^{k-2}\left(R(k-1-j)+\sum_{i=1}^{n_0-2k}\eta_-(n_0-k-i, i)\right. \\
& \left.\times R(k-1-j+i)\right){}^t\eta_+(j, n_0-k-1-j) \\
& -\eta_-(k-1, n_0-2k)V_+(n_0-k-1)+V_-(n_0-k-1)^t\eta_+(k-1, n_0-2k).
\end{aligned}
$$

By Theorem 3.1′ and $\mathrm{I}_{n-1}^{(k)}=0$,

$$
\begin{aligned}
\mathrm{I}''' = & \left\{-\sum_{j=0}^{k-2}\eta_-(j, n_0-k-1-j)\left(\sum_{i=0}^{n_0-2k-1} R(k-j+i)^t\eta_-(i, n_0-k-1-i)\right)\right\} \\
& \times{}^t\delta_+(n_0-k)+\delta_-(n_0-k) \\
& \times\left\{\sum_{j=0}^{k-2}\left(\sum_{i=0}^{n_0-2k-1}\eta_+(i, n_0-k-1-i)R(k-j+i)\right){}^t\eta_+(j, n_0-k-1-j)\right\}.
\end{aligned}
\tag{4.13}
$$

Next, applying Theorem 3.1′ to I′, we have

$$\mathrm{I}' = \delta_+(n_0-k)\tilde{\mathrm{I}}'{}^t\delta_-(n_0-k) + \delta_+(n_0-k)\widetilde{\mathrm{II}}' + \widetilde{\mathrm{III}}'{}^t\delta_+(n_0-k) + \mathrm{IV}',$$

where

$$\begin{aligned}
\tilde{\mathrm{I}}' = {} & R(k) + \sum_{i=1}^{n_0-2k} \eta_-(n_0-k-1-i, i)R(k+i) \\
& + \eta_-(k-1, n_0-2k)\, V_+(n_0-k-1)^t\delta_-(n_0-k) \\
& + \sum_{j=1}^{k-1} \left(R(k-j) + \sum_{i=1}^{n_0-2k} \eta_-(n_0-k-1-i, i) \right. \\
& \left. \times R(k-j+i) \right) {}^t\eta_-(n_0-k-1-j, j) \\
& - \sum_{j=0}^{k-2} \eta_-(j, n_0-k-1-j) \left(R(k-j) \right. \\
& \left. + \sum_{i=0}^{n_0-2k-1} R(k-j+i)^t\eta_-(i, n_0-k-i) \right),
\end{aligned}$$

$$\begin{aligned}
\widetilde{\mathrm{II}}' = {} & -\eta_-(k-1, n_0-2k)\, V_+(n_0-k-1) + V_-(n_0-k-1)^t\eta_+(k-1, n_0-2k) \\
& - \sum_{j=0}^{k-2} \eta_-(j, n_0-k-1-j) \\
& \times \left(R(k-1-j) + \sum_{i=1}^{n_0-2k} R(k-1-j+i)^t\eta_+(n_0-k-1-i, i) \right) \\
& + \sum_{j=0}^{k-2} \left(R(k-1-j) + \sum_{i=1}^{n_0-2k} \eta_-(n_0-k-1-i, i) \right. \\
& \left. \times R(k-1-j+i) \right) {}^t\eta_+(j, n_0-k-1-j),
\end{aligned}$$

$$\begin{aligned}
\widetilde{\mathrm{III}}' = {} & \sum_{j=0}^{n_0-2k-1} \eta_+(j, n_0-k-1-j)R(k+1+j) \\
& + \sum_{j=1}^{k-1} \left(\sum_{i=0}^{n_0-2k-1} \eta_+(i, n_0-k-1-i)R(k-j+i) \right) {}^t\eta_+(n_0-k-1-j, j),
\end{aligned}$$

$$\mathrm{IV}' = \sum_{j=0}^{k-2} \left(\sum_{i=0}^{n_0-2k-1} \eta_+(i, n_0-k-1-i)R(k-j+i) \right) {}^t\eta_+(j, n_0-k-1-j).$$

By Lemma 4.2(*ii*) and $\mathrm{II}_{n_0-1}^{(k)}=0$,

$$\begin{aligned}\tilde{\mathrm{I}}' &= \eta_-(k-1, n_0-2k)\Big\{R(n_0-k)\\ &\quad+\sum_{j=0}^{n_0-2k-2} R(1+j)^t\eta_-(j, n_0-k-1-j)\\ &\quad+V_+(n_0-k-1)^t\delta_-(n_0-k)\Big\}\\ &\quad+\mathrm{II}_{n_0-1}^{(k)}-V_+(n_0-k-1)^t\eta_-(n_0-2k-1, 1)\\ &=-V_+(n_0-k-1)^t\eta_-(n_0-2k-1, 1).\end{aligned}$$

Furthermore, since

$$\widetilde{\mathrm{II}}'=\mathrm{I}_{n_0-1}^{(k)}=0,$$

$$\begin{aligned}\mathrm{I}' &= -\delta_+(n_0-k)V_-(n_0-k-1)^t\eta_-(n_0-2k-1, k)^t\delta_+(n_0-k)\\ &\quad+\widetilde{\mathrm{III}}''^t\delta_+(n_0-k)+\widetilde{\mathrm{IV}}'. \qquad (4.14)\end{aligned}$$

Similarly to (4.13), we can apply $\mathrm{I}_{n_0-1}^{(k)}=\mathrm{III}_{n_0-1}^{(k)}=0$ to see that

$$\begin{aligned}\mathrm{II}' &= \delta_-(n_0-k)\eta_+(n_0-2k-1, k)V_+(n_0-k-1)^t\delta_-(n_0-k)\\ &\quad+\delta_-(n_0-k)\widetilde{\mathrm{III}}''+\widetilde{\mathrm{IV}}'', \qquad (4.15)\end{aligned}$$

where

$$\begin{aligned}\widetilde{\mathrm{III}}'' &= -\sum_{i=0}^{n_0-2k-1} R(k+i)^t\eta_-(i, n_0-k-1-i)\\ &\quad-\sum_{j=1}^{k-1}\eta_+(n_0-k-1-j, j)\left(\sum_{i=0}^{n_0-2k-1} R(k+1-j+i)^t\eta_-(i, n_0-k-1-i)\right),\\ \widetilde{\mathrm{IV}}'' &= -\sum_{j=0}^{k-2}\eta_-(j, n_0-k-1-j)\left(\sum_{i=0}^{n_0-2k-1} R(k-j+i)^t\eta_-(i, n_0-k-1-i)\right).\end{aligned}$$

Thus, by (4.12), (4.13), (4.14) and (4.15),

$$\begin{aligned}\mathrm{I}_{n_0+1}^{(k+1)} &= -\delta_-(n_0-k)(-\delta_+(n_0-k)V_-(n_0-k-1)^t\eta_-(n_0-2k-1, k)\\ &\quad+\eta_+(n_0-2k-1, k)V_+(n_0-k-1)^t\delta_-(n_0-k)\\ &\quad+\widetilde{\mathrm{III}}'+\widetilde{\mathrm{III}}'')^t\delta_+(n_0-k). \qquad (4.16)\end{aligned}$$

On the other hand, by using $\mathrm{IV}_{n_0-1}^{(k)}=0$, coming from $(*)_{n_0}$ and Step 5, we see that

$$
\begin{aligned}
\widetilde{\mathrm{III}}'+\widetilde{\mathrm{III}}''=\eta_+(n_0-2k-1,k)\Bigg(&R(n_0-k)\\
&+\sum_{j=0}^{n_0-k-2}R(1+j)'\eta_-(j,n_0-k-1-j)\Bigg)\\
&-\Bigg(R(n_0-k)+\sum_{j=0}^{n_0-k-2}\eta_+(j,n_0-k-1-j)\\
&\times R(1+j)\Bigg)'\eta_-(n_0-2k+1,k).
\end{aligned}
$$

Consequently, by substituting this into (4.16), it follows from Lemma 4.2 that $\mathrm{I}_{n_0+1}^{(k+1)}=0$. Q.E.D.

Step 12. (i) $\mathrm{II}_{n_0+1}^{(k+1)}=0$; (ii) $\mathrm{III}_{n_0+1}^{(k+1)}=0$.

Proof. We prove only (i), since (ii) is shown similarly. By applying $\mathrm{IV}_{n_0}^{(k)}=0$ (resp. $\mathrm{II}_{n_0}^{(k)}=0$) to the coefficient of $\delta_-(n_0-k)$ (resp. ${}'\eta_-(n_0-k-1,1)$) in $\mathrm{II}_{n_0+1}^{(k+1)}$, we see from Theorem 3.1′ that

$$
\begin{aligned}
\mathrm{III}_{n_0+1}^{(k+1)}=\ &V_-(n_0-k)'(\eta_-(n_0-2k-1,k+1)-\eta_-(n_0-k-1,1)\eta_-(n_0-2k,k))\\
&+(I-\delta_-(n_0-k)\delta_+(n_0-k))J_1-\delta_-(n_0-k)J_2\\
&-J_3{}'\eta_-(n_0-k-1,1)+J_4,
\end{aligned}
\tag{4.17}
$$

where

$$
\begin{aligned}
J_1=\ &R(k+1)+\sum_{i=1}^{n_0-2k-1}\eta_-(n_0-k-1-i,i)R(k+1+i)\\
&-\sum_{j=0}^{k-1}\eta_-(j,n_0-k-1-j)\Bigg(\sum_{i=0}^{n_0-2k-1}R(k-j+i)'\eta_-(i,n_0-k-i)\Bigg),
\end{aligned}
$$

$$
J_2=\sum_{j=1}^{k}\Bigg(\sum_{i=0}^{n_0-2k-1}\eta_+(i,n_0-k-i)R(k+1-j+i)\Bigg)'\eta_-(n_0-k-j,j),
$$

$$
\begin{aligned}
J_3=\ &-\sum_{j=0}^{k-1}\eta_-(j,n_0-k-j)\Bigg(\sum_{i=0}^{n_0-2k-1}R(k-j+i)'\eta_-(i,n_0-k-i)\Bigg)\\
&-\sum_{j=1}^{k-1}\Bigg(R(k-j)+\sum_{i=1}^{n_0-2k}\eta_-(n_0-k-i,i)R(k+1+i)\Bigg)'\eta_-(n_0-k-j,j)\\
&+\eta_-(k,n_0-2k)R(n_0-k),
\end{aligned}
$$

$$J_4 = \sum_{j=2}^{k} \left(R(k+1-j) + \sum_{i=1}^{n_0-2k-1} \eta_-(n_0-k-i, i) \right.$$

$$\left. \times R(k+1-j+i) \right) {}^t\eta_-(n_0-k-j, j).$$

Since

$$J_3 = \mathrm{II}_{n_0}^{(k)} - \left(R(k) + \sum_{i=1}^{n_0-2k} \eta_-(n_0-k-i, i) R(k+i) \right)$$

$$- V_-(n_0-k)^t\eta_-(n_0-2k, k) + \eta_-(k, n_0-k) R(n_0-k),$$

it follows from Theorem 3.1′ that

$$J_3 = -\delta_-(n_0-k) \left(\sum_{i=0}^{n_0-2k-2} \eta_+(i, n_0-k-1-i) R(k+1+i) \right)$$

$$- \left(R(k) + \sum_{i=1}^{n_0-2k-1} \eta_-(n_0-k-1-i, i) R(k+i) \right)$$

$$- V_-(n_0-k)^t\eta_-(n_0-2k, k),$$

and so

$$\delta_-(n_0-k) J_2 + J_3{}^t\eta_-(n_0-k-1, 1)$$

$$= -(I - \delta_-(n_0-k)\delta_+(n_0-k))$$

$$\times \left\{ R(k) + \sum_{i=1}^{n_0-2k-1} \eta_-(n_0-k-1-i, i) R(k+i) \right\} {}^t\eta_-(n_0-k-1, 1)$$

$$+ \delta_-(n_0-k)\delta_+(n_0-k) \sum_{j=2}^{k} \left\{ R(k+1-j) \right.$$

$$\left. + \sum_{i=1}^{n_0-2k-1} \eta_-(n_0-k-1-i, i) R(k+1-j+i) \right\}$$

$$\times {}^t\eta_-(n_0-k-j, j) + \delta_-(n_0-k)$$

$$\times \sum_{j=2}^{k} \left(\sum_{i=0}^{n_0-2k-2} \eta_+(n_0-k-1-i, i) R(k+2-j+i) \right) {}^t\eta_-(n_0-k-j, j)$$

$$- V_-(n_0-k)^t\eta_-(n_0-2k, k)^t\eta_-(n_0-k-1, 1). \tag{4.18}$$

Furthermore, we note from Theorem 3.1′ that

$$J_4 = \sum_{j=2}^{k} \left(R(k+1-j) + \sum_{i=1}^{n_0-2k-1} \eta_-(n_0-k-1-i, i) \times R(k+1-j+i) \right) {}^t\eta_-(n_0-k-j, j)$$

$$+ \delta_-(n_0-k) \sum_{j=2}^{k} \left(\sum_{i=0}^{n_0-2k-2} \eta_+(n_0-k-1-i, i) \times R(k+2-j+i) \right) {}^t\eta_-(n_0-k-j, j). \tag{4.19}$$

Therefore, by using Theorem 3.1′ and (4.9), we see from (4.17), (4.18) and (4.19) that

$$\mathrm{II}_{n_0+1}^{(k+1)} = (I - \delta_-(n_0-k)\delta_+(n_0-k))J_5, \tag{4.20}$$

where

$$J_5 = R(k+1) + \sum_{i=1}^{n_0-2k-2} \eta_-(n_0-k-1-i, i) R(k+1+i)$$

$$+ \sum_{j=1}^{k} \left(R(k+1-j) + \sum_{i=1}^{n_0-2k-2} \eta_-(n_0-k-1-i, i) \times R(k+1-j+i) \right) {}^t\eta_-(n_0-k-j, j)$$

$$- \sum_{j=0}^{k-1} \eta_-(j, n_0-k-1-j) \left(R(k-j)^t\delta_-(n_0-k) + \sum_{i=1}^{n_0-2k-1} R(k-j+i)^t\eta_-(i, n_0-k-i) \right)$$

$$+ \eta_-(k, n_0-2k-1) \left(R(n_0-k) + \sum_{i=1}^{k} R(n_0-k-i)^t\eta_-(n_0-k-i, i) \right)$$

$$+ V_-(n_0-k-1)^t\eta_-(n_0-2k-1, k+1).$$

We note from Lemma 4.2(*ii*) that

$$R(n_0-k) + \sum_{i=1}^{k} R(n_0-k-i)^t\eta_-(n_0-k-1-i, i)$$

$$= -V_+(n_0-k-1)^t\delta_+(n_0-k) - \sum_{i=0}^{n_0-2k-2} R(1+i)^t\eta_-(i, n_0-k-1-i).$$

Therefore, by using Theorem 3.1′, we get

$$
\begin{aligned}
J_5 = {} & R(k+1) + \sum_{i=1}^{n_0-2k-2} \eta_-(n_0-k-1-i, i) R(k+1+i) \\
& - V_-(n_0-k-1)\,{}^t\eta_-(n_0-2k-2, k+1) \\
& + \sum_{j=1}^{k} \left(R(k+1-j) + \sum_{i=1}^{n_0-2k-2} \eta_-(n_0-k-1-i, i) \right. \\
& \left. \times R(k+1-j+i) \right) {}^t\eta_-(n_0-k-1-j, j) \\
& - \sum_{j=0}^{k} \eta_-(j, n_0-k-1-j) \\
& \times \left(\sum_{i=0}^{n_0-2k-2} R(k-j+1+i)\,{}^t\eta_-(i, n_0-k-1-i) \right) + \left\{ \sum_{j=1}^{k} \left(R(k+1-j) \right. \right. \\
& \left. + \sum_{i=1}^{n_0-2k-2} \eta_-(n_0-k-1-i, i) R(k+1-j+i) \right) {}^t\eta_+(j-1, n_0-k-j) \\
& - \sum_{j=0}^{k-1} \eta_-(j, n_0-k-1-j) \left(R(k-j) \right. \\
& \left. + \sum_{i=1}^{n_0-2k-1} R(k-j+i)\,{}^t\eta_+(n_0-k-1-i, i) \right) \\
& + \eta_-(k, n_0-2k-1) \left(-V_+(n_0-k-1) \right. \\
& \left. + \sum_{i=1}^{k} R(n_0-k-i)\,{}^t\eta_+(i-1, n_0-k-i) \right) \\
& \left. + V_-(n_0-k-1)\,{}^t\eta_+(k, n_0-2k-1) \right\} {}^t\delta_-(n_0-k) \\
= {} & \mathrm{II}_{n_0}^{(k+1)} + \mathrm{I}_{n_0}^{(k+1)}\,{}^t\delta_-(n_0-k) \\
= {} & 0,
\end{aligned}
\tag{4.21}
$$

noting that $k+1 \leq N_0$. Consequently, by (4.20) and (4.21), we see that $\mathrm{II}_{n_0+1}^{(k+1)} = 0$. Q.E.D.

Finally, we will show $(*)''_{n_0, N_0}$, that is,

Step 13. $\mathrm{IV}^{(N_0)}_{n_0+1}=0.$

Proof. Since it follows from (4.8) and (4.9) that

$$V_+(N_0)+\sum_{j=0}^{N_0-1} V_+(N_0-1-j)^t\delta_-(N_0-j)^t\delta_+(N_0-j)$$

$$=V_-(N_0)+\sum_{j=0}^{N_0-1} \delta_-(N_0-j)\delta_+(N_0-j)V_-(N_0-1-j)$$

$$=R(0),$$

we see from (*ii*) in Step 6 that Step 13 holds. Q.E.D.

Thus we have completed the proof of Lemma 4.3. By noting Steps 1, 2 and 3 in the proof of Lemma 4.3, we have

Theorem 4.1. *For any* $n\in\mathbf{N}^*$,

(*i*) $V_+(n+1)=V_+(n)-\delta_+(n+1)V_-(n)^t\delta_+(n+1)$,
(*ii*) $V_+(n+1)=(I-\delta_+(n+1)\delta_-(n+1))V_+(n)$,
(*iii*) $V_-(n+1)=V_-(n)-\delta_-(n+1)V_+(n)^t\delta_-(n+1)$,
(*iv*) $V_-(n+1)=(I-\delta_-(n+1)\delta_+(n+1))V_-(n)$,
(*v*) $V_+(n)^t\delta_-(n+1)=\delta_+(n+1)V_-(n)$,
(*vi*) $V_+(n+1)^t\delta_-(n+1)=\delta_+(n+1)V_-(n+1)$.

By Theorems 3.1 and 4.1, we find that the parameters $\gamma_+(\cdot,*)$, $\gamma_-(\cdot,*)$, $V_+(\cdot)$ and $V_-(\cdot)$ in the KM_2O-Langevin equations (2.16) and (2.17) can be uniquely determined by $R(0)$, $\delta_+(\cdot)$ and $\delta_-(\cdot)$. We call the system $\{\gamma_+(n,k), \delta_+(m), V_+(m); m, n, k\in\mathbf{N}, n>k\}$ (resp. $\{\gamma_-(n,k), \delta_-(m), V_-(m); m, n, k\in\mathbf{N}, n>k\}$) the *forward* (resp. *backward*) *KM_2O-Langevin data* associated with X.

For future use in a modelling of time series, we will rewrite Lemma 4.2 into

Theorem 4.2. *For any* $n\in\mathbf{N}^*$,

$$(i)\quad \delta_+(n+1)=-\left(R(n+1)+\sum_{k=0}^{n-1}\gamma_+(n,k)R(k+1)\right)V_-(n)^{-1},$$

$$(ii)\quad \delta_-(n+1)=-\left({}^tR(n+1)+\sum_{k=0}^{n-1}\gamma_-(n,k)^tR(k+1)\right)V_+(n)^{-1}.$$

5. The Prediction Formula and Prediction Error

For each $m, n \in \mathbf{N}$, $m \geq n$, we define a *forward* (resp. *backward*) *prediction matrix* $P_+(m, n)$ (resp. $P_-(m, n)) \in M(d; \mathbf{R})$ and a *forward* (resp. *backward*) *prediction error matrix* $e_+(m, n)$ (resp. $e_-(m, n)) \in M(d; \mathbf{R})$ by

$$P_+(m, n) = E(X(m)^t I_+(n))\, V_+(n)^{-1/2}, \tag{5.1}$$

$$P_-(m, n) = E(X(-m)^t I_-(n))\, V_-(n)^{-1/2}, \tag{5.2}$$

$$e_+(m, n) = E((X(m) - P_{M_0^+(n)} X(m))^t (X(m) - P_{M_0^+(n)} X(m))), \tag{5.3}$$

$$e_-(m, n) = E((X(-m) - P_{M_0^-(n)} X(-m))^t (X(-m) - P_{M_0^-(n)} X(-m))). \tag{5.4}$$

Theorem 5.1. *For any $m, n \in \mathbf{N}$, $m \geq n$,*

$$(i) \quad X(m) = R(m) R(0)^{-1} X(0) + \sum_{k=1}^{m} P_+(m, k)\, V_+(k)^{-1/2} I_+(k),$$

$$(ii) \quad P_{M_0^+(n)} X(m) = R(m) R(0)^{-1} X(0) + \sum_{k=1}^{n} P_+(m, k)\, V_+(k)^{-1/2} I_+(k),$$

$$(iii) \quad e_+(m, n) = \sum_{k=n+1}^{m} P_+(m, k)^t P_+(m, k).$$

Proof. We put

$$W = \text{the left-hand side in } (i) - \text{the right-hand side in } (i).$$

By (2.20), we see that each component of W belongs to the linear hull of $\{I_{+j}(l); 1 \leq j \leq d, 0 \leq l \leq m\}$. It follows from (2.18) and (5.1) that $E(W^t I_+(l)) = 0$ $(0 \leq l \leq m)$. Hence, we find that $W = 0$, implying (i). (ii) follows from (2.18), (2.20) and (i). By noting (2.18) again, we see that (iii) follows from (i) and (ii). Q.E.D.

Similarly, we obtain

Theorem 5.2. *For any $m, n \in \mathbf{N}$, $m \geq n$,*

$$(i) \quad X(-m) = {}^t R(m) R(0)^{-1} X(0) + \sum_{k=1}^{m} P_-(m, k)\, V_-(k)^{-1/2} I_-(k),$$

(*ii*) $P_{M_0^-(n)}X(-m) = {}^tR(m)R(0)^{-1}X(0) + \sum_{k=1}^{n} P_-(m,k)V_-(k)^{-1/2}I_-(k)$,

(*iii*) $e_-(m,n) = \sum_{k=n+1}^{m} P_-(m,k)^t P_-(m,k)$.

We will show that the prediction matrices $P_+(\cdot,*)$ and $P_-(\cdot,*)$ can be also determined by $R(0)$, $\delta_+(\cdot)$ and $\delta_-(\cdot)$ through the following

Theorem 5.3. *For any* $m, n \in \mathbf{N}$, $m > n$,

(*i*) $P_+(n,n) = V_+(n)^{1/2}$,

$$P_+(m,n) = -\sum_{k=n}^{m-1} \gamma_+(m,k)P_+(k,n),$$

(*ii*) $P_-(n,n) = V_-(n)^{1/2}$,

$$P_-(m,n) = -\sum_{k=n}^{m-1} \gamma_-(m,k)P_-(k,n).$$

Proof. By (2.16), (2.18), (2.20) and (5.1), $P_+(n,n) = V_+(n)^{1/2}$. By substituting (2.16) with n replaced by m into the right-hand side of (5.1), we find from (2.18) and (2.20) that the second statement in (*i*) holds. (*ii*) is proved similarly. Q.E.D.

By using Theorem 4.1(*ii*), (*iv*), we see from Theorems 5.1(*iii*), 5.2(*iii*) and 5.3 that the *one-step prediction error matrices* $e_+(n+1,n)$ and $e_-(n+1,n)$ can be explicitly represented by $R(0)$, $\delta_+(l)$ and $\delta_-(l)$ $(1 \le l \le n+1)$ through the following

Theorem 5.4. *For any* $n \in \mathbf{N}$,

(*i*) $e_+(n+1,n) = (I - \delta_+(n+1)\delta_-(n+1)) \cdots (I - \delta_+(1)\delta_-(1))R(0)$,
(*ii*) $e_-(n+1,n) = (I - \delta_-(n+1)\delta_+(n+1)) \cdots (I - \delta_-(1)\delta_+(1))R(0)$.

6. A Construction Theorem

As a converse setting of Sections 2–5, we are given a system $\{V, \delta_+(n);\, n \in \mathbf{N}\}$ of members in $M(d;\mathbf{R})$ such that

$$V \text{ is a symmetric positive definite matrix,} \tag{6.1}$$

$$\delta_+(n) \in M(d;\mathbf{R}) \qquad (n \in \mathbf{N}). \tag{6.2}$$

Then we construct a triple $(V_+(1), \delta_-(1), V_-(1))$ by

$$\begin{aligned} V_+(1) &= V - \delta_+(1) V^t\delta_+(1), \\ \delta_-(1) &= V^t\delta_+(1) V^{-1}, \\ V_-(1) &= V - \delta_-(1) V^t\delta_-(1). \end{aligned} \tag{6.3}$$

In order to continue the construction of $(V_+(n), \delta_-(n), V_-(n))$ from $(V_+(n-1), \delta_-(n-1), V_-(n-1))$ $(n \in \mathbf{N})$,

$$\begin{aligned} V_+(n) &= V_+(n-1) - \delta_+(n) V_-(n-1)^t\delta_+(n), \\ \delta_-(n) V_+(n-1) &= V_-(n-1)^t\delta_+(n), \\ V_-(n) &= V_-(n-1) - \delta_-(n-1) V_+(n-1)^t\delta_-(n-1), \end{aligned} \tag{6.4}$$

we suppose that

$$V_+(n-1) \text{ are positive definite} \qquad (n \in \mathbf{N}), \tag{6.5}$$

where $V_+(0) = V_-(0) = V$.

Next, we construct a system $\{\gamma_+(m, n), \gamma_-(m, n); m, n \in \mathbf{N}^*, m > n\}$ of members in $M(d; \mathbf{R})$ according to the algorithm in Theorem 3.1.

Furthermore, we prepare any $\mathbf{R}^d$-valued stochastic process $I_+ = (I_+(n); n \in \mathbf{N}^*)$ on a probability space $(\Omega, \mathscr{B}, P)$ such that

$$E(I_+(n)) = 0 \qquad (n \in \mathbf{N}^*), \tag{6.6}$$

$$E(I_+(n)^t I_+(m)) = \delta_{nm} V_+(n) \qquad (n, m \in \mathbf{N}^*), \tag{6.7}$$

and then construct an $\mathbf{R}^d$-valued stochastic process $X_+ = (X(n); n \in \mathbf{N}^*)$ by

$$X(0) = I_+(0), \tag{6.8}$$

$$X(n) = -\sum_{k=0}^{n-1} \gamma_+(n, k) X(k) + I_+(n) \qquad (n \in \mathbf{N}). \tag{6.9}$$

Now we will show

Theorem 6.1. *X_+ is weakly stationary.*

For any $m, n \in \mathbf{N}^*$, we put

$$R(m, n) = E(X(m)^t X(n)). \tag{6.10}$$

Since

$$^tR(m, n) = R(n, m), \tag{6.11}$$

for the proof of Theorem 6.1, it suffices to show

$$R(m+l, n+l) = R(m, n) \qquad (m \geq n \geq 0, l \geq 1). \tag{6.12}$$

The proof is divided into 22 steps.

Step 1. $E(X(m)^t I_+(n)) = \delta_{mn} V_+(n)$ $(0 \leq m \leq n)$.

This follows immediately from (6.7), (6.8) and (6.9).

Step 2.

(*i*) $R(0,0) = R(1,1) = R(2,2) = V$,

(*ii*) $R(1,0) = R(2,1) = -\delta_+(1)V$,

(*iii*) $R(2,0) = \delta_+(1)\delta_+(1)V - \delta_+(2)V_-(1)$.

These are shown by a simple calculation.

Step 3.

(*i*) $\delta_-(n+1)V_+(n) = V_-(n)^t\delta_+(n+1) \qquad (n \in \mathbf{N}^*)$,

(*ii*) $\delta_-(n)V_+(n) = V_-(n)^t\delta_-(n) \qquad (n \in \mathbf{N})$.

Proof. (*i*) is included in (6.4) and (*ii*) is proved similarly to Step 7 in the proof of Lemma 4.3. Q.E.D.

Step 4.

(*i*) $$R(n, l-1) = -\sum_{k=0}^{n-1} \eta_+(k, n-k)R(k, l-1) \qquad (n \geq l \geq 1),$$

(*ii*) $$R(n,n) = -\sum_{k=0}^{n-1} \eta_+(k, n-k)R(k,n) + V_+(n) \qquad (n \geq 1),$$

(*iii*) $$V_+(n) = R(n,n) + \sum_{k=1}^{n-1} \eta_+(k-1, n-k)R(k,n) + \delta_+(n)\left(\sum_{k=1}^{n-1} \eta_-(n-k-1, k)R(k,n) + R(0,n)\right) \qquad (n \geq 2),$$

where $\{\eta_+(m,n), \eta_-(m,n); m \in \mathbf{N}^*, n \in \mathbf{N}\}$ *is defined by* (3.6) *and* (3.7).

Proof. By multiplying both sides of (6.9) by ${}^tX(l-1)$ (resp. ${}^tX(n)$) and then taking an expectation with respect to the probability P, we have (i) (resp. (ii)). (iii) follows from (ii) and the algorithm in Theorem 3.1′.

Q.E.D.

Step 5.

$$(i) \qquad R(n,0) = -\delta_+(n)V_-(n-1) - \sum_{k=1}^{n-1} \eta_+(k-1, n-k)R(k,0) \qquad (n \in \mathbf{N}),$$

$$(ii) \qquad V_-(n-1) = R(0,0) + \sum_{k=1}^{n-1} R(0,k){}^t\eta_-(n-1-k, k) \qquad (n \in \mathbf{N}),$$

$$(iii) \qquad R(n,0) = -\delta_+(n)V - \sum_{k=1}^{n-1} \eta_+(k, n-k)R(k,0) \qquad (n \in \mathbf{N}).$$

Proof. By a mathematical induction on n, we show (i) and (ii). By Step 2, (i) and (ii) for $n=1$ hold. Let us assume that (i) and (ii) hold for $n=n_0$. Since $V_-(n_0)$ is symmetric, by (6.4), (6.11) and (i) for $n=n_0$,

$$\begin{aligned} V_-(n_0) &= V_-(n_0-1) - V_-(n_0-1){}^t\delta_+(n_0){}^t\delta_-(n_0) \\ &= V_-(n_0-1) + \Bigg(R(0,n_0) \\ &\qquad + \sum_{k=1}^{n_0-1} R(0,k){}^t\eta_+(k-1, n_0-k)\Bigg){}^t\delta_-(n_0). \end{aligned}$$

On the other hand, by Theorem 3.1′,

$$\begin{aligned} &R(0,0) + \sum_{k=1}^{n_0} R(0,k){}^t\eta_-(n_0-k, k) \\ &\quad = R(0,0) + \sum_{k=1}^{n_0-1} R(0,k){}^t\eta_-(n_0-1-k, k) \\ &\quad + \Bigg(R(0,n_0) + \sum_{k=1}^{n_0-1} R(0,k){}^t\eta_+(k-1, n_0-k)\Bigg){}^t\delta_-(n_0). \end{aligned}$$

Hence, we have (ii) for $n=n_0+1$. Next, by (6.9) and Theorem 3.1′,

$$R(n_0+1, 0) = -\sum_{k=0}^{n_0} \eta_+(k, n_0+1-k)R(k,0)$$

$$= -\sum_{k=1}^{n_0} \eta_+(k-1, n_0+1-k)R(k,0)$$

$$- \delta_+(n_0+1)\left(R(0,0) + \sum_{k=1}^{n_0} \eta_-(n_0-k, k)R(k,0)\right),$$

which, together with (ii) for $n = n_0$, implies (i) for $n = n_0+1$. By substituting (ii) into (i), we see from Theorem 3.1′ that (iii) holds. Q.E.D.

Before proceeding to the next step, we put the following statements: for each $n \in \mathbf{N}^*$,

$$\sum_{k=0}^{n-1} \eta_+(k, n-k)R(k+1, 0) = \sum_{k=0}^{n-1} R(k+1, 0)'\eta_-(k, n-k), \qquad (6.13)_n$$

$$V_+(n) = R(0,0) + \sum_{k=1}^{n} R(k,0)'\eta_+(n-k, k), \qquad (6.14)_n$$

$$R(0,n) = -\delta_-(n)V_+(n-1) - \sum_{k=1}^{n-1} \eta_-(k-1, n-k)R(0,k), \qquad (6.15)_n$$

$$R(0,n) = -\delta_-(n)V - \sum_{k=1}^{n-1} \eta_-(k, n-k)R(0,k). \qquad (6.16)_n$$

Step 6. *For any fixed* $n \in \mathbf{N}$,

(i) *If* $(6.13)_n$ *holds, then* $(6.15)_{n+1}$ *holds.*
(ii) *If* $(6.14)_{n-1}$ *and* $(6.15)_n$ *hold, then* $(6.14)_n$ *holds.*
(iii) *If* $(6.14)_n$ *and* $(6.15)_n$ *hold, then* $(6.16)_n$ *holds.*

Proof. By (6.9) and Theorem 3.1′,

$$R(0, n+1) = -\sum_{k=0}^{n} R(0,k)'\eta_+(k, n+1-k)$$

$$= -\sum_{k=1}^{n} R(0,k)'\eta_+(k-1, n+1-k)$$

$$-\left(R(0,0) + \sum_{k=1}^{n} R(0,k)'\eta_-(n-k, k)\right)'\delta_+(n+1).$$

And so by (ii) in Step 5,

$$R(0,n+1)=-{}^t\left(\sum_{k=0}^{n-1}\eta_+(k,n-k)R(k+1,0)\right)-V_-(n)'\delta_+(n+1).$$

Therefore, we have (i). (ii) and (iii) are proved similarly as in the proof of (ii) and (iii) in Step 5. Q.E.D.

We will show $(6.13)_n$ and $(6.14)_n$ $(n\in\mathbf{N})$ by a mathematical induction on n in Step 7–Step 17. Immediately from (i) and (ii) in Step 2, $(6.13)_1$ and $(6.14)_1$ hold. For any fixed $n_0\in\{2,3,\dots\}$, let us assume that $(6.13)_l$ and $(6.14)_l$ hold for any $l\in\{1,2,\dots,n_0-1\}$.

Step 7. *If* $(6.13)_{n_0}$ *holds, then* $(6.14)_{n_0}$ *holds.*

This follows from (i) and (ii) in Step 6.

Step 8.

(i) *the left-hand side of* $(6.13)_{n_0}=-\sum_{k=1}^{n_0}C_{n_0}(k)'\delta_-(k)$,

(ii) *the right-hand side of* $(6.13)_{n_0}=\sum_{k=1}^{n_0}D_{n_0}(k)'\delta_-(k)$, *where*

$$C_{n_0}(k)=\sum_{l=k}^{n_0-1}\eta_+(l,n_0-l)\left(R(l+1-k,0)+\sum_{m=0}^{k-2}R(l-m,0)'\eta_+(m,k-1-m)\right)+\eta_+(k-1,n_0-k+1)V_+(k-1),\tag{6.17}$$

$$D_{n_0}(k)=R(n+1-k)+\sum_{m=0}^{k-2}R(n_0-m,0)'\eta_+(m,k-1-m).\tag{6.18}$$

Proof. Since it follows from (i) in Step 6 that $(6.15)_l$ $(1\le l\le n_0)$ hold, we get

the left-hand side of $(6.13)_{n_0}$

$$=-\sum_{k=1}^{n_0-1}\eta_+(k,n_0-k)\left(V_+(k)'\delta_-(k+1)+\sum_{l=1}^{k}R(l,0)'\eta_-(l-1,k+1-l)\right)-\delta_+(n_0)V'\delta_-(1).\tag{6.19}$$

By Theorem 3.1′,

$$\eta_-(l-1, k+1-l) = \delta_-(k+1-l) + \sum_{j=1}^{l-1} \delta_-(k+1-l+j)\eta_+(k-l, j), \quad (6.20)$$

$$\eta_-(k, n_0-k) = \delta_-(n_0-k) + \sum_{j=1}^{k} \delta_-(n_0-k+j)\eta_+(n_0-k-1, j). \quad (6.21)$$

By substituting (6.20) (resp. (6.21)) into (6.19) (resp. the right-hand side of $(6.13)_{n_0}$), we have (i) (resp. (ii)) in Step 8. Q.E.D.

Step 9. *For any* $k \in \{1, 2, \ldots, n_0\}$,

$$C_{n_0}(k) + D_{n_0}(k) = \sum_{l=0}^{k-1} \delta_+(n_0-l)E_{n_0}(l; k),$$

where for $l \in \{0, \ldots, k-3\}$,

$$\begin{aligned} E_{n_0}(l; k) = & \sum_{i=0}^{n_0-k-1} \eta_-(i, n_0-l-1-i)R(n_0-k-i, 0) \\ & + \eta_-(n_0-k, k-l-1)V_+(k-1) \\ & - \sum_{j=0}^{l} \left(R(l-j, 0) + \sum_{i=1}^{k-l-1} \eta_-(n_0-l-1-i, i) \right. \\ & \left. \times R(l-j+i, 0) \right) {}^t\eta_+(j, k-1-j) \\ & + \sum_{j=l+1}^{k-2} \left(\sum_{i=0}^{n_0-k-1} \eta_-(i, n_0-l-1-i) \right. \\ & \left. \times R(n_0-j-1-i, 0) \right) {}^t\eta_+(j, k-1-j), \end{aligned} \quad (6.22)$$

$$\begin{aligned} E_{n_0}(k-2; k) = & \sum_{i=0}^{n_0-k-1} \eta_-(i, n_0-k+1-i)R(n_0-k-i, 0) \\ & + \eta_-(n_0-k, 1)V_+(k-1) \\ & - \sum_{j=0}^{k-2} (R(k-2-j, 0) + \eta_-(n_0-k, 1) \\ & \times R(k-1-j, 0))^t\eta_+(j, k-1-j), \end{aligned} \quad (6.23)$$

$$E_{n_0}(k-1; k) = V_+(k-1) - \left(V + \sum_{j=0}^{k-2} R(k-1-j, 0)^t\eta_+(j, k-1-j) \right). \quad (6.24)$$

Proof. By (6.17),

$$C_{n_0}(k)=\eta_+(k-1, n_0-k+1)V_+(k-1)$$
$$+\sum_{l=k}^{n_0-1}\eta_+(l, n_0-l)R(l+1-k, 0)$$
$$+\sum_{m=0}^{k-2}\left(\sum_{l=k}^{n_0-1}\eta_+(l, n_0-l)R(l-m, 0)\right){}^t\eta_+(m, k-1-m). \tag{6.25}$$

Moreover, it follows from Theorem 3.1′ and (*iii*) in Step 5 that for any $m\in\{0, 1, \ldots, k-1\}$,

$$\sum_{l=k}^{n_0-1}\eta_+(l, n_0-l)R(l-m, 0)$$
$$=\sum_{j=0}^{m-1}\delta_+(n_0-j)\left(\sum_{i=0}^{n_0-k-1}\eta_-(i, n_0-1-j-i)R(n_0-1-m-i, 0)\right)$$
$$-R(n_0-m, 0)-\sum_{j=0}^{k-1-m}\eta_+(j, n_0-m-j)R(j, 0). \tag{6.26}$$

By substituting (6.26) into (6.25), we find from (6.18) that

$$C_{n_0}(k)=-D_{n_0}(k)+\eta_+(k-1, n_0-k+1)V_+(k-1)-\delta_+(n_0-k+1)V$$
$$+\sum_{m=0}^{k-2}\left\{\sum_{j=0}^{m-1}\delta_+(n_0-j)\left(\sum_{i=0}^{n_0-k-1}\eta_-(i, n_0-1-j-i)R(n_0-1-m-i, 0)\right)\right.$$
$$\left.-\sum_{j=0}^{k-1-m}\eta_+(j, n_0-m-j)R(j, 0)\right\}{}^t\eta_+(m, k-1-m)$$
$$+\sum_{j=0}^{m-1}\delta_+(n_0-j)\left(\sum_{i=0}^{n_0-k-1}\eta_-(i, n_0-1-j-i)R(n_0-k-i, 0)\right). \tag{6.27}$$

Further, it follows from Theorem 3.1′ that for any $m\in\{0, 1, \ldots, k-2\}$,

$$\eta_+(k-1, n_0-k+1)$$
$$=\delta_+(n_0-k+1)+\sum_{j=1}^{k-1}\delta_+(n_0-k+1+j)\eta_-(n_0-k, j), \tag{6.28}$$

$$\sum_{j=0}^{k-1-m} \eta_+(j, n_0-m-j)R(j,0)$$

$$= \sum_{j=0}^{k-1-m} \delta_+(n_0-m-j)\Bigg(R(j,0)$$

$$+ \sum_{i=j+1}^{k-1-m} \eta_-(n_0-m-i-1, i-j)R(i,0)\Bigg). \tag{6.29}$$

Thus, by substituting (6.28) and (6.29) into (6.27), we can see that Step 9 holds. Q.E.D.

Step 10. $E_{n_0}(k-1; k)=0$ *for any* $k \in \{1, \ldots, n_0\}$.

This follows from the assumption of mathematical induction that $(6.14)_{k-1}$ holds.

Step 11. $E_{n_0}(k-2; k)=0$ *for any* $k \in \{2, \ldots, n_0\}$.

Proof. By (6.23) and $(6.14)_{k-1}$,

$$E_{n_0}(k-2; k) = \sum_{i=0}^{n_0-k-1} \eta_-(i, n_0-k+1-i)R(n_0-k-i, 0)$$

$$+\eta_-(n_0-k, 1)V - \sum_{j=0}^{k-2} R(k-2-j, 0)^t \eta_+(j, k-1-j). \tag{6.30}$$

By Theorem 3.1′ and (*iii*) in Step 5,

$$\sum_{i=0}^{n_0-k-1} \eta_-(i, n_0-k+1-i)R(n_0-k-i, 0)$$

$$= -\delta_-(n_0-k+1)\delta_+(n_0-k)V$$

$$+ \sum_{i=0}^{n_0-k-3} \eta_-(i, n_0-k-1)R(n_0-k-2-i, 0)$$

$$+\delta_-(n_0-k)\Bigg(R(n_0-k-1, 0)$$

$$+ \sum_{i=1}^{n_0-k-2} \eta_+(n_0-k-1-i, i)R(n_0-k-1-i, 0)\Bigg).$$

By repeating the same procedure, we have

$$\sum_{i=0}^{n_0-k-1} \eta_-(i, n_0-k+1-i)R(n_0-k-i, 0)$$

$$= -\left(\sum_{i=1}^{n_0-k} \delta_-(i+1)\delta_+(i)\right) V$$

$$= (-\eta_-(n_0-k, 1)+\delta_-(1))V. \tag{6.31}$$

On the other hand, by Theorem 3.1′,

$$\sum_{j=0}^{k-2} R(k-2-j, 0)^t\eta_+(j, k-1-j)$$

$$= \left(R(k-2, 0)+V^t\delta_-(k-2)\right.$$

$$\left.+\sum_{j=1}^{k-3} R(j, 0)^t\eta_-(j, k-2-j)\right)^t\delta_+(k-1)$$

$$+\sum_{j=0}^{k-3} R(k-3-j, 0)^t\eta_+(j, k-2-j),$$

and so by $(6.16)_{k-2}$,

$$\sum_{j=0}^{k-2} R(k-2-j, 0)^t\eta_+(j, k-1-j) = \sum_{j=0}^{k-3} R(k-3-j, 0)^t\eta_+(j, k-2-j).$$

By repeating the same procedure,

$$\sum_{j=0}^{k-2} R(k-2-j, 0)^t\eta_+(j, k-1-j) = \delta_-(1)V. \tag{6.32}$$

Thus, by (6.30), (6.31) and (6.32), we have Step 11. Q.E.D.

Step 12. *For any* $l \in \{0, 1, \ldots, k-2\}$,

$$E_{n_0}(l; k) = \sum_{i=0}^{n_0-k} \eta_-(i, n_0-l-i)R(n_0-k-i, 0)$$

$$-\sum_{j=0}^{l}\left(R(l-j, 0)\right.$$

$$\left.+\sum_{i=1}^{k-l-2} \eta_-(n_0-l-i, i)R(l-j+i, 0)\right)^t\eta_+(j, k-1-j)$$

$$
\begin{aligned}
&- \sum_{j=l+1}^{k-2} \left(\sum_{i=j-l}^{k-l-2} \eta_-(n_0-l-1-i, i) \right. \\
&\left. \times R(i-j+l, 0) \right) {}^t\eta_+(j, k-1-j) \\
&+ \sum_{j=l+1}^{k-2} \left(\sum_{i=j-l}^{n_0-l-1} \eta_-(n_0-l-1-i, i) \right. \\
&\left. \times R(i-j+l, 0) \right) {}^t\eta_+(j, k-1-j).
\end{aligned}
$$

Proof. By substituting $(6.14)_{k-1}$ into (6.22), we have

$$
\begin{aligned}
E_{n_0}(l; k) = &\sum_{i=0}^{n_0-k} \eta_-(i, n_0-l-1-i) R(n_0-k-i, 0) \\
&- \sum_{j=0}^{l} \left(R(l-j, 0) + \sum_{i=1}^{k-l-2} \eta_-(n_0-l-1-i, i) \right. \\
&\left. \times R(l-j+i, 0) \right) {}^t\eta_+(j, k-1-j) \\
&+ \sum_{j=l+1}^{k-2} \left(\sum_{i=0}^{n_0-k} \eta_-(i, n_0-l-1-i) \right. \\
&\left. \times R(n_0-j-i-1, 0) \right) {}^t\eta_+(j, k-1-j).
\end{aligned} \tag{6.33}
$$

Since

$$
\begin{aligned}
&\sum_{i=j-l}^{k-l-2} \eta_-(n_0-l-1-i, i) R(i-j+l, 0) \\
&\quad + \sum_{i=0}^{n_0-k} \eta_-(i, n_0-l-1-i) R(n_0-j-i-1, 0) \\
&= \sum_{i=j-l}^{n_0-l-1} \eta_-(n_0-l-1-i, i) R(i-j+l, 0),
\end{aligned}
$$

it follows from (6.33) that Step 12 holds. Q.E.D.

For each $l, m \in \mathbf{N}$, put

$$
F(m, l) = \sum_{i=0}^{m} \eta_-(m-i, l+i) R(i, 0). \tag{6.34}
$$

Step 13.

(*i*) $F(m,1)=\delta_-(1)V,$

(*ii*) $F(1,l)=(\eta_-(1,l)-\delta_-(l+1)\delta_+(1))V,$

(*iii*) $F(m,l)=F(m-1,l)+\delta_-(m+l)\sum_{j=1}^{l-1}\delta_+(m+j)F(m-1,j).$

Proof. (*i*) and (*ii*) follow from (6.31) and Step 2, respectively. By using Theorem 3.1′ repeatedly,

$$\begin{aligned}F(m,l)&=F(m-1,l)+\delta_-(m+l)\left(\sum_{i=0}^{m-1}\eta_+(l+i-1,m-i)\right.\\&\qquad\left.\times R(i,0)+R(m,0)\right)\\&=F(m-1,l)+\delta_-(m+l)\sum_{j=1}^{l-1}\delta_+(m+j)F(m-1,j)\\&\qquad+\delta_-(m+l)\left(\sum_{j=0}^{m-1}\eta_+(j,m-j)R(j,0)+R(m,0)\right),\end{aligned}$$

which, together with (*iii*) in Step 5, implies (*iii*). Q.E.D.

For each $l\in\{0,\ldots,k-2\}$, put

$$G(l;k)=\sum_{i=0}^{l}R(l-i,0)^t\eta_+(i,k-1-i). \tag{6.35}$$

Step 14. *For any* $l\in\{0,\ldots,k-2\}$,

$$\begin{aligned}E_{n_0}(l;k)&=-G(l;k)-\sum_{i=1}^{k-l-2}\eta_-(n_0-l-1-i,i)G(l+i;k)\\&\quad+F(n_0-k,k-l-1)\\&\quad+\sum_{i=1}^{k-l-2}F(n_0-k+i,k-l-1-i)^t\eta_+(k-1-i,i).\end{aligned}$$

This follows from Step 12, (6.34) and (6.35).

Step 15. *For any $k \in \{2, \dots, n_0\}$ and any $l \in \{0, \dots, k-2\}$,*

$$E_{n_0}(l;\, k) = E_{n_0-1}(l;\, k) + \delta_-(n_0 - l - 1) \times \sum_{j=2}^{k-l-1} \delta_+(n_0 - l - j) E_{n_0-1}(l - 1 + j;\, k).$$

Proof. By (*iii*) in Step 13,

$$\begin{aligned}
F(n_0 - k, k - l - 1) &+ \sum_{i=1}^{k-l-2} F(n_0 - k + i, k - l - 1 - i)^t \eta_+(k - 1 - i, i) \\
&= F(n_0 - k - 1, k - l - 1) + \delta_-(1) V^t \eta_+(l + 1, k - l - 2) \\
&\quad + \sum_{i=1}^{k-l-3} F(n_0 - k + i - 1, k - l - 1 - i)^t \eta_+(k - 1 - i, i) \\
&\quad + \delta_-(n_0 - l - 1) \sum_{j=2}^{k-l-2} \delta_+(n_0 - l - j) \Bigg(F(n_0 - k - 1, k - l - j) \\
&\quad + \sum_{i=1}^{k-l-1-j} F(n_0 - k + i - 1, k - l - j - i)^t \eta_+(k - 1 - i, i) \Bigg) \\
&\quad + \delta_-(n_0 - l - 1) \delta_+(n_0 - k + 1) F(n_0 - k - 1, 1). \qquad (6.36)
\end{aligned}$$

On the other hand, by Theorem 3.1′,

$$\begin{aligned}
\eta_-(n_0 - l - 1 - i, i) &= \eta_-(n_0 - l - 2 - i, i) + \delta_-(n_0 - l - 1) \delta_+(n_0 - l - 1 - i) \\
&\quad + \delta_-(n_0 - l - 1) \sum_{j=1}^{i-1} \delta_+(n_0 - l - 1 - j) \eta_-(n_0 - l - 2 - i, j),
\end{aligned}$$

and so

$$\begin{aligned}
\sum_{i=1}^{k-l-2} & \eta_-(n_0 - l - 1 - i, i) G(l + i;\, k) \\
&= \sum_{i=1}^{k-l-2} \eta_-(n_0 - l - 2 - i, i) G(l + i;\, k) \\
&\quad + \delta_-(n_0 - l - 1) \delta_+(n_0 - k + 1) G(k - 2;\, k) \\
&\quad + \delta_-(n_0 - l - 1) \sum_{j=1}^{k-l-3} \delta_+(n_0 - l - 1 - j) \Bigg(G(l + j;\, k) \\
&\quad + \sum_{i=j+1}^{k-l-2} \eta_-(n_0 - l - 2 - i, j) G(l + i;\, k) \Bigg). \qquad (6.37)
\end{aligned}$$

Hence, by combining Step 14 with (6.36) and (6.37), we have Step 15.
Q.E.D.

Step 16. *For any* $k \in \{2, \dots, n_0\}$ *and any* $l \in \{0, \dots, k-2\}$,

$$E_{n_0}(l; k) = 0.$$

Proof. By Step 15, it suffices to show that $E_2(0; 2) = 0$, which is proved in Step 11. Q.E.D.

Step 17. *For any* $n \in \mathbf{N}$, $(6.13)_n$ *and* $(6.14)_n$ *hold.*

This follows from Step 8–Step 16.

Finally we will show (6.12) by a mathematical induction such that for any fixed $n_0 \in \{2, 3, \dots\}$, if

$$R(j, k) = R(j-k, 0) \qquad \text{for any } j, k \in \mathbf{N}^*, \ 0 \le k \le j \le n_0, \tag{6.38}$$

then

$$R(n_0+1, k) = R(n_0+1-k, 0) \qquad \text{for any } k \in \{0, \dots, n_0+1\}. \tag{6.39}$$

Step 18. *For any* $k \in \{1, \dots, n_0\}$,

$$R(n_0+1, k) = R(n_0, k-1) - \delta_+(n_0+1)H(k; n_0),$$

where

$$H(k; n) = R(0, k) + \sum_{j=0}^{n-1} \eta_-(j, n-j)R(n-k, j) \qquad (1 \le k \le n). \tag{6.40}$$

Proof. By multiplying both sides of (6.9) with n replaced by n_0+1 by ${}^tX(k)$ and then taking an expectation with respect to P, it follows from Step 1 and Theorem 3.1′ that

$$R(n_0+1, k) = -\delta_+(n_0+1)\left(R(0, k) + \sum_{j=1}^{n_0} \eta_-(n_0-j, j)R(j, k)\right)$$
$$- \delta_+(n_0)R(1, k) - \sum_{j=1}^{n_0-1} \eta_+(j, n_0-j)R(j+1, k).$$

And so by (6.11) and the assumption (6.38),

$$R(n_0+1, k) = -\delta_+(n_0+1)\left(R(0, k)+\sum_{j=1}^{n_0} \eta_-(n_0-j, j)R(n_0-k, n_0-j)\right)$$

$$-\delta_+(n_0)R(0, k-1)-\sum_{j=1}^{n_0-1} \eta_+(j, n_0-j)R(j, k-1)$$

$$= -\delta_+(n_0+1)H(k; n_0)-\delta_+(n_0)R(0, k-1)$$

$$-\sum_{j=1}^{n_0-1} \eta_+(j, n-j)R(j, k-1),$$

which, together with (i) in Step 4, yields Step 18.

Q.E.D.

Step 19. *$H(m; m)=0$ for any $m \in \mathbf{N}$.*

Proof. By Steps 6 and 17, we can apply $(6.16)_m$ to get Step 19.

Step 20. *For any $l \in \{1, \ldots, n_0\}$ and any $k \in \{1, \ldots, l-1\}$,*

$$H(k; l) = H(k; l-1)+\delta_-(l)\sum_{j=1}^{k-1} \delta_+(l-j)H(k-j; l-1-j).$$

Proof. By the assumption (6.38),

$$H(k; l) = R(0, k)+\sum_{j=0}^{l-k} \eta_-(j, l-j)R(l-k, j)$$

$$+\sum_{j=l-k+1}^{l-1} \eta_-(j, l-j)R(l-k, j)$$

$$= R(0, k)+\sum_{j=0}^{l-k} \eta_-(j, l-j)R(l-k-j, 0)$$

$$+\sum_{j=l-k+1}^{l-1} \eta_-(j, l-j)R(0, j-l+k).$$

By Theorem 3.1′,

$$H(k; l) = R(0, k)+\sum_{j=0}^{l-k-1} \eta_-(j, l-j-1)R(l-1-j, 0)$$

$$+\sum_{j=l-k}^{l-2} \eta_-(j, l-1-j)R(0, j-l+1+k)$$

$$+\delta_-(l)\left(R(l-k,0)+\sum_{j=1}^{l-k}\eta_+(l-j-1,j)R(l-k-j,0)\right.$$

$$\left.+\sum_{j=l-k+1}^{l-1}\eta_+(l-j-1,j)R(0,j-l+k)\right)$$

$$=H(k;l-1)+\delta_-(l)\left(R(l-k,0)+\sum_{j=1}^{l-k}\eta_+(l-j-1,j)R(l-k-j,0)\right.$$

$$\left.+\sum_{j=l-k+1}^{l-1}\eta_+(l-j-1,j)R(0,j-l+k)\right).$$

By using Theorem 3.1′ again,

$$H(k;l)=H(k;l-1)+\delta_-(l)\delta_+(l-1)H(k-1;l-1)$$

$$+\delta_-(l)\left(R(l-k,0)+\sum_{j=1}^{l-k}\eta_+(l-j-2,j)R(l-k-j,0)\right.$$

$$\left.+\sum_{j=l-k+1}^{l-2}\eta_+(l-j-2,j)R(0,j-l+k)\right).$$

The same repetition yields that

$$H(k;l)=H(k,l-1)+\delta_-(l)\sum_{j=1}^{k-1}\delta_+(l-j)H(k-j;l-1-j)$$

$$+\delta_-(l)\left(R(l-k,0)+\sum_{j=1}^{l-k}\eta_+(l-k-j,j)R(l-k-j,0)\right),$$

which, together with (*iii*) in Step 5, implies Step 20. Q.E.D.

Step 21. $R(n_0+1,k)=R(n_0,k-1)$ *for any* $k\in\{1,\dots,n_0\}$.

Proof. By Steps 2 and 19, we can apply a mathematical induction to see that $H(k;l)=0$ for any $l\in\{1,\dots,n_0\}$ and any $k\in\{1,\dots,l\}$. Hence Step 21 follows from Step 18. Q.E.D.

Step 22. $R(n_0+1,n_0+1)=R(0,0)$.

Proof. By multiplying both sides of (6.9) with n replaced by n_0+1 by ${}^tX(n_0+1)$ and then taking an expectation with respect to P, it follows from

Step 1 and the assumption (6.38) that

$$R(n_0+1, n_0+1) = -\delta_+(n_0+1)R(0, n_0+1) + V_+(n_0+1) + \sum_{j=1}^{n_0} \eta_+(j, n_0+1-j)R(0, n_0+1-j).$$

By Steps 6 and 17, we can substitute $(6.16)_{n_0+1}$ into the above to see that

$$R(n_0+1, n_0+1) = V_+(n_0+1) + \delta_+(n_0+1)\delta_-(n_0+1)V_+(n_0) - \sum_{j=1}^{n_0}\left(\eta_+(j, n_0+1-j) - \delta_+(n_0+1)\eta_-(n_0-j, j)\right)R(0, n_0+1-j),$$

and so by Theorems 3.1′ and 4.1(ii),

$$R(n_0+1, n_0+1) = V_+(n_0) - \sum_{j=1}^{n_0} \eta_+(j-1, n_0+1-j)R(0, n_0+1-j)$$

$$= V_+(n_0) - \sum_{j=1}^{n_0} \eta_+(n_0-j, j)R(0, j).$$

Therefore, by Step 17, we can apply $(6.14)_{n_0+1}$ to the above to see that $R(n_0+1, n_0+1) = R(0, 0)$. Q.E.D.

Thus, we have completed the proof of Theorem 6.1.

Added in proof. After this paper was prepared, the author learned from Doctor of Engineering H. Sakai that the algorithms (i), (ii) in Theorem 3.1, (ii), (iv) in Theorem 4.1 and (i), (ii) in Theorem 4.2 have been derived for the fitting of a multi-dimensional autoregressive model [N. Levinson, *J. Math. Phys.* **25** (1947), 261–278; J. Durbin, *Biometrika* **46** (1959), 306–316; P. Whittle, *Biometrika* **50** (1963), 129–134; R. A. Wiggins and E. A. Robinson, *J. Geophys. Res.* **70** (1965), 1885–1891]. This algorithm is called the Levinson–Durbin algorithm for the one-dimensional case and the Levinson–Whittle–Wiggins–Robinson algorithm for the multi-dimensional case. In particular, the fundamental relation in Lemma 4.3, and parameters $\delta_+(\cdot)$ and $\delta_-(\cdot)$ are called Burg's relation and partial autocorrelation coefficients, respectively. The role that autoregressive processes have in our theory of KM_2O-Langevin equations will be discussed in [27]. The author would like to thank Prof. Hideaki Sakai for his communication.

References

[1] L. de Branges, *Hilbert Spaces of Entire Functions*, Prentice-Hall, Englewood Cliffs, New Jersey, 1968.

[2] H. Dym and H. P. McKean, Jr., Application of de Branges spaces of integral functions to the prediction of stationary Gaussian processes, *Illinois J. Math.* **45** (1970) 299–343.

[3] H. Dym and H. P. McKean, Jr., *Gaussian Processes, Function Theory, and the Inverse Spectral Problem*, Academic Press, New York, 1976.

[4] A. N. Kolmogorov, Interpolation und extrapolation von stationären zufälligen folgen, *Dokl. Akad. Nauk SSSR* **5** (1941) 3–14.

[5] M. G. Krein, On a fundamental approximation problem in the theory of extrapolation and filtration of stationary random processes, *Dokl. Akad. Nauk SSSR* **94** (1954) 13–16.

[6] R. Kubo, Statistical mechanical theory of irreversible processes I, general theory and simple applications to magnetic and conduction problems, *J. Phys. Soc. Japan* **12** (1957) 570–586.

[7] T. Miyoshi, On (l, m)-string and $(\alpha, \beta, \gamma, \delta)$-Langevin equation associated with a stationary Gaussian process, *J. Fac. Sci. Univ. Tokyo, Sect. IA* **30** (1983) 139–190.

[8] T. Miyoshi, On an $\mathbf{R}^d$-valued stationary Gaussian process associated with (k, l, m)-string and $(\alpha, \beta, \gamma, \delta)$-Langevin equation, *J. Fac. Sci. Univ. Tokyo, Sect. IA* **31** (1984) 155–194.

[9] H. Mori, Transport, collective motion and Brownian motion, *Progr. Theor. Phys.* **33** (1965) 115–165.

[10] Y. Nakano and Y. Okabe, On a multi-dimensional $[\alpha, \beta, \gamma]$-Langevin equation, *Proc. Japan Acad.* **59** (1983) 171–173.

[11] Y. Okabe, Stationary Gaussian processes with Markovian property and M. Sato's hyperfunctions, *Jap. J. Math.* **41** (1973) 69–122.

[12] Y. Okabe, On a stationary Gaussian process with T-positivity and its associated Langevin equation and S-matrix, *J. Fac. Sci. Univ. Tokyo, Sect. IA* **26** (1979) 115–165.

[13] Y. Okabe, On a stochastic differential equation for a stationary Gaussian process with T-positivity and the fluctuation-dissipation theorem, *J. Fac. Sci. Univ. Tokyo, Sect. IA* **28** (1981) 169–213.

[14] Y. Okabe, On a stochastic differential equation for a stationary Gaussian process with finite multiple Markovian property and the fluctuation-dissipation theorem, *J. Fac. Sci. Univ. Tokyo, Sect. IA* **28** (1982) 793–804.

[15] Y. Okabe, On a Langevin equation, *Sugaku* **33** (1981) 306–324 (in Japanese).

[16] Y. Okabe, On a wave equation associated with prediction errors for a stationary Gaussian process, *Lecture Notes in Control and Information Sciences* **49** (1983) 215–226.

[17] Y. Okabe, A generalized fluctuation-dissipation theorem for the one-dimensional diffusion process, *Commun. Math. Phys.* **98** (1985) 449–468.

[18] Y. Okabe, On KMO-Langevin equations for stationary Gaussian process with T-positivity, *J. Fac. Sci. Univ. Tokyo, Sect. IA* **33** (1986) 1–56.

[19] Y. Okabe, On the theory of Brownian motion with the Alder-Wainwright effect, *J. Stat. Phys.* **45** (1986) 953–981.

[20] Y. Okabe, KMO-Langevin equation and fluctuation-dissipation theorem (I), *Hokkaido Math. J.* **15** (1986) 163–216.

[21] Y. Okabe, KMO-Langevin equation and fluctuation-dissipation theorem (II), *Hokkaido Math. J.* **15** (1986) 317–355.

[22] Y. Okabe, Stokes-Boussinesq-Langevin equation and fluctuation-dissipation theorem, *Prob. Theory and Math. Stat.*, vol. 2, Prohorov *et al.* (eds.), VNU Science Press, 1986, pp. 431–436.

[23] Y. Okabe, On long time tails of correlation functions for KMO-Langevin equations, to appear in *Proceedings of Fourth Japan-USSR symposium on probability theory*, Kyoto, July, 1986.
[24] Y. Okabe, On the theory of discrete KMO-Langevin equations with reflection positivity (I), to appear in *Hokkaido Math. J.*
[25] Y. Okabe, On the theory of discrete KMO-Langevin equations with reflection positivity (II), to be submitted to *Hokkaido Math. J.*
[26] Y. Okabe, On the theory of discrete KMO-Langevin equations with reflection positivity (III), in preparation.
[27] Y. Okabe, On stochastic difference equations for the multi-dimensional weakly stationary time series, in preparation.
[28] N. Wiener, *Extrapolation, interpolation, and smoothing of stationary time series*, Wiley, New York, 1949.

Bäcklund Transformations of Classical Orthogonal Polynomials

Kazuo Okamoto
Department of Mathematics
University of Tokyo
Tokyo, Japan

In this note we will see that the classical orthogonal polynomials, except for the Jacobi polynomials, yield a solution of the system of equations

$$t_0 = 1, \tag{1}$$

$$D^2 \log t_n = \frac{t_{n-1} t_{n+1}}{t_n^2} \qquad (n \geq 1), \tag{2}$$

D being a derivation. Although this fact might already be known, it is available to make a list of solutions of (1)–(2) expressed in terms of the classical orthogonal polynomials (see Sections 3–6). The system (1)–(2) will be called a *truncated Toda equation.*

By using the Darboux–Kametaka formulation of the Bäcklund transformation of the Toda equation, we will obtain the *Bäcklund transformation* of the truncated equation (1)–(2) (see Section 2).

ISBN 0-12-400466-0

We will study in Sections 7–8 nonlinear relations satisfied by the Jacobi polynomials. As for definitions and properties of the classical orthogonal polynomials, we quote [5].

1. Truncated Toda Equation

Let Ω be a domain in $\mathbf{C}$ and D a holomorphic derivation defined on the universal covering surface $\mathscr{U}(\Omega)$ of Ω. Given a holomorphic function f on $\mathscr{U}(\Omega)$, we define a family $\mathfrak{T}=\{t_n;\ n\in\mathbf{Z}_+\}$ of holomorphic functions by:

$$t_0=1, \tag{3}$$

$$t_1=f, \tag{4}$$

$$t_n=\begin{vmatrix} f & Df & \cdots & D^{n-1}f \\ Df & D^2f & \cdots & D^nf \\ & \cdots & \cdots & \\ D^{n-1}f & D^nf & \cdots & D^{2n-2}f \end{vmatrix} \qquad (n\geq 2). \tag{5}$$

Here $\mathbf{Z}_+$ denotes the set of non-negative integers. It is known ([1, 3, 4]) that

Theorem 1. *The family* $\mathfrak{T}$ *satisfies the truncated Toda equation* (1)–(2). *Conversely, under the condition* (4), *the unique solution of* (1)–(2) *is given by* (5).

In fact we can deduce from (5) the equation (2) by means of the Jacobi identity of determinants: see [2]. The right-hand side of (5) will be denoted by $\mathscr{J}(n;f;D)$, or simply by $\mathscr{J}(n;f)$ when there is no danger of confusion. By specifying the initial function f, we may write in the following as $\mathfrak{T}=\mathfrak{T}(f)$, $t_n=t_n(f)$ and so on. $\mathfrak{T}$ is called a *Darboux sequence* generated by f with respect to D; the expression

$$t_n(f)=\mathscr{J}(n;f;D) \tag{5$'$}$$

is the *Darboux formula* for a solution of the equation (1)–(2).

Another application of the Jacobi identity yields the following theorem:

Theorem 2. *For* $n\geq 1$,

$$\mathscr{J}(n-1;\,D^2f)\mathscr{J}(n+1;f)=\mathscr{J}(n;\,D^2f)\mathscr{J}(n;f)-\mathscr{J}(n;\,Df)^2. \tag{6}$$

We will call it the *Jacobi identity* for Darboux sequences.

2. Bäcklund Transformation

Let f and ϕ be holomorphic functions on $\mathscr{U}(\Omega)$ such that

$$D\phi + f = 0, \tag{7}$$

and let $\mathfrak{T}(f) = \{t_n(f)\}$ be a Darboux sequence generated by f with respect to a derivation D. We define a sequence $\Phi = \{\phi_n;\ n \in \mathbf{Z}_+\}$ of meromorphic functions by

$$\phi_0 = \phi, \tag{8}$$

$$\phi_{n+1} = (D + s_{n+1})\phi_n \qquad (n \geq 0), \tag{9}$$

where, for $t_n = t_n(f)$,

$$s_{n+1} = D \log \frac{t_n}{t_{n+1}}. \tag{10}$$

The following theorem is an immediate consequence of the Darboux-Kametaka formulation of the Bäcklund transformation of the Toda equation (cf. [1, 3, 4]).

Theorem 3. *If we set*

$$t'_{n+1} = \phi_n t_n, \tag{11}$$

then $\mathfrak{T}' = \{t'_n;\ n \in \mathbf{Z}_+\}$ $(t'_0 = 1)$ *is a Darboux sequence generated by* ϕ.

The correspondence thus obtained between the two Darboux sequences,

$$\mathfrak{T}(f) \cdots \to \mathfrak{T}(\phi),$$

is called a *Bäcklund transformation*, and the family Φ is called a *factor* of the transformation. By taking into consideration (5)′ we can rewrite (11) in the form

$$\mathscr{J}(n+1;\ \phi) = \phi_n \mathscr{J}(n;\ -D\phi). \tag{11$'$}$$

Remark 1. The condition

$$t'_0 = 1$$

follows from the truncated Toda equation for $\mathfrak{T}'$. In fact we have

$$\phi_1 = (D + s_1)\phi,$$

$$s_1 = -D \log f = -D^2\phi / D\phi,$$

and then, by virtue of (11),

$$t_2' = \phi \cdot D^2\phi - (D\phi)^2.$$

Remark 2. Given a Darboux sequence $\mathfrak{T}(f)$, set, besides (10),

$$r_n = D^2 \log t_n. \tag{12}$$

Then a factor Φ of the transformation satisfies, for $n \geq 1$, the relation

$$-D\phi_n = r_n \phi_{n-1}$$

(cf. [3, 4]).

3. Hermite Polynomials

The Hermite polynomials $H_n(x)$ are defined by the Rodrigues formula (see [5]):

$$H_n(x) = (-1)^n w(x)^{-1} \left(\frac{d}{dx}\right)^n w(x),$$

$$w(x) = \exp(-\tfrac{1}{2}x^2).$$

It is known ([4]) that they satisfy the nonlinear equation

$$\frac{d^2}{dx^2} \log H_n(x) = n \frac{H_{n-1}(x) H_{n+1}(x)}{H_n(x)^2} - n \qquad (n \geq 1), \tag{13}$$

which can be verified by virtue of the contiguity relations

$$\frac{d}{dx} H_n(x) = n H_{n-1}(x),$$

$$-\frac{d}{dx} H_n(x) + x H_n(x) = H_{n+1}(x).$$

We have from (13) a Darboux sequence generated by

$$f = x w(x)^{-1} \tag{14}$$

with respect to

$$D = \frac{d}{dx}. \tag{15}$$

Proposition 1. *For* $n \geq 2$,

$$\mathcal{J}(n; f) = n^! w(x)^{-n} H_n(x), \tag{16}$$

where we introduce the notation

$$n^! = 1!2! \cdots n!. \tag{17}$$

Since $\phi = -w(x)^{-1}$ is a solution of (7) with (14)-(15), we have a Bäcklund transformation from $\mathfrak{T}(f)$ to $\mathfrak{T}(\phi)$. On the other hand, by using the truncated Toda equation, we can show the

Proposition 2. *For* $n \geq 0$,

$$\mathcal{J}(n+1; \phi) = (-1)^{n+1} n^! w(x)^{-n-1}. \tag{18}$$

It follows from (16) and (18) that a factor of the transformation is given by

$$\phi_n = \frac{(-1)^{n+1}}{w(x) H_n(x)}.$$

Moreover, since $Df = (1+x^2) w(x)^{-1}$, we deduce from the Jacobi identity (6) the following proposition.

Proposition 3. *For* $n \geq 0$,

$$\mathcal{J}(n; -(1+x^2) w(x)^{-1}) = (-1)^n n^! w(x)^{-n} \sum_{k=0}^{n} k! H_k(x)^2.$$

4. Tchebichef Polynomials

In this section we put

$$x = \cos\theta, \qquad D = -(1-x^2)^{1/2} \frac{d}{dx} = \frac{d}{d\theta}, \qquad w(\theta) = \frac{1}{\sin\theta}.$$

We can show

Proposition 4. *For* $n \geq 0$,

$$\mathcal{J}(n; \cot\theta) = ((n-1)^!)^2 w(\theta)^{n^2} \cos(n\theta),$$

$$\mathcal{J}(n; 2\cot\theta(1+\cot^2\theta)) = (n^!)^2 w(\theta)^{(n+1)^2} \sin(n+1)\theta.$$

Recall that the Tchebichef polynomials of the first kind, $T_n(x)$, are defined by

$$\cos(n\theta) = T_n(\cos\theta),$$

while the polynomials $U_n(x)$, given by

$$\sin(n+1)\theta = U_n(\cos\theta)\sin\theta,$$

are the Tchebichef polynomials of the second kind. Since

$$\left(\frac{d}{d\theta}\right)^2 \cot\theta = 2\cot\theta(1+\cot^2\theta),$$

we obtain from the truncated Toda equation and the Jacobi identity the following proposition.

Proposition 5. *For* $n \geq 0$,

$$\mathcal{J}(n; -d/d\theta \cot\theta) = (n^!(n-1)^!)^2 w(\theta)^{n(n+1)}.$$

Hence a factor of the Bäcklund transformation from $\mathfrak{T}((d/d\theta)^2 \cot\theta)$ to $\mathfrak{T}(-d/d\theta \cot\theta)$ is of the form

$$\phi_n = \frac{(n+1)!}{\sin(n+1)\theta} w(\theta)^{n+1},$$

and for the transformation from $\mathfrak{T}(-d/d\theta \cot\theta)$ to $\mathfrak{T}(\cot\theta)$ we have

$$\phi_n = n!\cos((n+1)\theta)w(\theta)^{n+1}.$$

We can compute an explicit form of the Bäcklund transformation from $\mathfrak{T}(\cot\theta)$ to $\mathfrak{T}(\phi)$, where

$$\phi = \log w = -\log\sin\theta.$$

In fact we obtain from (6) and (11) the

Proposition 6. *For* $n \geq 1$,

$$\mathcal{J}(n; \phi) = (n-1)^!(n-2)^!\left(\phi - \sum_{k=1}^{n-1} \frac{1}{k}\cos^2 k\theta\right) w(\theta)^{n(n-1)}.$$

A factor of the transformation is given by

$$\phi_n = n!\left(\phi - \sum_{k=1}^{n} \frac{1}{k}\cos^2 k\theta\right)\frac{w(\theta)^n}{\cos(n\theta)}.$$

5. Laguerre Polynomials

Let $L_n^\alpha(x)$ denote the Laguerre polynomials. We have for $n \in \mathbf{Z}_+$

$$L_n^\alpha(x) = \frac{1}{n!} w(x)^{-1} \left(\frac{d}{dx}\right)^n w(x) x^n,$$

$$= \frac{\Gamma(\alpha+1+n)}{\Gamma(\alpha+1)n!} F(-n, \alpha+1; x),$$

where $F(a, c; x)$ is the confluent hypergeometric function and

$$w(x) = x^\alpha e^{-x}.$$

Consider a sequence of functions of the form

$$t_n = n^! \theta(n; \alpha)^{-1} x^{-n(n+1+\alpha)} L_n^\alpha(x),$$

where we use the notation (17) and set

$$\theta(n; \alpha) = \begin{cases} \dfrac{1}{\Gamma(\alpha+1) \cdots \Gamma(\alpha+n)} & n \geq 1, \\ 1 & n = 0, \\ \Gamma(\alpha) \cdots \Gamma(\alpha+n+1) & n \leq -1. \end{cases}$$

Proposition 7. *$\mathfrak{T} = \{t_n;\ n \in \mathbf{Z}_+\}$ is a Darboux sequence generated by*

$$f_\alpha = \Gamma(\alpha+1) x^{-\alpha-2} L_1^\alpha(x)$$

$$= \frac{\Gamma(\alpha+2)}{x^{\alpha+2}} - \frac{\Gamma(\alpha+1)}{x^{\alpha+1}}, \tag{19}$$

with respect to $D = d/dx$.

Since

$$Df_{\alpha-1} + f_\alpha = 0, \tag{20}$$

we have from (6) the quadratic relation of the Laguerre polynomials:

$$(n+1) L_{n-1}^{\alpha+1}(x) L_{n+1}^{\alpha-1}(x) - (\alpha+n) L_n^{\alpha+1}(x) L_n^{\alpha-1}(x) + \alpha L_n^\alpha(x)^2 = 0.$$

The equation (20) defines the Bäcklund transformation from $\mathfrak{T}(f_\alpha)$ to $\mathfrak{T}(f_{\alpha-1})$ whose factor is

$$\phi_n = (n+1)! \Gamma(\alpha) x^{-n-1\alpha} L_{n+1}^{\alpha-1}(x) / L_n^\alpha(x).$$

6. Gegenbauer Polynomials

In this section we set

$$D = w(x)^{-1}\frac{d}{dx},$$

$$w(x) = (1-x^2)^{-1/2}.$$

Proposition 8. *For $n \in \mathbf{Z}_+$,*

$$\mathcal{J}(n; f) = n^! \theta(n; 2\lambda - 1) w^{n(2\lambda+n)} C_n^\lambda(x),$$

where $C_n^\lambda(x)$ are the Gegenbauer polynomials and

$$f = \Gamma(2\lambda+1) x w(x)^{2\lambda+1}.$$

Moreover by putting

$$\phi = -\Gamma(2\lambda) w(x)^{2\lambda},$$

we have

$$D\phi = -f,$$

$$D^2\phi = \Gamma(2\lambda+1)(2\lambda - (2\lambda+1)w(x)^2) w(x)^{2\lambda};$$

hence it follows that

Proposition 9. *For $n \geq 1$,*

$$\mathcal{J}(n; \phi) = (-1)^n (n-1)^! \theta(n; 2\lambda-1)^{-1} w(x)^{n(2\lambda+n-1)},$$

$$\mathcal{J}(n; D^2\phi) = (-1)^n n^! \theta(n+1: 2\lambda - 1)^{-1} w(x)^{n(2\lambda+n+1)} \sum_{k=0}^{n} \frac{k!}{\Gamma(2\lambda+k)} C_k^\lambda(x)^2.$$

In the case $\lambda = \frac{1}{2}$, $C_n^\lambda(x)$ reduce to the Legendre polynomials $P_n(x)$.

7. Hypergeometric Functions

Under the assumption that none of a, b, $c-a$, $c-b$ are integers, we consider the hypergeometric functions

$$F_n = F\left(a-n, b+n, c; \frac{1-x}{2}\right). \tag{21}$$

We set

$$D = w(x)^{-1}\, d/dx, \tag{22}$$

$$w(x) = (1-x^2)^{-1/2},$$

$$\kappa = \tfrac{1}{2}(b-a+1), \qquad \chi = \tfrac{1}{2}(a+b-2c+1), \tag{23}$$

$$t_n = \theta(n) w(x)^{(n-a)(n+b)} F_n,$$

$$\theta(n) = \frac{\theta(n;\kappa)\theta(n;\kappa-1)}{\theta(n;b)\theta(n;b-c)\theta(n;c-a)\theta(n;-a)}.$$

The following theorem can be verified by computation.

Theorem 4. *We have, for* $n \in \mathbf{Z}$,

$$D^2 \log t_n + \chi_n(a,b,c)\frac{d}{dx}\log t_n = \frac{t_{n-1}t_{n+1}}{t_n} - \omega_n(a,b,c;x)w(x)^2,$$

where

$$\chi_n(a,b,c) = \frac{(\kappa+a)\chi}{(n+\kappa)(n+\kappa-1)}.$$

$$\omega_n(a,b,c;x) = (b+n)(a-n)\chi\frac{(\kappa+a-1)x+\chi}{(n+\kappa)(n+\kappa-1)}.$$

We omit the details of computation (see [4]).

Corollary. *If* $\chi = 0$, *then the functions* t_n *yield a solution of the Toda equation.*

8. Jacobi Polynomials

We apply the results of the preceding section to the Jacobi polynomials

$$P_n^{(\alpha,\beta)}(x) = \binom{n+\alpha}{n} F\left(-n, n+\alpha+\beta+1, \alpha+1; \frac{1-x}{2}\right),$$

that is, the case in (21)

$$a = 0, \qquad b = \alpha+\beta+1, \qquad c = \alpha+1.$$

From (23) we have

$$\kappa=\tfrac{1}{2}(\alpha+\beta)+1, \qquad \chi=\tfrac{1}{2}(\beta-\alpha);$$

moreover we set

$$p(n)=\frac{(n+\kappa)(n+\kappa-1)}{(n+2\kappa-1)(n+\alpha)(n+\beta)(n+1)},$$

$$\chi_n=\frac{(\kappa-1)\chi}{(n+\kappa)(n+\kappa-1)},$$

$$\omega'_n(x)=-n(n+2\kappa-1)\frac{\chi(\chi-x)}{(n+\kappa)(n+\kappa-1)},$$

$$t_n=\theta(n)w(x)^{n(n+2\kappa-1)}P_n^{(\alpha,\beta)}(x),$$

$$\theta(n)=\frac{\theta(n;\kappa)\theta(n;\kappa-1)}{\theta(n;2\kappa-1)\theta(n;\beta)\theta(n;\alpha+1)n!},$$

$$\omega_n(x)=-n(n+2\kappa-1)\chi_n x+\omega_n.$$

Proposition 10. *The Jacobi polynomials $P_n=P_n^{(\alpha,\beta)}(x)$ satisfy the equalities*

$$p(n)w(x)^{-2}\left[D\log P_n+\chi_n\frac{d}{dx}\log P_n\right] =\frac{P_{n-1}P_{n+1}}{P_n}-p(n)[n(n+2\kappa-1)+\omega'_n(x)],$$

where D is the derivation (22).

It follows that

Theorem 5. *For $n\geq 1$,*

$$D^2\log t_n+\chi_n\frac{d}{dx}\log t_n=\frac{t_{n-1}t_{n+1}}{t_n^2}-\frac{\omega_n(x)}{1-x^2}. \tag{24}$$

The equation (24) is reduced to the truncated Toda equation, if $\chi=0$. This case occurs when

$\alpha=\beta=\lambda-\tfrac{1}{2}$ (*Gegenbauer*),

$\alpha=\beta=-\tfrac{1}{2}$ (*Tchebichef*),

$\alpha=\beta=0$ (*Legendre*).

For example, as for the Legendre polynomials, we obtain

$$t_n = n^!(n-1)^! w(x)^{n(n+1)} P_n(x)$$
$$= \mathscr{J}(n; xw(x)^2),$$

$$xw(x)^2 = \frac{x}{1-x^2}.$$

References

[1] G. Darboux, *Leçons sur la Théorie Générale des Surfaces*, tII, Chelsea, 1972.
[2] F. R. Gantmacher, *The Theory of Matrices*, Chelsea, 1959.
[3] Y. Kametaka, Hypergeometric solutions of Toda equation, *RIMS Kokyuroku* **554** (1985) 26–46.
[4] K. Okamoto, Sur les échelles associées aux fonctions spéciales et l'équation de Toda, *J. Fac. Sci. Univ. Tokyo* **34** (1987) 709–740.
[5] G. Szegö, *Orthogonal polynomials*, Amer. Math. Soc. Colloq. Publ., 1959.

A Deformation of Dirichlet's Class Number Formula

Takashi Ono
Department of Mathematics
The Johns Hopkins University
Baltimore, Maryland

1. Dirichlet's Formula

Let l be a prime of the form $l = 4N+1$, ζ be a primitive lth root of unity, $k = \mathbf{Q}(\sqrt{l})$ and $K = \mathbf{Q}(\zeta)$. As is well known, the real quadratic field k is contained in the lth cyclotomic field K. Consider two sets of integers[1]

$$A = \{a \in \mathbf{Z};\quad 1 \le a \le l-1,\quad \lambda(a) = +1\},$$

$$B = \{b \in \mathbf{Z};\quad 1 \le b \le l-1,\quad \lambda(b) = -1\},$$

and put

$$F_A(t) = \prod_{a \in A} (1 - t\zeta^a), \qquad F_B(t) = \prod_{b \in B} (1 - t\zeta^b). \tag{1}$$

[1] $\lambda(x) = (x/l)$ denotes the Legendre symbol.

ISBN 0-12-400466-0

Then we have

$$F_A(t)F_B(t)=N_{K/\mathbf{Q}}(1-t\zeta)=t^{l-1}+\cdots+t+1,$$

$$F_A(t)=N_{K/k}(1-t\zeta)=\sum_{m=0}^{2N} c_m t^m \in \mathfrak{o}_k[t], \tag{2}$$

$$F_B(t)=\sum_{m=0}^{2N} d_m t^m \in \mathfrak{o}_k[t],$$

where $\mathfrak{o}_k$ denotes the ring of integers of k. Note that $d_m=c'_m$, the conjugate of c_m with respect to the quadratic extension $k/\mathbf{Q}$. By a simple algebraic consideration, we see that $F_A(t)$, $F_B(t)$ are reciprocal polynomials, i.e.,

$$c_\nu=c_{2N-\nu},\ d_\nu=d_{2N-\nu},\qquad 0\le\nu\le 2N. \tag{3}$$

Clearly, we have

$$F_A(1)F_B(1)=l. \tag{4}$$

On the other hand, we know that

$$F_B(1)/F_A(1)=\varepsilon^{2h}\qquad\text{(Dirichlet's formula)}, \tag{5}$$

where h is the class number of k and ε is the fundamental unit of k. From (1), (2), (3), (4), (5), we get

$$\sqrt{l}\varepsilon^h=F_B(1)=\sum_{m=0}^{2N} d_m=2\sum_{m=0}^{N-1} d_m+d_N. \tag{6}$$

In this paper, we shall obtain a recurrence relation for d_m in (6) (Theorem 1, (14)). Such a relation facilitates the calculation of ε^h, sometimes even h itself.[2] We shall also obtain an upper bound for ε^h which is smaller than such a bound of Hua.

2. A Recurrence Relation

Let a_n, $1\le n\le M$, be complex numbers. Put

$$F(t)=\prod_{n=1}^{M}(1-ta_n)=\sum_{m=0}^{M} b_m t^m,\qquad b_m=\frac{F^{(m)}(0)}{m!}. \tag{7}$$

Call $G(t)$ the logarithmic derivative of $F(t)$:

$$G(t)=\frac{F'(t)}{F(t)}=-\sum_{n=1}^{M}\frac{a_n}{1-ta_n}.$$

[2] Compare pp. 398–409, H. Hasse, *Vorlesungen über Zahlentheorie*, Springer, 1950.

Then we have

$$G^{(\nu)}(t) = -\nu!\sum_{n=1}^{M}\frac{a_n^{\nu+1}}{(1-ta_n)^{\nu+1}},$$

and hence

$$G^{(\nu)}(0) = -\nu!\,s_{\nu+1} \qquad \text{with } s_{\nu+1} = \sum_{n=1}^{M} a_n^{\nu+1}.$$

Taking the higher derivatives of both sides of

$$F'(t) = F(t)G(t),$$

we get

$$F^{(m)}(t) = \sum_{\nu=0}^{m-1}\binom{m-1}{\nu}F^{(m-1-\nu)}(t)G^{(\nu)}(t),$$

and hence

$$F^{(m)}(0) = -\sum_{\nu=0}^{m-1}\binom{m-1}{\nu}F^{(m-1-\nu)}(0)\,\nu!\,s_{\nu+1}.$$

In view of (7), we have

$$mb_m = -\sum_{\nu=0}^{m-1} b_{m-1-\nu}s_{\nu+1},$$

or

$$mb_m = -\sum_{\mu=1}^{m} b_{m-\mu}s_\mu, \qquad 1 \le m \le M, \qquad s_\mu = \sum_{n=1}^{M} a_n^\mu. \tag{8}$$

3. A Deformation of Dirichlet's Formula

Applying the formula (8) to polynomials (1), we get

$$\begin{aligned} mc_m &= -\sum_{\mu=1}^{m} c_{m-\mu}\sigma_\mu, \qquad \sigma_\mu = \sum_{a\in A}\zeta^{a\mu}, \qquad 1\le m\le 2N,\\ md_m &= -\sum_{\mu=1}^{m} d_{m-\mu}\tau_\mu, \qquad \tau_\mu = \sum_{b\in B}\zeta^{b\mu}, \qquad 1\le m\le 2N. \end{aligned} \tag{9}$$

Now, we have

$$\sigma_\mu + \tau_\mu = \sum_{x\in(\mathbf{Z}/l\mathbf{Z})^\times} \zeta^{x\mu} = \left(\sum_{x\in\mathbf{Z}/l\mathbf{Z}} \zeta^{x\mu}\right) - 1 = -1 \tag{10}$$

since $\mu \geq 1$. On the other hand, we have

$$\sigma_\mu - \tau_\mu = \sum_{x\in(\mathbf{Z}/l\mathbf{Z})^\times} \lambda(x)\zeta^{x\mu} = \lambda(\mu)\sqrt{l} \qquad \text{(Gauss's theorem).} \tag{11}$$

Hence, from (10), (11), we have

$$2\tau_\mu = -(1+\lambda(\mu)\sqrt{l}). \tag{12}$$

Substituting (12) in (9), we have

$$\begin{aligned} md_m &= \sum_{\mu=1}^{m} \left(\frac{1+\lambda(\mu)\sqrt{l}}{2}\right) d_{m-\mu} \\ &= \sum_{\mu=1}^{m} \left(\frac{1-\lambda(\mu)}{2} + \lambda(\mu)\omega\right) d_{m-\mu}, \end{aligned}$$

where $\omega = (1+\sqrt{l})/2$.

If we put $\alpha_\mu = (1-\lambda(\mu))/2 + \lambda(\mu)\omega$, then we have

$$\alpha_\mu = \begin{cases} \omega & \text{if } \lambda(\mu) = +1, \\ \omega' & \text{if } \lambda(\mu) = -1, \end{cases}$$

and hence

$$md_m = \sum_{\mu=1}^{m} \alpha_\mu d_{m-\mu}, \qquad 1 \leq m \leq \frac{l-1}{2}.$$

In view of (6), changing μ, m for ν, n, respectively, we can summarize our argument as follows:

Theorem 1. *Let l be a prime such that $l \equiv 1 \pmod 4$. Put $N = (l-1)/4$, $\omega = (1+\sqrt{l})/2$, $\omega' = (1-\sqrt{l})/2$ and*

$$\alpha_\nu = \begin{cases} \omega & \text{if } \lambda(\nu) = +1, \\ \omega' & \text{if } \lambda(\nu) = -1, \end{cases}$$

where $\lambda(n) = (\nu/l)$, the Legendre character. Let $k = \mathbf{Q}(\sqrt{l})$ be the real quadratic field and h, ε be, respectively, the class number and the fundamental unit of k. Then we have the relation

$$\sqrt{l}\varepsilon^h = 2\sum_{n=0}^{N-1} d_n + d_N, \tag{13}$$

where d_n, which are integers of k, are determined by the recurrence relation

$$nd_n = \sum_{\nu=1}^{n} \alpha_\nu d_{n-\nu}, \qquad d_0 = 1,\ d_1 = \omega, \qquad 1 \le n \le N. \tag{14}$$

Here are some illustrations.

(*i*) $l = 5$. $N = 1$, $\omega = (1+\sqrt{5})/2 = \varepsilon$.

$$\sqrt{5}\varepsilon^h = 2d_0 + d_1 = 2 + \omega = \sqrt{5}\omega.$$

Hence $h = 1$.

(*ii*) $l = 17$. $N = 4$, $\omega = (1+\sqrt{17})/2$, $\omega^2 = 4 + \omega$. Since $17 = 4^2 + 1$, we see immediately that $\varepsilon = 4 + \sqrt{17} = 3 + 2\omega$.

ν	1	2	3	4
$\lambda(\nu)$	1	1	-1	1
α_ν	ω	ω	ω'	ω
d_ν	ω	$2+\omega$	$3+\omega$	$1+2\omega$

$$2d_2 = \alpha_1 d_1 + \alpha_2 d_0 = \omega^2 + \omega = 4 + 2\omega \Rightarrow d_2 = (5+\sqrt{17})/2 = 2 + \omega,$$

$$3d_3 = \alpha_1 d_2 + \alpha_2 d_1 + \alpha_3 d_0 = \omega(2+\omega) + \omega^2 + \omega' = 9 + 3\omega \Rightarrow d_3 = 3 + \omega,$$

$$4d_4 = \alpha_1 d_3 + \alpha_2 d_2 + \alpha_3 d_1 + \alpha_4 d_0 = \omega(3+\omega) + \omega(2+\omega) + \omega'\omega + \omega$$

$$= 4 + 8\omega \Rightarrow d_4 = 1 + 2\omega.$$

$$\sqrt{17}\varepsilon^h = \sqrt{17}(4+\sqrt{17})^h = 2(d_0 + d_1 + d_2 + d_3) + d_4$$

$$= 2(1 + \omega + (2+\omega) + (3+\omega)) + 1 + 2\omega = 13 + 8\omega$$

$$= 17 + 4\sqrt{17} = \sqrt{17}(4+\sqrt{17}) = \sqrt{17}\varepsilon.$$

Hence $h = 1$.

(*iii*) $l = 29$. $N = 7$, $\omega = (1+\sqrt{29})/2$, $\omega^2 = 7 + \omega$. Since $29 = 5^2 + 4$, we see immediately that $\varepsilon = (5+\sqrt{29})/2 = 2 + \omega$.

	1	2	3	4	5	6	7
$\lambda(\nu)$	1	-1	-1	1	1	1	1
α_ν	ω	ω'	ω'	ω	ω	ω	ω
d_ν	ω	4	$\omega - 2$	$1 - \omega$	-1	$1+\omega$	$4+\omega$

$$2d_2 = \alpha_1 d_1 + \alpha_2 d_0 = \omega^2 + \omega' = 8 \Rightarrow d_2 = 4,$$

$$3d_3 = \alpha_1 d_2 + \alpha_2 d_1 + \alpha_3 d_0 = \omega \cdot 4 + \omega'\omega + \omega' = 3\omega - 6 \Rightarrow d_3 = \omega - 2,$$

$$4d_4 = \alpha_1 d_3 + \alpha_2 d_2 + \alpha_3 d_1 + \alpha_4 d_0 = \omega(\omega - 2) + 4\omega' + \omega'\omega + \omega$$
$$= 4 - 4\omega \Rightarrow d_4 = 1 - \omega,$$

$$5d_5 = \alpha_1 d_4 + \alpha_2 d_3 + \alpha_3 d_2 + \alpha_4 d_1 + \alpha_5 d_0$$
$$= \omega(1 - \omega) + \omega'(\omega - 2) + 4\omega' + \omega^2 + \omega = -5 \Rightarrow d_5 = -1,$$

$$6d_6 = \alpha_1 d_5 + \alpha_2 d_4 + \alpha_3 d_3 + \alpha_4 d_2 + \alpha_5 d_1 + \alpha_6 d_0$$
$$= -\omega + \omega'(1 - \omega) + \omega'(\omega - 2) + \omega \cdot 4 + \omega^2 + \omega = 6 + 6\omega \Rightarrow d_6 = 1 + \omega,$$

$$7d_7 = \alpha_1 d_6 + \alpha_2 d_5 + \alpha_3 d_4 + \alpha_4 d_3 + \alpha_5 d_2 + \alpha_6 d_1 + \alpha_7 d_0$$
$$= \omega(1 + \omega) - \omega' + \omega'(1 - \omega) + \omega(\omega - 2) + \omega \cdot 4 + \omega^2 + \omega$$
$$= 28 + 7\omega \Rightarrow d_7 = 4 + \omega.$$

$$\sqrt{29}\varepsilon^h = \sqrt{29}(2 + \omega)^h = 2(d_0 + d_1 + d_2 + d_3 + d_4 + d_5 + d_6) + d_7 = 5\omega + 12$$
$$= \sqrt{29}(2 + \omega).$$

Hence $h = 1$.

4. An Upper Bound of ε^h

For a complex number x and a natural number n, we put

$$(x)_n = x(x+1)\cdots(x+n-1).$$

We shall also put $(x)_0 = 1$. By induction on integers $N \geq 0$, one verifies easily that

Lemma 1.

$$\sum_{n=0}^{N} \frac{(x)_n}{n!} = \frac{(x+1)_N}{N!}.$$

Lemma 2. *The notation being as in Theorem* 1, *we have*

$$|d_n| \leq \frac{(\omega)_n}{n!}, \qquad n \geq 0.$$

Proof (by induction on n). If $n=0$ or 1, the assertion is obvious. Assume that the inequality is true for n. Then, by (14) and Lemma 1, we have

$$|d_{n+1}| \leq \frac{\omega}{n+1} \sum_{\nu=0}^{n} |d_\nu| \leq \frac{\omega}{n+1} \sum_{\nu=0}^{n} \frac{(\omega)_n}{n!}$$

$$= \frac{\omega}{n+1} \frac{(\omega+1)_n}{n!} = \frac{(\omega)_{n+1}}{(n+1)!}, \qquad \text{Q.E.D.}[3]$$

Theorem 2. *The notation being as in Theorem* 1, *we have*

$$\sqrt{l}\varepsilon^h < 2\frac{(\omega+1)_N}{N!} = 2(1+\omega)\left(1+\frac{\omega}{2}\right)\cdots\left(1+\frac{\omega}{N}\right).$$

Proof. By Lemma 2 and Lemma 1, we have

$$\sqrt{l}\varepsilon^h \leq 2\sum_{n=0}^{N-1} |d_n| + |d_N| < 2\sum_{n=0}^{N} |d_n|$$

$$\leq 2\sum_{n=0}^{N} \frac{(\omega)_n}{n!} = 2\frac{(\omega+1)_N}{N!}, \qquad \text{Q.E.D.}$$

5. Comparison with Hua's Upper Bound

Hua[4] obtained an upper bound for $h \log \varepsilon$ for any real quadratic field of discriminant Δ:

$$h \log \varepsilon < \sqrt{\Delta}(1+\log\sqrt{\Delta}). \tag{15}$$

By some elementary (but long) considerations, one verifies that

$$2(1+\omega)\left(1+\frac{\omega}{2}\right)\cdots\left(1+\frac{\omega}{N}\right) < \sqrt{l}\exp(\sqrt{l}(1+\log\sqrt{l})), \tag{16}$$

where l, N, and ω are the same as in Theorem 1. The inequality (16) shows that our upper bound for ε^h obtained by Theorem 2 is smaller than the one derived from (15). To make a long story short, consider the simplest case: $l=5$. First of all, it is easy to check (16) for this case. Next, Theorem

[3] Note that, when $\lambda(\nu)=-1$, we have $|\alpha_\nu| = |\omega'| = \omega - 1 \leq \omega$.

[4] L.-K. Hua, On the least solution of Pell's equation, *Bull. Amer. Math. Soc.* **48** (1942) 731–735.

2 implies that $h=1$ because

$$h<\frac{\log(1+\omega)+\log 2-\log\sqrt{l}}{\log \varepsilon}\approx 1.76814415.$$

On the other hand, Hua's formula (15) implies that

$$h<\frac{\sqrt{l}(1+\log\sqrt{l}))}{\log \varepsilon}\approx 8.38606629,$$

and we only know that $h\leq 7$.[5]

6. A Parting Remark

It is tempting to collect as many prime numbers of the form $l=4N+1$ as possible for which the following inequality holds:

$$2\prod_{n=1}^{N}\left(1+\frac{\omega}{n}\right)\leq\sqrt{l}\varepsilon^3,$$

because by Theorem 2 we have $h=1$ for these primes.

[5] Note that h must be odd since $l\equiv 1 \pmod 4$.

A Method of Harmonic Analysis on Semisimple Symmetric Spaces

Toshio Oshima
Department of Mathematics
University of Tokyo
Tokyo, Japan

0. Introduction

A homogeneous space $X = G/H$ of a connected Lie group G is called a *symmetric space* if there exists an involutive automorphism σ of G such that H is an open subgroup of the fixed point group of σ.

A connected Lie group G' is a symmetric space: Putting $G = G' \times G'$, G' is a homogeneous space of G by the map $G \times G' \ni ((g_1, g_2), x) \mapsto g_1 x g_2^{-1}$ and the isotropy group of the identity element equals $\Delta G' = \{(g, g) \in G; g \in G'\}$, which is the fixed-point group of the involution $(g_1, g_2) \mapsto (g_2, g_1)$ of G. We call this typical example a *group case.*

In this note we consider the case where G is a connected real linear semisimple Lie group. Then the symmetric space is called a *semisimple symmetric space.* Moreover if σ is a Cartan involution, the group H is a maximal compact subgroup of G and the symmetric space is called a *Riemannian symmetric space of non-compact type*, which has been studied well from several points of view.

Algebraic Analysis, Volume II

ISBN 0-12-400466-0

Hereafter we assume that G is a connected real linear semisimple Lie group. Then the symmetric space $X = G/H$ admits an invariant measure and we have a unitary representation of G on $L^2(X)$. The most fundamental problem on the harmonic analysis on the semisimple symmetric space is to give an explicit decomposition of $L^2(X)$ into irreducible unitary representations of G, which we call a *Plancherel formula*. In the group case, this was accomplished by the great work due to Harish-Chandra.

In [O2] I proposed an approach to obtain the Plancherel formula. It works well for the most continuous spectra on $L^2(X)$ with respect to the ring $D(X)$ of the invariant differential operators on X. But the method is not strong enough or not easy to analyze the discrete spectra, which we call *discrete series* for X. On the other hand, by using Flensted-Jensen's duality method, we now have sufficient information for the discrete series ([FJ1], [MO1], [Ma2], [V], [MO2]). Combining this with the results explained in [O2], we can obtain a Plancherel formula for $L^2(X)$. In this note, I will explain some new aspects concerning the Plancherel formula which were not clear in [O2].

1. Notation

Let K be a σ-stable maximal compact subgroup of G, θ be the corresponding Cartan involution of both G and the Lie algebra $\mathfrak{g}$, $\mathfrak{g} = \mathfrak{h} + \mathfrak{q}$ (resp. $\mathfrak{k} + \mathfrak{p}$) be the decompositions of $\mathfrak{g}$ with respect to $+1$ and -1 eigenspaces for σ (resp. θ). Fix maximal abelian subspaces $\mathfrak{a}$ of $\mathfrak{p} \cap \mathfrak{q}$, $\mathfrak{a}_\mathfrak{p}$ of $\mathfrak{p}$, $\mathfrak{j}$ of $\mathfrak{q}$ and $\tilde{\mathfrak{j}}$ of $\mathfrak{g}$ with $\mathfrak{a}_\mathfrak{p} \supset \mathfrak{a}$, $\mathfrak{j} \supset \mathfrak{a}$ and $\tilde{\mathfrak{j}} \supset \mathfrak{a}_\mathfrak{p} \cup \mathfrak{j}$ and put $\mathfrak{t} = \mathfrak{j} \cap \mathfrak{k}$. For a real vector space V we denote the dual space by V^* and the complexification of V by V_c. We identify $\mathfrak{g}_c$ and $\mathfrak{g}_c^*$ by the Killing form $\langle\ ,\ \rangle$ of $\mathfrak{g}$. Let $\Sigma(\mathfrak{a})$, $\Sigma(\mathfrak{a}_\mathfrak{p})$, $\Sigma(\mathfrak{j})$ and $\Sigma(\tilde{\mathfrak{j}})$ be the root systems corresponding to the respective pairs $(\mathfrak{g}, \mathfrak{a})$, $(\mathfrak{g}, \mathfrak{a}_\mathfrak{p})$, $(\mathfrak{g}_c, \mathfrak{j}_c)$ and $(\mathfrak{g}_c, \tilde{\mathfrak{j}})$. The corresponding Weyl groups are denoted by $W(\mathfrak{a})$, $W(\mathfrak{a}_\mathfrak{p})$, $W(\mathfrak{j})$ and $W(\tilde{\mathfrak{j}})$. We fix respective compatible positive systems $\Sigma(\mathfrak{a})^+$, $\Sigma(\mathfrak{a}_\mathfrak{p})^+$, $\Sigma(\mathfrak{j})^+$ and $\Sigma(\tilde{\mathfrak{j}})^+$. Put $\rho = \frac{1}{2}\sum_{\alpha \in \Sigma(\tilde{\mathfrak{j}})^+} \alpha$. Let $\Psi(\mathfrak{a}) = \{\alpha_1, \ldots, \alpha_l\}$ and $\Psi(\mathfrak{j}) = \{\tilde{\alpha}_1, \ldots, \tilde{\alpha}_{l'}\}$ be the fundamental system of $\Sigma(\mathfrak{a})^+$ and $\Sigma(\mathfrak{j})^+$, respectively. Here l is called the *split rank* of X and l' is called the *rank* of X.

For a subgroup G' of G and a subalgebra $\mathfrak{b}$ of $\mathfrak{g}$, we denote by $Z_{G'}(\mathfrak{b})$ (resp. $N_{G'}(\mathfrak{b})$) the centralizer (resp. normalizer) of $\mathfrak{b}$ in G' and by Lie(G') the Lie algebra of G'. Let $\mathfrak{n}$ be the maximal nilpotent subalgebra of $\mathfrak{g}$ corresponding to $\Sigma(\mathfrak{a}_\mathfrak{p})^+$ and put $A = \exp(\mathfrak{a})$, $A_\mathfrak{p} = \exp(\mathfrak{a}_\mathfrak{p})$ and $M = Z_K(\mathfrak{a}_\mathfrak{p})$. Then the group $P = MA_\mathfrak{p}N$ is a minimal parabolic subgroup of G.

Let P_σ denote the parabolic subgroup of G with the Langlands decomposition $P_\sigma = M_\sigma A_\sigma N_\sigma$ such that $M_\sigma A_\sigma = Z_G(\mathfrak{a})$ and, moreover, such that the Lie algebra $\mathfrak{n}_\sigma$ of N_σ is spanned by the root spaces $\mathfrak{g}(\mathfrak{a};\alpha)$ for the roots α in $\Sigma(\mathfrak{a})^+$. Put $\mathfrak{m}_\sigma = \mathrm{Lie}(M_\sigma)$ and $\mathfrak{a}_\sigma = \mathrm{Lie}(A_\sigma)$. Let $\mathfrak{g}(\sigma)$ be the Lie algebra spanned by the root spaces $\mathfrak{g}(\mathfrak{a}_\mathfrak{p};\lambda)$ for the roots $\lambda \in \Sigma(\mathfrak{a}_\mathfrak{p})^+$ with $\lambda|\mathfrak{a} = 0$, and let $\mathfrak{m}(\sigma)$ be the centralizer of $\mathfrak{g}(\sigma)$ in $\mathfrak{m}_\sigma$. Let $G(\sigma)$ and $M(\sigma)_0$ be the respective analytic subgroups of G, and put $M(\sigma) = M(\sigma)_0 \mathrm{Ad}_G^{-1}(\mathrm{Ad}_G(K) \cap \exp\sqrt{-1}\mathfrak{a}_\mathfrak{p})$. Then $\mathfrak{m}_\sigma$ is the direct sum of $\mathfrak{m}(\sigma)$ and $\mathfrak{g}(\sigma)$ and moreover we have $M_\sigma \subset M(\sigma)$, $G(\sigma) \subset H$ and $M_\sigma = M(\sigma)G(\sigma)$.

Let $W(\mathfrak{a}; H)$ be the subgroup of $W(\mathfrak{a})$ whose elements have representatives in $K \cap H$. Put $r = \#(W(\mathfrak{a}; H)\backslash W(\mathfrak{a}))$ and $W(\mathfrak{a}; H)\backslash W(\mathfrak{a}) = \{w_1, \ldots, w_r\}$. For every $w \in W(\mathfrak{a}; H)\backslash W(\mathfrak{a})$ we fix a representative of w in $W(\mathfrak{a})$ and also that of w in K and denote them by the same symbol w. We can choose the representative w so that $\mathrm{Ad}(w)\mathfrak{a}_\mathfrak{p} = \mathfrak{a}_\mathfrak{p}$, $\mathrm{Ad}(w)\mathfrak{j} = \mathfrak{j}$, $\mathfrak{m}(\sigma) \cap \mathrm{Ad}(w)^{-1}\mathfrak{h} = \mathfrak{m}(\sigma) \cap \mathfrak{h}$ and $w(\Sigma(\mathfrak{j})_\theta^+) = \Sigma(\mathfrak{j})_\theta^+$. Here $\Sigma(\mathfrak{j})_\theta = \{\alpha \in \Sigma(\mathfrak{j});\ \alpha|\mathfrak{a} = 0\}$ and $\Sigma(\mathfrak{j})_\theta^+ = \Sigma(\mathfrak{j})_\theta \cap \Sigma(\mathfrak{j})^+$. We may assume $w_1 = e$. The restricted root system for the reductive symmetric pair $(\mathfrak{m}(\sigma), \mathfrak{m}(\sigma) \cap \mathfrak{h})$ is identified with $\Sigma(\mathfrak{j})_\theta$.

We will use the standard notation **N**, **Z**, **R** and **C**. Here **N** means the set of non-negative integers. Moreover, for a manifold M, $\mathscr{A}(M)$ and $\mathscr{B}(M)$ denote the space of real analytic functions on M and that of hyperfunctions on M, respectively.

2. Smooth Imbedding

We define a compact G-manifold with finitely many G-orbits and every open G-orbit isomorphic to X. Put $I = \{1, \ldots, r\}$ and $\hat{X} = G \times \mathbf{R}^l \times I$. Let $x = (g, t, i)$ be an element of $\hat{X}$. Put $\mathrm{sgn}\, x = \mathrm{sgn}\, t \in \{-1, 0, 1\}^l$, where $\mathrm{sgn}\, t = (\mathrm{sgn}\, t_1, \ldots, \mathrm{sgn}\, t_l)$ with $t = (t_1, \ldots, t_l) \in \mathbf{R}^l$. Moreover, put $\Theta_x = \{\alpha_j;\ t_j \neq 0\} \subset \Psi(\mathfrak{a})$ and denote by W_x the subgroup of $W(\mathfrak{a})$ generated by the reflections with respect to $\alpha \in \Theta_x$. Let P_x be the closure of $P_\sigma W_x P_\sigma$ in G, $P_x = M_x A_x N_x$ be its Langlands decomposition with $A_x \subset A_\sigma$ and $\mathfrak{p}_x = \mathfrak{m}_x + \mathfrak{a}_x + \mathfrak{n}_x$ be the corresponding decomposition of the Lie algebra. Let $\{H_1, \ldots, H_l\}$ be the dual basis of $\{\alpha_1, \ldots, \alpha_l\}$. Define a closed subgroup $P(x)$ of G by

$$P(x) = (M_x \cap w_i^{-1} H w_i) A_x N_x,$$

and put

$$a(x) = a(t) = \exp\Big(-\sum_{t_j \neq 0} H_j \log|t_j|\Big) \in A_x.$$

Definition 2.1. We define that two elements $x=(g,t,i)$ and $x'=(g',t',i')$ of $\hat{X}$ are equivalent if and only if the following three conditions hold:

(*i*) $\operatorname{sgn} x=\operatorname{sgn} x'$,
(*ii*) $W(\mathfrak{a};H)w_iW_x=W(\mathfrak{a};H)w_{i'}W_{x'}$,
(*iii*) $ga(x)P(x)=g'a(x')(M_x\cap w_{i'}^{-1}Hw_i)A_xN_x$.

This in fact defines an equivalence relation in $\hat{X}$, and the quotient space of $\hat{X}$ by this equivalence relation is denoted by $\tilde{X}$. Let $\pi:\hat{X}\to\tilde{X}$ be the natural projection. The action of $g_0\in G$ on $\tilde{X}$ is defined by $g_0\pi(g,x,i)=\pi(g_0g,x,i)$.

Theorem 2.2. (*i*) *We can define a real analytic structure on* $\tilde{X}$ *compatible with the quotient topology, and then* $\tilde{X}$ *is a connected compact real analytic manifold without boundaries.*

(*ii*) *The action of G on* $\tilde{X}$ *is analytic, and the G-orbital decomposition is of normal crossing type in the following sense: For any point* $p\in\tilde{X}$, *there exists a local coordinate system* $(t,x)=(t_1,\ldots,t_k,x_1,\ldots,x_n)$ *in a neighborhood of p such that the points* (t,x) *and* (t',x') *are in the same G-orbit if* $\operatorname{sgn} t_j=\operatorname{sgn} t'_j$ *for* $j=1,\ldots,k$.

(*iii*) *The orbit* $G\pi(x)$ *for an element* $x\in\hat{X}$ *is isomorphic to* $G/P(x)$. *There are just* 2^l *open orbits and they are isomorphic to X. The number of the compact orbits equals r and they are contained in the closure of every open orbit.*

(*iv*) *We identify X with the open G-orbit* $G\pi(e,(1,\ldots,1),1)$. *Then any invariant differential operator on X has an analytic extension to an invariant differential operator on* $\tilde{X}$.

This result was first obtained by [O1] for a Riemannian symmetric space of non-compact type (cf. [Sc]). The proof of the above theorem is similar as in [O1] and given in [O4]. A little different type of imbedding was given in [OS] for some series of semisimple symmetric spaces.

3. Asymptotic Expansion

The Plancherel formula is almost equivalent to a simultaneous spectral decomposition of $L^2(X)$ with respect to the ring $D(X)$ of invariant differential operators on X. Therefore it is important to study the simultaneous eigenspace of $D(X)$.

We identify the ring $D(\tilde{X})$ of invariant differential operators on $\tilde{X}$ with $D(X)$ by Theorem 2.2(*iv*). Let J be a finite-codimensional ideal of $D(\tilde{X})$ and consider the space $\mathscr{B}(X;\mathscr{M})$ of the hyperfunction solutions of the system

$$\mathscr{M}\colon Du=0 \qquad (\forall D\in J)$$

on X. It is proved in [O3] that the system $\mathscr{M}$ has regular singularities along every boundary component of X (i.e., G-orbit contained in the boundary of X in $\tilde{X}$). We can prove that every solution $u\in\mathscr{B}(X;\mathscr{M})$ has the asymptotic expansion

$$u(ka(t))\sim\sum_{\nu=1}^{m}\sum_{\alpha\in\mathbf{N}^l} a_{\nu,\alpha}(k)\phi_\nu(\log t_1,\ldots,\log t_l)t^{\lambda_\nu+\alpha}, \tag{3.1}$$

with $k\in K$ and $t\in(0,\infty)^l$ when t tends to 0. Here $a_{\nu,\alpha}\in\mathscr{B}(K)$, $t^{\lambda_\nu}=t_1^{\lambda_{\nu,1}}\cdots t_l^{\lambda_{\nu,l}}$ and ϕ_ν are non-zero homogeneous polynomials and do not depend on u. Since $a_{\nu,\alpha}$ are right $Z_{K\cap H}(\mathfrak{a})$-invariant, we regard $a_{\nu,\alpha}$ as elements of $\mathscr{B}(B)$ by putting $B=K/Z_{K\cap H}(\mathfrak{a})$. We can choose λ_ν so that for a given open subset U of B and a given number ν, if $a_{\mu,\beta}|_U=0$ for any (μ,β) satisfying $\lambda_\nu=\lambda_\mu+\beta$, then $a_{\nu,\alpha}|_U=0$ $(\forall\alpha)$. The numbers λ_ν are determined by $\mathscr{M}$ and called the *characteristic exponents* of $\mathscr{M}$.

The meaning of (3.1) is as follows: For any $N\in\mathbf{N}$ we can choose $m\in\mathbf{N}$ so that the $\mathscr{B}(K)$-valued function

$$(0,\infty)\ni t\mapsto\left(\prod_{j=1}^{l}t_j^{-N}\right)\left(u(ka(t))\right.$$
$$\left.-\sum_{\nu=1}^{m}\sum_{\alpha\in\{0,\ldots,M\}^l}a_{\nu,\alpha}(k)\phi_\nu(\log t_1,\ldots,\log t_l)t^{\lambda_\nu+\alpha}\right)$$

uniformly converges to 0 in the topology of $\mathscr{B}(K)$ when t converges to 0.

Now we fix an element k_0 of K. Then the point $k_0a(t)H\in X$ converges to a point $p=\pi(k_0,0,e)$ in a distinguished boundary B, which is isomorphic to $K/Z_{K\cap H}(\mathfrak{a})$. In fact, we have the following identification:

$$\begin{array}{ccc} K/Z_{K\cap H}(\mathfrak{a})\times(0,1)^l\ni(k,t) & \mapsto & \bar{k}a(t)H\in X\\ \downarrow & & \downarrow\\ K/Z_{K\cap H}(\mathfrak{a})\times(-1,1)^l\ni(k,t) & \mapsto & \pi(\bar{k},t,e)\in\tilde{X}\end{array}$$

Here $\bar{k}\in K$ is a representative of $k\in K/Z_{K\cap H}(\mathfrak{a})$, all the maps above are diffeomorphisms onto open subsets, and B is identified with $K/Z_{K\cap H}(\mathfrak{a})$. Then the expansion has the following properties:

If $a_{\nu,0}(k)=0$ for $\nu=1,\ldots,m$ in a neighborhood of p in B, then $u=0$ on X.

If there exists a neighborhood U of p in B such that $a_{\nu,0}(k)$ are real analytic in U for $\nu=1,\ldots,m$, then there exist a neighborhood U' of k_0 in B and a positive number ε such that $a_{\nu,\alpha}(k)$ are real analytic in U and $\sum_\alpha a_{\nu,\alpha}(k)t^\alpha$ converge to real analytic functions on $U'\times(-\varepsilon,\varepsilon)^l$ and both sides of (3.1) are equal. In this case, we say that u is *ideally analytic* at p, which was named by M. Sato.

If u satisfies a system of differential equations $\mathcal{N}$ with

$$\mathrm{SS}\,\mathcal{N}\cap T_p^*\tilde{X}\subset T_p^*B,$$

then u is ideally analytic at p.

The expansion (3.1) for K-finite functions was studied by [HC] and [CM] in the group case and by [Ba] and [O3] in a general case. Also in the group case, the expansion was obtained by [W] assuming a certain growth condition for v.

Fix ν. If $a_{\nu',0}(k)=0$ in a neighborhood U of p in B for any ν' satisfying

$$\left(\tfrac{1}{2}(\lambda_{\nu,1}-\lambda_{\nu',1}),\ldots,\tfrac{1}{2}(\lambda_{\nu,l}-\lambda_{\nu',l})\right)\in\mathbf{N}^l-\{0\}$$

or

$$\lambda_{\nu'}=\lambda_\nu \qquad \text{and} \qquad \deg\phi_{\nu'}>\deg\phi_\nu,$$

then the correspondence which maps u to $a_{\nu,0}$ defines a $\mathfrak{g}$-equivariant map to a certain subspace (described by differential equations) of local sections over U of a certain line bundle over the distinguished boundary B, which decomposes with multiplicity free into a finite direct sum of the spaces of local sections over U belonging to principal series for X. This boundary value map is studied in [O3]. If $U=B$, then the map is G-equivariant.

Fix an open subset U of B and for the expression (3.1), put $\bar{\Lambda}=\{(\lambda_\nu,\deg\phi_\nu);\ a_{\nu,0}|_U\neq 0\}$ and $\Lambda=\{(\lambda,k)\in\bar{\Lambda};\ \{(\lambda',k')\in\bar{\Lambda};\ (1)\ (\mathrm{Re}(\lambda_1-\lambda_1'),\ldots,\mathrm{Re}(\lambda_l-\lambda_l'))\in[0,\infty)^l-\{0\}$, or (2) $\lambda=\lambda'$ and $k'>k\}=\varnothing\}$. We call Λ the *set of leading exponents* of u with respect to U. Also for the point p we say that Λ is the set of leading exponents of u at p by choosing U a sufficiently small neighborhood of p.

There are r distinguished boundaries. Changing $a(t)$ by $a(t)w_i^{-1}$ in the left-hand side of (3.1), the same argument as above is valid for any distinguished boundary.

4. Asymptotic Behavior of Spherical Functions

Let $\hat{K}$ denote the set of equivalence classes of finite-dimensional irreducible representations of K and $C_\delta^\infty(X)$ denote the space of K-finite C^∞-functions on X of type δ. For an element $\lambda \in \mathfrak{j}_c^*$, an algebra homomorphism χ_λ of $D(X)$ to $\mathbf{C}$ is defined through the Harish-Chandra isomorphism $D(X) \simeq I(W(\mathfrak{j}))$, where $I(W(\mathfrak{j}))$ is the algebra of $W(\mathfrak{j})$-invariant elements in the symmetric algebra of $\mathfrak{j}_c$. Put

$$\mathscr{B}(X; \mathscr{M}_\lambda) = \{u \in \mathscr{B}(X);\ Du = \chi_\lambda(D)u\ (\forall D \in D(X))\},$$

$$\mathscr{A}_\delta(X; \mathscr{M}_\lambda) = C_\delta^\infty(X) \cap \mathscr{B}(X; \mathscr{M}_\lambda),$$

and

$$\mathscr{A}_K(X; \mathscr{M}_\lambda) = \bigoplus_{\delta \in \hat{K}} \mathscr{A}_\delta(X; \mathscr{M}_\lambda).$$

Since $\mathscr{A}_K(X; \mathscr{M}_\lambda) = \mathscr{A}_K(X; \mathscr{M}_{w\lambda})$ for any $w \in W(\mathfrak{j})$, we may assume

$$\operatorname{Re}\langle \lambda, \alpha \rangle \geq 0 \qquad (\forall \alpha \in \Sigma(\mathfrak{j})^+) \tag{4.1}$$

and

there exists a subset Θ of $\Psi(\mathfrak{j})$ such that

$$\Sigma(\mathfrak{j}) \cap \lambda^\perp = \sum_{\alpha \in \Theta} \mathbf{C}\alpha \cap \Sigma(\mathfrak{j}). \tag{4.2}$$

For any $v \in \mathscr{B}(X; \mathscr{M}_\lambda)$, we can choose λ_ν in the expression (3.1) so that

$$\{\lambda_\nu;\ \nu = 1, \ldots, m\} = \{(\langle \rho - w\lambda, H_1 \rangle, \ldots, \langle \rho - w\lambda, H_l \rangle);\ w \in W(\mathfrak{j})\}. \tag{4.3}$$

We call an element ψ of $\mathscr{A}_K(X; \mathscr{M}_\lambda)$ a *spherical function* on X. Then ψ is ideally analytic at every boundary point of X in $\tilde{X}$. Let $\Lambda_i(\psi)$ be the set of leading exponents of ψ on the distinguished boundary $B_i = \pi(G \times \{0\} \times \{i\})$. Since the invariant measure on X asymptotically behaves like

$$t_1^{-2\langle \rho, H_1 \rangle - 1} \cdots t_l^{-2\langle \rho, H_l \rangle - 1}\, dt_1 \cdots dt_l\, dx_1 \cdots dx_n$$

near the distinguished boundary (which is defined by $t_1 = \cdots = t_l = 0$), we can easily show that the following two conditions are equivalent:

$$\psi \in \mathscr{A}_K(X; \mathscr{M}_\lambda) \cap L^p(X). \tag{4.4}$$

If $(\lambda, m) \in \bigcup_{i=1}^r \Lambda_i(\psi)$, then $\operatorname{Re}\left(\lambda_j - \frac{2}{p}\langle \rho, H_j \rangle\right) \geq 0 \qquad$ for $j = 1, \ldots, l.$

(4.5)

If $\psi \in L^{2+\varepsilon}(X)$ for any $\varepsilon > 0$, we say that ψ is tempered. We can prove that any spectrum of $D(X)$ in $L^2(X)$ corresponds to a $(\mathfrak{g}, K)$-module realized in $\mathscr{A}_K(X; \mathscr{M}_\lambda) \cap \bigcap_{\varepsilon>0} L^{2+\varepsilon}(X)$. Especially, discrete series for X correspond to $(\mathfrak{g}, K)$-modules $L^2_K(X; \mathscr{M}_\lambda) = \mathscr{A}_K(X; \mathscr{M}_\lambda) \cap L^2(X)$. By using the boundary value maps for suitable boundary components of X, the problem to get tempered spherical functions is reduced to the study of discrete series for some semisimple symmetric spaces. (In fact, if ψ is tempered and not in $L^2(X)$ and if ψ generates an irreducible $(\mathfrak{g}, K)$-module, then, for example, we may assume that there exist a positive number k and a leading exponent (λ, m) in a suitable $\Lambda_i(\psi)$ such that $\operatorname{Re} \lambda_j = \langle \rho, H_j \rangle$ for $j \leq k$ and $\operatorname{Re} \lambda_j > \langle \rho, H_j \rangle$ for $j > k$. Then the $(\mathfrak{g}, K)$-module is imbedded to a unitary representation space induced from a representation of a parabolic subgroup of G which is parametrized modulo 1-dimensional representations by the discrete series for a semisimple symmetric space with split rank $l-k$. The imbedding is given by a boundary value map for the boundary component defined by $t_1 = \cdots = t_k = 0$.)

To study the asymptotic behavior of ψ, we use Flensted-Jensen's duality method. Let $\mathfrak{g}^d$, $\mathfrak{k}^d$ and $\mathfrak{h}^d$ be the subalgebras of $\mathfrak{g}_c$ defined by

$$\mathfrak{g}^d = \mathfrak{k} \cap \mathfrak{h} + \sqrt{-1}(\mathfrak{k} \cap \mathfrak{q}) + \sqrt{-1}(\mathfrak{p} \cap \mathfrak{h}) + \mathfrak{p} \cap \mathfrak{q},$$
$$\mathfrak{k}^d = \mathfrak{k} \cap \mathfrak{h} + \sqrt{-1}(\mathfrak{p} \cap \mathfrak{h}), \qquad \mathfrak{h}^d = \mathfrak{k} \cap \mathfrak{h} + \sqrt{-1}(\mathfrak{k} \cap \mathfrak{q}).$$

Let G^d, K^d and H^d be the respective analytic subgroups of the complexification G_c of G. The Riemannian symmetric space $X^r = G^d / K^d$ is called the *non-compact Riemannian form* of X. Then $D(X^r)$ is naturally isomorphic to $D(X)$ and $\delta \in \hat{K}$ corresponds to an equivalence class of a finite-dimensional irreducible representation of H^d through the holomorphic representation of the complexification K_c of K in G_c, which we denote by the same symbol δ. Let $C^\infty_\delta(X^r)$ denote the space of H^d-finite functions on X^d of type δ and put

$$\mathscr{A}(X^r; \mathscr{M}_\lambda) = \{u \in \mathscr{A}(X^r);\ Du = \chi_\lambda(D)u\ (\forall D \in D(X^r))\},$$
$$\mathscr{A}_\delta(X^r; \mathscr{M}_\lambda) = \mathscr{A}(X^r; \mathscr{M}_\lambda) \cap C^\infty_\delta(X^r)$$

and

$$\mathscr{A}_K(X^r; \mathscr{M}_\lambda) = \bigoplus_{\delta \in \hat{K}} \mathscr{A}_\delta(X^r; \mathscr{M}_\lambda).$$

Then there exists a bijective $\mathfrak{g}$-homomorphism

$$\eta : \mathscr{A}_K(X; \mathscr{M}_\lambda) \rightrightarrows \mathscr{A}_K(X^r; \mathscr{M}_\lambda), \tag{4.6}$$

which we call the *Flensted-Jensen isomorphism* ([FJ1]). Since $\psi\in\mathcal{A}_K(X;\mathcal{M}_\lambda)$ is K-finite, $\psi(ka)$ $(k\in K, a\in A)$ can be extended holomorphically to a function $\tilde{\psi}(ka)$ $(k\in K_c,\ a\in A)$ and then $\eta(\psi)|_{H^dA}=\tilde{\psi}|_{H^dA}$. Here we remark that $G=KAH$ and $G^d=H^dAK^d$.

Since X^r is a Riemannian symmetric space, there is only one distinguished boundary B^d which is isomorphic to G^d/P^d. Here P^d is the minimal parabolic subgroup of G^d with the Langlands decomposition $P^d=M^dA_\mathfrak{p}^dN^d$ such that Lie(N^d) is the linear span of the root spaces in $\mathfrak{g}^d$ for the roots in $\Sigma(\mathfrak{j})^+$, Lie$(A_\mathfrak{p}^d)=\mathfrak{a}_\mathfrak{p}^d$ and moreover $M^d=Z_{K^d}(\mathfrak{a}_\mathfrak{p}^d)$ by denoting $\mathfrak{a}_\mathfrak{p}^d=\sqrt{-1}\mathfrak{t}+\mathfrak{a}$.

For any non-zero $u\in\mathcal{A}(X^r;\mathcal{M}_\lambda)$, the set of leading exponent of u on the distinguished boundary B^d contains the exponent

$$((\langle\rho-\lambda,\tilde{H}_1\rangle,\ldots,\langle\rho-\lambda,\tilde{H}_{l'}\rangle),m),$$

where $\{\tilde{H}_1,\ldots,\tilde{H}_{l'}\}$ is the dual basis of $\{\tilde{\alpha}_1,\ldots,\tilde{\alpha}_{l'}\}=\Psi(\mathfrak{j})$ and $2m$ equals the number of the roots in a reduced root system with the Weyl group $W_\lambda=\{w\in W(\mathfrak{j});\ w\lambda=\lambda\}$.

The boundary value map β_λ of $\mathcal{A}(X^r;\mathcal{M}_\lambda)$ for B^d with respect to this leading exponent gives the G^d-isomorphism

$$\beta_\lambda:\mathcal{A}(X^r;\mathcal{M}_\lambda)\xrightarrow{\sim}\mathcal{B}(B^d;L_\lambda),$$

where

$$\mathcal{B}(B^d;L_\lambda)=\{f\in\mathcal{B}(G^d);f(gman)=f(g)a^{\lambda-\rho}\quad\text{for } g\in G^d,\ m\in M^d,\ a\in A_\mathfrak{p}^d,\ n\in N^d\}.$$

The inverse map is a constant multiple of the Poisson transformation

$$\begin{array}{cccc}\mathcal{P}_\lambda:\mathcal{B}(B^d;L_\lambda) & \rightarrow & & \mathcal{A}(X^r;\mathcal{M}_\lambda)\\ \cup\!\!\!\!\!\!\!\!\text{ } & & & \cup\!\!\!\!\!\!\!\!\text{ }\\ f & \mapsto & (\mathcal{P}_\lambda f)(g)= & \displaystyle\int_{K^d}f(gk)\,dk\end{array}$$

(cf. [K-]). Denoting

$$\mathcal{B}_\delta(B^d;L_\lambda)=\{f\in\mathcal{B}(B^d;L_\lambda);f\text{ is }H^d\text{-finite of type }\delta\}$$

and

$$\mathcal{B}_K(B^d;L_\lambda)=\bigoplus_{\delta\in\hat{K}}\mathcal{B}_\delta(B^d;L_\lambda),$$

we have the $\mathfrak{g}$-isomorphism

$$\beta_\lambda \circ \eta : \mathscr{A}_K(X; \mathscr{M}_\lambda) \xrightarrow{\sim} \mathscr{B}_K(B^d; L_\lambda). \tag{4.7}$$

Definition 4.1. For $\psi \in \mathscr{A}_K(X; \mathscr{M}_\lambda)$, we put

$$\mathrm{FBI}_\lambda(\psi) = \mathrm{supp}\, \beta_\lambda \circ \eta(\psi).$$

We remark that $\mathrm{FBI}_\lambda(\psi)$ is an H^d-invariant closed subset of G^d/P^d and that the structure of the double coset decomposition $H^d \backslash G^d / P^d$ is studied in [Ma1]. Especially, $\#(H^d \backslash G^d / P^d) < \infty$ and there are r open cosets $O_i = \{H^d w_i P^d;\ i = 1, \ldots, r\}$.

Since $a(t) \in G^d \cap G$, the asymptotic behavior of $\psi(ka(t)w_i^{-1})$ $(k \in K)$ when $t \to 0$ coincides with that of $(\eta \circ \psi)(kw_i a(t))$ $(k \in H^d)$. The latter is determined by the set of leading exponents of $\eta \circ \psi$ on O_i. On the other hand, the relation between boundary values of an element $u \in \mathscr{A}(X^r; \mathscr{M}_\lambda)$ with respect to several exponents are described by using intertwining operators between class-1 principal series for G^d. Studying the intertwining operators precisely and combining the above argument, we finally obtain the following result ([O4]):

Theorem 4.2. *(i)* $\psi \in L^p(X) \cap \mathscr{A}_K(X; \mathscr{M}_\lambda)$ *if and only if*

$$\mathrm{FBI}_\lambda(\psi) Cl(P^d w^{-1} P^d) \cap \left(\bigcup_{i=1}^{r} O_i \right) = \varnothing$$

for any $w \in W(\mathfrak{j})$ *satisfying*

$$\left(\mathrm{Re}\left\langle \left(1 - \frac{2}{p}\right)\rho - w\lambda, H_1 \right\rangle, \ldots, \mathrm{Re}\left\langle \left(1 - \frac{2}{p}\right)\rho - w\lambda, H_l \right\rangle \right) \notin (0, \infty)^l.$$

(ii) $\psi \in \mathscr{A}_K(X; \mathscr{M}_\lambda)$ *is tempered if and only if*

$$\mathrm{FBI}_\lambda(\psi) Cl(P^d w^{-1} P^d) \cap \left(\bigcup_{i=1}^{r} O_i \right) = \varnothing$$

for any $w \in W(\mathfrak{j})$ *satisfying*

$$(\mathrm{Re}\langle w\lambda, H_1 \rangle, \ldots, \mathrm{Re}\langle w\lambda, H_l \rangle) \notin (-\infty, 0]^l.$$

5. Discrete Series

Using the isomorphism (4.7) and Theorem 4.2, the problem to get an L^2-spherical function is reduced to get an H^d-finite hyperfunction in

$\mathscr{B}(B^d; L_\lambda)$ with a suitable support property. By an analysis on the structure of $H^d\backslash G^d/P^d$, we can conclude that a non-empty H^d-invariant subset V of G^d/P^d which satisfies the condition in Theorem 4.2(i) by setting $p=2$ and replacing $\mathrm{FBI}_\lambda(\psi)$ by V only exists when rank $X=\mathrm{rank}(K/K\cap H)$ and moreover V should be a union of closed H^d-orbits in G^d/P^d.

Theorem 5.1. *The discrete series for X is non-empty if and only if*

$$\operatorname{rank} X=\operatorname{rank}(K/K\cap H). \tag{5.1}$$

It is proved in [FJ1] that the discrete series for X is non-empty if (5.1) holds, and the necessity of the condition (5.1) is proved in [MO1].

To parametrize the discrete series for X we prepare some notation. Hereafter we assume the condition (5.1). Let $\mathfrak{a}'$ be a maximal abelian subspace of $\sqrt{-1}(\mathfrak{k}\cap\mathfrak{q})$ and let Σ denote the root system for the pair $(\mathfrak{g}^d, \mathfrak{a}')$ and fix a positive system Σ^+ of Σ. Put $M'=Z_{G^d}(\mathfrak{a}')$, $A'=\exp\mathfrak{a}'$, $\mathfrak{n}^+=\sum_{\alpha\in\Sigma^+}\mathfrak{g}(\mathfrak{a}';\alpha)$, $N^+=\exp\mathfrak{n}^+$, $\rho'=\sum_{\alpha\in\Sigma^+}\frac{1}{2}m_\alpha\alpha$ and $\rho_t'=\frac{1}{2}\sum_{\alpha\in\Sigma^+}m_\alpha^+\alpha$. Here for a root $\alpha\in\Sigma^+$, $\mathfrak{g}^d(\mathfrak{a}';\alpha)$ is the corresponding root space in $\mathfrak{g}^d$, $m_\alpha=\dim\mathfrak{g}^d(\mathfrak{a}';\alpha)$ and $m_\alpha^+=\dim(\mathfrak{g}^d(\mathfrak{a}';\alpha)\cap\mathfrak{h}^d)$. Then $P'=M'A'N'$ is a minimal parabolic subgroup of G^d and we denote by P_c' the complexification of P' in G_c.

Let $L_{K/M\cap H}$ be the lattice in $(\mathfrak{a}')^*$ generated by the highest weights of finite-dimensional representations of K with a nontrivial $(K\cap H)$-fixed vector. We fix $x\in K^d$ such that $\mathrm{Ad}(x)(\sqrt{-1}\mathfrak{t}+\mathfrak{a})=\mathfrak{a}'$. Using $\mathrm{Ad}(x)$ we can identify $\mathfrak{j}_c^*$ with $(\mathfrak{a}')_c^*$. We may assume $\Sigma(\mathfrak{j})^+$ corresponds to $\Sigma(\mathfrak{a}')^+$ by changing x if necessary and for $\lambda\in\mathfrak{j}_c^*$ we denote by λ' the corresponding element of $(\mathfrak{a}')_c^*$ and put $\mu_\lambda=\lambda'+\rho'-2\rho_t'$.

Put $\mathscr{C}=\{\lambda\in\mathfrak{j}_c^*;\ \mathrm{Re}\langle\lambda,\alpha\rangle>0\ (\forall\alpha\in\Sigma(\mathfrak{j})^+)\}$. For an element $\lambda\in\mathfrak{j}_c^*$ with $\mu_\lambda\in L_{K/K\cap H}$ we can define a holomorphic line bundle $\tilde{L}_\lambda$ over G_c/P_c' whose section f over an open subset U of G_c/P_c' satisfies

$$f(gman)=f(g)a^{\lambda'-\rho'}\qquad(g\in U, m\in M', a\in A', n\in N').$$

We can choose elements $v_1=e, v_2,\dots, v_m$ of $Z_{K^d}(\mathfrak{a}')$ such that $\{H^dv_jP'; j=1,\dots,m\}$ is the set of all the closed H^d-P' double cosets in G^d with $H^dv_iP'\neq H^dv_jP'$ if $i\neq j$. Put $V_j=K_cv_jP_c'$. We identify V_j with compact algebraic submanifolds of G_c/P_c' and put

$$B_\lambda^j=H_{V_j}^n(G_c/P_c;\ \mathscr{O}(\tilde{L}_\lambda)), \tag{5.2}$$

where n is the complex codimension of V_j in G_c/P_c, which does not depend on j. Here $\mathcal{O}(\tilde{L}_\lambda)$ means a sheaf of sections of $\tilde{L}_\lambda$ in the sense of Zariski topology.

Put $L^2_K(X; \mathcal{M}_\lambda) = L^2(X) \cap \mathcal{A}_K(X; \mathcal{M}_\lambda)$. Suppose $\psi \in L^2_K(X; \mathcal{M}_\lambda)$. Then

$$\operatorname{supp} \beta_\lambda \circ \eta(\psi) \subset \bigcup_{j=1}^{m} H^d v_j P'.$$

By using the action of K_c we can extend $\beta_\lambda \circ \eta(\psi)$ to an element of $\bigoplus_j B^j_\lambda$ and we have the following theorem.

Theorem 5.2. *(i) Let $\lambda \in \bar{\mathscr{C}}$.* If $L^2_K(X; \mathcal{M}_\lambda) \neq \{0\}$, *then $\lambda \in \mathscr{C}$ and $\mu_\lambda \in L_{K/K\cap H}$ and we have the $\mathfrak{g}$-isomorphism*

$$L^2_K(X; \mathcal{M}_\lambda) \simeq \bigoplus_{j=1}^{m} B^j_\lambda.$$

(ii) If $\lambda \in \mathscr{C}$ and $\mu_\lambda \in L_{K/K\cap H}$, then B^j_λ is $\{0\}$ or an irreducible $\mathfrak{g}$-module.

Theorem 5.2(i) is given in [MO1]. A much simpler proof is shown in [Ma2] by using Theorem 4.2(i). Theorem 5.2(ii) is given for a generic λ and also many cases are proved by using a theory of $\mathscr{D}$-modules. The other singular cases are proved in [V] checking case by case.

Now we examine the condition for $B^j_\lambda \neq \{0\}$. By changing the choice of the positive system Σ^+ of Σ, we may assume $j = 1$.

Theorem 5.3. *Suppose $\lambda \in \bar{\mathscr{C}}$ and $\mu_\lambda \in L_{K/K\cap H}$.*

(i) *Put $m^-_\alpha = m_\alpha - m^+_\alpha$ and $m^0_\alpha = m^+_\alpha - m^-_\alpha$ for $\alpha \in \Sigma$. Then $B^1_\lambda \neq \{0\}$ if and only if the following (5.3) holds*

Let $\{\beta_1, \ldots, \beta_k\}$ be a sequence of roots in Σ^+ satisfying the following two conditions. Then $\langle \mu_\lambda, \beta_k \rangle \geq 0$. (5.3)

β_i is a simple root in the set $\{\alpha \in \Sigma^+; \langle \alpha, \beta_1 \rangle = \cdots = \langle \alpha, \beta_{i-1} \rangle = 0\}$. (5.4)

Put $n_i = \sum_{\alpha \in \Sigma \cap (\beta_i + \mathbf{Z}\beta_1 + \cdots + \mathbf{Z}\beta_{i-1})} m^0_\alpha$. Then $n_i < m_{\beta_i}$ for $i = 1, \ldots, k-1$ *and $n_k = m_{\beta_k}$.* (5.5)

(ii) *Put $S = \mathfrak{p}_c/\mathfrak{p}_c \cap \operatorname{Lie}(P'_c)$ and let $S^{(j)}$ be the j-th symmetric algebra of S. Let $S^{(j)}(\lambda)$ be the vector bundle over $K_c/K_c \cap P'_c$ associated to the* *holomorphic representation π_λ^j of $K_c \cap P'_c$ which is defined by the condition*

$$\pi_\lambda^j(man) = a^{2\rho'_t-\lambda'}\,\mathrm{Ad}(man) \qquad \text{for } m \in M' \cap K_c,\ a \in A' \cap K_c,\ n \in N' \cap K_c.$$

Then for any $\delta \in \hat{K}$, we have

$$[B_\lambda^1 : \delta] = \sum_{\nu=1}^{\infty} \sum_{i=1}^{\infty} (-1)^i [H^i(K_c/K_c \cap P'_c;\ S^{(\nu)}(\lambda)) : \delta].$$

The necessity of the condition (5.3) for $B_\lambda^1 \neq \{0\}$ is given in [Ma2]. The sufficiency will be given in [MO2]. Theorem 5.3 (*ii*) is proved by a standard argument if we admit the vanishing of the cohomology groups of the right-hand side of (5.2) except the n-th degree. The vanishing theorem was first proved by [BB] in the case when P'_c is a Borel subgroup of G_c. A vanishing theorem which can be applied to our case was proved by Kashiwara by using b-functions (cf. [Ka]). A proof similar to [BB] is given in [Bi].

References

[Ba] E. P. van den Ban, Invariant differential operators on a semisimple symmetric space and finite multiplicities in a Plancherel formula, preprint.

[Bi] F. Bien, Spherical $\mathcal{D}$-module and representation of reductive Lie groups, Ph.D. dissertation, M.I.T., Cambridge, Massachusetts, 1986.

[BB] A. A. Beilinson and J. Bernstein, Localization de $\mathfrak{g}$-modules, *C.R. Acad. Sci. Paris* **292** (1981) 15–18.

[CM] W. Casselman and D. Miličić, Asymptotic behavior of matrix coefficients of admissible representations, *Duke Math. J.* **49** (1982) 869–930.

[De] P. Delorme, Injection de modules sphériques pour les espaces symétriques réductifs dans certaines représentations induites, *Non-Commutative Harmonic Analysis and Lie Groups. Proceedings 1985*, Lect. Notes in Math., Springer, **1243** (1987) 108–135.

[FJ1] M. Flensted-Jensen, Discrete series for semisimple symmetric spaces, *Ann. of Math.* **111** (1980) 253–311.

[FJ2] M. Flensted-Jensen, *Analysis on Non-Riemannian Symmetric Spaces*, Regional conference series in mathematics, No. 61, A.M.S., Providence, 1986.

[HC] Harish-Chandra, Some results on differential equations (unpublished 1960), *Collected Papers*, Vol. 3, Springer, 1984, pp. 7–56.

[Ka] M. Kashiwara, The universal Verma module and the b-function, *Adv. Studies in Pure Math.* **6** (1984) 67–81.

[K-] M. Kashiwara, A. Kowata, K. Minemura, K. Okamoto, T. Oshima and M. Tanaka, Eigenfunctions of invariant differential operators on a symmetric space, *Ann. of Math.* **107** (1978) 1–39.

[KO] M. Kashiwara and T. Oshima, Systems of differential operators with regular singularities and their boundary value problems, *Ann. of Math.* **106** (1977) 145–200.

[Ma1] T. Matsuki, The orbits of affine symmetric spaces under the action of minimal parabolic subgroups, *J. Math. Soc. Japan* **31** (1979) 331–357.

[Ma2] T. Matsuki, A description of discrete series for semisimple symmetric spaces II, *Advanced Studies in Pure Math.* **14** (1988) 531–540.

[MO1] T. Matsuki and T. Oshima, A description of discrete series for semisimple symmetric spaces, *Advanced Studies in Pure Math.* **4** (1984) 331–390.

[MO2] T. Matsuki and T. Oshima, A description of discrete series for semisimple symmetric spaces III, to appear.

[O1] T. Oshima, A realization of Riemannian symmetric spaces, *J. Math. Soc. Japan* **30** (1978) 117–132.

[O2] T. Oshima, Fourier analysis on semisimple symmetric spaces, *Non-Commutative Harmonic Analysis. Proceedings, 1980*, Lect. Notes in Math., Springer, **880** (1981) 357–369.

[O3] T. Oshima, Boundary value problems for systems of linear partial differential equations with regular singularities, *Advanced Studies in Pure Math.* **4** (1984) 391–432.

[O4] T. Oshima, A realization of semisimple symmetric spaces and construction of boundary value maps, *Advanced Studies in Pure Math.* **14** (1988) 603–650.

[O5] T. Oshima, Asymptotic behavior of spherical functions on semisimple symmetric spaces, *Advanced Studies in Pure Math.* **14** (1988) 561–601.

[OS] T. Oshima and J. Sekiguchi, Eigenspaces of invariant differential operators on an affine symmetric space, *Invent. Math.* **57** (1980) 1–81.

[Sc] H. Schlichtkrull, *Hyperfunctions and Harmonic Analysis on Symmetric Spaces*, Birkhäuser, Boston, 1984.

[V] D. Vogan, Irreducibility of certain series of representation for semisimple symmetric spaces, to appear in *Advanced Studies in Pure Math.* **14**.

[W] N. Wallach, Asymptotic expansion of generalized matrix entries of representations of real reductive groups, Lect. Notes in Math., Springer, **1024** (1983) 287–369.

A Note on Ehrenpreis' Fundamental Principle on a Symmetric Space

Toshio Oshima
Department of Mathematics
University of Tokyo
Tokyo, Japan

Yutaka Saburi
Chiba Jr. College
Ichikawa, Japan

Masato Wakayama
Fukuyama University
Fukuyama, Japan

0. Introduction

Ehrenpreis' fundamental principle (on a Euclidean space) was first announced in Ehrenpreis [1], and proved independently by Palamodov [1] and Ehrenpreis [2] on several function spaces, such as distributions, infinitely differentiable functions or holomorphic functions on a convex domain in a Euclidean space. On the other hand, in the case of hyperfunctions on a convex domain in $\mathbf{R}^l$, it was proved by Kaneko [1] and Oshima [1], whose methods are different from each other. It is as follows.

Algebraic Analysis, Volume II

ISBN 0-12-400466-0

Let $A(D_x)$ be an $m_1 \times m_0$ matrix whose entries are in $\mathbf{C}[D_x]$, the ring of linear differential operators on a Euclidean space $\mathbf{R}^l$ with constant coefficients. Then we can find a multiplicity variety $\{(d_k, V_k)\}_k$ for ${}^tA(\zeta)$, which is a set of finite pairs of algebraic varieties V_k in $\mathbf{C}^l$ and row vectors d_k of length m_0 whose elements are in $\mathbf{C}[\zeta][D_\zeta]$, the ring of linear holomorphic differential operators on $\mathbf{C}^l$ with polynomial coefficients. Let E denote a certain function space of $\mathbf{C}[D_x]$-module. Then every element u in the kernel of the map $A(D_x): E^{m_0} \to E^{m_1}$ can be represented in the form

$$u(x) = \sum_k \int_{V_k} d_k^* \exp\langle\sqrt{-1}x, \zeta\rangle \, d\mu_k(\zeta),$$

where each μ_k is a measure with the support in V_k which satisfies some growth condition at infinity corresponding to E. The integral converges in the topology of E.

Ehrenpreis' fundamental principle is, in fact, fundamental in the general theory of systems of linear differential equations with constant coefficients and, for example, the solvability of the systems or regularity of their solutions are reduced to the principle. Many more interesting applications are given in Ehrenpreis [2].

Our aim is to extend Ehrenpreis' fundamental principle on Euclidean spaces to that on symmetric spaces. In Section 1 we introduce distributions and hyperfunctions of exponential type on a Riemannian symmetric space and give a theorem of Paley–Wiener type for the dual spaces of these function spaces. In Section 2 we announce our current result on Ehrenpreis' fundamental principle on these function spaces, which claims that the principle is valid on the function spaces if the symmetric space is a Cartesian product of Riemannian symmetric spaces of Euclidean type or non-compact type whose rank equals one. In Section 3 we mention briefly the crucial point of the proof of our Ehrenpreis' fundamental principle. The complete proof will be given in Oshima–Saburi–Wakayama [1], [2].

The study of Fourier transformations of functions on a Riemannian symmetric space was developed by Harish-Chandra [1], Helgason [1], Trombi–Varadarajan [1], Kostant [1], Eguchi [1] and others. On the other hand, a Poisson integral representation of a simultaneous eigenfunction of all invariant differential operators on a symmetric space is studied by Kashiwara–Kowata–Minemura–Okamoto–Oshima–Tanaka [1], Oshima–Sekiguchi [1] and others. Ehrenpreis' fundamental principle on symmetric spaces is a generalization of the Poisson integral representation and therefore our method gives it another proof.

1. A Paley–Wiener Theorem

In this section we determine a symmetric space G/K, on which we formulate Ehrenpreis' fundamental principle. Then we introduce two (test) function spaces on G/K and give their Fourier–Laplace transformation and a theorem of Paley–Weiner type.

1.1. Symmetric Space

We will use the standard notations **Z**, **R** and **C** for the ring of integers, the field of real numbers and the field of complex numbers, respectively. The set of non-negative integers is denoted by **N**. For an **R**-vector space V, we denote its complexification by $V_{\mathbf{C}}$. Let G be a connected real reductive linear Lie group and K its maximal compact subgroup. Let $\mathfrak{g}$ and $\mathfrak{k}$ be the Lie algebras of G and K, respectively. Let $\mathfrak{g}=\mathfrak{k}+\mathfrak{p}$ be a Cartan decomposition of $\mathfrak{g}$ and $\mathfrak{g}=\mathfrak{k}+\mathfrak{a}+\mathfrak{n}$ an Iwasawa decomposition of $\mathfrak{g}$. We denote by $\mathfrak{g}^*$ and $\mathfrak{a}^*$ the dual spaces of $\mathfrak{g}$ and $\mathfrak{a}$, respectively. We fix an $\mathrm{Ad}(G)$ invariant bilinear form $\langle\ ,\ \rangle$ on $\mathfrak{g}$ which is positive definite on $\mathfrak{p}$. Let A and N denote the analytic subgroups of G corresponding to $\mathfrak{a}$ and $\mathfrak{n}$, respectively. Let M^* be the normalizer of A in K, M the centralizer of A in K and W the factor group M^*/M, the (little) Weyl group. The Weyl group W has the natural adjoint action on $\mathfrak{a}$. W also acts on $\mathfrak{a}^*$ and $\mathfrak{a}^*_{\mathbf{C}}$. Let $\mathfrak{h}$ be a Cartan subalgebra of $\mathfrak{g}$ containing $\mathfrak{a}$. We choose a positive system P^+ of roots of $(\mathfrak{g}_{\mathbf{C}}, \mathfrak{h}_{\mathbf{C}})$ so that $\{\alpha|_{\mathfrak{a}};\ \alpha\in P^+\}$ is the positive system of restricted roots on $\mathfrak{a}$ corresponding to $\mathfrak{n}$. We put $\rho=(1/2)(\sum_{\alpha\in P^+}\alpha)|_{\mathfrak{a}_{\mathbf{C}}}$.

The homogeneous space G/K is a Riemannian symmetric space and the dimension l of $\mathfrak{a}$ is called the rank of G/K or real rank of G. There is a real analytic diffeomorphism:

$$\begin{array}{ccc} \exp: \mathfrak{p} & \rightrightarrows & G/K \\ \Downarrow & & \Downarrow \\ X & \mapsto & (\exp X)K. \end{array}$$

Our aim is to extend Ehrenpreis' fundamental principle on the Euclidean spaces to that on the above symmetric space.

1.2. Function Spaces $\mathscr{C}_*(G/K)$ and $\mathscr{A}_*(G/K)$

For $x\in G/K$, we put

$$|x|=\langle \exp^{-1}x, \exp^{-1}x\rangle^{1/2}.$$

Let $\{X_1, \ldots, X_m\}$ be a basis of $\mathfrak{g}$. X_j determine right G-invariant vector fields on G:

$$(X_j f)(g) = \frac{d}{dt} f(\exp(-tX_j)g)|_{t=0} \qquad (g \in G),$$

where f is a C^∞ function on G. X_j act also on functions on G/K which are identified with right K-invariant functions on G. We put $X = (X_1, \ldots, X_m)$ and for $\alpha \in \mathbf{N}^m$, we define a differential operator X^α on G/K by

$$X^\alpha = X_1^{\alpha_1} \cdots X_m^{\alpha_m}.$$

We introduce the following two function spaces on G/K:

$$\mathscr{C}_*(G/K) = \{\varphi \in C^\infty(G/K);\ \text{for } \forall \alpha \in \mathbf{N}^m \text{ and } \forall r \geq 0$$
$$\|\varphi\|_{\alpha,r} = \sup_{x \in G/K} |(X^\alpha \varphi)(x)|\, e^{r|x|} < \infty\},$$

$$\mathscr{A}_*(G/K) = \{\varphi \in C^\infty(G/K);\ \text{for } \forall J \in \tilde{\mathcal{O}}(\mathfrak{g}_{\mathbf{C}}^*) \text{ and } \forall r \geq 0$$
$$\|\varphi\|_{J,r} = \sup_{x \in G/K} |(J(X)\varphi)(x)|\, e^{r|x|} < \infty\},$$

where

$$\tilde{\mathcal{O}}(\mathfrak{g}_{\mathbf{C}}^*) = \{J \in \mathcal{O}(\mathfrak{g}_{\mathbf{C}}^*);\ J \text{ is of infra-exponential type}\}$$

and the differential operator $J(X)$ of infinite order is defined through the Taylor expansion of J at the origin with the basis $\{X_1, \ldots, X_m\}$. Here, for a finite-dimensional $\mathbf{C}$-vector space V, we denoted by $\mathcal{O}(V)$ the space of entire holomorphic functions on V and a function $f(x)$ on V is said to be of *infra-exponential type* if it satisfies the growth condition

$$\lim_{|x| \to \infty} f(x)\, e^{-\varepsilon|x|} = 0 \qquad \text{for any } \varepsilon > 0,$$

where $|x|$ denotes the norm of $x \in V$. $\mathscr{A}_*(G/K)$ is a subspace of $\mathscr{C}_*(G/K)$, the elements f of $\mathscr{A}_*(G/K)$ are real analytic on G/K. $\mathscr{C}_*(G/K)$ is a so-called FS space. We denote the dual spaces of $\mathscr{C}_*(G/K)$ (resp. $\mathscr{A}_*(G/K)$) by $\mathscr{C}'_*(G/K)$ (resp. $\mathscr{A}'_*(G/K)$) and call its element a *distribution* (resp. *hyperfunction*) *of exponential type* on G/K.

1.3. Fourier–Laplace Transformation

For $g \in G$ there are unique elements $\kappa(g) \in K$, $H(g) \in \mathfrak{a}$ and $n(g) \in N$ such that

$$g = \kappa(g) \exp(H(g)) n(g).$$

The Fourier–Laplace transform $\mathscr{F}\varphi$ of $\varphi \in C_*(G/K)$ is a function on $\mathfrak{a}_{\mathbf{C}}^* \times K/M$ and is defined by

$$\mathscr{F}\varphi(\lambda; kM) = \int_G \varphi(x)\, e^{(\sqrt{-1}\,\lambda - \rho)(H(x^{-1}k))}\, dx \qquad ((\lambda, kM) \in \mathfrak{a}_{\mathbf{C}}^* \times K/M).$$

1.4. Function Spaces $\mathscr{Z}_*^{\mathscr{C}}(\mathfrak{a}_{\mathbf{C}}^* \times K/M)_W$ and $\mathscr{Z}_*^{\mathscr{A}}(\mathfrak{a}_{\mathbf{C}}^* \times K/M)_W$

We introduce the following two function spaces on $\mathfrak{a}_{\mathbf{C}}^* \times K/M$:

$$\mathscr{Z}_*^{\mathscr{C}}(\mathfrak{a}_{\mathbf{C}}^* \times K/M) = \{\Phi \in C^\infty(\mathfrak{a}_{\mathbf{C}}^* \times K/M);\ \Phi \text{ is holomorphic in } \lambda \in \mathfrak{a}_{\mathbf{C}}^* \text{ and } \|\Phi\|^{r',r} < \infty \text{ for } \forall r' \in \mathbf{N} \text{ and } \forall r \geq 0\}$$

and

$$\mathscr{Z}_*^{\mathscr{A}}(\mathfrak{a}_{\mathbf{C}}^* \times K/M) = \{\Phi \in \mathscr{Z}_*^{\mathscr{C}}(\mathfrak{a}_{\mathbf{C}}^* \times K/M);\ \|\Phi\|^{J,r} < \infty \text{ for } \forall J = (J_1, J_2) \in \tilde{\mathscr{O}}(\mathfrak{a}_{\mathbf{C}}^*) \times \tilde{\tilde{\mathscr{O}}}(\mathbf{C}) \text{ and } \forall r \geq 0\}$$

where, denoting by $\Delta_{K/M}$ the Laplacian on K/M, we put

$$\|\Phi\|^{r',r} = \sup_{\substack{|\mathrm{Im}\,\lambda| \leq r \\ k \in K}} |\Delta_{K/M}^{r'} \Phi(\lambda; kM)|(1 + |\mathrm{Re}\,\lambda|)^{r'},$$

$$\|\Phi\|^{J,r} = \sup_{\substack{|\mathrm{Im}\,\lambda| \leq r \\ k \in K}} |J_1(\lambda) J_2(\Delta_{K/M}) \Phi(\lambda; kM)|,$$

and $\tilde{\tilde{\mathscr{O}}}(\mathbf{C}) = \{J \in \mathscr{O}(\mathbf{C});\ J(z^2) \text{ is of infra-exponential type}\}$. Let V be a subset in $\mathfrak{a}_{\mathbf{C}}^*$. For $\Phi \in C(V \times K/M)$ we define its Poisson integral $\check{\Phi}$ by

$$\check{\Phi}(\lambda; x) = \int_K \Phi(\lambda; kM)\, e^{-(\sqrt{-1}\,\lambda + \rho)(H(x^{-1}k))}\, dk \qquad ((\lambda, x) \in V \times G).$$

Then we define the following closed subspaces $\mathscr{Z}_*^{\mathscr{C}}(\mathfrak{a}_{\mathbf{C}}^* \times K/M)_W$ and $\mathscr{Z}_*^{\mathscr{A}}(\mathfrak{a}_{\mathbf{C}}^* \times K/M)_W$ of $\mathscr{Z}_*^{\mathscr{C}}(\mathfrak{a}_{\mathbf{C}}^* \times K/M)$ and $\mathscr{Z}_*^{\mathscr{A}}(\mathfrak{a}_{\mathbf{C}}^* \times K/M)$, respectively:

$$\mathscr{Z}_*^{\mathscr{C}}(\mathfrak{a}_{\mathbf{C}}^* \times K/M)_W = \{\Phi \in \mathscr{Z}_*^{\mathscr{C}}(\mathfrak{a}_{\mathbf{C}}^* \times K/M);\ \check{\Phi}(w\lambda; x) = \check{\Phi}(\lambda\ x) \text{ for } \forall w \in W \text{ and } \forall (\lambda, x) \in \mathfrak{a}_{\mathbf{C}}^* \times G\},$$

$$\mathscr{Z}_*^{\mathscr{A}}(\mathfrak{a}_{\mathbf{C}}^* \times K/M)_W = \mathscr{Z}_*^{\mathscr{A}}(\mathfrak{a}_{\mathbf{C}}^* \times K/M) \cap \mathscr{Z}_*^{\mathscr{C}}(\mathfrak{a}_{\mathbf{C}}^* \times K/M)_W.$$

1.5. A Paley–Wiener Theorem

Theorem 1. *Let G be a connected real reductive linear Lie group and K its maximal compact subgroup. Then we have the following linear topological* *isomorphisms*:

$$\mathcal{F}: \mathcal{C}_*(G/K) \rightrightarrows \mathcal{Z}_*^{\mathcal{C}}(\mathfrak{a}_{\mathbf{C}}^* \times K/M)_W.$$

$$\mathcal{F}: \mathcal{A}_*(G/K) \rightrightarrows \mathcal{Z}_*^{\mathcal{A}}(\mathfrak{a}_{\mathbf{C}}^* \times K/M)_W.$$

Under a natural normalization of the Lebesgue measure $d\lambda$ *on* $\mathfrak{a}_{\mathbf{C}}^*$, *the inverse* $\mathcal{F}^{-1}$ *of* $\mathcal{F}$ *is given by*

$$\mathcal{F}^{-1}\Phi(xK) = \frac{1}{\# W}\int_{\mathfrak{a}^*} \check{\Phi}(\lambda; x)\frac{d\lambda}{|c(\lambda)|^2} \qquad (x \in G),$$

where $c(\lambda)$ *is Harish-Chandra's c-function for the class* 1 *principal series of* G *with respect to* K.

We can get the proof of the theorem thanks to Theorem 4.1.1 in Eguchi [1].

Remark. The distributions of exponential type on a Euclidean space were studied in Hasumi [1], Morimoto [1], Sebastiaõ e Silva [1] and others. As to the theory of Fourier transformation of hyperfunctions on a Euclidean space, there are works by Sato [1], Kawai [1], Morimoto [1], Zharinov [1], [2], Nagamachi [1], Saburi [1], Kaneko [2] and others. Our proof of the above Paley–Wiener theorem for $\mathcal{A}_*(G/K)$ is due to the work of Kaneko [2] in addition to that of Eguchi [1].

2. Ehrenpreis' Fundamental Principle

In this section we present the formulation of our Ehrenpreis' fundamental principle on a symmetric space.

2.1. Invariant Differential Operators

We denote by $\mathbf{D}(G/K)$ the ring of G-invariant differential operators on G/K. $\mathbf{D}(G/K)$ is commutative and there exists a generator system $\{\Delta_1, \ldots, \Delta_l\}$ of $\mathbf{D}(G/K)$:

$$\mathbf{D}(G/K) = \mathbf{C}[\Delta_1, \ldots, \Delta_l].$$

We put $D = (\Delta_1, \ldots, \Delta_l)$. For $P \in \mathbf{C}[D]$, we put

$$P(\lambda) = P(D)\, e^{(\sqrt{-1}\lambda - \rho)H(x^{-1})}\big|_{x=e} \qquad (\lambda \in \mathfrak{a}_{\mathbf{C}}^*).$$

For $\varphi \in \mathscr{C}_*(G/K)$ the following equality holds:

$$\mathscr{F}(P(D)\varphi)(\lambda; kM) = P(\lambda)F\varphi(\lambda; kM).$$

2.2. Presentation of the Problem

Let $\mathscr{M}$ be a finitely generated $\mathbf{D}(G/K)(=\mathbf{C}[D])$-module. For a ring R and positive integers p and q we denote by $\mathrm{Mat}(p, q; R)$ the set of matrices of size $p \times q$ with the entries in R. Under this notation, we take a free resolution of $\mathscr{M}$:

$$0 \leftarrow \mathscr{M} \leftarrow \mathbf{C}[D]^{m_0} \xleftarrow{{}^tA^0(D)} \mathbf{C}[D]^{m_1} \xleftarrow{{}^tA^1(D)} \mathbf{C}[D]^{m_2} \leftarrow \cdots,$$

where $A^i(D) = [A^i_{jk}(D)] \in \mathrm{Mat}(m_{i+1}, m_i; \mathbf{C}[D])$.

Let E denote $\mathscr{C}'_*(G/K)$ or $\mathscr{A}'_*(G/K)$. Then we get naturally the following complex:

$$0 \to \mathrm{Ker}(A(D)) \to E^{m_0} \xrightarrow{A^0(D)} E^{m_1} \xrightarrow{A^1(D)} E^{m_2} \to \cdots. \tag{$*$}$$

We put $A = A^0$.

Our aim is to represent the elements in $\mathrm{Ker}(A(D))$ by Fourier–Laplace integrals (Ehrenpreis' fundamental principle) and to show the exactness of this complex.

We denote the dual space of E by E' and put $Z_W = \mathscr{F}(E')$. To present our problem we take the Fourier–Laplace transformation of the dual complex of $(*)$:

$$0 \leftarrow Z_W^{m_0}/{}^tA(\lambda)Z_W^{m_1} \leftarrow Z_W^{m_0} \xleftarrow{{}^tA^0(\lambda)} Z_W^{m_1} \xleftarrow{{}^tA^1(\lambda)} Z_W^{m_2} \leftarrow \cdots,$$

where $A^i(\lambda) = [A^i_{jk}(\lambda)] \in \mathrm{Mat}(m_{i+1}, m_i; \mathbf{C}[\lambda])$. We put

$$V_A = \mathrm{supp}(\mathscr{O}^{m_0}_{\mathfrak{a}^*_{\mathbf{C}}}/{}^tA(\lambda)\mathscr{O}^{m_1}_{\mathfrak{a}^*_{\mathbf{C}}}) = \{\lambda \in \mathfrak{a}^*_{\mathbf{C}};\ \mathrm{rank}\, A(\lambda) < m_0\},$$

where $\mathscr{O}_{\mathfrak{a}^*_{\mathbf{C}}}$ denotes the sheaf of holomorphic functions on $\mathfrak{a}^*_{\mathbf{C}}$. There exists a multiplicity variety (with a Noetherian operator) $\{(d_n(\lambda, \partial_\lambda), V_n)\}_{1 \le n \le N}$ for $A(\lambda)$ (Ehrenpreis [2], Palamodov [1] and Björk [1]), where

$$\partial_\lambda = \left(\frac{\partial}{\partial \lambda_1}, \ldots, \frac{\partial}{\partial \lambda_l}\right),$$

$$d_n(\lambda, \partial_\lambda) = (d^1_n(\lambda, \partial_\lambda), \ldots, d^{m_0}_n(\lambda, \partial_\lambda)) \in (\mathbf{C}[\lambda][\partial_\lambda])^{m_0}$$

and V_n are algebraic varieties in $\mathfrak{a}^*_{\mathbf{C}}$. A multiplicity variety

$\{(d_n(\lambda,\partial_\lambda),V_n)\}_{1\le n\le N}$ for $A(\lambda)$ is characterized as follows:

$$V_A=V_1\cup\cdots\cup V_N \quad \text{and, for } Q(\lambda)\in C[\lambda]^{m_0},$$

$$Q(\lambda)\in {}^tA(\lambda)C[\lambda]^{m_1}\Leftrightarrow\left(\sum_{m=1}^{m_0} d_n^m(\lambda,\partial_\lambda)Q_m(\lambda)\right)\Bigg|_{V_n}=0 \quad \text{for } n=1,\ldots,N.$$

Then our conjecture on Ehrenpreis' fundamental principle on a symmetric space G/K is stated as follows:

$$\mathrm{FP}(\mathscr{C}'_*(G/K);\mathbf{D}(G/K)):$$

$$\{u\in\mathscr{C}'_*(G/K)^{m_0};\ A(D)u=0\}$$

$$=\Bigg\{\sum_{n=1}^{N}\int_{V_n\times K/M} d_n(\lambda,\partial_\lambda)^*(\Delta_{K/M})^r(1+|\lambda|)^r\,e(\lambda;x^{-1}k)\,d\mu_n(\lambda;kM);$$

$$d\mu_n \text{ are bounded measures on } V_n\times K/M \text{ and}$$

$$\mathrm{supp}(\mu_n)\subset(V_n\cap(\mathfrak{a}^*\times\sqrt{-1}\mathfrak{a}_r^*))\times K/M \text{ for some } r\ge 0\Bigg\},$$

where $d_n(\lambda,\partial_\lambda)^*$ denote the adjoint operators of $d_n(\lambda,\partial_\lambda)$, $\mathfrak{a}_r^*=\{H\in\mathfrak{a}^*;\ \langle H,H\rangle^{1/2}\le r\}$ and we put $e(\lambda;g)=e^{(\sqrt{-1}\lambda-\rho)H(g)}$ for $(\lambda,g)\in\mathfrak{a}_{\mathbf{C}}^*\times G$.

$$\mathrm{FP}(\mathscr{A}'_*(G/K);\mathbf{D}(G/K)):$$

$$\{u\in\mathscr{A}'_*(G/K)^{m_0};\ A(D)u=0\}$$

$$=\Bigg\{\sum_{n=0}^{N}\int_{V_n\times K/M} d_n(\lambda,\partial_\lambda)^*J_1(\lambda)J_2(\Delta_{K/M})\,e(\lambda;x^{-1}k)\,d\mu_n(\lambda;kM);$$

$$J_1\in\tilde{\mathscr{O}}(\mathfrak{a}_{\mathbf{C}}^*),\ J_2\in\tilde{\tilde{\mathscr{O}}}(\mathbf{C}),\ d\mu_n \text{ are bounded measures on } V_n\times K/M$$

$$\text{and } \mathrm{supp}(\mu_n)\subset(V_n\cap(\mathfrak{a}^*\times\sqrt{-1}\mathfrak{a}_r^*))\times K/M \text{ for some } r\ge 0\Bigg\}.$$

2.3. Current Result

Theorem 2 (Ehrenpreis' fundamental principle on a symmetric space). *Let G be a Cartesian product of connected real reductive linear Lie groups of real rank* 1 *and K its maximal compact subgroup, then* $\mathrm{FP}(\mathscr{C}'_*(G/K);\mathbf{D}(G/K))$ *and* $\mathrm{FP}(\mathscr{A}'_*(G/K);\mathbf{D}(G/K))$ *hold.*

Corollary 1. *If G/K satisfies the assumption in Theorem* 1, *then $\mathscr{C}'_*(G/K)$ and $\mathscr{A}'_*(G/K)$ are injective $\mathbf{D}(G/K)$-modules*:

$$\mathscr{C}'_*(G/K)^{m_0} \xrightarrow{A^0(D)} \mathscr{C}'_*(G/K)^{m_1} \xrightarrow{A^1(D)} \mathscr{C}'_*(G/K)^{m_2} \qquad \text{(exact)},$$

$$\mathscr{A}'_*(G/K)^{m_0} \xrightarrow{A^0(D)} \mathscr{A}'_*(G/K)^{m_1} \xrightarrow{A^1(D)} \mathscr{A}'_*(G/K)^{m_2} \qquad \text{(exact)}.$$

In particular, for a nonzero $P(D) \in \mathbf{D}(G/K)$, we have

$$\mathscr{C}'_*(G/K) \xrightarrow{P(D)} \mathscr{C}'_*(G/K) \to 0 \qquad \text{(exact)},$$

$$\mathscr{A}'_*(G/K) \xrightarrow{P(D)} \mathscr{A}'_*(G/K) \to 0 \qquad \text{(exact)}.$$

The solvability of a single equation easily follows from a theorem of Paley–Wiener type and therefore the above maps defined by $P(D)$ are surjective under the assumption that G is a connected reductive linear group. It was already proved in other function spaces by Helgason [1] and Eguchi [2] (cf. Helgason [3]).

An operator $A(D) \in \mathrm{Mat}(m_1, m_0; \mathbf{D}(G/K))$ is said to be *hypoelliptic*, if $V_A \cap \mathfrak{a}^*$ is compact. For hypoelliptic operators we have the following

Corollary 2. *Suppose that G/K satisfies the assumption in Theorem* 2 *and $A(D) \in \mathrm{Mat}(m_1, m_0; \mathbf{D}(G/K))$ is hypoelliptic. If $u \in \mathscr{C}'_*(G/K)^{m_0}$ (or $\in \mathscr{A}'_*(G/K)^{m_0}$) satisfies the equation $A(D)u = 0$, then u is real analytic. In particular, if G is commutative, then u is extended as an entire holomorphic function.*

Example. Let G/K be the Euclidean space $\mathbf{R}^{1+l}$, $x = (x_0, \ldots, x_l)$ its coordinate system and

$$A(D) = \frac{\partial}{\partial x_0} - \sum_{i=1}^{l} \frac{\partial^2}{\partial x_i^2}.$$

If a locally summable function u on $\mathbf{R}^{1+l}$ satisfies the conditions

$$A(D)u = 0 \text{ and } |u(x)| \le r\, e^{r|x|} (x \in \mathbf{R}^{1+l}) \text{ for some } r > 0$$

in the distribution sense, then u is extended as an entire holomorphic function on $\mathbf{C}^{1+l}$.

3. Crucial Point in the Proof of Theorem 2

A crucial point in the proof of Theorem 2 is to show that the quotient space $Z_W^{m_0}/{}^tA(\lambda)Z_W^{m_1}$ is topologically isomorphic to a certain function space on $V_A\times K/M$. In this section we define the function spaces and give the isomorphism between the two spaces. We treat mainly the case of distributions of exponential type.

3.1. Function Space $\mathscr{L}_*^{\mathscr{C}}(V_A\times K/M)_W$

We define a function space on $V_A\times K/M$:

$\mathscr{L}_*^{\mathscr{C}}(V_A\times K/M)_W=\{\Psi=(\Psi_1,\dots,\Psi_N);\ \Psi_n$ are elements in $C(V_n\times K/M)$,

respectively, with the growth condition $C(RD)$

and there exists $\Phi\in C^\infty(\mathfrak{a}_{\mathbf{C}}^*\times K/M)^{m_0}$ which

satisfies $C(\mathscr{O})$, $C(W)$ and $d\Phi=\Psi\}$,

where $d=\{(d_n, V_n)\}_{1\le n\le N}$ is a Noetherian operator for $A(\lambda)$:

$$d\Phi=(d_1\Phi|_{V_1},\dots,d_N\Phi|_{V_N}),$$

and the conditions $C(\mathscr{O})$, $C(W)$, $C(RD)$ are as follows:

$C(\mathscr{O})$: Φ_m are holomorphic in $\lambda\in\mathfrak{a}_{\mathbf{C}}^*$.

$C(W)$: $\check{\Phi}(w\lambda;x)=\check{\Phi}(\lambda;x)$ for $\forall w\in W$ and $\forall(\lambda,x)\in\mathfrak{a}_{\mathbf{C}}^*\times G$,

$C(RD)$: $\|\Psi\|^{r',r}<\infty$ for $\forall r'\in\mathbf{N}$ and $\forall r\ge 0$, where

$$\|\Psi\|^{r',r}=\sum_n \sup_{\substack{(\lambda,k)\in V_n\times K\\ |\mathrm{Im}\,\lambda|\le r}}(1+|\lambda|)^{r'}|\Delta_{K/M}^{r'}\Psi_n(\lambda;kM)|.$$

The claim we have to show is that a Noetherian operator for $A(\lambda)$ gives the topological isomorphism

$$\mathscr{L}_*^{\mathscr{C}}(\mathfrak{a}_{\mathbf{C}}^*\times K/M)_W^{m_0}/{}^tA(\lambda)\mathscr{L}_*^{\mathscr{C}}(\mathfrak{a}_{\mathbf{C}}^*\times K/M)_W^{m_1}\overset{d}{\rightarrow}\mathscr{L}_*^{\mathscr{C}}(V_A\times K/M)_W \tag{2*}$$

3.2. Fourier Expansions of Functions in $C^\infty(K/M)$ and $\mathscr{A}(G/K)$

To construct the inverse of the above map we need some results about the Fourier expansion of a function on K/M. Let $\hat{K}$ denote the set of equivalence classes of irreducible unitary representations of K. For $\delta \in \hat{K}$ we take a Hilbert space V_δ which realizes a representation of K belonging to δ. The dimensions $d(\delta)$ of V_δ are finite and there exists a constant $C>0$ such that $d(\delta) \leq C(1+|\delta|)^C$. Here we fix a norm on the $\mathfrak{t}_{\mathbb{C}}^*$, the complexification of the dual space $\mathfrak{t}^*$ of the Cartan subalgebra $\mathfrak{t}$ of $\mathfrak{k}$, and $|\delta|$ denotes the norm of the highest weight of the representation belonging to $\delta \in \hat{K}$. We put

$$V_\delta^M = \{v \in V_\delta;\ \delta(m)v = v \text{ for } \forall m \in M\},$$
$$l(\delta) = \dim V_\delta^M \text{ and } \hat{K}_M = \{\delta \in \hat{K};\ l(\delta) > 0\}.$$

For each $\delta \in \hat{K}$, we take an orthonormal basis $\{v_1, \ldots, v_{d(\delta)}\}$ so that $\{v_1, \ldots, v_{l(\delta)}\} \subset V_\delta^M$ and put

$$f_{ij}^\delta(kM) = (v_j, \delta(k)v_i) \qquad (1 \leq i \leq l(\delta),\ 1 \leq j \leq d(\delta)),$$
$$f^\delta = [f_{ij}^\delta]_{\substack{1 \leq i \leq l(\delta) \\ 1 \leq j \leq d(\delta)}},$$

where (,) denotes the inner product of V_δ. Then the following set constitutes a complete orthonormal basis of $L^2(K/M)$:

$$\{d(\delta)^{1/2} f_{ij};\ \delta \in \hat{K}_M,\ 1 \leq i \leq l(\delta) \text{ and } 1 \leq j \leq d(\delta)\}.$$

For $\Phi \in L^2(K/M)$, and $\delta \in \hat{K}_M$, we define

$$\mathcal{F}_{K/M}^\delta \Phi = \Phi^\delta = \int_K \Phi(kM)\overline{f^\delta(kM)}\,dk \qquad \text{and } \mathcal{F}_{K/M}\Phi = \{\Phi^\delta\}_{\delta \in \hat{K}_M},$$

where $\overline{f^\delta(kM)}$ denote the complex conjugates of $f^\delta(kM)$. We call $\mathcal{F}_{K/M}\Phi$ the Fourier coefficients of Φ. The Fourier coefficients of C^∞ functions on K/M are characterized as follows:

$$\Phi \in C^\infty(K/M) \Leftrightarrow |||\mathcal{F}_{K/M}\Phi|||^{r'} = \sup_{\delta \in \hat{K}_M} |\Phi^\delta|(1+|\delta|)^{r'} < \infty \qquad \text{for } \forall r' \geq 0.$$

There is also a characterization of the Fourier coefficients of real analytic functions on K/M:

$$\Phi \in \mathcal{A}(K/M) \Leftrightarrow |||\mathcal{F}_{K/M}\Phi|||^{J} = \sup_{\delta \in \hat{K}_M} |\Phi^\delta|\,|J(\delta)| < \infty$$
$$\text{for } \forall J \in \tilde{C}(\mathfrak{t}_{\mathbb{C}}^*), \text{ where } \tilde{C}(\mathfrak{t}_{\mathbb{C}}^*)$$
$$= \{J \in C(\mathfrak{t}_{\mathbb{C}}^*);\ J \text{ is of infra-exponential type}\}.$$

3.3. Function Spaces $H^{\mathscr{C}}(\mathfrak{a}_{\mathbf{C}}^*)_W$ and $H^{\mathscr{C}}(V_A)_W$

Using the above characterization of Fourier expansions of the functions on K/M, we can realize the function space on $\mathfrak{a}_{\mathbf{C}}^* \times K/M$, $\mathscr{Z}_*^{\mathscr{C}}(\mathfrak{a}_{\mathbf{C}}^* \times K/M)_W$, as (a Cartesian product of) that on $\mathfrak{a}_{\mathbf{C}}^*$. We can also realize the function space on $V_A \times K/M$, $\mathscr{Z}_*^{\mathscr{C}}(V_A \times K/M)_W$, as (a Cartesian product of) that on V_A.

We first start with the characterization of the Fourier coefficients of the functions in $\mathscr{Z}_*^{\mathscr{C}}(\mathfrak{a}_{\mathbf{C}}^* \times K/M)_W$. Let V be a subset in $\mathfrak{a}_{\mathbf{C}}^*$. For $\Phi \in C(V \times K/M)$ and $\delta \in \hat{K}_M$, we define

$$\mathscr{F}_{K/M}^{\delta}\Phi(\lambda) = \Phi^{\delta}(\lambda) = \int_K \Phi(\lambda : kM)\overline{f^{\delta}(kM)}\, dk \qquad (\lambda \in V)$$

and

$$\mathscr{F}_{K/M}\Phi = \{\Phi^{\delta}\} = \{\Phi^{\delta}\}_{\delta \in \hat{K}_M}.$$

We can identify Φ with $\mathscr{F}_{K/M}\Phi$. Then we have the following characterization of the Fourier coefficients of the functions in $\mathscr{Z}_*^{\mathscr{C}}(\mathfrak{a}_{\mathbf{C}}^* \times K/M)_W$:

$$\mathscr{Z}_*^{\mathscr{C}}(\mathfrak{a}_{\mathbf{C}}^* \times K/M)_W \overset{\mathscr{F}_{K/M}}{\rightrightarrows} H^{\mathscr{C}}(\mathfrak{a}_{\mathbf{C}}^*)_W = \{\{\Phi^{\delta}\};$$
$$\Phi^{\delta} \text{ satisfy the conditions } \mathrm{F}(\mathscr{O}), \mathrm{F}(W) \text{ and } \mathrm{F}(RD)\}.$$

Here the conditions $\mathrm{F}(\mathscr{O})$, $\mathrm{F}(W)$ and $\mathrm{F}(RD)$ are as follows:

$\mathrm{F}(\mathscr{O})$: $\Phi^{\delta} \in \mathrm{Mat}(l(\delta), d(\delta); \mathscr{O}(\mathfrak{a}_{\mathbf{C}}^*))$,

$\mathrm{F}(W)$: For any $\delta \in \hat{K}_M Q_{\check{\delta}}^{-1}\Phi \in \mathscr{O}(\mathfrak{a}_{\mathbf{C}}^*)$ and

$$Q_{\check{\delta}}^{-1}(w\lambda)\Phi^{\delta}(w\lambda) = Q_{\check{\delta}}^{-1}(\lambda)\Phi^{\delta}(\lambda) \qquad \text{for } \forall w \in W \text{ and } \forall \lambda \in \mathfrak{a}_{\mathbf{C}}^*,$$

where $\check{\delta}$ denotes the contragradient representation of δ, and $Q_{\check{\delta}} \in \mathrm{Mat}(l(\delta), l(\delta); \mathbf{C}[\lambda])$ is the matrix defined in Kostant [1] (also cf. Kostant [2] and Helgason [2]),

$$\mathrm{F}(RD): |||\{\Phi^{\delta}\}|||^{r',r} = \sup_{|\mathrm{Im}\,\lambda| \le r} \sup_{\delta \in \hat{K}_m} |\Phi^{\delta}(\lambda)|(1+|\delta|)^{r'}(1+|\lambda|)^{r'} < \infty$$
$$\text{for } \forall r' \ge 0 \text{ and } \forall r \ge 0,$$

with

$$|\Phi^{\delta}(\lambda)| = \max_{\substack{1 \le i \le l(\delta) \\ 1 \le j \le d(\delta)}} |\Phi_{ij}^{\delta}(\lambda)|.$$

Next we go on to the characterization of the Fourier coefficients of the functions in $\mathscr{Z}_*^{\mathscr{C}}(V_A \times K/M)_W$. For $\Psi = (\Psi_1, \ldots, \Psi_N) \in \mathscr{Z}_*^{\mathscr{C}}(V_A \times K/M)_W$, we define

$$\mathscr{F}^{\delta}_{K/M}\Psi = \Psi^{\delta} = (\Psi^{\delta}_1, \ldots, \Psi^{\delta}_N) \text{ and } \mathscr{F}_{K/M}\Psi = \{\Psi^{\delta}\}.$$

We also define the Fourier coefficients of a function Φ in $C^{\infty}(\mathfrak{a}^*_{\mathbf{C}} \times K/M)^{m_0}$:

$$\mathscr{F}^{\delta}_{K/M}\Phi = \Phi^{\delta} = (\Phi^{\delta}_1, \ldots, \Phi^{\delta}_{m_0}) \text{ and } \mathscr{F}_{K/M}\Phi = \{\Phi^{\delta}\}.$$

For $\Phi = \{\Phi^{\delta}\} \in \prod_{\delta \in \hat{K}_M} \mathrm{Mat}(l(\delta), d(\delta); \mathcal{O}(\mathfrak{a}^*_{\mathbf{C}}))^{m_0}$ and a Noetherian operator d for $A(\lambda)$ we define

$$d\Phi = \{d\Phi^{\delta}\} \in \prod_{\delta \in \hat{K}_M} \mathrm{Mat}(l(\delta), d(\delta); \mathcal{O}(\mathfrak{a}^*_{\mathbf{C}}))^{N},$$

where for $\Omega = (\Omega^1, \ldots, \Omega^{m_0}) \in \mathrm{Mat}(l(\delta), d(\delta); \mathcal{O}(\mathfrak{a}^*_{\mathbf{C}}))^{m_0}$ we put

$$d\Omega = (d_1\Omega, \ldots, d_N\Omega) \in \mathrm{Mat}(l(\delta), d(\delta); \mathcal{O}(\mathfrak{a}^*_{\mathbf{C}}))^{N}$$

with

$$d_n\Omega = \left[\sum_{m=1}^{m_0} d^m_n \Omega^m_{ij}|_{V_n} \right]_{\substack{1 \le i \le l(\delta) \\ 1 \le j \le d(\delta)}}.$$

Then $\mathscr{F}_{K/M}$ commute with the Noetherian operator d for $A(\lambda)$:

$$d\mathscr{F}_{K/M}\Phi = \mathscr{F}_{K/M}\, d\Phi \qquad (\Phi \in C^{\infty}(\mathfrak{a}^*_{\mathbf{C}} \times K/M)^{m_0}).$$

Now we define a function space on V_A:

$$H^{\mathscr{C}}(V_A)_W = \{\{\Psi^{\delta} = (\Psi^{\delta}_1, \ldots, \Psi^{\delta}_N)\};\ \Psi^{\delta}_n \in \mathrm{Mat}(l(\delta), d(\delta); C(V_n)),$$

there exist $\Phi^{\delta} \in \mathrm{Mat}(l(\delta), d(\delta); \mathcal{O}(\mathfrak{a}^*_{\mathbf{C}}))^{m_0}$ such that the conditions CF(W), CF(RD) and $d\Phi^{\delta} = \Psi^{\delta}$ hold},

where the conditions CF(W) and CF(RD) are as follows:

CF(W): For $\forall \delta \in \hat{K}_M$

$$Q^{-1}_{\check{\delta}}\Phi \in \mathrm{Mat}(l(\delta), d(\delta); \mathcal{O}(\mathfrak{a}^*_{\mathbf{C}})),$$

$$Q^{-1}_{\check{\delta}}(w\lambda)\Phi^{\delta}(w\lambda) = Q^{-1}_{\check{\delta}}(\lambda)\Phi^{\delta}(\lambda) \qquad \text{for } \forall w \in W \text{ and } \forall \lambda \in \mathfrak{a}^*_{\mathbf{C}},$$

CF(RD): $|||\{\Psi^{\delta}\}|||^{r',r}$

$$= \sum_n \sup_{\substack{\lambda \in U \\ |\mathrm{Im}\,\lambda| \le r}} \sup_{\delta \in \hat{K}_m} |\Psi^{\delta}_n(\lambda)|(1+|\delta|)^{r'}(1+|\lambda|)^{r'} < \infty$$

for $\forall r' \ge 0$ and $\forall r \ge 0$.

Then we have the following topological isomorphism:

$$\mathscr{Z}_*^{\mathscr{C}}(V_A\times K/M)_W \xrightarrow[\sim]{\mathscr{F}_{K/M}} H^{\mathscr{C}}(V_A)_W.$$

3.4. Construction of d^{-1}

By virtue of the study in the preceding section, we have the following commutative diagram:

$$\begin{array}{ccc} \mathscr{Z}_*^{\mathscr{C}}(\mathfrak{a}_{\mathbf{C}}^*\times K/M)_W^{m_0}/{}^tA(\lambda)\mathscr{Z}_*^{\mathscr{C}}(\mathfrak{a}_{\mathbf{C}}^*\times K/M)_W^{m_1} & \xrightarrow{d} & \mathscr{Z}_*^{\mathscr{C}}(V_A\times K/M)_W \\ \mathscr{F}_{K/M}\ \wr\downarrow & & \wr\downarrow\ \mathscr{F}_{K/M} \\ H^{\mathscr{C}}(\mathfrak{a}_{\mathbf{C}}^*)_W^{m_0}/{}^tA(\lambda)H^{\mathscr{C}}(\mathfrak{a}_{\mathbf{C}}^*)_W^{m_1} & \xrightarrow[d]{} & H^{\mathscr{C}}(V_A)_W \end{array}$$

Here for $\Phi=\{\Phi^\delta=(\Phi^{\delta,1},\ldots,\Phi^{\delta,m_1})\}\in H^c(\mathfrak{a}_{\mathbf{C}}^*)_W^{m_1}$ we define

$${}^tA(\lambda)\Phi=\{({}^tA(\lambda)\Phi^\delta)^1,\ldots,({}^tA(\lambda)\Phi^\delta)^{m_0}\},$$

with

$$({}^tA(\lambda)\Phi^\delta)^m=\left[\sum_{\mu=1}^{m_1} a_{\mu m}(\lambda)\Phi_{ij}^{\delta,\mu}\right]_{\substack{1\le i\le l(\delta)\\ 1\le j\le d(\delta)}}.$$

Therefore to construct d^{-1} of (2∗), it is sufficient to construct d^{-1} of the diagram:

$$H^c(\mathfrak{a}_{\mathbf{C}}^*)_W^{m_0}/{}^tA(\lambda)H^c(\mathfrak{a}_{\mathbf{C}}^*)_W^{m_1} \overset{d}{\to} H^c(V_A)_W. \tag{3∗}$$

Our method to construct d^{-1} of (3∗) follows that in Ehrenpreis [2]. Here we briefly explain it. Suppose G satisfies the hypothesis in Theorem 2. Then for $\Psi\in H^c(V_A)_W$, by a technique similar to Ehrenpreis [2], we can construct $\Omega=\{\Omega^\delta\}\in\prod_{\delta\in\hat{K}_M}\mathrm{Mat}(l(\delta),d(\delta);\mathscr{O}(\mathfrak{a}_{\mathbf{C}}^*))^{m_0}$ such that the following conditions hold:

$$d\Omega=\Psi,$$

$$\mathrm{CF}(W1)\colon Q_\delta^{-1}\Omega^\delta\in\mathrm{Mat}(l(\delta),d(\delta);\mathscr{O}(\mathfrak{a}_{\mathbf{C}}^*))^{m_0},$$

and for any $r', r\ge 0$, there exists some constant $C>0$ such that

$$|||\Omega|||^{r',r}\le C|||\Psi|||^{C,C},$$

where the constant C depends only on r', r and $A(\lambda)$. Yet, in general, the above Ω does not satisfy the condition:

$$\mathrm{CF}(W2)\colon Q_\delta^{-1}(w\lambda)\Omega^\delta(w\lambda)=Q_\delta^{-1}(\lambda)\Omega^\delta(\lambda)$$

for $\forall w\in W$, $\forall\lambda\in\mathfrak{a}_{\mathbf{C}}^*$ and $\forall\delta\in\hat{K}_M$.

Therefore we put

$$\Phi^{\delta}(\lambda)=\frac{1}{\# W}\sum_{w\in W} Q_{\check{\delta}}(\lambda)Q_{\check{\delta}}^{-1}(w\lambda)\Phi^{\delta}(w\lambda) \qquad (\lambda\in\mathfrak{a}_{\mathbf{C}}^{*},\ \delta\in\hat{K}_M).$$

Then $\Phi=\{\Phi^{\delta}\}$ satisfies the condition CF($W2$) in addition to all the above ones which Ω satisfies. Hence Φ is what we required:

$$\Phi+{}^{t}A(\lambda)H^{\mathscr{C}}(\mathfrak{a}_{\mathbf{C}}^{*})_{W}^{m_1}=d^{-1}\Psi.$$

In the case of hyperfunctions of exponential type we can also prove Theorem 2 by the same way as in the case of distributions of exponential type if we replace the conditions C(RD), F(RD) and CF(RD) by the following conditions C(ED), F(ED) and CF(ED), respectively:

$$\mathrm{C}(ED):\|\Psi\|^{J,r}=\sum_{n}\sup_{\substack{(\lambda,k)\in V_n\times K\\ |\mathrm{Im}\,\lambda|}}|J_1(\lambda)J_2(\Delta_{K/M})\Psi_n(\lambda;kM)|<\infty$$

for $\forall J=(J_1,J_2)\in\tilde{\mathscr{O}}(\mathfrak{a}_{\mathbf{C}}^{*})\times\tilde{\tilde{\mathscr{O}}}(\mathbf{C})$ and $\forall r\geq 0$.

$$\mathrm{F}(ED):\ |||\{\Psi^{\delta}\}|||^{J,r}=\sup_{|\mathrm{Im}\,\lambda|\leq r}\ \sup_{\delta\in\hat{K}_M}|\Phi^{\delta}(\lambda)J_2(\delta)J_1(\lambda)|<\infty$$

for $\forall J=(J_1,J_2)\in\tilde{\mathscr{O}}(\mathfrak{a}_{\mathbf{C}}^{*})\times\tilde{C}(\mathfrak{t}_{\mathbf{C}}^{*})$ and $\forall r\geq 0$.

$$\mathrm{CF}(ED):|||\{\Psi^{\delta}\}|||^{J,r}=\sum_{n}\sup_{\substack{\lambda\in V_n\\ |\mathrm{Im}\,\lambda|\leq r}}\ \sup_{\delta\in\hat{K}_M}|\Psi^{\delta}(\lambda)J_2(\delta)J_1(\lambda)|<\infty$$

for $\forall J=(J_1,J_2)\in\tilde{\mathscr{O}}(\mathfrak{a}_{\mathbf{C}}^{*})\times\tilde{C}(\mathfrak{t}_{\mathbf{C}}^{*})$ and $\forall r\geq 0$.

Remark. In Theorem 2 we made a hypothesis that G is a Cartesian product of connected real reductive linear Lie groups of real rank 1. If G is of this type, then $l(\delta)=1$ for any $\delta\in\hat{K}_M$. Hence $Q_{\check{\delta}}^{-1}(\lambda)$ are scalar in this case. If $Q_{\check{\delta}}^{-1}(\lambda)$ are not diagonalizable matrices in Mat($l(\delta)$, $d(\delta)$; $\mathscr{O}(\mathfrak{a}_{\mathbf{C}}^{*})$), then it is difficult to construct the above Ω. The crucial point in the construction is to control the growth condition for Ω^{δ} when $|\delta|$ tends to infinity retaining the conditions $d\Omega=\Psi$ and CF($W1$). This is the reason why we need the hypothesis in Theorem 1. Hence we can show that our Ehrenpreis' fundamental principle holds for K-finite distributions and hyperfunctions of exponential type on G/K under the assumption that G is an arbitrary connected real reductive linear Lie group.

References

J.-E. Björk, [1] *Rings of Differential Operators*, North-Holland, Amsterdam, 1979.

M. Eguchi, [1] Asymptotic expansions of Eisenstein integrals and Fourier transform on symmetric spaces, *J. Funct. Anal.* **34** (1979) 167–216.

[2] An application of topological Paley–Wiener theorems to invariant differential equations on symmetric spaces, *Analyse Harmonique sur les Groupes de Lie II*, Lect. Notes in Math., Springer, **739** (1979) 193–206.

L. Ehrenpreis, [1] A fundamental principle for systems of linear differential equations with constant coefficients and some applications, *Proc. Int. symp. on linear spaces, Jerusalem*, 1960.

[2] *Fourier Analysis in Several Complex Variables*, Wiley–Interscience, New York, 1970.

M. Flensted-Jensen, [1] *Analysis on non-Riemannian symmetric spaces*, Regional conference series in math. 61, AMS, 1986.

I. M. Gel'fand and G. E. Shilov, [1] *Generalized Functions 2*, Academic Press, New York, 1968.

Harish-Chandra, [1] Spherical functions on a semisimple Lie group I, *Amer. J. Math.* **80** (1958) 241–310.

[2] Spherical functions on a semisimple Lie group II, *Amer. J. Math.* **80** (1958) 553–613.

[3] Invariant eigendistributions on a semisimple Lie group, *Trans. Amer. Math. Soc.* **119** (1965) 457–508.

M. Hashizume, A. Kowata, K. Minemura and K. Okamoto, [1] An integral representation of an eigenfunction of the Laplacian on the Euclidean space, *Hiroshima Math. J.* **2** (1972) 535–545.

M. Hashizume, K. Minemura and K. Okamoto, Harmonic functions on hermitian hyperbolic spaces, *Hiroshima Math. J.* **3** (1973) 81–108.

M. Hasumi, [1] Note on the n-dimensional tempered ultra-distributions, *Tôhoku Math. J.* **13** (1961) 94–104.

S. Helgason, [1] The surjectivity of invariant differential operators on symmetric spaces I, *Ann. of Math.* **98** (1973) 451–479.

[2] A duality for symmetric spaces with applications to group representations, II. Differential equations and eigenspace representations, *Advan. Math.* **22** (1976) 187–219.

[3] Invariant differential equations on homogeneous manifolds, *Bull. Amer. Math. Soc.* **83** (1977) 751–774.

[4] *Differential Geometry, Lie Groups, and Symmetric Spaces*, Academic Press, New York, 1978.

[5] *Groups and Geometric Analysis*, Academic Press, New York, 1984.

T. Inoue, K. Okamoto and M. Tanaka, [1] An integral representation of an eigenfunction of invariant differential operators on a symmetric space, *Hiroshima Math. J.* **4** (1974) 413–419.

M. Kashiwara, A. Kowata, K. Minemura, K. Okamoto, T. Oshima and M. Tanaka, [1] Eigenfunctions of invariant differential operators on a symmetric space, *Ann. Math.* **107** (1978) 1–39.

A. Kaneko, [1] Fundamental principle and extension of solutions of linear differential equations with constant coefficients, in *Hyperfunctions and Pseudo-Differential Equations*, Lect. Notes in Math., **287**, Springer, Berlin, 1973, pp. 122–134.

[2] *Introduction to Hyperfunctions II*, Press Univ. Tokyo, Tokyo, 1982 (in Japanese).

T. Kawai, [1] On the theory of Fourier hyperfunctions and its application to partial differential equations with constant coefficients, *J. Fac. Sci. Univ. Tokyo Sect. IA* **17** (1970) 467–517.

B. Kostant, [1] On the existence and irreducibility of certain series of representations, *Bull. Amer. Math. Soc.* **75** (1969) 627–642.

[2] On the existence and irreducibility of certain series of representations, in *Lie Groups and Their Representations* (I. M. Gel'fand, ed.), Halsted, New York, 1975, pp. 231–329.

M. Morimoto, [1] *Fourier transform and hyperfunctions*, Sem. Notes in Math. **2** (1978), Sophia Univ. (in Japanese).

S. Nagamachi, [1] The theory of vector valued Fourier hyperfunctions of mixed type I, II, *Publ. RIMS* **17** (1981) 25-93.

T. Oshima, [1] A proof of Ehrenpreis' fundamental principle in hyperfunctions, *Proc. Japan Acad.* **50** (1974) 16-18.

T. Oshima, Y. Saburi and M. Wakayama, [1] Paley-Wiener theorems on a symmetric space and its application, in preparation. [2] Ehrenpreis' fundamental principle on a symmetric space, in preparation.

T. Oshima and J. Sekiguchi, [1] Eigenspaces of invariant differential operators on an affine symmetric space, *Invent. Math.* **57** (1980) 1-81.

V. P. Palamodov, [1] *Linear Differential Operators with Constant Coefficients*, Nauk, Moscow, 1967, (in Russian). English edition, Springer-Verlag, Berlin, 1970.

Y. Saburi, [1] Fundamental properties of modified Fourier hyperfunctions, *Tokyo J. Math.* **8** (1985) 231-273.

M. Sato, [1] Theory of hyperfunctions, *Sûgaku* **10** (1958) 1-27 (in Japanese).

H. Schlichtkrull, [1] *Hyperfunctions and Harmonic Analysis on Symmetric Spaces*, Birkhauser, Boston, 1984.

J. Sebastiaõ e Silva, [1] Les fonctions analytiques comme ultra-distributions dans le calcul opérationnel, *Math. Ann.* **136** (1958) 58-96.

M. Sugiura, [1] Fourier series of smooth functions on a compact Lie group, *Osaka J. Math.* **8** (1971) 33-47.

P. C. Trombi and V. S. Varadarajan, [1] Spherical transforms on semisimple Lie groups, *Ann. Math.* **94** (1971) 246-303.

N. R. Wallach, [1] Asymptotic expansion of generalized matrix entries of representations of reductive groups, in *Lie Group Representations I*, Lect. Notes in Math. **1024**, Springer, 1983, pp. 287-369.

[2] Kostant's P^γ and Q^γ matrices and intertwining operators, in *Proceedings of the symposium in pure mathematics of the American Math. Soc.*, Vol. XXVI. "Harmonic analysis on homogeneous spaces" ed. by Calvin C. Moore, 1973, pp. 269-273.

V. V. Zharinov, [1] The Laplace transform of Fourier hyperfunctions and other similar cases of analytic functionals I, *Teoret. Mat. Fiz.* **33** (1977) 291-309.

[2] The Laplace transform of Fourier hyperfunctions and other similar cases of analytic functionals II, *Teoret. Mat. Fiz.* **37** (1978) 12-29.

Resurgence, Quantized Canonical Transformations, and Multi-Instanton Expansions*

F. Pham
Department of Mathematics
Université de Nice
Nice, France

Quantum field theory and quantum mechanics often deal with divergent series, e.g., the so-called "perturbation expansions" (expansions in powers of the coupling constant) or the "semi-classical expansions" (expansions in powers of the Planck constant). The problems we are interested in deal with obstructions to Borel summability of such expansions. The works of Balian and Bloch [BB], followed by other physicists, have shown that such obstructions can be read in the classical mechanics of the problem, provided one takes into account the complex trajectories. The main clue in that connection is the existence of "instantons," or trajectories giving a finite action integral in an infinite time. While physicists such as Voros [V]

* A joint work with B. Candelpergher and C. Nosmas.
Reference [*]
B. Candelpergher, C. Nosmas, and F. Pham, *Résurgence et développements semi-classiques* (Séminaire semi-classique, Nice 1985–87), to appear.

Algebraic Analysis, Volume II

ISBN 0-12-400466-0

developed the ideas of Balian and Bloch, a powerful mathematical machinery was invented independently by Ecalle to study and master the obstructions to Borel summability of divergent series met in all kinds of mathematical problems. Because of its formal analogies with usual differential calculus, he named his machinery *alien differential calculus.* The general framework in which this machinery operates is the theory of *resurgent functions.* Among many other achievements, Ecalle's theory succeeded in proving the conjectures of Voros in [V] (cf. [E2] for a very short sketch; a more detailed version, based on Ecalle's scratchpads, will appear in [*]).

From the beginning of my acquaintance with Ecalle's theory, I have been intrigued by the question of its connection with Sato's *microdifferential calculus.* Although both theories can be said to have a common starting point (the local study of obstructions to analytic continuation in one variable, i.e., microfunctions of one variable) they seem to develop in completely different directions, Ecalle's point of view being one of *global* analytic continuation in *one* variable, whereas Sato (following Maslov [M], Egorov [Eg], Hörmander [H]) insists on covariance with respect to transformations which are defined *microlocally*, and involve several variables.

In relation with the above-mentioned physical problems, what puzzled me is that despite its success in proving Voros' conjectures, Ecalle's theory seemed to miss some aspects (e.g., the so-called uniform approximations of "semi-classical" physicists [BM]) where the microlocal point of view gives deep insight.

The main novelty of the present work is to make *quantized canonical transformations*, a notion named after Sato's "quantized contact transformations," enter (yet shyly) the world of resurgence. After having sketched some of the basic notions of resurgence theory (in an unorthodox fashion, more suited to our purpose), we shall use quantized canonical transformations to study a conjecture of Zinn-Justin [Z2], [Z3] on the building of *multi-instantons expansions* in quantum mechanics [Z1], reducing his conjecture to very natural "resurgence conjectures." This might be a first, very special example of a—yet to be created—theory of "alien differential calculus in several variables," covariant under canonical transformations.

Part 1 of the present article is a summary of the first chapters of [*]; part 2 will be reproduced in [*] as a last chapter.

1. Resurgence (A Radar Approach)

A very convenient feature of resurgence theory is the freedom to formulate any statement in either one of the two following "models," related to each other by Laplace transformation: the "multiplicative model," where the variable is designated by a Latin letter (say x), and the "convolutive model," where the corresponding Greek letter is used (say, ξ).

Our way of exposition will differ somewhat from that of Ecalle: instead of starting with the concept of resurgent function, we shall introduce first what we call *extended resurgent functions*, a concept inspired by the ideas of Balian and Bloch. This might seem a superficial difference of presentation, but is no longer so in the "relative" case (resurgent functions depending on parameters), when "confluence of singularities" occurs.

1.1 Sectorial Germs at Infinity and Laplace Transformation

Let S^1 be the circle at infinity in $\mathbf{C}$ (the set of "radar directions," to borrow an image of Voros). The direction of $0z$, for $z \in \mathbf{C}^* = \mathbf{C} - 0$, will be denoted by $z\infty \in S^1$. For $\alpha \in S^1$, $\hat{\alpha}$ will be the complex number of modulus 1 such that $\hat{\alpha}\infty = \alpha$. $\langle\ ,\ \rangle$ will be the euclidean scalar product in $\mathbf{C} = \mathbf{R}^2$.

1.1.1. Sectorial Neighbourhoods of Infinity

Let $A \subset S^1$ be an open arc (i.e., connected open set). An open set $U \subset \mathbf{C}$ is called a *sectorial neighbourhood of infinity of direction A* (resp. *codirection A*) if the following condition is satisfied:

(nbhd$^\infty(A)$) $\forall A' \Subset A, \exists x \in U, xA' \subset U,$
where xA' is the sector of summit x and direction A';
resp.
($\check{\text{n}}$bhd$^\infty(A)$) $\forall \alpha \in A, \exists a \in \mathbf{R}, P_\alpha(a) \subset U,$
where $P_\alpha(a)$ is the half-plane
$P_\alpha(a) := \{\xi \in \mathbf{C} \mid \langle \hat{\alpha}, \xi \rangle < -a\}.$

Remark. Let A be a "small" arc, i.e., an arc of length $< \pi$.
The *copolar* of A is defined by

$$\check{A} := \bigcup_{\alpha \in A} \{\beta \in \mathbf{S}^1 \mid \langle \hat{\alpha}, \hat{\beta} \rangle < 0\}.$$

Then

$$\check{\text{n}}\text{bhd}^{\infty}(A) = \text{nbhd}^{\infty}(\check{A}).$$

1.1.2. Sectorial Germs at Infinity

Sectorial germs at infinity are defined in an obvious way as germs of functions which are analytic in sectorial neighbourhoods of infinity: taking the inductive limit of $\mathcal{O}(U)$ when $U \in \text{nbhd}^{\infty}(A)$ (resp. $U \in \check{\text{n}}\text{bhd}^{\infty}(A)$) one thus gets

$\mathcal{O}^{\infty}(A)$ (sectorial germs at infinity of *direction* A), resp.
$\check{\mathcal{O}}^{\infty}(A)$ (sectorial germs at infinity of *codirection* A)

Sectorial germs of exponential type at infinity

$\mathscr{E}(A) :=$ the elements of $\mathcal{O}^{\infty}(A)$ which are *of exponential type in the direction* A (cf. [Bo]).
$\mathscr{E}_{-\infty}(A) :=$ the elements of $\mathcal{O}^{\infty}(A)$ which are *of exponential type* $-\infty$ (i.e., decrease quicker than any exponential) *in the direction A.*

1.1.3. The Laplace Isomorphism

Theorem. *Let A be a small arc, and A* be its complex conjugate. One has a canonical isomorphism (Laplace transformation)*

$$\mathscr{L} : \check{\mathcal{O}}^{\infty}(A)/\mathcal{O}(\mathbf{C}) \to \mathscr{E}(A^*)/\mathscr{E}_{-\infty}(A^*).$$

Idea of the proof. Let $\Phi \in \mathcal{O}(U)$, $U \in \check{\text{n}}\text{bhd}^{\infty}(A)$. For every $A' \Subset A$, U contains the closure of some $z\check{A}'$;
let $\gamma := -\partial(z\check{A}')$ (oriented boundary).

Construction of the mapping $\mathscr{L}$: Following an idea of Ecalle (private communication, July 1987), one first proves the

Key-lemma. *Let γ be a finite union of half-lines, and let Φ be holomorphic in a neighbourhood of γ. Then, by adding to Φ a suitable entire function, one can transform it into a function Φ' whose modulus is bounded on γ by any given function.*

Taking Φ' to be bounded on γ by an exponential, one can thus define

$$(\mathscr{L}_{\gamma}\Phi')(x) := \int_{\gamma} e^{-x\xi}\Phi'(\xi)\, d\xi.$$

Of course $\mathcal{L}_\gamma \Phi' \in \mathcal{E}(A'^*)$ and, using Cauchy's theorem (which for entire Φ' allows any kind of deformation of γ inside a bounded region), one checks that modulo $\mathcal{E}_{-\infty}(A'^*)$

(i) It only depends on the class of Φ' (i.e., the class of Φ) modulo $\mathcal{O}(\mathbf{C})$.
(ii) It does not depend on γ (apply the key-lemma to the union of the integration paths under consideration.)

Proof that $\mathcal{L}$ is an isomorphism: The inverse mapping can be defined by

$$(\bar{\mathcal{L}}_z \Phi)(\xi) := -\frac{1}{2\pi i} \int_{z\alpha} e^{x\xi} \Phi(x)\, dx,$$

where $\alpha \in A^*$, and z is so chosen that the half-line $z\alpha$ lies in the domain of holomorphy of Φ. Checking that this defines an inverse of $\mathcal{L}$ (in the quotient spaces we consider) follows from classical arguments (of the kind to be found in [Bo], for instance). It is interesting to notice that $\bar{\mathcal{L}}_z \Phi$ belongs to $\mathcal{E}(\check{A})$, and that the inclusion $\mathcal{E}(\check{A}) \subset \check{\mathcal{O}}^\infty(A)$ induces an *isomorphism*

$$\mathcal{E}(\check{A}) / \mathcal{E}(\check{A}) \cap \mathcal{O}(\mathbf{C}) \to \check{\mathcal{O}}^\infty(A) / \mathcal{O}(\mathbf{C}).$$

1.1.4. Remark

A section of the sheaf $\mathcal{E}/\mathcal{E}_{-\infty}$ over an arc A^* of length $> \pi$ does not in general come from a global section $\Phi \in \mathcal{E}(A^*)$. But when it does, such a Φ is unique. This comes from the well-known fact that a sectorial germ at infinity cannot be exponentially decreasing along an arc of length $> \pi$, unless it is zero.

1.2. Algebras of Resurgent Functions and Resurgent Symbols

1.2.1. Endless Continuation of Analytic Functions

Resurgence theory starts from the hypothesis that all the functions considered in the convolutive model (functions of ξ) are "everywhere continuable without cuts"—or "endlessly continuable," as we shall say for short. This means multivalued analytic functions such that any of their germs can be continued along any straight path, with the only exception being a discrete set of obstacles which can be bypassed in succession on the right or on the left, at will (cf. [E1]; a slightly stronger definition is proposed in [*]). This property can be shown to be preserved by convolution, and this allows one to define various "*convolution algebras*," which will be denoted by **boldface** letters. It will be understood that the corresponding "multiplicative algebras," deduced from them by the Laplace isomorphism when it can be defined, are denoted by the corresponding plainface letters.

1.2.2. Extended Resurgent Functions

In Section 1.0 we considered convolution algebras of *sectorial germs at infinity*, forming sheaves on the circle S^1, namely

$$\mathcal{O}^\infty, \qquad \check{\mathcal{O}}^\infty, \qquad \check{\mathcal{O}}^\infty/\mathcal{O}(\mathbf{C}).$$

Adding the "endless continuation" hypothesis, one gets the subsheaves

$$\mathcal{P}^\infty, \qquad \check{\mathcal{P}}^\infty, \qquad \check{\mathcal{P}}^\infty/\mathcal{O}(\mathbf{C});$$

the latter is our sheave of *extended resurgent functions*, which will be denoted by $\hat{\mathcal{R}}$

1.2.3. Resurgent Functions

For any point $\omega \in \mathbf{C}$, sectorial neighbourhoods of ω can be defined in an obvious way; the following notations, copied on the $\omega = \infty$ case, are self-explanatory:

$\mathrm{nbhd}^\omega(A)$	$\check{\mathrm{n}}\mathrm{bhd}^\omega(A)$
$\mathcal{O}^\omega$	$\check{\mathcal{O}}^\omega$ (sectorial germs at ω)
$\mathcal{P}^\omega$	$\check{\mathcal{P}}^\omega$ (endlessly continuable sectorial germs at ω).

$\mathcal{R}^\omega := \check{\mathcal{P}}^\omega/\mathcal{O}_\omega$, where $\mathcal{O}_\omega$ is the space of germs of analytic functions at ω, will be called the sheaf of *resurgent microfunctions* at ω. Resurgent microfunctions at 0 will simply be called *resurgent functions*, and we shall write $\mathcal{R}$ short for $\mathcal{R}^0$. Since the endless continuation hypothesis allows us to turn endlessly around ω, these sheaves are locally constant, and can be conveniently identified with their spaces of *multivalued* sections on the circle (the word "sheaf" never appears in the papers of Ecalle). When the continuation along the circle is *single-valued*, we shall say that the microfunction is *simply ramified*; this implies that it can be written (in a unique way) as an infinite linear combination of $\delta^{(k)}(\xi - \omega)$, $k \in \mathbf{Z}$ (k-th derivative of δ-function). The *variation* homomorphism

$$\mathrm{var}: \mathcal{R}^\omega \to \mathcal{P}^\omega$$

is defined by associating to a resurgent microfunction $[\Phi]_\omega$ (class of Φ

modulo $\mathcal{O}_\omega$) the function $\Phi_{\omega+} - \Phi_{\omega-}$, where $\Phi_{\omega+}$ (resp. $\Phi_{\omega-}$) means the element of $\mathscr{P}^\omega$ deduced from $\Phi \in \mathscr{P}^\omega$ by bypassing ω on the right (resp. on the left). Simply ramified microfunctions are those for which the variation is holomorphic at ω.

Beware. Our sheaf $\hat{\mathscr{R}}$ of extended resurgent functions is *not* locally constant, since the endless continuation hypothesis does not prevent singularities from accumulating at infinity in the same sectorial neighbourhood.

1.2.4. Decomposition of Extended Resurgent Functions into "Resurgent Symbols"

Let $\Phi \in \check{\mathscr{P}}^\infty(A)$, $\{\Phi\}$ = class of Φ mod. $\mathcal{O}(\mathbf{C}) \in \hat{\mathscr{R}}(A)$. Choosing $\alpha \in A$, one can suppose that Φ is holomorphic in $\mathbf{C} - \Gamma_\alpha$, where Γ_α is a discrete union of "cuts" (half-lines) parallel to α. Continuation of Φ across Γ_α, from the right (resp. from the left), is possible except for a discrete set of points $\omega \in \Gamma_\alpha$, the singular points "seen from the right (resp. left)" in the direction α. For any $\omega \in \mathbf{C}$ one defines in an obvious way a canonical "singularity homomorphism"

$$\mathscr{s}^{\omega}_{\alpha+} (\text{resp. } \mathscr{s}^{\omega}_{\alpha-})\colon \hat{\mathscr{R}}(A) \to \mathscr{R}^\omega(A),$$

which is non-zero only if ω is a singular point on the side considered. (N.B. The origin of each cut is seen from both sides, and $\mathscr{s}^{\omega}_{\alpha+}\{\Phi\} = \mathscr{s}^{\omega}_{\alpha-}\{\Phi\}$ in that case.) The collection of all these homomorphisms will be denoted by

$$\dot{\mathscr{s}}_{\alpha+} (\text{resp. } \dot{\mathscr{s}}_{\alpha-})\colon \hat{\mathscr{R}}(A) \to \dot{\mathscr{R}}(A),$$

where $\dot{\mathscr{R}}(A)$ is the space of *resurgent symbols* in the codirection A:

Definition. A *resurgent symbol* in the codirection A is a collection

$$\dot{\varphi} = (\varphi^\omega \in \mathscr{R}^\omega(A))_{\omega \in \mathbf{C}},$$

where $\varphi^\omega \neq 0$ only for $\omega \in \Omega$, a discrete subset of $\mathbf{C}$ called the *support* of $\dot{\varphi}$, such that $\mathbf{C} - \Omega\alpha \in \check{\text{n}}\text{bhd}^\infty(A)$ $(\Omega\alpha := \bigcup_{\omega \in \Omega} \omega\alpha)$.

Theorem. *The "decomposition homomorphism"* $\dot{\mathscr{s}}_{\alpha+}$ (resp. $\dot{\mathscr{s}}_{\alpha-}$) *is an isomorphism* (*of convolutive algebras*).

This is a kind of "Mittag-Leffler theorem" for resurgent functions (reconstruction of a function from its singularities); for the proof, cf. [*].

For $\{\Phi\} \in \hat{\dot{\mathscr{R}}}(A)$, the support of $\dot{\jmath}_{\alpha+}\{\Phi\}$ (resp. $\dot{\jmath}_{\alpha-}\{\Phi\}$) will be called the *right* [resp. *left*] *singular support* of $\{\Phi\}$ in the direction α. An extended resurgent function is called *pure* when its singular support consists of one single point: denoting by $\dot{\mathscr{R}}^{\omega}$ the space of *elementary resurgent symbols* (i.e. res. symbols with only one component) at ω, we thus define

$$\hat{\dot{\mathscr{R}}}^{\omega}_{\alpha+} := (\dot{\jmath}_{\alpha+})^{-1}(\dot{\mathscr{R}}^{\omega}) \quad \text{(pure on the right, in the direction } \alpha)$$
$$\hat{\dot{\mathscr{R}}}^{\omega}_{\alpha-} := (\dot{\jmath}_{\alpha-})^{-1}(\dot{\mathscr{R}}^{\omega}) \quad \text{(pure on the left, in the direction } \alpha).$$

1.2.5. Resurgent Functions in the Multiplicative Model

The convolutive algebras $\hat{\dot{\mathscr{R}}}$, $\mathscr{R}$, $\dot{\mathscr{R}}$ are transformed by the Laplace isomorphism (0.2 or its local version) into multiplicative algebras $\hat{\dot{\mathcal{R}}}$, $\mathcal{R}$, $\dot{\mathcal{R}}$. The algebra $\dot{\mathcal{R}}$ (of *resurgent symbols in the multiplicative model*) can be seen as an algebra of formal (eventually infinite) linear combinations

$$\dot{\varphi} = \sum_{\omega} \varphi_{\omega}\, e^{-\omega x},$$

where $\varphi_{\omega} \in \mathcal{R}$ (algebra of *resurgent functions in the multiplicative model*).

The special case of *simply ramified* resurgent functions is worth mentioning: since the k-th derivative of $\delta(\xi)$ has x^k as its Laplace transform, simply ramified elements of $\mathcal{R}$ are *formal power series*, and the corresponding (pure) extended resurgent functions can be seen as *summations* of these power series, *up to functions of exponential type* $-\infty$ (in Ecalle's recent terminology [E3], one speaks of "pre-summations").

1.2.6. Summable Resurgent Functions

What one would like to call "summable" resurgent functions should form—in the multiplicative model—an algebra of *true functions* (not classes modulo $\mathscr{E}_{-\infty}$). Imposing exponential growth conditions in the convolutive model allows one to define such a subalgebra of $\hat{\dot{\mathscr{R}}}$ (cf. [*]). But this "naïve" definition excludes many interesting cases which one would obviously like to call "summable."

A more general definition of summability, based on a refinement of remark 1.0.3, is proposed by Ecalle in [E3].

1.3. Alien Derivatives

1.3.1. A Sectorial Point of View

Let $A \subset S^1$ be a given arc. *Connection automorphisms* are defined for $\alpha \in A$ by

$$\boldsymbol{\sigma}_{(\alpha)} := \dot{\jmath}_{\alpha+} \circ (\dot{\jmath}_{\alpha-})^{-1} : \dot{\mathscr{R}}(A) \to \dot{\mathscr{R}}(A).$$

Denoting by $<_\alpha$ the order relation $\omega <_\alpha \omega' \Leftrightarrow \omega' \in \omega\alpha, \omega' \neq \omega$, we have already mentioned the fact that $\mathscr{s}^{\omega}_{\alpha+}\{\Phi\} = \mathscr{s}^{\omega}_{\alpha-}\{\Phi\}$ when ω is minimal for $<_\alpha$ in the singular support of $\{\Phi\}$. More precisely, $\boldsymbol{\sigma}_{(\alpha)} \simeq \mathbf{1}$ in $\dot{\mathscr{R}}(A)$ in the following sense: consider the iterated action of $\boldsymbol{\sigma}_{(\alpha)} - \mathbf{1}$ on a given $\dot{\varphi} \in \dot{\mathscr{R}}(A)$; there exists a discrete set $\Omega \subset \mathbf{C}$, containing the supports of all the symbols thus obtained, such that each iteration of $\boldsymbol{\sigma}_{(\alpha)} - \mathbf{1}$ "kills" the initial part of the symbol on which it acts (where "initial part" means forgetting the components whose support is not minimal for $<_\alpha$). This allows us to define the formal logarithm

$$\dot{\Delta}_{(\alpha)} := \log \boldsymbol{\sigma}_{(\alpha)},$$

which is a derivation of the convolution algebra $\dot{\mathscr{R}}(A)$, called the *alien derivation of resurgent symbols.*

Decomposing the resurgent symbols into their elementary components, the action of $\dot{\Delta}_{(\alpha)}$ decomposes into a collection of linear mappings

$$\dot{\Delta}_\eta : \dot{\mathscr{R}}^{\omega}(A) \to \dot{\mathscr{R}}^{\omega+\eta}(A),$$

where $\eta \in \mathbf{C}$ is such that $\eta\infty = \alpha$, and $\dot{\Delta}_\eta$ also depends on ω but in a trivial way (obvious translations).

An explicit formula can be given for $\dot{\Delta}_\eta \boldsymbol{\varphi}^{\omega} (\boldsymbol{\varphi}^{\omega} \in \dot{\mathscr{R}}^{\omega}(A))$: take $\operatorname{var} \boldsymbol{\varphi}^{\omega}$, and try to continue it analytically along the direction $\alpha = \eta\infty$; if $\omega + \eta$ is the first singular point met, $\dot{\Delta}_\eta \boldsymbol{\varphi}^{\omega}$ is the microfunction at ω defined by this analytic continuation; if other singular points have been met first, one has several ways of bypassing them before getting at $\omega + \eta$, and $\dot{\Delta}_\eta \boldsymbol{\varphi}^{\omega}$ is given by a suitably weighted "mean" of these analytic continuations.

Composing the linear mappings $\dot{\Delta}_\eta$ with the obvious isomorphisms (translations in ξ)

$$\dot{\mathscr{R}}^{\omega} (\text{resp. } \dot{\mathscr{R}}^{\omega+\eta}) \approx \dot{\mathscr{R}}^{0} = \mathscr{R},$$

one gets the alien derivations of resurgent functions:

$$\Delta_\eta : \mathscr{R}(A) \to \mathscr{R}(A),$$

which are indexed by all the η's such that $\eta\infty \in A$.

Given a $\boldsymbol{\varphi} \in \mathscr{R}(A)$, the set of all η's such that $\Delta_\eta \boldsymbol{\varphi} \neq 0$ is a discrete subset of the sector $0A$, called the *derivated singular support* of $\boldsymbol{\varphi}$ (in the A direction).

1.3.2. Ecalle's Point of View

Since $\mathscr{R}$ is locally constant, it can be made constant by lifting it to the universal covering of the circle. This amounts to choosing a "base point" α_0 in S^1 (say, the real positive direction), and continuing every $\mathscr{R}_\alpha$ $(\alpha \in S^1)$ to $\mathscr{R}_{\alpha_0}$ along a path on the circle S. The space of multivalued sections of our sheaf $\mathscr{R}$, canonically isomorphic to $\mathscr{R}_{\alpha_0}$, is what Ecalle denotes $\mathscr{R}$. Alien derivations of resurgent functions,

$$\Delta_\eta : \mathscr{R} \to \mathscr{R},$$

must now be indexed by $\eta \in \tilde{\mathbf{C}}^*$ (universal covering of $\mathbf{C}^* = \mathbf{C} - 0$).

A remarkable *resurgence subalgebra* (i.e., subalgebra stable under alien derivations) of $\mathscr{R}$, is $\mathscr{R}^{(1)}$, the algebra of "*simple resurgent functions,*" i.e., the resurgent functions which are simply ramified at 0 as well as all their iterated alien derivatives. Restricted to $\mathscr{R}^{(1)}$, Ecalle's alien derivations may be indexed by $\eta \in \mathbf{C}^*$, forgetting about universal coverings.

1.4. Resurgent Functions Depending on Parameters

Our "extended resurgent functions" are easily made to "depend analytically on parameters;" the definition is left to the reader, with a warning: in making the discrete sets of "accessible singularities" (in the definition [*] of endless continuability) depend analytically on the parameters, we have to allow for possible "confluence" of singular points. In other words, considered in the product space $\mathbf{C}$ (the ξ-plane) times the parameter space, the singularity sets are analytic hypersurfaces, "relatively discrete" but not necessarily smooth with respect to the projection on the parameter space. Those values of the parameters for which confluence of singularities occurs are called *turning points.* Outside the turning points, it is straightforward to define what "analytic dependence on parameters" means for a *resurgent symbol.* But *it is not true* that given an extended resurgent function depending analytically on parameters q, the resurgent symbol defined by its (right or left) decomposition (in some direction α) depends analytically on q outside the turning points: "Stokes discontinuities" occur for those values of q for which two non-colinear "cuts" $\omega\alpha$, $\omega\alpha'$ become colinear. This phenomenon can also be detected in the "absolute" case (without parameters), moving the "radar direction" α: this is "Stokes phenomenon," well known for solutions of ordinary linear differential equations. Stokes discontinuities are thus given by the "connection automorphisms" of Section 1.2.1.

2. Quantized Canonical Transformations and Multi-Instanton Expansions

Consider the Schrödinger operator in one dimension

$$\mathscr{H} = -x^{-2}\frac{d^2}{dq^2} + V(q),$$

where q is the "space" variable, and V is the "potential" function, which we shall assume to be polynomial, increasing to $+\infty$ for $q \to \pm\infty$; the parameter $x = 1/\hbar$ is the inverse of the Planck constant.

Our problem is to study the spectrum of $\mathcal{H}$ as a function of the parameter x.

In the special case of the harmonic oscillator (Section 1: $V(q) = q^2$) a "quantized canonical transformation" will allow us to solve the differential equation

$$(\mathcal{H} - x^{-1}E)\Psi(q, x) = 0 \tag{0}$$

(with E an arbitrary parameter) and check that only for *odd integer values* of E is the solution square integrable (exponentially decreasing for $q \to \pm\infty$).

In Section 2.2, a similar method will allow us *locally* to solve equation (0) for potentials "locally looking like the harmonic oscillator" ($V(q) \sim (q - q_0)^2$). But of course, we can no longer hope that constant values of E will ensure square integrability of the solutions: E will have to be a function of x, hopefully an *extended resurgent function* of x.

Taking V to be an even function with two quadratic minima (e.g. $V = (q^2 - q_0^2)^2/2$), Section 2.3 will show that very natural "resurgence conjectures" about the data of Section 2.2 imply a conjecture of *Zinn-Justin* [Z2, Z3], which gives a constructive characterization of the $E(x)$'s as the solutions of an "implicit function" equation.

"Constructive" means that we can make out an algorithm for building—to any given order—the decomposition of $E(x)$ into resurgent symbols, *i.e.*, the "multi-instantons expansions" of Zinn-Justin.

The "correspondence principle." The starting point of our study will be the *classical* Hamiltonian

$$H = -p^2 + V(q),$$

which we deduce from $\mathcal{H}$ through the correspondence principle

$$p \leftrightarrow -x^{-1}\frac{d}{dq}.$$

[Traditionally, the correspondence principle reads $p \leftrightarrow (i/\hbar)\, d/dq$, so that our p is (i) times the usual "momentum" p: the reason for our unusual convention is that we like to write $\Psi \simeq e^{-xS(q)}$ instead of $\Psi \simeq e^{iS(q)/\hbar}$, so that our $S(q)$—the "support" of the resurgent function Ψ—is (i) times the usual (Hamilton–Jacobi) action function.]

2.1. The Harmonic Oscillator

Our treatment of the harmonic oscillator will bear some analogy with that of Bargmann [B]. But we shall move more freely in the complex domain (cf. 2.1.2).

2.1.1. *A Quantized Canonical Transformation*

With our convention for the correspondence principle, the classical Hamiltonian reads

$$H = -p^2 + q^2 = -2\hat{p}\hat{q},$$

where

$$\begin{cases} \hat{p} = \dfrac{p-q}{\sqrt{2}}, \\ \hat{q} = \dfrac{p+q}{\sqrt{2}}, \end{cases} \qquad \text{so that conversely} \qquad \begin{cases} p = \dfrac{\hat{p}+\hat{q}}{\sqrt{2}}, \\ q = \dfrac{\hat{q}-\hat{p}}{\sqrt{2}}. \end{cases} \tag{1}$$

The transformation (1) is canonical, with generating function

$$S(q, \hat{q}) = \sqrt{2}q\hat{q} - \frac{q^2}{2} - \frac{\hat{q}^2}{2} \tag{1$'$}$$

(in the sense that $dS = p\,dq - \hat{p}\,d\hat{q}$).

Consider the following transformation $\mathscr{L}_{S,\gamma}$ on functions of two variables q, x:

$$\Psi(q, x) = (\mathscr{L}_{S,\gamma}\Phi)(q, x) := \int_\gamma e^{-xS(q,\hat{q})}\Phi(\hat{q}, x)\,d\hat{q} \tag{2}$$

(conditions on the integration path γ will be made precise in Section 2.1.2). Derivating under the integration symbol, and formally integrating by parts, one checks that the operators

$$q, \mathbf{p} := -x^{-1}\partial_q \qquad \text{(acting on } \Psi)$$

$$\hat{q}, \hat{\mathbf{p}} := -x^{-1}\partial_{\hat{q}} \qquad \text{(acting on } \Phi)$$

transform into each other as follows:

$$\begin{cases} \mathbf{p}\mathscr{L}_{S,\gamma} = \mathscr{L}_{S,\gamma}\left(\dfrac{\hat{\mathbf{p}}+\hat{q}}{\sqrt{2}}\right), \\ q\mathscr{L}_{S,\gamma} = \mathscr{L}_{S,\gamma}\left(\dfrac{\hat{q}-\hat{\mathbf{p}}}{\sqrt{2}}\right). \end{cases} \tag{1$''$}$$

In other words, $\mathscr{L}_{S,\gamma}$ is a "quantization" of the canonical transformation (1).

It follows that $\mathscr{H}\mathscr{L}_{S,\gamma}=\mathscr{L}_{S,\gamma}\hat{\mathscr{H}}$, where $\mathscr{H}=-x^{-2}\partial_q^2+V(q)$ is the Schrödinger operator, and $\hat{\mathscr{H}}=-2\hat{q}\hat{\mathbf{p}}-x^{-1}=-x^{-1}(2\hat{q}\partial_{\hat{q}}+1)$.

The Schrödinger equation (0) thus becomes

$$(\hat{q}\partial_{\hat{q}}-s)\Phi=0, \qquad \text{with } s=\frac{E-1}{2}. \tag{3}$$

It has the solution

$$\Phi=\hat{q}^s \text{ (times an arbitrary function of } x). \tag{3)'$$

Notice that for non-integral s this solution is ramified around $\hat{q}=0$.

2.1.2. Steepest Descent, and "Pure" Solutions of the Schrödinger Equation

Considered as a function of $\hat{q}$ (depending on the parameter q), the generating function S has a quadratic critical point at $\hat{q}=\sqrt{2}q$. To give a meaning to the integral (2), with Φ as in (3)$'$ and for *real positive x*, one just has to choose the (endless) integration path γ in such a way that near infinity it satisfies a "descent condition" for the function $-\text{Re } S$ (so that the integrand decreases exponentially). If this is done in such a way that γ doesn't depend on q (or depends continuously on q), the integral $\mathscr{L}_{\gamma,S}\Phi$ will indeed allow being derivated under the integration symbol, integrated by parts, etc., *defining an entire function of q which will be a solution of the Schrödinger equation.*

By Cauchy's theorem, this solution $\mathscr{L}_{\gamma,S}\Phi$ will only depend on the homology class of γ subject to the descent condition, *on the Riemann surface* of $\log \hat{q}$. Denoting by $\hat{R}$ the canonical automorphism of this Riemann surface ($\hat{R}$ = one anti-clockwise turn), one easily sees that the homology space of such paths is a *2-dimensional* $\mathbf{C}(\hat{R})$*-vector space.* Equivalently, it can be considered a *1-dimensional* $\mathbf{C}(\hat{R}^{1/2})$*-vector space*, where

$$\hat{R}^{1/2}=\tfrac{1}{2}\text{ anti-clockwise turn.}$$

Figure 1 illustrates this by showing a typical generator γ_0, and some remarkable elements deduced from it.

$$\gamma_1=\hat{R}^{1/2}\gamma_0,$$

$$\gamma'=\gamma_0+\gamma_1=(1+\hat{R}^{1/2})\gamma_0,$$

$$\gamma''=(\hat{R}^{1/2}+\hat{R})\gamma_0, \text{ etc.}$$

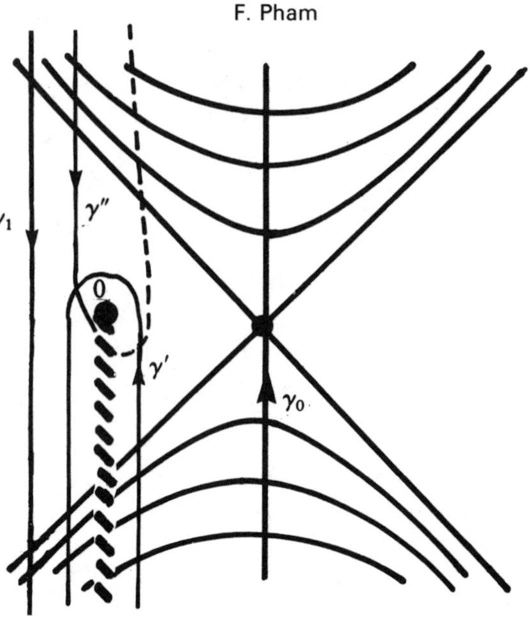

FIGURE 1 A "cut" has been drawn (shaded line) to define a "first sheet."

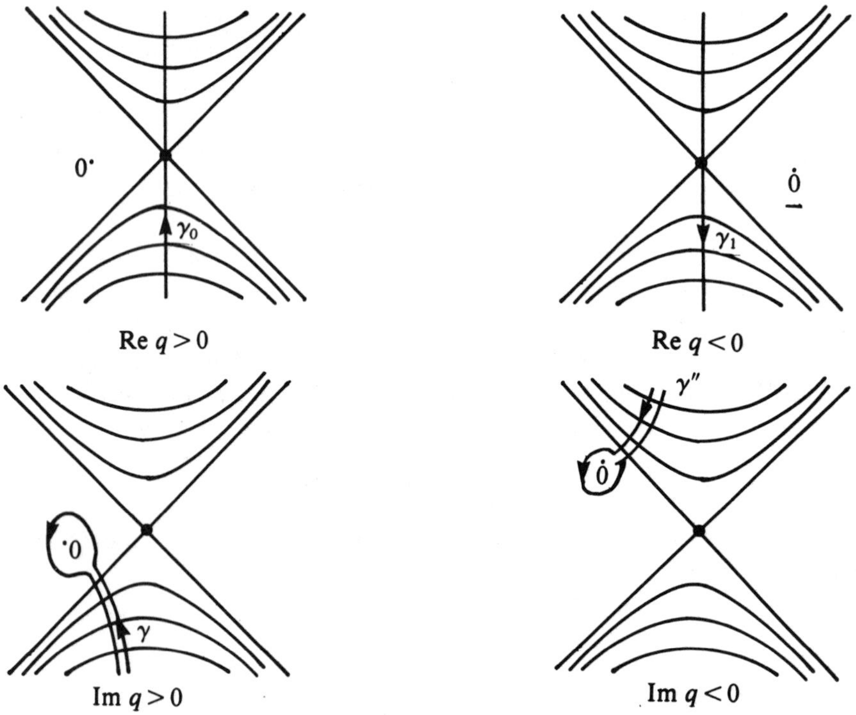

FIGURE 2

As shown in Fig. 2, each of these elements has the following remarkable property:

For Re $q>0$ [resp. Re $q<0$], γ_0 [resp. γ_1] is homologous to a *steepest descent path through the saddlepoint*;
For Im $q>0$ [resp. Im $q<0$], γ' [resp. γ''] is homologous to a *steepest descent loop around the origin.*

Taking $\xi = S(q, \hat{q})$ as the new integration variable in (2), one thus sees that for any q in the corresponding half-space, the corresponding integral (2) is a *pure resurgent function of* x (holomorphically depending on q), with support

$\xi = S(q, \sqrt{2}q) = q^2/2$	for $\mathscr{L}_{\gamma_0, s}\Phi$	(when Re $q>0$)
	and $\mathscr{L}_{\gamma_1, s}\Phi$	(when Re $q<0$)
$\xi = S(q, 0) = -q^2/2$	for $\mathscr{L}_{\gamma', s}\Phi$	(when Im $q>0$)
	and $\mathscr{L}_{\gamma'', s}\Phi$	(when Im $q<0$):

Remark. If $s \in \mathbf{N}$, $\mathscr{L}_{\gamma', s}\Phi = 0$, so that all the above constructed solutions are multiples of each other. One gets another linearly independent solution of the Schrödinger equation by replacing the loop γ' (for instance) by "half of it" (a path with *one* end at the origin)—this also works for any s with Re $s > -1$.

2.1.3. Elementary Symbols and Their Resurgence Equations

Decomposing the integration path γ into "steepest descent paths and loops," one gets the decomposition of $\mathscr{L}_{\gamma, s}\Phi$ into elementary symbols (in the real positive direction). An elementary symbol

$$\Psi^0 \text{ with support } q^2/2, \text{ resp.}$$

$$\Psi_0 \text{ with support } -q^2/2$$

can be defined by integrating over a *local* (i.e., truncated) steepest descent path through the saddlepoint (say, oriented upwards) (*resp.* integrating over a local steepest descent loop around 0 (say, oriented counter-clockwise)).

One thus gets, for instance,

$$\Psi^0 = -\mathfrak{z}^{q^2/2}\mathscr{L}_{\gamma_1,S}\Phi$$

$$\Psi_0 = \mathfrak{z}^{-q^2/2}\mathscr{L}_{\gamma',S}\Phi = \mathfrak{z}^{-q^2/2}\mathscr{L}_{\gamma'',S}\Phi,$$

say, for Re $q < 0$ (such a restriction ensures that each of the above symbols depends holomorphically on q).

Either of these elementary symbols can be identified with an *asymptotic expansion* which is easily computed from its integral representation. It has the following form:

$$\Psi^0 = \frac{\sqrt{2}\, e^{-xq^2/2}}{\sqrt{x}} (\sqrt{2}\, q)^s \sum_{k=0}^{\infty} a_k x^{-k} q^{-2k},$$

$$\Psi_0 = \frac{2\, e^{xq^2/2}}{(\sqrt{2}\, xq)^{s+1}\, e^{i\pi s}\Gamma(-s)} \sum_{k=0}^{\infty} b_k x^{-k} q^{-2k},$$

where a_k, b_k are polynomials in s, of degree $\leq 2k$.

By a purely geometrical reasoning (deformation of integration paths) one can show that these elementary symbols verify the following resurgence equations for Re $q < 0$:

$$\dot{\Delta}_{e^{i\pi}q^2}\boldsymbol{\psi}^0 = -\boldsymbol{\psi}_0,$$

$$\dot{\Delta}_{q^2}\boldsymbol{\psi}_0 = (1 - e^{-2i\pi s})\boldsymbol{\psi}^0.$$

2.1.4. Quantization Condition

For q real > 0, $\mathscr{L}_{\gamma_0,S}\Phi$ is pure and can be characterized as the unique solution (up to a factor) which vanishes exponentially when $q \to +\infty$.

But this solution is no longer pure for q real < 0, since its "right decomposition" is then given by

$$\mathscr{L}_{\gamma_0,S}\Phi = \underset{\text{(pure on the right)}}{\mathscr{L}_{\gamma',S}\Phi} - \underset{\text{(pure)}}{\mathscr{L}_{\gamma_1,S}\Phi}.$$

The condition

$$s \in \mathbf{N} \qquad (\Leftrightarrow \mathscr{L}_{\gamma',S}\Phi = 0)$$

is therefore necessary and sufficient for $\mathscr{L}_{\gamma_0,S}\Phi$ to vanish exponentially when $q \to -\infty$: this is the well-known "quantization condition" for the harmonic oscillator.

2.2. Local Resurgence in the Bottom of a Well

We now want to study *locally* the more general case of a potential function with a quadratic minimum (say, at 0)

$$V \in \mathbf{C}\{q\}, \qquad V \sim q^2.$$

2.2.1. A Quantized Canonical Transformation

The classical Hamiltonian reads

$$\begin{aligned} H(q, p) &= -p^2 + V(q) = (-p + U(q))(p + U(q)), \\ U(q) &= \sqrt{V(q)}, \qquad U \sim q. \end{aligned} \tag{4}$$

Introducing new canonical coordinates $\hat{p}$, $\hat{q}$ such that

$$\begin{cases} p = U(q) \ \Leftrightarrow \hat{p} = 0, \\ p = -U(q) \Leftrightarrow \hat{q} = 0, \end{cases} \tag{5}$$

the classical Hamiltonian now reads

$$\hat{H}(\hat{q}, \hat{p}) = u(\hat{q}, \hat{p})\hat{q}\hat{p} \qquad (u \in \mathbf{C}\{\hat{q}, \hat{p}\}, u(0, 0) \neq 0). \tag{6}$$

"*Quantizing*" the canonical transformation, we formally transform the Schrödinger operator $\mathscr{H}$ into a *microdifferential operator*

$$\hat{\mathscr{H}} = \hat{\mathscr{H}}(q, x^{-1}, x^{-1}\partial_q),$$

having $\hat{H}$ for its principal (zeroth order) symbol.

[N.B. The new coordinates have been allowed to remove their hats.]

Dictionary. What we mean by a microdifferential operator can be summarized as follows:

"*Microdifferential symbols*": $\sum_{k=0}^{\infty} a_k(q, p)x^{-k}$, with $a_k \in \mathbf{C}\{q, p\}$, having a common polyradius of convergence ρ, satisfying the "Borel" convergence condition

$$\sum_{k=0}^{\infty} \|a_{k+1}\|_\rho \frac{\xi^k}{k!} \in \mathbf{C}\{\xi\}.$$

Any microdifferential symbol defines a microdifferential operator

$$P = \sum_{k=0}^{\infty} a_k(q, -x^{-1}\partial_q)x^{-k},$$

with the convention that the $-x^{-1}\partial_q$ are written *on the right* of the q's.

The function $a_0(q, p)$ is denoted by $\sigma(P)$, and called the *zeroth-order symbol* of P. In particular $\sigma(-x^{-1}\partial_q)=p$.

The ring of all such microdifferential operators will be denoted by $\mathscr{D}_{(x)}$.

Up to the notation (x instead of ∂_ξ), these are just Sato's *microdifferential operators* of order ≤ 0 in two variables (q, ξ) in the codirection $d\xi$, *whose total symbol doesn't depend on* ξ.

By the same conventions, our "quantized canonical transformations" are the restrictions to $\mathscr{D}_{(x)}$ of Sato's "quantized contact transformations."

Comment. By this dictionary, our Section 2.1 yielded a complete description of the so-called "regular analytic interactions of simple holonomic systems" (cf. [KKO]). But reducing it to such a description would be misleading: on the one hand, our study in Section 2.1 was *global*, and our transformations dealt with *functions*, not just differential equations; besides, even limited to a (micro-)*local* study, our approach will now lead us *beyond* the framework of holonomic systems[1] (the module $\mathscr{D}_{(x)}q^{s(x)}$ of Section 2.2 will *not* be holonomic, since the "monodromy exponent" s will no longer be a constant).

Notations. Besides the notation $\mathscr{D}_{(x)}$ for our ring of microdifferential operators, we shall use the notation $\mathscr{O}_{(x)}$ [resp. $\mathbf{C}_{(x)}$] for the (commutative) ring of "*analytic symbols*"—subring of elements of $\mathscr{D}_{(x)}$ whose coefficients a_k depend only on q (resp. are constant).

2.2.2. Solving the "Canonically Transformed" Schrödinger Equation

We are now aiming at an algebraic study of the microdifferential equation

$$P\Phi=0, \tag{7}$$

where $P\in\mathscr{D}_{(x)}$ is a microdifferential operator such that

$$\sigma(P)=u(q,p)qp \quad (u \text{ invertible}). \tag{8}$$

Let us introduce the following space of *meromorphic* microdifferential operators:

$$\mathscr{D}'_{(x)} \qquad (\text{resp. } \mathscr{O}'_{(x)}),$$

[1] Cf. [P4], where I first met—with astonishment—an example of this phenomenon.

consisting of those operators which can be deduced from elements of $\mathcal{D}_{(x)}$ (resp. $\mathcal{O}_{(x)}$) by the formal substitution $x^{-1} \mapsto q^{-1}x^{-1}$ (so that the coefficient of x^{-k} may have a pole of order $\leq k$). This space is a right $\mathcal{D}_{(x)}$-module [resp. a commutative ring].

Division lemma. *With P as in* (8), *any $A \in \mathcal{D}'_{(x)}$ can be written $A = QP + r$,*

$$Q \in \mathcal{D}'_{(x)}, r \in \mathcal{O}'_{(x)} \qquad (r \text{ uniquely determined}).$$

This is a meromorphic version of a well-known result [SKK]: the same statement in $\mathcal{D}_{(x)}$, $\mathcal{O}_{(x)}$, with P satisfying $\sigma(P) = up$ (u invertible) (cf. also [P1] 2^{d} Part, Section 3). Applying this lemma to $A = x^{-1}\partial_q$, and noticing that in that case the "remainder" term r is divisible by x^{-1}, let us separate in it the simple pole terms:

$$\begin{aligned} r(q, x) &:= x^{-1}\frac{s(x)}{q} + x^{-1}b'(q, x), \\ s(x) &= \sum_{k=0}^{\infty} s_k x^{-k} \in \mathbf{C}_{(x)}, \end{aligned} \tag{9}$$

where $b' \in \mathcal{O}'_{(x)}$ has no simple pole terms, and can thus be written $b' = \partial_q b(q, x)$, $b \in \mathcal{O}'_{(x)}$.

We thus get the following

Proposition. *The microdifferential equation* (7) *is equivalent to*

$$\partial_q \Phi / \Phi = \frac{s(x)}{q} + \partial_q b(q, x), \tag{7$'$}$$

which can be formally solved by

$$\begin{aligned} \Phi &= a(q, x) q^{s(x)}, \\ a(q, x) &= \exp b(q, x). \end{aligned} \tag{10}$$

Notice that the *monodromy exponent*

$$s = s_0 + s_1 x^{-1} + s_2 x^{-2} + \cdots \in \mathbf{C}_{(x)} \tag{11}$$

is uniquely determined by P, whereas $a \in \mathcal{O}'_{(x)}$ is determined up to a factor in $\mathbf{C}_{(x)}$.

Remark. Expressions like $a(q, x)q^{s(x)}$ can be considered as a canonical way of writing any element of $\mathcal{D}_{(x)}q^{s(x)}$, the $\mathcal{D}_{(x)}$—module generated by $q^{s(x)}$—making $a(q, x)$ run over $\mathcal{O}'_{(x)}$ in general, or in $\mathcal{O}_{(x)}$ in the special case $s(x) = s \in \mathbf{N}$.

Addendum: Dependence on the parameter E. Suppose P depends *linearly* on a parameter E, its zeroth-order symbol $\sigma(P)$ being independent of E. Then the "remainder" term r in the division lemma depends on E in the following fashion:

$$r = \sum_{k=0}^{\infty} r_k(E)x^{-k}, \qquad r_k(E) = \text{polynomial of degree} \leq k.$$

The dependence on E in the Proposition is therefore the following:

$$s(x) = \sum_{k=0}^{\infty} s_k(E)x^{-k},$$

$$b(q, x) = \sum_{k=0}^{\infty} b_k(q, E)x^{-k},$$

with s_k and b_k polynomials in E of degree $\leq k+1$.

Remark. In the case we are interested in ($P = \hat{\mathscr{H}} - Ex^{-1}$ of Section 2.2.1), one has $s_0(E) = (E-1)/2$.

2.2.3. "Local" Steepest Descent and "Confluent Symbols"

Sections 2.2.1, 2.2.2 consisted of purely algebraic manipulations of abstract microdifferential operators. We now want to make them enter the realm of *analysis*, giving a meaning to the integral

$$\Psi(q, x) = \int_{\gamma} e^{-xS(q,\hat{q})}\Phi(\hat{q}, x)\, d\hat{q}, \tag{12}$$

where $S(q, \hat{q})$ is the generating function of the canonical transformation in Section 2.2.1, and Φ is a solution of $(\hat{\mathscr{H}} - x^{-1}E)\Phi = 0$.

Since our hypotheses are just local, we can no longer consider γ as an endless path. It must be a "truncated" path: we are only concerned with its behaviour in the vicinity of $\hat{q} = 0$ and of the saddlepoint $\hat{q} = \hat{q}(q)$—the topological situation being the same, locally, as in Section 2.1.2.

It follows that $\Psi(q, x)$ must not be considered here as a true function, but as some kind of equivalence class, a "confluent symbol."

Dictionary: Confluent symbols. Let $\mathscr{S}$ be a germ of plane analytic curve at the origin of $\mathbf{C}^2 \ni (q, \xi)$, transverse to the ξ-axis. For any direction α in the ξ-plane, let $\mathscr{S}\alpha$ be the union in $\mathbf{C}^2$ of all half-lines $\{q\} \times \xi\alpha$, $(q, \xi) \in \mathscr{S}$.

A confluent symbol with support in $\mathscr{S}$ is,

(1) *in the convolutive model,*
the class, modulo $\mathbf{C}\{q, \xi\}$, of the germ at $(0, 0)$ of a function analytic outside $\mathscr{S}\alpha$, continuable near $(0, 0)$ to the universal covering of $\mathbf{C}^2\backslash\mathscr{S}$;

(2) *in the multiplicative model,*
a truncated Laplace transform of a confluent symbol in the first sense.

In short, a confluent symbol is the *local* analogue of an extended resurgent function in x, holomorphically depending on q.

Remark. Such objects played an important role in Kashiwara–Kawai [KK], as far as the convolutive model is concerned. Their Laplace transforms were considered in [P3].

Proposition. *The integral* (12), *with* Φ *given by Proposition* 2.2.2, *can be considered as defining a confluent symbol with support* $\operatorname{Supp} \Psi \subset \{\xi = S^0(q)\} \cup \{\xi = S_0(q)\}$, *where*

$$S_0(q) = S(q, 0)$$

$$S^0(q) = S(q, \hat{q}(q)) \qquad \text{(where } \hat{q}(q) \text{ is the saddlepoint).}$$

The dependence on γ *of this confluent symbol, and its "purity" properties, can be described exactly as in Section* 2.1.2.

Sketch of the proof. In Proposition 2.2.2, the expression (10) which "defined" Φ was given no meaning but *formal.* Now we can define it as the Laplace transform of a "microfunction" $\boldsymbol{\phi}(q, \xi) \in \mathcal{O}(\tilde{U}^*)/\mathcal{O}(\tilde{U})$, where $\tilde{U}$ is the universal covering of

$$U = \{(q, \xi) \in \mathbf{C}^2 | 0 < |q| < \delta, |\xi| < \eta |q|\},$$

(where δ, η are conveniently chosen positive numbers), whereas $\tilde{U}^*$ is the universal covering of $U^* = U\backslash\{\xi = 0\}$. This being done, consider the double integral

$$\boldsymbol{\psi}(q, \xi) = \frac{-1}{2\pi i} \int_{\gamma\times\beta} \frac{\boldsymbol{\phi}(\hat{q}, \hat{\xi})}{\xi - \hat{\xi} - S(q, \hat{q})} \, d\hat{q} \wedge d\hat{\xi}, \tag{13}$$

where γ is the integration path in (12), and β a path as represented on Fig. 3.

Our claim is that (13) defines a confluent symbol with the above-mentioned support (in the convolutive model). Its Laplace transform is what we mean by (12).

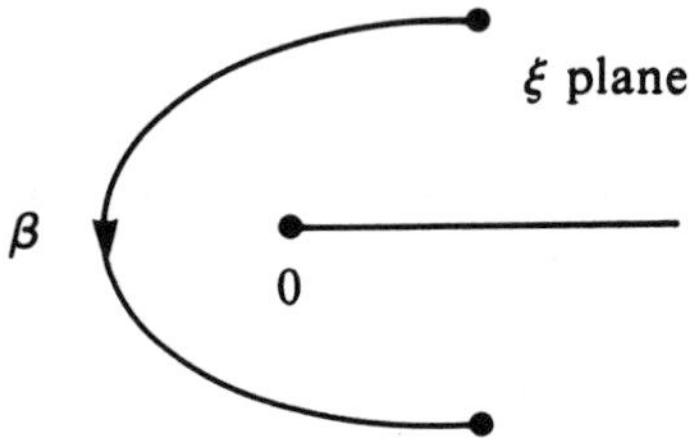

FIGURE 3

2.2.4. Asymptotic Expansions, and Local Resurgence Equations

These can be computed for the decompositions of confluent symbols into elementary symbols, exactly as in the case of the harmonic oscillator (cf. 2.1.3).

The results look exactly the same, with

$$\left.\begin{matrix} q^2/2 \\ -q^2/2 \\ s \end{matrix}\right\} \text{replaced by} \left\{\begin{matrix} S^0(q) \\ S_0(q) \\ s(x) \text{ (the "monodromy exponent")} \end{matrix}\right.$$

2.2.5. Comparison of Confluent Solutions

Lemma. *If two confluent solutions Ψ_1, Ψ_2 of the Schrödinger equation are* pure (*for some q*) with the same support, *they coincide up to normalization, i.e.* $\Psi_1(q, x) = C(x)\Psi_2(q, x)$ (*this result holds for confluent solutions of any differential equation having "simple characteristics" at generic points of its support*).

This easy lemma, combined with the comparison of asymptotic expansions, allows us to see to what extent the confluent solutions constructed in Sections 2.2.2 and 2.2.3 depend on the choice of the canonical transformation. We shall consider two cases.

First comparison lemma. *Let Ψ_1, Ψ_2 be constructed from two canonical transformations both satisfying condition* (5), *and with the same integration path γ.*

Then

$$\Psi_1(q, x) = C(x)\Psi_2(q, x), \tag{14}$$

where $C(x)$ is an invertible element of $\mathbf{C}_{(x)}$.

Second comparison lemma. *What if we "turn condition* (5) *upside-down"? Replace* (5) *by*

$$\begin{aligned} p = U(q) &\Leftrightarrow \hat{q} = 0, \\ p = -U(q) &\Leftrightarrow \hat{p} = 0. \end{aligned} \tag{$\tilde{5}$}$$

Then the monodromy exponent $s(x)$ *is replaced by* $\tilde{s}(x) = -s(x) - 1$.

If Ψ *and* $\tilde{\Psi}$ *are confluent symbols constructed according to* (5) *and* ($\tilde{5}$) *respectively, with integration paths chosen so as to give the same support (pure for some value of* q*), then*

$$\Psi = \frac{x^{-s-1/2}}{\Gamma(-s)} C(x)\tilde{\Psi}, \tag{15}$$

where $C(x)$ *is an invertible element of* $\mathbf{C}_{(x)}$.

Addendum: How does $C(x)$ *in formulas* (14) *and* (15) *depend on the parameter* E? Let us make the same hypothesis as in the Addendum to Proposition 2.2.2. Then

$$C(x, E) = \exp A(x, E), \qquad A(x, E) = \sum_{k=0}^{\infty} A_k(E) x^{-k},$$

with A_k a *polynomial in E of degree* $\leq k+1$.

2.3. Quantization of the Symmetric Double Well

As a typical example of how the local considerations of Section 2.2 work in a global context, we shall study the case of the symmetric double well

$$V(q) = \left(\frac{q^2 - q_0^2}{2} \right)^2$$

(or, more generally, any even function with the same shape, having two quadratic minima).

2.3.1. *Local Resurgence Equations*

In the classical limit at zero energy, two remarkable solutions of the Hamilton–Jacobi equations are

$$S^0(q) = \int_{q_0}^{q} \sqrt{V(q')}\, dq', \qquad \text{where } \sqrt{V(q')} = \frac{q^2 - q_0^2}{2},$$

and

$$\tilde{S}^0(q) = \int_{-q_0}^{q} -\sqrt{V(q')}\, dq' = S^0(-q).$$

Each of these determines, up to normalization, a formal solution of the Schrödinger equation (0)[2]

$$\Psi^0(q, x) = \varphi^0(q, x)\, e^{-xS^0(q)},$$

and

$$\tilde{\Psi}^0(q, x) = \Psi^0(-q, x),$$

where φ^0 is a formal power series of x^{-1}, with coefficients analytic in q on the universal covering of $\mathbf{C}\backslash\{-q_0, q_0\}$. It will be convenient to consider them rather as holomorphic functions of q in the cut plane $\operatorname{Arg}(q-q_0) \neq 0$, $\operatorname{Arg}(q+q_0) \neq \pi$.

From the results of Section 2.2, one easily deduces the following

Theorem. *For* $q \simeq q_0$ $(q \neq q_0)$, the elementary symbol Ψ^0 [*resp.* $\tilde{\Psi}^0$] is "locally resurgent" (in the sense commented hereafter) and satisfies local resurgence equations

$$-\dot{\Delta}_{e^{i\pi}z}\Psi^0 = \frac{x^{-s-1/2}}{\Gamma(-s)} C(x)\, e^{\omega x}\tilde{\Psi}^0, \qquad (16)$$
$$\dot{\Delta}_z\tilde{\Psi}^0 = \Gamma(s+1)x^{s+1/2}\tilde{C}(x)\, e^{-\omega x}\Psi^0,$$

where $z = 2S^0(q)$ *is the "singularity shift"* $(z = S^0 - (-S^0))$,

$$\omega = S^0(q) + \tilde{S}^0(q) = \tfrac{1}{2}\oint \sqrt{V(q)}\, dq$$

is the "half-period" (*independent of* q), *and* $s = s(x) \in \mathbf{C}_{(x)}$ *is the "monodromy exponent" of Section* 2.2 (*don't forget it also depends on* E). $C(x)$ *and* $\tilde{C}(x)$ *are elements of* $\mathbf{C}_{(x)}$ *depending on* E *as indicated in* 2.2.5 (*addendum to formula* (15)).

Notice that since $\tilde{\Psi}^0$ *is related to* Ψ^0 *by symmetry, the coefficients in the above resurgence equations do not depend on the normalization of* Ψ^0.

Comment. The elementary symbol Ψ^0 is defined only for $q \neq q_0$, so that in the ξ variable, the radius of convergence of $\psi^0(q, \xi)$ (its Borel transform)

[2] Don't forget that the energy is $x^{-1}E$, with E an arbitrary parameter.

is expected to vanish when $q \to q_0$. What we mean here by "local resurgence" is the following property: there exists a disc of *fixed* radius around $S^0(q)$ such that for every q close enough to q_0 ($q \neq q_0$), $\psi^0(q, \cdot)$ has no singularity in that disc outside the points $S^0(q)$ and $-S^0(q)$ (i.e., analytic continuation is possible along every path which avoids these two points).

Remark. Our proof of the theorem is constructive, in the sense that the coefficients of the formal power series $s(x)$, $C(x)$, $\tilde{C}(x)$, can be constructed to any finite order by a finite algorithm. A possible procedure is the following.

First step: Construct a canonical transformation. A convenient choice for the case of $V(q) = ([q^2 - 1]/2)^2$ is for instance

$$S(q, \hat{q}) \text{ (generating function)} = q\hat{q} - \frac{q^3}{6} + \frac{q}{2} - \tfrac{2}{3}(1+\hat{q})^{3/2} + \tfrac{1}{3}. \qquad (17)$$

Second step: Solve the canonically transformed Schrödinger equation. This is like solving transport equations—a problem for which physicists have efficient methods. Notice that the above chosen canonical transformation transforms the Schrödinger equation into another *differential* (not micro-differential!) equation, namely

$$[(1+\hat{q})^{1/2}(2\hat{q}\,\partial_{\hat{q}} + 1) + \tfrac{1}{2}(1+\hat{q})^{-1/2}\hat{q} - E + x^{-1}(\hat{q}\,\partial_{\hat{q}}^2 + \partial_{\hat{q}})]\Phi(\hat{q}, x) = 0 \quad (18)$$

a globally defined, although multivalued, second-order differential equation; continuation to the "second sheet" in $\hat{q}$ corresponds to q going from a neighbourhood of q_0 to a neighbourhood of $-q_0$.

Third step. Call Ψ a confluent symbol defined near q_0 by an integral of the type (12) (Section 2.2.3). Call $\tilde{\Psi}$ the confluent symbol near $-q_0$ deduced from Ψ by changing q into $-q$; $\tilde{\Psi}$ can be continued analytically to a neighbourhood of q_0 (just change sheets for the integration variable $\hat{q}$!).

Near q_0, the asymptotic expansions of either of these confluent symbols on either branch of their supports can be computed to any degree of accuracy (as double power series in $q - q_0$, $x^{-1}/(q-q_0)^2$). Comparing them gives the coefficients of the series $C(x)$, $\tilde{C}(x)$ to any desired degree of accuracy.

2.3.2. Global Resurgence Conjectures and the Zinn-Justin Quantization Condition

Resurgence conjecture 0. Everything in the above theorem 2.3.1 is (globally) resurgent in x; more precisely, $s(x)$, $C(x)$, $\tilde{C}(x)$ are resurgent functions of

x, with derivated singular support in $2\omega\mathbf{Z}$; $\Psi^0(q, x)$ can be normalized in such a way that it is resurgent in x (and of course analytic in q), with derivated singular support in $\{\xi = -S^0(q) + 2\omega\mathbf{Z}\}$.

The next conjecture deals with the dependence of $\Psi^0(q, x) = \Psi^0(q, x; E)$ on the parameter E. We shall denote by $\Psi(q, x; E)$ the extended resurgent function of x, *pure on the right* for $q > q_0$, having $\Psi^0(q, x; E)$ for its symbol (in the real positive direction).

Resurgence conjecture 1. The normalization of Ψ^0 in resurgence conjecture 0 can be so chosen that substitution in $\Psi(q, x; E)$ of any extended resurgent function $E = E(x)$ gives an *extended resurgent function* $\Psi(q, x; E(x))$—of course, no longer pure for $q > q_0$, but with singular support $S^0(q) + \Omega$, $\Omega = \text{sing supp } E(x)$.

Notice that any function thus obtained is exponentially decreasing for $q \to +\infty$.

Theorem. *A function* $\Psi(q, x; E(x))$ *of the type considered in resurgence conjecture* 1 *will be an* even (*resp.* odd) *function of* q (*and therefore an eigenfunction of the Schrödinger operator*) iff $E(x)$ *satisfies the following* "quantization condition"

$$\frac{x^{-s(x,E)-1/2}}{\Gamma(-s(x, E))} C(x, E)\, e^{\omega x} = \pm 1, \qquad (19)_\pm$$

(+1 *in the even case*, −1 *in the odd case*), *where* $C(x, E)$ *is the coefficient in the first resurgence equation* (16); *more precisely,* $C(x, E)$ *and* $s(x, E)$ *must be considered here as* extended, pure on the right *resurgent functions.*

Idea of the proof. Ψ will have the same parity property as its decomposition in the region $-q_0 < q < q_0$, which is easily deduced from its decomposition in the region $q > q_0$, thanks to the first resurgence equation (16).

Resolution of equation $(19)_\pm$ (after Zinn-Justin [Z2][Z3]). For $x \gg 0$, $e^{\omega x}$ is large, so that $-s(x, E)$ must be close to a pole of the Gamma function:

$$s(x, E) \simeq N \in \mathbf{N};$$

since $s(x, E) = (E-1)/2 + O(x^{-1})$, this implies the "Bohr-Sommerfeld quantization rule" $E \simeq 2N + 1$. Zinn-Justin noticed that equations of the form $(19)_{\pm}$ admit for every $N \in \mathbf{N}$ unique *formal series solutions*

$$\dot{E}^{(N_{\pm})}(x) = \sum_{n=0}^{\infty} \varepsilon_n^{(N_{\pm})} \lambda^n, \qquad \lambda = \sqrt{x}\, e^{-\omega x}, \qquad (20)_{\pm}$$

where $\varepsilon_n^{(N_{\pm})} \in \mathbf{C}[[x^{-1}]][\ln x]_{n-1}$ (polynomial of degree $\leq n-1$ in $\ln x$), with $\varepsilon_0^{(N_{\pm})} = 2N+1+O(x^{-1})$.

Resurgence conjecture 2. *$\dot{E}^{(N_{\pm})}(x)$ is a resurgent symbol*, and the corresponding extended resurgent function $E^{(N_{\pm})}(x)$, having $\dot{E}^{(N_{\pm})}(x)$ for its *right-decomposition*, solves equation $(19)_{\pm}$.

2.3.3. Concluding Comments

Notice that no canonical transformation was involved in the *statement* of our theorems 2.3.1, 2.3.2, nor in our "resurgence conjectures," which could presumably be proved in purely "Ecalle" fashion (constructing the relevant functions as infinite sums of "resurgence monomials," etc.). But even if this is so, we feel that our "quantized canonical transformations" deserve better understanding in relation with resurgence properties. When $S(q, \hat{q})$ is chosen as in eq. (17), it is most likely that the integral (12) can be given as concrete a meaning as it was given in Section 1 (in contradistinction with the sophistications of Section 2.2), with a *true function of* $\hat{q}$ and x as the integrand, and with an endless integration path satisfying a *global* descent condition on the Riemann surface of S.

Investigating this example in greater detail might give us hints about questions which are beyond the present scope of Ecalle's theory, such as *what is a "quantized canonical transformation of resurgent functions?" How do alien derivatives behave under such transformations?*

References

[B] V. Bargmann, *Comm. Pure Appl. Math.* **14** (1961).

[BB] R. Balian and C. Bloch, Solution of the Schrödinger equation in terms of classical paths. *Ann. of Physics* **85** (1974).

[BM] M. Berry and K. Mount, Semiclassical approximations in wave mechanics. *Rep. Progr. Phys.* **35** (1972).

[Bo] R. Boas, *Entire functions.* New York, Academic Press (1952).

[E1] J. Ecalle, *Les fonctions résurgentes,* tome 3 (Publ. math. Université de Paris-Sud 81-06).

[E2] J. Ecalle, Singularités irrégulières et résurgence multiple. In *Cinq applications des fonctions résurgentes* (preprint 84T 62, Orsay).

[E3] J. Ecalle, L'accélération des fonctions résurgentes (survol) (manuscrit, September 1987).

[Eg] I. V. Egorov, On canonical transformations of pseudo-differential equations. *Usp. Mat. Nauk* **25** (1969).

[H] L. Hörmander, Fourier integral operators. *Acta Math.* **127** (1971).

[K] M. Kashiwara, Microlocal calculus. In *Mathematical Problems in Theoretical Physics*, Lecture Notes in Physics **39** (1975).

[KK] M. Kashiwara and T. Kawai, On holonomic systems of microdifferential equations III. *Publ. R.I.M.S., Kyoto Univ.* **17**, 3 (1981).

[KKO] M. Kashiwara, T. Kawai, and T. Oshima, A study of Feynman integrals by micro-differential equations. *Commun. Math. Physics* **60** (1978).

[M] V. Maslov, *Theory of perturbations and asymptotic methods.* Moscow State University (1965).

[P1] F. Pham, *Singularités des systèmes différentiels de Gauss-Manin.* Progress in Math. **2**, Birkhäuser (1980).

[P2] F. Pham, Calcul microdifférentiel complexe et méthode semi-classique. R.C.P. no. 25, I.R.M.A. Strasbourg (1983).

[P3] F. Pham, Transformées de Laplace des microsolutions de systémes holonomes. *L'enseignement mathématique* **30** (1984).

[P4] F. Pham, Exercice semi-classique. In Actes du colloque. "*Méthodes semiclassiques en mécanique quantique*," Hellfer, Robert, & Sjöstrand eds. (1985).

[SKK] M. Sato, T. Kawai, and M. Kashiwara, *Microfunctions and pseudodifferential equations*, Lecture Notes in Mathematics **287** (1973).

[V] A. Voros, The return of the quartic oscillator (the complex WKB method). *Ann. Inst. H. Poincaré* **29**, 3 (1983).

[Z1] J. Zinn-Justin, The principles of instanton calculus. *Les Houches, Lecture Notes 1982.* North-Holland, Amsterdam (1982).

[Z2] J. Zinn-Justin, Multi-instanton contributions in Quantum Mechanics (II). *Nuclear Physics* **B218** (1983).

[Z3] J. Zinn-Justin, Instantons in Quantum Mechanics: numerical evidence for a conjecture. *J. Math. Phys.* **25** (3) (1984).

[*] B. Candelpergher, C. Nosmas, and F. Pham, *Résurgence et développements semi-classiques* (Seminaire semi-classique, Nice 1985–87), to appear.

Quantization of Extended Vortices and sDiff $\mathbf{R}^3$

Mario Rasetti
Dipartimento di Fisica
Politecnico di Torino
Torino, Italy

and

Tullio Regge
Dipartimento di Fisica Teorica
Università di Torino
Torino, Italy

1. Introduction

There is no general structure theorem on the set of unitary irreducible representations of sDiff $\mathbf{R}^3$. These representations have some physical interest in that they are related to the theory of quantum vortices in an incompressible superfluid. In [1, 2] we have discussed a particular set of induced representations and of course the broader set of representations obtained from this set by repeated tensor products. Such representations describe infinitesimal vortices only and account for macroscopic vortices in the semiclassical limit.

ISBN 0-12-400466-0

This picture is satisfactory from the physical point of view and it may be relevant to the study of the dispersion spectrum in the intermediate region between the phonon–roton branches and the macroscopic vortices. However it leaves entirely open the question of the existence of unitary representations which correspond directly to macroscopic vortices [3, 4]. We have conjectured that the irreducible unitary representations of the group sDiff $\mathbf{R}^3$ are the quantum counterpart of the isotopy classes of knots and that the Casimir operators play the role of knot invariants.

2. sDiff $\mathbf{R}^3$ as Phase Space for Vortices

A map $\Gamma: S^1 \to \mathbf{R}^3$, modulo reparametrizations in Diff S^1, will be referred to as a *classical vortex.* Explicitly we can represent it by the functions $x^i(\sigma)$, $i = 1, \ldots, 3$. We define then

$$\mathbf{A}(\mathbf{x}) = \frac{k}{4\pi} \int_\Gamma \frac{1}{|\mathbf{x}-\mathbf{y}|}\, d\mathbf{y}, \qquad \mathbf{v}(\mathbf{x}) = \operatorname{curl} \mathbf{A}. \tag{2.1}$$

We refer to $\mathbf{A}(\mathbf{x})$ and $\mathbf{v}(\mathbf{x})$ as the velocity field induced by the vortex. The constant k is given, in the case of superfluid helium, by $k = Nh/m$ where h is Planck's constant, m the mass of the helium atom and $N \in \mathbf{Z}$.

The Lagrangian of the vortex is then given by

$$\mathscr{L} = \tfrac{1}{3} k\rho \int_\Gamma \varepsilon_{ijl} x^i\, dx^j \frac{\partial x^l}{\partial t} - \frac{1}{2}\rho \int_{\mathbf{R}^3} \mathbf{v}^2\, d\mathbf{x}, \tag{2.2}$$

where $\rho = m/\Omega$ is the mass density of the fluid, Ω denoting the specific volume per atom.

From (2.2) one can develop a canonical formalism [1, 2] with the help of Dirac's theory of constrained Hamiltonian systems [5, 6]. The momentum density $P_i(\sigma)$ satisfies indeed the constraint

$$P_k(\sigma) = \tfrac{1}{3} k\rho\varepsilon_{ijk} x^j \frac{\partial x^k}{\partial \sigma}\,. \tag{2.3}$$

In [1, 2] we have shown that these constraints can be eliminated by describing the system entirely in terms of the integrated functionals:

$$J(\mathbf{b}, \Gamma) = \int_\Gamma \mathbf{B}(\mathbf{x}) \cdot d\mathbf{x}. \tag{2.4}$$

Here $\mathbf{B}(\mathbf{x})$ is a classical vector field whereas $\mathbf{b}(\mathbf{x}) = \text{curl } \mathbf{B}(\mathbf{x})$. In (2.4) the vectors $\mathbf{b}$ and $\mathbf{B}$ are used as labels for the dynamical variables on phase space. Obviously the $J(\mathbf{b}, \Gamma)$, hereafter referred to as *currents*, do not depend on the parametrization of Γ. The Poisson brackets (P.B.) are then calculated from (2.4) and the Dirac commutation relations derived in [1, 2].

We obtain the P.B.

$$\{J(\mathbf{b}), J(\mathbf{c})\} = -\frac{1}{k\rho} J(\text{curl}(\mathbf{b} \times \mathbf{c})). \tag{2.5}$$

This suggests the quantum commutator

$$[J(\mathbf{b}), J(\mathbf{c})] = \frac{i\hbar}{k\rho} J(\text{curl}(\mathbf{b} \times \mathbf{c})) = i\frac{\Omega}{2\pi} J(\mathbf{c}\nabla \cdot \mathbf{b} - \mathbf{b}\nabla \cdot \mathbf{c}). \tag{2.6}$$

In the sequel it will be convenient to introduce a particular set of currents, which shall be denoted $J(\mathbf{q}, n, \Gamma)$ or simply $J(\mathbf{q}, n)$, corresponding to the choice $\mathbf{B}(\mathbf{x}) \cdot d\mathbf{x} = dx^n \exp(i\mathbf{q} \cdot \mathbf{x})$, where $n, \mathbf{q}$ are polarization and Fourier mode labels respectively. With this notation the Hamiltonian is then given by

$$H = \frac{k^2\rho}{16\pi^3} \int \frac{d\mathbf{q}}{q^2} \sum_{n=1}^{3} J(\mathbf{q}, n) J(-\mathbf{q}, n). \tag{2.7}$$

H is thus a quadratic element in the enveloping algebra of the infinite-dimensional Lie algebra $\mathfrak{A}$ generated by the $J(\mathbf{b}, \Gamma)$ or equivalently by the $J(\mathbf{q}, n)$.

Upon denoting by $\mathfrak{A}(k)$ the subset spanned by all the $J(\mathbf{b})$ where $\mathbf{b}$ is a homogeneous polynomial in $\mathbf{x}$ of degree k, one can see from (2.5), (2.6) that $\mathfrak{A}(1)$ is a finite Lie subalgebra of $\mathfrak{A}$ isomorphic to the Lie algebra $SL(3, \mathbf{R})$. In general however the set of elements $\mathfrak{A}(k)$ with $k \leq p$ forms a subalgebra only if $p = 1$ since

$$[\mathfrak{A}(k), \mathfrak{A}(n)] \subset \mathfrak{A}(k+n-1). \tag{2.8}$$

On the other hand the set

$$\bigcup_{k=p}^{\infty} \mathfrak{A}(k) \equiv \mathfrak{U}(p) \tag{2.9}$$

is a subalgebra of $\mathfrak{A}$. Moreover, if $p \geq 1$, $\mathfrak{U}(p)$ is an invariant subalgebra of $\mathfrak{U}(1)$ and one can define the factor algebra

$$\mathfrak{T}(p) \equiv \mathfrak{U}(1)/\mathfrak{U}(p+1). \tag{2.10}$$

We denote with $\mathfrak{T}(p, q)$ the subset of elements of $\mathfrak{T}(p)$ where $\mathbf{b}$ is of degree q in $\mathbf{x}$. Clearly $\mathfrak{T}(p, q) = \emptyset$ for $q > p$. $\mathfrak{T}(p)$ forms a finite Lie algebra and moreover

$$[\mathfrak{T}(p, q), \mathfrak{T}(p, m)] \subset \mathfrak{T}(p, m+q-1). \tag{2.11}$$

Let $\mathfrak{K}(p)$ be the subgroup of sDiff $\mathbf{R}^3$ leaving the origin at $\mathbf{x} = 0$ and the derivatives of a generic function in C^∞ fixed up to the order p. Then $\mathfrak{T}(p)$ generate the group $\mathfrak{K}(0)/\mathfrak{K}(p)$.

With these definitions in mind it is easy to check that a representation for $J(\mathbf{b})$ is given by

$$\mathfrak{D}: J(\mathbf{b}) \to \frac{\Omega}{2\pi i} b^k \frac{\partial}{\partial x^k}, \tag{2.12}$$

and that it corresponds to the induced representations by volume-preserving diffeomorphisms acting on elements of $\mathscr{L}^2(\mathbf{R}^3)$. The representation $\mathfrak{D}$ is a particular case of the set $\mathscr{D}^p$ given by

$$\mathscr{D}^p: J(\mathbf{b}) \to \frac{\Omega}{2\pi i} \left\{ b^k \frac{\partial}{\partial x^k} + \sum_{n=1}^{p} \frac{1}{n!} \frac{\partial^n b^i}{\partial x^I} T_i^I \right\}, \tag{2.13}$$

where $I = \{i_1, i_2, \ldots, i_n\}$ modulo permutations. The commutation relations of the T_i^I are given by

$$[T_i^I, T_k^K] = \sum_p \{\delta_i^p T_k^{I \cup (K \setminus p)} - \delta_k^p T_i^{(I \setminus p) \cup K}\}. \tag{2.14}$$

Since we consider measure-preserving diffeomorphisms only, the T_i^I need to be traceless.

This representation can be formally obtained from (2.12) by tensoring it with itself as

$$J(\mathbf{b}) = \frac{\Omega}{2\pi i} b^a(\mathbf{x}) \frac{\partial}{\partial x^a} + \frac{\Omega}{2\pi i} b^a(\mathbf{y}) \frac{\partial}{\partial y^a}, \tag{2.15}$$

where $\mathbf{y} = \mathbf{x} + \boldsymbol{\xi}$ and $\boldsymbol{\xi}$ is nilpotent of order $p+1$. Physically these Lie algebras, as discussed in [1, 2], represent vortices which have an infinitesimal size but in which the moments up to the order p can still be considered as meaningful dynamical variables. For instance, the case $p = 1$ introduces essentially an intrinsic spin of the vortex.

3. Vortices and Knots

The dynamical description of vortices in an incompressible superfluid given above is adequate in many physical applications but it still leaves several open questions. The first unsatisfactory feature is well known and related to the logarithmic divergence already present at the classical level in the Hamiltonian (2.7); this divergence is formally similar to the one appearing in the self-impedance of a thin conducting loop [7] and it can be controlled by a standard regularization procedure [2].

The second unsolved problem concerns the existence of unitary irreducible representations which correspond to macroscopic vortices. These representations could be easily constructed should we know a quasi-invariant measure on the space of loops under the group sDiff $\mathbf{R}^3$ of volume preserving diffeomorphisms in $\mathbf{R}^3$.

To the best of our knowledge no such measure is known and a convenient representation of macroscopic vortices can only be obtained in the present framework as a particular limit in the low-lying energy states in the second quantized theory of the infinitesimal vortex [1, 2].

Physically such a limit leads to a picture where the quantum vortex is represented as a giant bound state of many elementary vortices constrained on a differentiable disk whose boundary is the macroscopic vortex and such that the momentum of each individual vortex lies along the normal to the surface.

A somewhat similar picture was proposed by R. Feynman many years ago [8].

Here we prefer to discuss without any pretence of rigour the possible connection between unitary irreducible representations (irreps) of sDiff $\mathbf{R}^3$ and the theory of knot invariants.

It appears from the previous discussion that the set $\{J(\mathbf{b}, \Gamma), \text{all } \mathbf{b}\}$ gives a complete classical physical description of the vortex Γ.

In the quantum domain the $J(\mathbf{b}, \Gamma)$ do not commute and this introduces an uncertainty in the shape of the vortex. As already discussed in [1, 2], it is for instance possible to define with absolute sharpness the projection of the vortex on an arbitrary plane (say the x, y plane) and in this case the area of the projection is, in the appropriate units, the momentum of the vortex along the z axis.

In this case we expect that the global position of the vortex along the z axis should be totally undetermined. At first sight it would appear that also the z coordinate of each point of the vortex should be completely undetermined since it does not commute with the x, y coordinates of the projection.

This is not the case as we shall show in the sequel. The x, y projection $\Gamma_\perp$ plays the role of a maximal abelian subalgebra $\mathfrak{H}$ of $\mathfrak{A}$ and it is generated by the subset of $J(\mathbf{b}, \Gamma)$ given by

$$B_1 = \partial_y V(x, y), \qquad B_2 = -\partial_x V(x, y), \qquad B_3 = 0, \tag{3.1}$$

that is, $b^1 = 0$, $b^2 = 0$, $b^3 = \Delta V$. The choice $V = \frac{1}{2}\ln[(x-\xi)^2 + (y-\eta)^2]$ yields for $J(\mathbf{b}, \Gamma)$ the winding number $N(\xi, \eta, \Gamma_\perp)$ of $\Gamma_\perp$ around the point ξ, η and it is reasonable to assume that the knowledge of $N(\xi, \eta, \Gamma_\perp)$ for all ξ, η conveys the complete information about $\Gamma_\perp$.

This discussion can be supplemented by looking at the appropriate ascending sequence of subalgebras $\mathfrak{H}^p$ of degree p of the enveloping algebra $\mathfrak{P}(\mathfrak{A})$ of $\mathfrak{A}$ which commute with $\mathfrak{H}$ and among themselves.

As an example in $\mathfrak{H}^2$ we look at the integrals of the kind

$$J(\mathcal{S}, \Gamma_\perp) = \int_{\Gamma\times\Gamma} \chi_{\mathcal{S}}(x, y)\delta(x - x')\delta(y - y')(z - z')(dx \wedge dy' - dy \wedge dx'), \tag{3.2}$$

where $\chi_{\mathcal{S}}$ denotes the characteristic function of the set $\mathcal{S} \subset \mathbf{R}^2$. $J(\mathcal{S}, \Gamma_\perp)$ is then seen to be expressible as the sum

$$J(\mathcal{S}, \Gamma) = \sum_{P \in \mathcal{S}} \delta z_P, \tag{3.3}$$

where the δz_P is the level difference at the self-crossing of $\Gamma_\perp$ and the sum is carried only on the points P contained in the set $\mathcal{S}$. In this way it is possible to deduce from the set $\{J(\mathcal{S}, \Gamma_\perp)$, all $\mathcal{S}\}$ the value of all the level differences δz_P. If $\Gamma_\perp$ is sharply known and it has only simple self-crossings then this additional information allows the complete reconstruction of Dehn's diagram and the computation of knot invariants of Γ considered as a knot in $\mathbf{R}^3$.

We conjecture that the extra ambiguities arising in the case of multiple and degenerated self-crossings can be likewise resolved by computing suitable elements in $\mathfrak{H}^p$, $p > 2$.

This conjecture makes all the more interesting and pressing the construction of irreps corresponding to truly extended vortices, if any, since all these observables do not have a precise meaning either in the irreps generated by (2.13) or in their tensor products, including irreps obtained via second quantization.

Should the conjecture be resolved affirmatively, then the Casimir operators would play the role of knot invariants and quite possibly they would be directly related to them. In general they are not expected to be expressible as polynomials in $\mathfrak{A}$ and this implies that their definition in terms of integrals of the kind (3.2) requires an indefinite number of iterated integrations over Γ.

4. Conclusions

The quantum theory of vortices appears as directly related to the mathematical problem of constructing and classifying all the irreducible representations of sDiff $\mathbf{R}^3$. In turn this problem is related to the construction of a quasi-invariant measure on the space of loops under sDiff $\mathbf{R}^3$. The present non-relativistic theory of vortices is a particular limiting case of the Kalb–Ramond model [9] of a relativistic string in interaction with a pseudovector field. Vortices in $\mathbf{R}^3$ present therefore in a simplified version many of the still unsolved questions of the string model.

The question of the existence of a suitable loop measure is a common theme and a central feature of a variety of problems on the borderline between physics and mathematics.

Finally, the notion that the classifications of knots and links according to their topological invariants may have a quantum analog in the irreps of sDiff $\mathbf{R}^3$ is also interesting and deserves further attention.

Acknowledgement

We are indebted to D. Sharp and S. Sternberg for many interesting discussions.

References

[1] M. Rasetti and T. Regge, *Physica* **A80** (1975) 217.

[2] M. Rasetti and T. Regge, Quantum vortices, in *Highlights of Condensed Matter Theory*, (F. Bassani *et al.*, eds.), Ed. Compositori, Bologna, 1985.

[3] G. A. Goldin, R. Menikoff, and D. H. Sharp, *Diffeomorphism Groups and Quantized Vortex Filaments*, Los Alamos Nat. Lab., Theor. Div. Preprint LA-UR-86-3723.

[4] G. A. Goldin, R. Menikoff, and D. H. Sharp, Diffeomorphism Groups, Coadjoint Orbits and the Quantization of Classical Fluids, and Quantized Vortex Filaments in Incompressible Fluids, in *The Physics of Phase Spaces* (Y. S. Kim and W. W. Zachary, eds.), Lecture Notes in Physics No. 278, Springer-Verlag, New York, 1987.

[5] P. A. M. Dirac, *Lectures on Quantum Mechanics*, Belfer Graduate School of Science, Yeshiva University, New York, 1965.

[6] A. J. Hanson, T. Regge, and C. Teitelboim, *Constrained Hamiltonian Systems*, Accad. Naz. dei Lincei, Roma, 1976.

[7] F. Linder, *J. Am. Inst. Electr. Eng.* **46** (1927) 614.

[8] R. P. Feynman, *Progress in Low Temperature Physics*, Vol. 1, Chapter II (C. G. Goerter, ed.), North-Holland, Amsterdam, 1955.

[9] F. Lund and T. Regge, *Phys. Rev.* **D14** (1976) 1524.

Moduli Space for Fuchsian Groups

Kyoji Saito
Research Institute for Mathematical Sciences
Kyoto University
Kyoto, Japan

We attach to the moduli space for Fuchsian groups (= Fricke moduli space) an S^1-bundle structure and a complex structure. This gives another approach to the Teichmüller theory. The motivation for this work is found in conformal field theory.

ISBN 0-12-400466-0

1. Introduction

1.1.

First we explain briefly a motivation from physics. The conformal field theory has been initiated by Belavin, Polyakov and Zamolodchikov [10]. In their original work, the field operators depend on complex parameters z, which run over rational curves. The theory has been extended to non-rational curves, whose coordinate patchings depend on the moduli of the curves. Hence one encounters a problem: the relation between coordinates and moduli for curves of higher genus ([9], [17], [21–22]).

Beilinson, Manin and Schectman [8] approached this problem by introducing an infinite-dimensional fiber space over the moduli of curves. It consists of all *formal local coordinates* for curves. On this infinite-dimensional space, the τ-function in the sense of M. Sato [39, 40] (cf. [14]) was calculated by [26] (cf. [1]).

The present work has a motivation to understand these works by proposing a finite-dimensional space instead of the infinite-dimensional one. Namely, by considering all *global coordinates* for curves instead of infinitesimal coordinates, one obtains a real 3-dimensional fiber space over the moduli, which we shall investigate in detail in the present work. Actually the space is very classical. The basic ideas for the space have roots in the work of Fricke [20], and we call this space the *Fricke moduli.*

1.2.

To explain more details of this approach, we recall Poincaré and Klein's uniformization of Riemann surfaces ([37]).

Except for $\mathbf{P}^1(\mathbf{C})$, $\mathbf{C}$, $\mathbf{C}^*$ or an elliptic curve, a Riemann surface X admits a holomorphic universal covering (= uniformization) map:

$$\rho : \mathbf{D} \longrightarrow X,$$

from the unit disc $\mathbf{D} := \{z \in \mathbf{C} : |z| < 1\}$. We propose to use this coordinate z of the disc $\mathbf{D}$ for the parameter in the conformal theory, since z has both local and global nature for the Riemann surface X.

A uniformization map ρ for X is determined by two data: (i). The image $\rho(0) \in X$ of $0 \in \mathbf{D}$ in X, and (ii). The angle 2θ of change $z' = \exp(2i\theta)z$ for $2\theta \in PS^1 := S^1/\pm 1$. Hence the set $\{\rho$: uniformization map for $X\}$ forms a real 3-dimensional variety as a PS^1-bundle over X, denoted by

$$\pi : PS^1(X) \longrightarrow X.$$

Then there exists a universal uniformization map for X

$$u: PS^1(X)\times \mathbf{D} \longrightarrow X$$

in the sense that for each fixed $\rho \in PS^1(X)$, $u(\rho, *): \mathbf{D} \to X$ is a uniformization map for X such that

(i) $u(\rho, 0) = \pi(\rho)$,
(ii) $u(\rho\cdot \exp(2i\theta), z) = u(\rho, \exp(2i\theta)z)$.

Here $\exp(2i\theta) \in PSO(2) := SO(2)/\pm 1$ acts on fibers of $PS^1(X)$ as a principal bundle. In fact, the double covering $S^1(X)$ of $PS^1(X)$ as a principal $SO(2)$-bundle (cf. (1.3) Remark 1) is constructed in terms of Fuchsian groups in the following (1.3). (May skip to (1.4).)

1.3.

To agree with the later formulation of this paper, we use the upper half complex plane $\mathbf{H} := \{z \in \mathbf{C}: \mathrm{Im}(z) > 0\}$ instead of the disc $\mathbf{D}$.

Recall the facts $\mathrm{Aut}(\mathbf{H}) = PSL(2, \mathbf{R})$ and $H \simeq PSL(2, \mathbf{R})/PSO(2) \simeq SL(2, \mathbf{R})/SO(2)$, where $SO(2)$ = the isotropy at $i \in \mathbf{H}$. Let $\Gamma \subset SL(2, \mathbf{R})$ be a Fuchsian group such that $X \simeq \Gamma\backslash\mathbf{H} \simeq \Gamma\backslash SL(2, \mathbf{R})/SO(2)$ (biholomorph).

Using these notations, the S^1-bundle discussed in (1.2) is given by

$$S^1(X) := \Gamma\backslash SL(2, \mathbf{R}).$$

The maps u and π in (1.2) are lifted to $S^1(X)$ by the same notions as:

$$\pi: S^1(X) := \Gamma\backslash SL(2, \mathbf{R}) \to X := \Gamma\backslash SL(2, \mathbf{R})/SO(2),$$

$$\Gamma\backslash g \mapsto \Gamma\backslash g/SO(2),$$

$$u: S^1(X)\times \mathbf{H} := \Gamma\backslash SL(2, \mathbf{R}) \times SL(2, \mathbf{R})/SO(2) \to X := \Gamma\backslash SL(2, \mathbf{R})/SO(2),$$

$$\Gamma\backslash g\times h/SO(2) \mapsto \Gamma\backslash g\cdot h/SO(2).$$

The properties (i) and (ii) for π and u are verified immediately.

Remark 1. By using $PSL(2,\mathbf{R})$, we have

$$PS^1(X) = \Gamma\backslash PSL(2, \mathbf{R}).$$

Instead, we use the double cover $SL(2, \mathbf{R})$ as a building block in the present paper, due to (3.3) Assertion. (cf. (5.1) Note 1.)

Remark 2. The $PS^1(X)$ is embedded to a unit circle bundle in the tangent bundle of X by the Poincaré metric (cf. (2.5)).

1.4.

The universal map u for a single Riemann surface X in (1.2-3) will be extended to the family $\mathscr{X}_g$ of Riemann surfaces:

$$\Phi : \mathscr{X}_g \longrightarrow \mathscr{T}_g,$$

over a finite covering $\mathscr{T}_g$ of the Teichmuller space T_g of genus g (cf. (5.3) Remark). Namely, we shall describe a S^1-bundle Π and the universal map U in (13.1):

$$\Pi : S^1(\mathscr{X}_g) \longrightarrow \mathscr{X}_g,$$

$$U : S^1(\mathscr{X}_g) \times \mathbf{H} \longrightarrow \mathscr{X}_g,$$

whose restriction over a point $t \in \mathscr{T}_g$ are the maps u_t and π_t for a Riemann surface $X_t := \Phi^{-1}(t)$ as described in (1.2-3).

In other words, the space $S^1(\mathscr{X}_g)$ is the double covering of the parameter space for all uniformizations for all Riemann surfaces of genus g. As the case for a single X, the space $S^1(\mathscr{X}_g)$ is constructed in terms of Fuchsian groups, i.e., by the Fricke moduli space in this paper. (See (5.3).)

1.5.

The theory for discrete subgroups of $PSL(2, \mathbf{R})$ and $SL(2, \mathbf{R})$ is quite different from that for other Lie groups, due to its rich connection with the theory of Riemann surfaces.

Fricke [20] has initiated the study of the moduli of Riemann surfaces using Fuchsian groups. His work was followed by several authors ([19, 35, 27]). In the mean time, the theory of moduli has been developed rather differently in Teichmüller theory and in Kleinian group theory (for instance cf. [30]). However, in the present work, we continue working in Fuchsian groups to retain the idea of Fricke.

Along this line, we construct directly the spaces $\mathscr{X}_g$ and $\mathscr{T}_g$ with their complex structures and the unit circle bundle $S^1(\mathscr{X}_g)$ embedded in the relative tangent space $T(\mathscr{X}_g/\mathscr{T}_g)$, where the main tool for the construction is Weil's representation space R_0 (cf. [48-50]).

As a result, our approach gives another construction and proof of the complex structure on the Teichmüller spaces. Namely, a key step in Ahlfors and Bers's theory for Teichmüller spaces is the analytic dependence of the normalized solution w_μ of the Beltrami equation $\partial w/\partial \bar{z} = \mu \cdot \partial w/\partial z$ on the coefficients μ [5]. This step is replaced in this paper by a simultaneous trivialization of some cohomology classes (cf. (8.5) *Lemma**).

(For another construction of the complex structure, cf. [46]).

1.6.

As mentioned above, in preparing the paper, the author found that a series of works of A. Weil [48–50] treated the same object in general. Namely, the small as well as infinitesimal deformations for discrete subgroups of Lie groups were studied. We recall and use them in an essential way in the present paper (cf. § 4).

We also remark on the close relation of our approach with the cohomology theory on symmetric domains, developed by Matsushima and Murakami [33, 34] (see (7.1)), (cf. also [18], [15], [42], [29]).

1.7.

Finally the author would like to point out a strong similarity between the conformal field theory and the theory of period maps for primitive forms ([36, 38]), which inspired him to do this work. The mathematical common roots for both theories should yet be studied.

So, motivated anew by conformal field theory and influenced by the works of Weil, the present paper is a detailed account of a construction of the Fricke moduli space with the above-mentioned complex structures. Each step is elementary and classical, compared with far developed theories in differential and algebraic geometry and Teichmüller spaces, still the moduli of Fuchsian groups seems to form a beautiful world, which demands finer understanding yet.

1.8.

The contents of the present paper are as follows. Sections 2 and 3 are devoted to preparations on $SL(2, \mathbf{R})$ and Fuchsian groups, which may be skipped or returned to when necessary. Weil's result on representation space is recalled in Section 4. The Fricke moduli space as a real manifold is introduced in Section 5, whose infinitesimal structure is studied in Section 6. The complex structures on the moduli are introduced in Section 7. The integrability of them are shown in Section 9 and Section 10. The key lemma on simultaneous trivialization of cohomology classes is given in (8.5). The action $SL(2, \mathbf{R})$ on the moduli space, which we call the gauge transformation, is studied in Sections 11–12. The universal uniformization map U is described in Section 13, which might be understandable immediately after Section 5 so far as the complex structure is not concerned.

1.9.

The author expresses his gratitude to Professors Oikawa, Koizumi and Kusunoki for guidance to the literatures on Fuchsian groups and Fricke moduli and to Professor Murakami for noticing the work of Weil [50] and also for helpful discussions.

He expresses his gratitude to Dr. and Mrs. G. Kato for the correction of the English text. He expresses also thanks to the members of the seminar in Kyoto for discussions and interests.

2. $SL(2, \mathbf{R})$ and $PSL(2, \mathbf{R})$

We fix some notations and concepts on $SL(2, \mathbf{R})$ for later use.

2.1.

Let $\mathbf{P}^1(\mathbf{R})(\mathbf{P}^1(\mathbf{C}))$ be the real (complex) projective space for column vectors $\left[\begin{smallmatrix} u \\ v \end{smallmatrix}\right]$, whose inhomogeneous coordinate will be denoted by $z := u/v$. $\mathbf{P}^1(\mathbf{C})$ decomposes into three orbits: the upper half plane $\mathbf{H} := \{z \in \mathbf{C}: \operatorname{Im}(z) > 0\}$, the lower half plane $\bar{\mathbf{H}}$, and $\mathbf{P}^1(\mathbf{R})$ by the projective linear action of $PSL(2, \mathbf{R}) \simeq SL(2, \mathbf{R})/\{\pm 1\}$. The points $z = \pm i$ are fixed points of $PSO(2)$ in each $\mathbf{H}$ and $\bar{\mathbf{H}}$, since $u^2 + v^2 = (u + iv)(u - iv)$. So we obtain:

$$\mathbf{H} \simeq PSL(2, \mathbf{R})/PSO(2) \simeq SL(2, \mathbf{R})/SO(2), \qquad z = g(i) \leftarrow g/SO(2).$$

The action of $g \in SL(2, \mathbf{R})$ on $z \in \mathbf{H}$ from the left is given by the fractional linear transformation $g(z) = (az + b)/(cz + d)$ for $g = \left[\begin{smallmatrix} a & b \\ c & d \end{smallmatrix}\right]$.

2.2. Adjoint Actions of $SL(2, \mathbf{R})$

Let $\mathfrak{sl}(2, \mathbf{R}) := \{g \in M(2, \mathbf{R}): \operatorname{tr}(g) = 0\}$ (resp. $\mathfrak{so}(2) := \{g \in M(2, \mathbf{R}): \operatorname{tr}(g) = 0, {}^t g = -g\}$) be the Lie algebra for $SL(2, \mathbf{R})$ $(PSL(2, \mathbf{R}))$ (resp. $SO(2)$).

The adjoint action $A \cdot \operatorname{Ad}(g)$ of $g \in SL(2, \mathbf{R})$ on $A \in SL(2, \mathbf{R})$ is defined by the map $\operatorname{Ad}(g): A \mapsto g^{-1}Ag$. The adjoint representation $u \cdot \operatorname{ad}(g) = g^{-1}ug$ of $g \in SL(2, \mathbf{R})$ on $u \in \mathfrak{sl}(2, \mathbf{R})$ is the differential $d\operatorname{Ad}(g)_*$ of the map $\operatorname{Ad}(g)$ at the identity in $SL(2, \mathbf{R})$.

We denote by Θ the sheaf of germs of holomorphic vector fields on $\mathbf{H}$. Hence $\Gamma(\mathbf{H}, \Theta)$ means the space of global holomorphic vector fields on $\mathbf{H}$. The action of $g \in SL(2, \mathbf{R})$ on $\mathbf{H}$ induces an action on $\Gamma(\mathbf{H}, \Theta)$ by pull-back $(dg_*)^{-1}$. For convenience we shall call it also the adjoint action on $\Gamma(\mathbf{H}, \Theta)$, denoted by the same $\operatorname{ad}(g)$ as a right action. This should not cause confusion.

Therefore, for

$$g=\begin{bmatrix} a & b \\ c & d \end{bmatrix}\in SL(2,\mathbf{R}) \quad \text{and} \quad v=\varphi(z)\frac{\partial}{\partial z}\in\Gamma(\mathbf{H},\Theta),$$

$$v\cdot \mathrm{ad}(g):=\varphi(g(z))\left(\frac{\partial g}{\partial z}\right)^{-1}\frac{\partial}{\partial z}=\varphi(g(z))(cz+d)^2\frac{\partial}{\partial z}. \tag{2.2.1}$$

The adjoint actions have associativity:

$$v\cdot \mathrm{ad}(g_1g_2)=(v\cdot \mathrm{ad}(g_1))\cdot \mathrm{ad}(g_2) \qquad \text{for } g_1, g_2\in SL(2,\mathbf{R}). \tag{2.2.2}$$

2.3.

For an element $g\in SL(2,\mathbf{R})$, let us define a linear map:

$$\delta_g:\mathfrak{sl}(2,\mathbf{R})\to\mathfrak{sl}(2,\mathbf{R}), \qquad v\mapsto v-v\cdot \mathrm{ad}(g). \tag{2.3.1}$$

Assertion. *Let a, $b\in SL(2, \mathbf{R})$ be hyperbolic elements (i.e., $|\mathrm{tr}(a)|$, $|\mathrm{tr}(b)|>2$). Then the following are equivalent.*

(*i*) $[a,b]=0$, (*ii*) $\ker(\delta_a)=\ker(\delta_b)$, (*iii*) $\mathrm{Im}(\delta_a)=\mathrm{Im}(\delta_b)$.

2.4. Representation of $\mathfrak{sl}(2,\mathbf{R})$ in $\Gamma(\mathscr{H},\Theta)$

An element u of $\mathfrak{sl}(2,\mathbf{R})$ may be regarded as a right $SL(2,\mathbf{R})$ invariant vector field on $SL(2,\mathbf{R})$. Hence we define $\iota(u)$ by the direct image $d\tilde{\pi}_*(-u)$ for the natural projection $\tilde{\pi}:SL(2,\mathbf{R})\to\mathbf{H}$.

$$\iota(u):=d\tilde{\pi}_*(-u)\in\Gamma(\mathbf{H},\Theta). \tag{2.4.1}$$

An explicit calculation $\iota(u)=(d/dt\,\exp(-ut)z)|_{t=0}\,d/dz$ shows that

$$\iota(u)=[-1,z]u\begin{bmatrix} z \\ 1\end{bmatrix}\frac{d}{dz}. \tag{2.4.2}$$

Assertion. 1. *The map $\iota:\mathfrak{sl}(2,\mathbf{R})\to\Gamma(\mathbf{H},\Theta)$ is an $SL(2,\mathbf{R})$ equivariant injective Lie algebra homomorphism.*

2. *The map ι induces the following commutative diagram:*

$$\begin{array}{ccccccccc} 0 & \longrightarrow & \mathfrak{so}(2) & \longrightarrow & \mathfrak{sl}(2,\mathbf{R}) & \longrightarrow & \mathfrak{sl}(2,\mathbf{R})/\mathfrak{so}(2) & \longrightarrow & 0 \\ & & \downarrow{\scriptstyle\iota} & & \downarrow{\scriptstyle\iota} & & \downarrow{\scriptstyle\iota} & & \\ 0 & \longrightarrow & \Gamma(\mathbf{H},\Theta)(-[i]) & \longrightarrow & \Gamma(\mathbf{H},\Theta) & \longrightarrow & T_i(\mathbf{H}) & \longrightarrow & 0, \end{array} \tag{2.4.3}$$

where $\Gamma(\mathbf{H},\Theta(-[i])):=\{$holomorphic vector fields on $\mathbf{H}$ which is zero at $z=i\}$, and $T_i(\mathbf{H}):=$ the tangent space of $\mathbf{H}$ at $z=i$.

Proof. 1. Let

$$H=\begin{bmatrix}1 & 0\\ 0 & -1\end{bmatrix},\qquad X=\begin{bmatrix}0 & 1\\ 0 & 0\end{bmatrix}\quad\text{and}\quad Y=\begin{bmatrix}0 & 0\\ 1 & 0\end{bmatrix}$$

be the basis of $\mathfrak{sl}(2,\mathbf{R})$. Their representations by ι are

$$\iota(H)=-2z\frac{d}{dz},\qquad \iota(X)=-\frac{d}{dz},\qquad \text{and } \iota(Y)=z^2\frac{d}{dz},$$

which are vector fields of degree 2 real polynomial coefficients, which satisfy the same relations as H, X and Y. Thus ι is homomorphic.

The equivariance of ι follows from a relation

$$[-1,z]g^{-1}ug\begin{bmatrix}z\\ 1\end{bmatrix}=[-1,g(z)]u\begin{bmatrix}g(z)\\ 1\end{bmatrix}\cdot(cz+d)^2$$

for $g=\left[\begin{smallmatrix}a & b\\ c & d\end{smallmatrix}\right]\in SL(2,\mathbf{R})$ and $u\in\mathfrak{sl}(2,\mathbf{R})$.

2. One has

$$\iota(u)|_{z=i}=(-\beta-\gamma+i(-\alpha+\delta))\frac{d}{dz}\quad\text{for } u=\begin{bmatrix}\alpha & \beta\\ \gamma & \delta\end{bmatrix}\in\mathfrak{sl}(2,\mathbf{R}).$$

Thus (*i*) $\iota(u)|_{z=0}=0\Leftrightarrow\beta+\gamma=0$ and $\alpha=\delta=0\Leftrightarrow u\in\mathfrak{so}(2)$, and (*ii*) $\iota(\mathfrak{sl}(2,\mathbf{R}))|_{z=i}=(\mathbf{R}+i\mathbf{R})\,d/dz$, which spans the tangent space $T_i(\mathbf{H})$ of $\mathbf{H}$ at $z=i$.

Note. The definitions of ad (2.2.1) and ι (2.4.1) using pull-backs of vector fields are made to coordinate with the definition of the action of Γ on forms on $\mathbf{H}$ by pull-backs (cf. (8.4)).

2.5. The Circle Bundle over H

The projection $\pi: PSL(2,\mathbf{R})\to\mathbf{H}(g\mapsto g(i))$ is a principal fiber bundle with the fiber $PSO(2)$. Fix a tangent vector $v_0=d/dz\in T_i(\mathbf{H})$ at $z=i$.

Assertion. *The correspondence $g\in PSL(2,\mathbf{R})\mapsto dg_*(v_0)\in T_{g(i)}(\mathbf{H})$ (= the tangent space of $\mathbf{H}$ at $z=g(i)$), is an embedding of $PSL(2,\mathbf{R})$ into $T(\mathbf{H})$ as a unit circle bundle. The Hermitian metric associated to the unit circle bundle is the* Poincaré *metric of* $\mathbf{H}$.

To see this, let us represent any element $g\in SL(2,\mathbf{R})$ uniquely as

$$g=\begin{bmatrix}1 & x\\ 0 & 1\end{bmatrix}\begin{bmatrix}\sqrt{y} & 0\\ 0 & 1/\sqrt{y}\end{bmatrix}\begin{bmatrix}\cos(\theta) & \sin(\theta)\\ -\sin(\theta) & \cos(\theta)\end{bmatrix},\tag{2.5.1}$$

for $x\in\mathbf{R}$, $y\in\mathbf{R}_{>0}$, and $\theta\in S^1$. Then $g(i)=x+yi$ gives the coordinate for the base $\mathbf{H}$. The tangent vector $dg_*(v^0)$ is calculated as

$$dg_*(v_0) = \frac{dg}{dz}(i)\frac{d}{dz} = (ci+d)^{-2}\frac{d}{dz} = y\,e^{2i\theta}\frac{d}{dz}. \tag{2.5.2}$$

This particularly implies that the map $g \mapsto (g(i), dg_*(v_0))$ defines the embedding of $PSL(2, \mathbf{R})$ into the tangent bundle $T(\mathbf{H})$. Furthermore, the formula implies that the metric form associated to the unit circle is the Poincaré metric form

$$\left(\frac{dz}{y\,e^{2i\theta}}\right) \otimes \left(\frac{\overline{dz}}{y\,e^{2i\theta}}\right) = \frac{dz \otimes \overline{dz}}{y^2}. \tag{2.5.3}$$

2.6.

Any automorphism of $PSL(2, \mathbf{R})$ is given by an adjoint action,

$$\mathrm{Aut}(PSL(2, \mathbf{R})) = \mathrm{Ad}(GL(2, \mathbf{R})). \tag{2.6.1}$$

Here

$$\begin{aligned} \mathrm{Ad}(GL(2, \mathbf{R})) &\simeq PGL(2, \mathbf{R}) = PGL_+(2, \mathbf{R}) \cup PGL_-(2, \mathbf{R}) \\ &= PSL(2, \mathbf{R}) \cup \left(PSL(2, \mathbf{R}) \cdot \mathrm{Ad}\begin{bmatrix} 1 & 0 \\ 0 & -1 \end{bmatrix}\right). \end{aligned}$$

The element $\left[\begin{smallmatrix} 1 & 0 \\ 0 & -1 \end{smallmatrix}\right]$ induces the map $z \mapsto -z$ interchanging the upper and lower half places.

Let $\beta \in \mathrm{Aut}(PSL(2, \mathbf{R}))$ and $d\beta \in \mathrm{Aut}(\mathfrak{sl}(2, \mathbf{R}))$ be its differential. Then the following diagram is commutative as a result of (2.6.1).

$$\begin{array}{ccc} \mathrm{ad}: \mathfrak{sl}(2, \mathbf{R}) \times \mathrm{ad}(SL(2, \mathbf{R})) & \longrightarrow & \mathfrak{sl}(2, \mathbf{R}) \\ \Big\downarrow {\scriptstyle d\beta \times \beta} & & \Big\downarrow {\scriptstyle d\beta} \\ \mathrm{ad}: \mathfrak{sl}(2, \mathbf{R}) \times \mathrm{ad}(SL(2, \mathbf{R})) & \longrightarrow & \mathfrak{sl}(2, \mathbf{R}). \end{array} \tag{2.6.2}$$

2.7.

Let $p(t) \in M(n)$ be a family of $n \times n$ matrices for $t \in (-1, 1)$. It is obvious that

$$\frac{d}{dt}\det(p(t)) = \mathrm{tr}\left(p(t)^{-1}\frac{d}{dt}p(t)\right) \cdot \det(p(t)). \tag{2.7.1}$$

Particularly if $p(t) \in SL(2, \mathbf{R})$, then $p(t)^{-1}\, dp/dt(t) \in \mathfrak{sl}(2, \mathbf{R})$.

For $\beta \in \mathrm{Aut}(PSL(2, \mathbf{R}))$, using (2.6.1) we have,

$$(\beta p(t))^{-1}\left(\frac{d\beta p}{dt}(t)\right) = d\beta\left(p(t)^{-1}\frac{dp}{dt}(t)\right). \tag{2.7.2}$$

3. Fuchsian Groups

Some definitions and results concerning Fuchsian groups are recalled. For details, refer to the literature in the reference.

3.1.

It is well known that a subgroup Γ of $SL(2, \mathbf{R})$ is discrete if and only if its action on $\mathbf{H}$ is properly discontinuous.

By a *Fuchsian group of the first kind*, we mean a discrete subgroup Γ of $SL(2, \mathbf{R})$ (or of $PSL(2, \mathbf{R})$) such that the quotient $\Gamma \backslash \mathbf{H}$ is completed to a compact Riemann surface by adding a finite number of cusp points. It is equivalent to say that Γ is a discrete subgroup of $SL(2, \mathbf{R})$ with $\mathrm{vol}(\Gamma \backslash \mathbf{H}) < \infty$ (Siegel). Generators and relations for such groups are well known [20].

3.2.

This paper is concerned mainly with the following case. By a *Fuchsian surface group*, we mean a Fuchsian group Γ of the first kind such that (i) the action of Γ on $\mathbf{H}$ is fixed-point free, and (ii) $\Gamma \backslash \mathbf{H}$ is compact. This is the case when the quotient $\Gamma \backslash \mathbf{H}$ is a smooth compact Riemann surface, and Γ is isomorphic to the fundamental group of $\Gamma \backslash \mathbf{H}$.

The following facts are well known.

Assertion. *Let Γ be a Fuchsian surface group. Then*

(i) *Except for the neutral element, any element of Γ is hyperbolic, i.e., it is a matrix in $SL(2, \mathbf{R})$ with $|\mathrm{trace}| > 2$.*

(ii) *As an abstract group, Γ is represented as*

$$\left\langle a_1, \ldots, a_g, b_1, \ldots, b_g \colon \prod_{i=1}^{g} (a_i b_i a_i^{-1} b_i^{-1}) = 1 \right\rangle, \tag{3.2.1}$$

for an integer $g \geq 2$, where g is the genus of $\Gamma \backslash \mathbf{H}$.

The abstract group given by (3.2.1) will also be referred to as an (abstract) *Fuchsian surface group of genus g.*

3.3.

The following fact asked by Siegel [44] and answered by Bers [11] gives a finer structure on the moduli (cf. (5.1) Note 1).

Assertion. *Let $\Gamma \subset PSL(2, \mathbf{R})$ be a Fuchsian surface group with a system of generators $a_1, \ldots, a_g, b_1, \ldots, b_g$ satisfying the relation in (3.2.1). Then representatives A_i, B_i of a_i, b_i $(i = 1, \ldots, g)$ in $SL(2, \mathbf{R})$ also satisfy*

$$\prod_{i=1}^{g} (A_i B_i A_i^{-1} B_i^{-1}) = 1. \tag{3.3.1}$$

4. Representation Space $R_0(\Gamma, G)$

Weil's representation space $R(\Gamma, G)$ and its basic properties are recalled [48, 49]. Here G denotes a Lie group and Γ denotes an abstract group with discrete topology. After Section 5, G will be $PSL(2, \mathbf{R})$ or $SL(2, \mathbf{R})$ and Γ will be a Fuchsian group.

4.1.

Let G and Γ be as above. Let us define:

$$R(\Gamma, G) := \text{the set of all homomorphism of } \Gamma \text{ into } G. \tag{4.1.1}$$

The set $R(\Gamma, G)$ is embedded in the product space $\prod_{\gamma \in \Gamma} G_\gamma$, where G_γ is the same as G for every $\gamma \in \Gamma$, so that it carries naturally the induced topology. If Γ is generated by a finite set A, then $R(\Gamma, G)$ is a closed real analytic subset of $\prod_{\gamma \in A} G_\gamma$ defined by the equations given by the relations among the elements of A. Put

$$R_0(\Gamma, G) := \{r \in R(\Gamma, G) \colon r(\Gamma) \text{ is a discrete subgroup of } G \text{ isomorphic to } \Gamma \text{ such that } G/r(\Gamma) \text{ is compact}\}. \tag{4.1.2}$$

It was conjectured by A. Selberg [41] and proven by A. Weil [48, 49] that $R_0(\Gamma, G)$ is open in $R(\Gamma, G)$. Let us recall the precise statement.

Theorem (Weil). *Let G be a connected Lie group. Let Γ be a discrete group, and r_0 an injective representation of Γ into G, such that $r_0(\Gamma)$ is discrete in G with compact quotient space $G/r_0(\Gamma)$. Then there are a neighbourhood $\mathfrak{U}$ of r_0 in the space $R(\Gamma, G)$ of all representations of Γ in G, a compact subset K of G and, for every $g \in G$, a neighbourhood W_g of g in G, and a neighbourhood $\mathfrak{U}_g$ of r_0 in $\mathfrak{U}$, such that $G = r(\Gamma) \circ K$ for all $r \in \mathfrak{U}$ and that the union of the sets $r^{-1}(W_g)$, for all $r \in \mathfrak{U}_g$, is a finite set for all $g \in G$.*

4.2. The Action of $\mathrm{Aut}(\Gamma)\times\mathrm{Aut}(G)$ on $R(\Gamma, G)$

Let $\mathrm{Aut}(\Gamma)$ (resp. $\mathrm{Aut}(G)$) be the group of automorphisms (resp. continuous automorphisms) of Γ (resp. G). Then we naturally have an action of $\mathrm{Aut}(\Gamma)$ from the left and an action of $\mathrm{Aut}(G)$ from the right on $R_0(\Gamma, G)$ by composition of homomorphism:

$$\begin{aligned}\mathrm{Aut}(\Gamma)\times R(\Gamma, G)\times\mathrm{Aut}(G)&\to R(\Gamma, G)\\ p\times r\times q&\mapsto q\circ r\circ p\end{aligned}\tag{4.2.1}$$

We have the following simple consequences.

(*i*) The actions of $\mathrm{Aut}(\Gamma)$ and $\mathrm{Aut}(G)$ commute, i.e., $(q\circ r)\circ p= q\circ(r\circ p)$.

(*ii*) $\mathrm{Ad}(\gamma)\circ r=r\circ\mathrm{Ad}(r(\gamma))$ for $\forall\gamma\in\Gamma$ and $\forall r\in R(\Gamma, G)$.

(*iii*) The action of $\mathrm{Aut}(\Gamma)$ on $R_0(\Gamma, G)$ is fixed-point free. (Since $r\in R_0(\Gamma, G)$ is faithful, $r(\gamma)=r\circ p(\gamma)$ implies $\gamma=p(\gamma)$ $\forall\gamma\in\Gamma$.)

(*iv*) The action of $\mathrm{Aut}(\Gamma)$ on $R_0(\Gamma, G)$ is proper.

(It is enough to show that for any $r\in R_0(\Gamma, G)$, there is a neighbourhood $\mathfrak{U}$ of r such that the set $\{p\in\mathrm{Aut}(\Gamma)\colon p(\mathfrak{U})\cap\mathfrak{U}\neq\emptyset\}$ is finite. One chooses $\mathfrak{U}$ as follows. Let A be a finite subset of Γ which generates Γ. Let us apply the Theorem (4.1) for r so that for each g of A let W_g be a neighbourhood of $r(g)$ in G and let $\mathfrak{U}_g$ be a neighbourhood of r in $R(\Gamma, G)$ as described in the Theorem. Then put $\mathfrak{U}:=\bigcap_{g\in A}\mathfrak{U}_g\cap\bigcap_{g\in A} W_g$.)

Note. Detailed discussions on the topology of the $R(\Gamma, G)$ and related topics with proofs independent of Weil, are found in Macbeath and Singerman [31, 32] and Harvey [25].

5. The Spaces $S^1(\mathscr{X}(\Gamma))$, $\mathscr{X}(\Gamma)$ and $\mathscr{T}(\Gamma)$

Real varieties $S^1(\mathscr{X}(\Gamma))$, $\mathscr{X}(\Gamma)$ and $\mathscr{T}(\Gamma)$ are introduced in (5.3).

5.1.

We apply Weil's representation space $R(\Gamma, G)$ for abstract Fuchsian groups Γ in this section. As for the group G, we employ $SL(2,\mathbf{R})$ (cf. (3.3) Assertion, and Note 1 below).

The following fact may have been known since the time of Fricke. The proof given here is similar to that in [2] (cf. also [25]).

Assertion. *For a Fuchsian surface group* Γ *of genus* g, *the set* $R_0(\Gamma, SL(2,\mathbf{R}))$ *is a smooth real analytic space of dimension* $6g-3$.

Proof. $R_0(\Gamma, SL(2,\mathbf{R}))$ is realized as an open subset of

$$R(\Gamma, SL(2,\mathbf{R})) = \left\{(A_1,\ldots,A_g,B_1,\ldots,B_g)\in \prod_{i=1}^{2g} SL(2,\mathbf{R}): R(A,B)=1\right\}.$$

Here $R(A,B) := \prod_{i=1}^{g}(A_i\cdot B_i\cdot A_i^{-1}\cdot B_i^{-1})$ is the matrix presentation of the fundamental relation so that $R(A,B)=1$ is the defining matrix equation for the variety $R(\Gamma, SL(2,\mathbf{R}))$. The smoothness of $R_0(\Gamma, G)$ at a point ρ is shown by the implicit function theorem: i.e., by the surjectivity of the differential dR_*: $T_\rho(\prod_{i=1}^{2g} SL(2,\mathbf{R}))\to T_{R(\rho)}(SL(2,\mathbf{R}))$ of the map $R(A,B)$ at the point ρ. Each tangent space is identified with $\prod_{i=1}^{2g}\mathfrak{sl}(2,\mathbf{R})$ and $\mathfrak{sl}(2,\mathbf{R})$ regarded as right invariant vector fields. Then for $(\underline{\xi},\underline{\eta}) = (\xi_i,\eta_i)_{i=1,\ldots,g}\in \prod_{i=1}^{2g}\mathfrak{sl}(2,\mathbf{R})$, the value $dR_*(\underline{\xi},\underline{\eta})\in\mathfrak{sl}(2,\mathbf{R})$ is calculated as

$$\sum_{i=1}^{g}\left(\prod_{j=1}^{i-1}(A_jB_jA_j^{-1}B_j^{-1})\right)A_iB_i(\delta_{A_i}(\eta_i)-\delta_{B_i}(\xi_i))A_i^{-1}B_i^{-1}\left(\prod_{k=i+1}^{g}(A_kB_kA_k^{-1}B_k^{-1})\right)$$

at a point $\rho=(A_i,B_i)_{i=1,\ldots,g}\in\prod_{i=1}^{2g} SL(2,\mathbf{R})$, where the notation δ is introduced in (2.3). At the point $\rho\in R_0(\Gamma, SL(2,\mathbf{R}))$, the elements $\{A_i, B_i\colon i=1,\ldots,g\}$ are hyperbolic and none are commutative. Then the assertion (2.3) implies that the difference map $\delta A(\eta)-\delta B(\xi)\colon \mathfrak{sl}(2,\mathbf{R})^2\to\mathfrak{sl}(2,\mathbf{R})$ is surjective. And so is dR_* at ρ.

Note 1. One may develop a parallel theory for $R_0(\Gamma, G)$ for any cyclic covering group G of $PSL(2,\mathbf{R})$. Then $R_0(\Gamma, G)$ is a finite covering space of $R_0(\Gamma, PSL(2,\mathbf{R}))$ and the $PSO(2)$-bundle structure is replaced by its cyclic covering group bundle (cf. (13.1) Note).

In this paper, we shall not be concerned with this fine structure and discuss only the $SO(2)$-bundle for the case $G=SL(2,\mathbf{R})$ without loss of generality so far as infinitesimal structure is concerned.

5.2.

The following are also elementary or reduced to (4.1).

Assertion. (*i*) *The action of* $\mathrm{Aut}(SL(2,\mathbf{R}))$ *on* $R_0(\Gamma, SL(2,\mathbf{R}))$ *from the right is proper and fixed-point free.*

(*ii*) *The actions of* $SL(2,\mathbf{R})$ *and* $SO(2,\mathbf{R})$ *on* $R_0(\Gamma, SL(2,\mathbf{R}))$ *have real analytic sections.*

Proof. (*i*) Any element of $\mathrm{Aut}(SL(2,\mathbf{R}))$ is given by an adjoint action for some $g \in GL(2,\mathbf{R})$. Then $\rho \cdot \mathrm{ad}(g) = \rho \Leftrightarrow g$ is commutative with $\Gamma_\rho := \rho(\Gamma) \Leftrightarrow g \in \mathrm{Center}(GL(2,\mathbf{R})) \Leftrightarrow \mathrm{Ad}(g) = 1$.

The properness of the action follows from (4.1) Theorem.

(*ii*) There is a standard construction of local sections [2, 11].

As for a section for the action of $SL(2,\mathbf{R})$, take $R_0(\Gamma, SL(2,\mathbf{R})) \cap \{(A_i, B_i) \in \prod_{i=1}^{2g} SL(2,\mathbf{R})$: A_1 is hyperbolic whose repelling and attracting fixed points are 0 and ∞ respectively and the fixed points of B_1 are two real numbers, the product of which has the value $\pm 1\}$.

As for a section for the action of $SO(2)$, take $R_0(\Gamma, SL(2,\mathbf{R})) \cap \{(A_i, B_i) \in \prod_{i=1}^{2g} SL(2,\mathbf{R})$: A_1 is hyperbolic whose repelling fixed point is $0\}$.

5.3.

As consequences of the assertions in (5.1), (5.2) and the commutativity (4.2)(*i*), we define five real analytic manifolds:

$$S^1(\tilde{\mathscr{X}}(\Gamma)) := R_0(\Gamma, SL(2,\mathbf{R})), \qquad S^1(\mathscr{X}(\Gamma)) := \Gamma \backslash R_0(\Gamma, SL(2,\mathbf{R})),$$

$$\tilde{\mathscr{X}}(\Gamma) := R_0(\Gamma, SL(2,\mathbf{R}))/SO(2), \qquad \mathscr{X}(\Gamma) := \Gamma \backslash R_0(\Gamma, SL(2,\mathbf{R}))/SO(2),$$

$$\mathscr{T}(\Gamma) := R_0(\Gamma, SL(2,\mathbf{R}))/SL(2,\mathbf{R}) = \Gamma \backslash R_0(\Gamma, SL(2,\mathbf{R}))/SL(2,\mathbf{R}).$$

The equality in the last line follows from the relation (4.2)(*ii*). A justification of these names will be clarified in the Remark below.

By definition, one has natural maps, which are named in the following commutative diagram:

$$\begin{array}{ccccc} & S^1(\tilde{\mathscr{X}}(\Gamma)) & \xrightarrow{d\mathscr{R}_*} & S^1(\mathscr{X}(\Gamma)) & \\ & \downarrow \tilde{\Pi} & & \downarrow \Pi & \\ \tilde{\Omega} & \tilde{\mathscr{X}}(\Gamma) & \xrightarrow{\mathscr{R}} & \mathscr{X}(\Gamma) & \Omega \\ & \downarrow \tilde{\Phi} & & \downarrow \Phi & \\ & \mathscr{T}(\Gamma) & = & \mathscr{T}(\Gamma), & \end{array} \tag{5.3.1}$$

where

(*i*) $d\mathcal{R}_*$ and $\mathcal{R}$ are covering maps by the adjoint action of Γ,

(*ii*) $\tilde{\Pi}$ and Π (resp. $\tilde{\Omega} := \tilde{\Phi} \circ \tilde{\Pi}$ and $\Omega := \Phi \circ \Pi$) are principal fiber bundles with the structure group SO(2) (resp. SL(2, **R**)),

(*iii*) The group $\mathrm{Aut}(\Gamma)/\Gamma$ is acting properly and discontinuously on the spaces $\mathcal{T}(\Gamma)$, $\mathcal{X}(\Gamma)$ and $S^1(\mathcal{X}(\Gamma))$. The projections Ω, Π and Φ are equivariant to this action, due to the commutativity (4.2)(*i*).

Remark. For each $\rho \in R_0(\Gamma, SL(2, \mathbf{R}))$, let us associate a pair $(\rho\Gamma\backslash\mathbf{H}, (a_1, \ldots, a_g, b_1, \ldots, b_g))$ of a Riemann surface and a canonical generator system of its fundamental group. This is a map from $R_0(\Gamma, SL(2, \mathbf{R}))$ to the Teichmüller space T_g for Riemann surfaces of genus g. The map is factored through $\mathcal{T}(\Gamma)$. The induced map $\mathcal{T}(\Gamma) \to T_g$ is a finite covering by the action of $\mathbf{Z}_2^{2g-1}$. Hence if we assume the complex structure on T_g due to Weil [47] (cf. [4, 13]), $\mathcal{T}(\Gamma)$ also carries that structure.

As stated in the introduction, independent of this Teichmüller space theory, we construct the complex structures on $\mathcal{X}(\Gamma)$ and $\mathcal{T}(\Gamma)$ in Sections 7–10 of this paper. Further, $S^1(\mathcal{X}(\Gamma))$ is identified with a unit circle bundle in the relative complex tangent bundle of $\mathcal{X}(\Gamma)/\mathcal{T}(\Gamma)$ which corresponds to the Poincaré metric of the fibers.

6. The Real Tangent Spaces

The real tangent vectors of the space $S^1\mathcal{X}(\Gamma)$ is described in terms of the cocycles of certain group cohomology (cf. (6.2)). For Teichmüller space T_g, a corresponding statement is found in Earle and Eells [16]. Both works are influenced by the work of Weil [50].

6.1. Notations for Group Cohomology

Let M be a left K-module for an algebra K and let a group Γ act on M from the right, denoted by ad. A mapping $z: \Gamma \to M$ is a cocycle if it satisfies the following cocycle condition:

$$z(\gamma_1\gamma_2) = z(\gamma_1)\cdot \mathrm{ad}(\gamma_2) + z(\gamma_2) \qquad \text{for } \gamma_1, \gamma_2 \in \Gamma. \tag{6.1.1}$$

The set of all cocycles over M, denoted by $\mathcal{Z}^1(\Gamma, M)$, forms naturally a left K-module. The coboundary map $\delta: M \to \mathcal{Z}^1(\Gamma, M)$ is given by

$$\delta(m)(\gamma) := m - m\cdot \mathrm{ad}(\gamma), \tag{6.1.2}$$

and the first cohomology group as a left K-module is defined by

$$H^1(\Gamma, M) := \mathcal{Z}^1(\Gamma, M)/\delta M. \tag{6.1.3}$$

Let us fix a point $\rho \in S^1(\mathscr{X}(\Gamma))$ for a Fuchsian surface group Γ and take M to be $\mathfrak{sl}(2, \mathbf{R})$ on which Γ acts by the adjoint action $\mathrm{ad}_\rho := \mathrm{ad} \circ \rho$. Instead of $\mathscr{Z}^1(\Gamma, M)$, δ, $H^1(\Gamma, M)$, etc., we use notations $\mathscr{Z}^1_\rho(\Gamma, M)$, δ_ρ, $H^1_\rho(\Gamma, M)$, etc. to stress their dependence on ρ.

6.2.

The (real) tangent space of a (real) variety A at a point $a \in A$ will be denoted by $T_a(A)$.

We define a map e. For a tangent vector $v \in T_\rho(S^1(\mathscr{X}(\Gamma)))$ at ρ, let $\rho(t) \in S^1(\mathscr{X}(\Gamma))$ for $t \in (-1, 1)$ be a differentiable path with $\rho(0) = \rho$ and $d\rho/dt|_{t=0} = v$. Then for $\gamma \in \Gamma$, the 2×2 matrix

$$\left((\rho(t)(\gamma))^{-1} \frac{d\rho(t)(\gamma)}{dt} \right)\Bigg|_{t=0}$$

depends only on ρ, v and γ, and it belongs to $\mathfrak{sl}(2, \mathbf{R})$ ((2.7.1)). Denoting it by $\rho(\gamma)^{-1}\partial_v\rho(\gamma)$, we define a map $e_\rho(v): \Gamma \to \mathfrak{sl}(2, \mathbf{R})$, by

$$e_\rho(v)(\gamma) := \rho(\gamma)^{-1}\partial_v\rho(\gamma). \tag{6.2.1}$$

Assertion. i) *The map $e = e_\rho$ (6.2.1) defines an isomorphism of real vector spaces*:

$$e: T_\rho(S^1(\mathscr{X}(\Gamma))) \simeq \mathscr{Z}^1_\rho(\Gamma, \mathfrak{sl}(2, \mathbf{R})) \tag{6.2.2}$$

ii) *The map e induces the isomorphisms of the subspaces*:

$$e: T_\rho(\rho \cdot \mathrm{Ad}(SL(2, \mathbf{R}))) \simeq \delta_\rho(\mathfrak{sl}(2, \mathbf{R})), \tag{6.2.3}$$

and

$$e: T_\rho(\rho \cdot \mathrm{Ad}(SO(2))) \simeq \delta_\rho(\mathfrak{so}(2)). \tag{6.2.4}$$

Here we denote by $\rho \cdot \mathrm{Ad}(SL(2, \mathbf{R}))$ and $\rho \cdot \mathrm{Ad}(SO(2))$ the orbits of ρ.

Proof. i) Let us check that $e(v)$ is a cocycle.

$$\begin{aligned} e(v)(\gamma_1\gamma_2) &= \rho(\gamma_1\gamma_2)^{-1}\partial_v\rho(\gamma_1\gamma_2) \\ &= \rho(\gamma_2)^{-1}\rho(\gamma_1)^{-1}((\partial_v\rho(\gamma_1))\rho(\gamma_2) + \rho(\gamma_1)\partial_v\rho(\gamma_2)) \\ &= e(v)(\gamma_1) \cdot \mathrm{ad}_\rho(\gamma_2) + e(v)(\gamma_2). \end{aligned}$$

Injectivity of e. Since $v \neq 0$ implies that for at least one γ of the generators of Γ, $\partial_v\rho(\gamma) \neq 0$, then $e(v)(\gamma) \neq 0$.

Surjectivity of e. For any cocycle z and a sequence $a_1, \ldots, a_n$ of Γ, applying the cocycle condition (6.1.1) repeatedly, one gets

$$z(a_1 \cdots a_n) = \sum_{i=1}^{n} z(a_i) \cdot \mathrm{ad}_\rho(a_{i+1} \cdots a_n)$$

$$= \sum_{i=1}^{n} \rho(a_{i+1} \cdots a_n)^{-1} z(a_i) \rho(a_{i+1} \cdots a_n). \quad (6.2.5)$$

If the product $a_1 \cdots a_n$ is the identity of Γ, then the left-hand side of (6.2.5) is zero. Multiplying (6.2.5) by $\rho(a_1 \cdots a_n)$, one obtains the relation

$$\sum_{i=1}^{n} \rho(a_1 a_2 \cdots a_i) z(a_i) \rho(a_{i+1} \cdots a_n) = 0. \quad (6.2.6)$$

Here we have expressed the relation as the product of matrixes.

Let the group Γ be generated by a set A with a set B of relations given by the words of the letters of A. Then a cocycle z is uniquely determined by the data $(z(a))_{a \in A}$ due to (6.2.5). Conversely, a given system $(z(a))_{a \in A} \in \prod_{a \in A} \mathfrak{sl}(2, \mathbf{R})$, defines a cocycle if and only if they satisfy the relation (6.2.6) coming from the words of B.

For a given cocycle $z \in \mathscr{Z}^1_\rho(\Gamma, \mathfrak{sl}(2, \mathbf{R}))$, let us denote by v the tangent vector of the product space $\prod_{a \in A} GL(2, \mathbf{R})$ at the point $(\rho(a))_{a \in A}$ defined by the relation

$$\partial_v \rho(a) := \rho(a) z(a), \quad (6.2.7)$$

for $a \in A$. The fact that $\rho^{-1}(a) \partial_v \rho(a) \in \mathfrak{sl}(2, \mathbf{R})$ for $a \in A$ implies that v is tangent to the subspace $\prod_{a \in A} SL(2, \mathbf{R})$ ((2.7.1)).

Let us regard $R = a_1 \cdots a_n$ for $R \in B$ as a matrix-valued real function on the product space $\prod_{a \in A} SL(2, \mathbf{R})$, whose level set defines the $S^1(\mathscr{X}(\Gamma))$. Using the Leibnitz rule, (6.2.6) and (6.2.7) imply

$$\partial_v \rho(R) = \sum_{i=1}^{n} \rho(a_1 a_2 \cdots a_{i-1})(\partial_v \rho(a_i)) \rho(a_{i+1} \cdots a_n) = 0.$$

This implies that v is tangent to the level set of R and hence $v \in T_\rho(S^1 \mathscr{X}(\Gamma))$.

ii) Consider an element $u \in \mathfrak{sl}(2, \mathbf{R})$ and a point $\rho \in S^1 \mathscr{X}(\Gamma)$. Let v be the tangent vector of the orbit $\rho \cdot \mathrm{Ad}(\exp(tu))$ for $t \in \mathbf{R}$ at ρ. We have $\partial_v \rho = d/dt(\exp(-tu)\rho \exp(tu))|_{t=0} = -u\rho + \rho u$ and therefore

$$e(v) = \rho^{-1} \partial_v \rho = u - \rho^{-1} u \rho = \delta_\rho(u).$$

This proves (6.2.3) and (6.2.4).

6.3.

The formula (6.2.3) is reformulated as follows.

Assertion. *The map $\imath$ defined by*

$$\imath := e^{-1} \circ \delta : \mathfrak{sl}(2, \mathbf{R}) \to \Gamma(S^1(\mathscr{X}(\Gamma)), T(S^1(\mathscr{X}(\Gamma)))), \tag{6.3.1}$$

is a Lie algebra homomorphism equivariant to the action of $SL(2, \mathbf{R})$.

Proof. The image $\imath(X)$ for $X \in \mathfrak{sl}(2, \mathbf{R})$ is a real analytic vector field on $S^1(\mathscr{X})$, since

$$\partial_{\imath(X)}\rho(\gamma) = \rho(\gamma)(\delta_\rho(X)(\gamma)) = \rho(\gamma) \cdot X - X \cdot \rho(\gamma) \tag{6.3.2}$$

depends analytically on $\rho \in S^1(\mathscr{X}(\Gamma))$ for $\gamma \in \Gamma$.

Using (6.3.2), we have

$$\begin{aligned}\partial_{\imath(Y)}\partial_{\imath(X)}\rho(\gamma) &= \partial_{\imath(Y)}(\rho(\gamma) \cdot X - X \cdot \rho(\gamma))\\ &= (\rho(\gamma) \cdot Y - Y \cdot \rho(\gamma))X - X(\rho(\gamma) \cdot Y - Y \cdot \rho(\gamma)),\end{aligned}$$

and therefore

$$[\partial_{\imath(Y)}, \partial_{\imath(X)}]\rho(\gamma) = \rho(\gamma) \cdot [Y, X] - [Y, X] \cdot \rho(\gamma).$$

This means that $\imath$ is a homomorphism.

Using (6.3.2), the equivariance of $\imath$ is shown by the equality

$$\partial_{\imath(X \cdot \mathrm{Ad}(g))}(\rho \cdot \mathrm{Ad}(g))(\gamma) = (\partial_{\imath(X)}\rho(\gamma)) \cdot \mathrm{Ad}(g).$$

6.4.

Corollary. *For every $\rho \in S^1(\mathscr{X}(\Gamma))$, one has the isomorphism e_ρ of the short exact sequences*

$$\begin{array}{ccccccccc} 0 \longrightarrow & T_{\Pi(\rho)}(X_{\Omega(\rho)}) & \longrightarrow & T_{\Pi(\rho)}(\mathscr{X}(\Gamma)) & \longrightarrow & T_{\Omega(\rho)}(\mathscr{T}(\Gamma)) & \longrightarrow 0 \\ & \downarrow e_\rho & & \downarrow e_\rho & & \searrow e_\rho & \\ 0 \longrightarrow & \mathfrak{sl}(2, \mathbf{R})/\mathfrak{so}(2) & \longrightarrow & \mathscr{Z}^1_\rho(\Gamma, \mathfrak{sl}(2, \mathbf{R}))/\delta_\rho\, \mathfrak{so}(2) & \longrightarrow & H^1_\rho(\Gamma, \mathfrak{sl}(2, \mathbf{R})) & \longrightarrow 0 \end{array} \tag{6.4.1}$$

Here $X_{\Omega(\rho)} := \Phi^{-1}\Omega(\rho)$ is the real two-dimensional submanifold in $\mathscr{X}$.

Proof. The differentials of $\tilde{\Omega}$ and $\tilde{\Pi}$ at the point $\rho \in S^1(\tilde{\mathscr{X}}(\Gamma))$ induce the following short exact sequences:

$$0 \longrightarrow T_\rho(\rho\cdot \mathrm{Ad}(SL(2,\mathbf{R}))) \longrightarrow T_\rho(S^1(\tilde{\mathscr{X}}(\Gamma))) \xrightarrow{d\tilde{\Omega}_*} T_{\tilde{\Omega}(\rho)}(\mathscr{T}(\Gamma)) \longrightarrow 0, \tag{6.4.2}$$

$$0 \longrightarrow T_\rho(\rho\cdot \mathrm{Ad}(SO(2))) \longrightarrow T_\rho(S^1(\tilde{\mathscr{X}}(\Gamma))) \xrightarrow{d\tilde{\Pi}_*} T_{\Omega(\rho)}(\tilde{\mathscr{X}}(\Gamma)) \longrightarrow 0. \tag{6.4.3}$$

These, combined with the isomorphisms e of (6.2) Assertion, induce the isomorphism e_ρ (6.4.1).

6.5.

The action of $\mathrm{Aut}(\Gamma)\times \mathrm{Aut}(PSL(2,\mathbf{R}))$ on the tangent spaces is described as below.

Assertion. 1. *Let $\alpha\times\beta$ be an element of* $\mathrm{Aut}(\Gamma)\times\mathrm{Aut}(PGL(2,\mathbf{R}))$. *For a cocycle $z \in \mathscr{Z}^1_\rho(\Gamma, \mathfrak{sl}(2,\mathbf{R}))$ at $\rho\in S^1(\mathscr{X}(\Gamma))$, the composition $d\beta\circ z\circ\alpha$ defines an element of $\mathscr{Z}^1_{\alpha\rho\beta}(\Gamma,\mathfrak{sl}(2,\mathbf{R}))$ so that one gets an isomorphism*:

$$\mathscr{Z}^1_\rho(\Gamma,\mathfrak{sl}(2,\mathbf{R})) \underset{\alpha\times d\beta}{\simeq} \mathscr{Z}^1_{\alpha\rho\beta}(\Gamma,\mathfrak{sl}(2,\mathbf{R})). \tag{6.5.1}$$

Here $d\beta$ is defined as in (2.6) *an automorphism of* $\mathfrak{sl}(2,\mathbf{R})$.

2. *The isomorphism* (6.5.1) *is compatible with the coboundary map δ* (6.1.2) *and the isomorphism e* (6.2.1) *in the sense that the following diagrams are commutative*:

$$\begin{array}{ccc} \mathfrak{sl}(2,\mathbf{R}) & \underset{d\beta}{\simeq} & \mathfrak{sl}(2,\mathbf{R}) \\ \Big\downarrow \delta_\rho & & \Big\downarrow \delta_{\alpha\rho\beta} \\ \mathscr{Z}^1_\rho(\Gamma,\mathfrak{sl}(2,\mathbf{R})) & \underset{\alpha\times d\beta}{\simeq} & \mathscr{Z}^1_{\alpha\rho\beta}(\Gamma,\mathfrak{sl}(2,\mathbf{R})), \end{array} \tag{6.5.2}$$

$$\begin{array}{ccc} T_\rho(S^1(\mathscr{X}(\Gamma)) & \underset{d(\alpha\times\beta)}{\simeq} & T_{\alpha\rho\beta}(S^1(\mathscr{X}(\Gamma)) \\ \Big\downarrow e & & \Big\downarrow e \\ \mathscr{Z}^1_\rho(\Gamma,\mathfrak{sl}(2,\mathbf{R})) & \underset{\alpha\times d\beta}{\simeq} & \mathscr{Z}^1_{\alpha\rho\beta}(\Gamma,\mathfrak{sl}(2,\mathbf{R})). \end{array} \tag{6.5.3}$$

Proof. Recalling (2.6.2), one calculates each case as below.

1.

$$d\beta\circ z\circ\alpha(\gamma_1\circ\gamma_2)=d\beta\circ z(\alpha(\gamma_1)\cdot\alpha(\gamma_2))$$
$$=d\beta(\alpha(\gamma_1)\,\mathrm{ad}_\rho(\alpha(\gamma_2))+z(\alpha(\gamma_2)))$$
$$=d\beta(z(\alpha(\gamma_1))\,\mathrm{ad}_\rho(\alpha(\gamma_2)))+d\beta(z(\alpha(\gamma_2)))$$
$$=d\beta(z(\alpha(\gamma_1)))\cdot\beta(\mathrm{ad}_{\beta\rho\alpha}(\alpha(\gamma_2)))+d\beta(z(\alpha(\gamma_2))). \tag{6.5.2}$$

2.

$$\delta_{\alpha\times\beta}(d\beta(X))(\gamma)=d\beta(X)-d\beta(X)\cdot\mathrm{ad}(\beta(\rho(\alpha(\gamma))))$$
$$=d\beta(X-X\cdot\mathrm{ad}(\rho(\alpha(\gamma))))$$
$$=(\alpha\times d\beta)(X-X\cdot\mathrm{ad}(\rho))(\gamma)=\alpha\times d\beta(\delta_\rho(X)(\gamma)). \tag{6.5.3}$$

For $v\in T_\rho(S^1(\tilde{\mathscr{X}}(\Gamma)))$ take a path $p(t)\in S^1(\tilde{\mathscr{X}}(\Gamma))(t\in(-1,1))$ such that $p(0)=\rho$ and $dp/dt(0)=v$. Recalling (2.7.2), we calculate as follows:

$$((\alpha\times d\beta)e(v))(\gamma)=d\beta(e(v)(\alpha(\gamma)))=d\beta\left(p(t,\alpha(\gamma))^{-1}\frac{dp}{dt}(t,\alpha(\gamma))\Big|_{t=0}\right)$$
$$=(\beta p(t,\alpha(\gamma))^{-1}\frac{d\beta p}{dt}(t,\alpha(\gamma))|_{t=0}=e(d(\alpha\times\beta)_*(v)).$$

7. The Complex Structures on the Tangent Spaces

Complex structures on the real tangent spaces of $\mathscr{T}(\Gamma)$ and $\mathscr{X}(\Gamma)$ are introduced by use of $H^1_\rho(\Gamma,\Gamma(\mathbf{H},\Theta))$ in (7.2). This may be regarded as a particular case of the Kodaira-Spencer theory [29].

7.1.

Recall the embedding $\iota\colon\mathfrak{sl}(2,\mathbf{R})\to\Gamma(\mathbf{H},\Theta)$ (2.4), which is equivariant to the adjoint action of $\Gamma_\rho:=\rho(\Gamma)$. This induces the following commutative diagram.

$$\begin{array}{ccc}\mathfrak{sl}(2,\mathbf{R}) & \xrightarrow{\ \iota\ } & \Gamma(\mathbf{H},\Theta)\\ \downarrow{\scriptstyle\delta_\rho} & & \downarrow{\scriptstyle\delta_\rho}\\ \mathscr{Z}^1_\rho(\Gamma,\mathfrak{sl}(2,\mathbf{R})) & \xrightarrow{\ \iota\ } & \mathscr{Z}^1_\rho(\Gamma,\Gamma(\mathbf{H},\Theta)).\end{array} \tag{7.1.1}$$

Assertion. *The above map induces the isomorphism*

$$H^1_\rho(\Gamma, \mathfrak{sl}(2, \mathbf{R})) \simeq H^1_\rho(\Gamma, \Gamma(\mathbf{H}, \Theta)) \tag{7.1.2}$$

as real vector spaces of rank $6g-6$.

Proof. Since a cocycle z is determined by $(z(a))_{a\in A}$ for $A=\{a_1, \ldots, a_g, b_1, \ldots, b_g\}$ with a single relation of (6.2.6) for the relation $\Pi a_i b_i a_i^{-1} b_i^{-1}=1$, one has $\dim_{\mathbf{R}} \mathscr{Z}^1_\rho(\Gamma, \mathfrak{sl}(2, \mathbf{R}))=3\cdot 2g-3$ and hence $\dim_{\mathbf{R}} H^1_\rho(\Gamma, \mathbf{sl}(2, \mathbf{R}))=6g-6$.

The proof of bijectivity we give here is an indirect one, using harmonic analysis due to Matsushima and Murakami [33, 34] (cf. Earle and Eells [16]).

First we show an isomorphism $H^1_\rho(\Gamma, \Gamma(\mathbf{H}, \Theta)) \simeq H^1(\Gamma_\rho \backslash \mathbf{H}, \Theta)$. For this purpose, we construct a complex $C^*(\Theta)$, whose cohomology group naturally coincides with both sides. The k-dimensional nerve of the complex is equal to $(\Gamma)^{k+1}=\Gamma\times\cdots\times\Gamma$ ($k+1$ copies). For a nerve $(\gamma_0, \ldots, \gamma_k)\in(\Gamma)^{k+1}$, we associate the module $\Gamma(\mathbf{H}_{\gamma_0,\ldots,\gamma_k}, \Theta)$, and so

$$C^k(\Theta) := \prod_{\gamma_0,\ldots,\gamma_k} \Gamma(\mathbf{H}_{\gamma_0,\ldots,\gamma_k}, \Theta).$$

Here $\mathbf{H}_{\gamma_0,\ldots,\gamma_k} :=$ the fiber product of the maps $\gamma_i\colon \mathbf{H}\to\mathbf{H}$ ($i=0, \ldots, k$) over the space $\mathbf{H}$. Note that $\mathbf{H}_{\gamma_0,\ldots,\gamma_k}$ are biholomorphic to $\mathbf{H}$, since the projection $p_i: \mathbf{H}_{\gamma_0,\ldots,\gamma_k} \to \mathbf{H}_{\gamma_0,\cdot,\hat{\gamma}_i,\cdot,\gamma_k}$ ($i=0, \ldots, k$) are so. The coboundary map δ is defined for $c\in C^k(\Theta)$ as usual by

$$(\delta c)(\gamma_0, \ldots, \gamma_{k+1}) := \sum_{j=0}^{k+1} (-1)^{j+1}(dp_j)_*^{-1} c(\gamma_0, \ldots, \hat{\gamma}_j, \ldots, \gamma_{k+1}).$$

On one side, the complex $C^*(\Theta)$ is naturally identified with the homogeneous chain complex for the cohomology $H^*(\Gamma, \Gamma(\mathbf{H}, \Theta))$. On the other hand $C^*(\Theta)$ may be regarded as a chain complex for a Cech cohomology for the Stein covering $\{\mathbf{H}_\gamma, \gamma\in\Gamma\}$ of $\rho(\Gamma)\backslash\mathbf{H}$, which is a simple covering in the sense of Weil. Hence the cohomology of $C^*(\Theta)$ naturally coincides with $H^*(\rho(\Gamma)\backslash\mathbf{H}, \Theta)$.

In Matsushima and Murakami [33, 34] the isomorphism $H^1_\rho(\Gamma, sl(2, \mathbf{R})) \simeq H^1(\Gamma_\rho\backslash\mathbf{H}, \Theta)$ is shown in a general setting for discrete groups in semisimple Lie groups. These altogether prove the Assertion.

Let us denote by E the composition of $\iota\circ e$ of e (6.2.1) with ι,

$$E: T_\rho S^1(\mathscr{X}(\Gamma)) \to \mathscr{Z}^1_\rho(\Gamma, \Gamma(\boldsymbol{H}, \Theta)). \tag{7.1.3}$$

We shall refer to E as a *cocycle map* at the point ρ.

Corollary. (*i*) *For each* $\rho \in S^1(\mathscr{X}(\Gamma))$, *the E induces isomorphisms:*

$$\begin{array}{ccccccccc} 0 \longrightarrow & T_{\Pi(\rho)}(X_{\Omega(\rho)}) & \longrightarrow & T_{\Pi(\rho)}(\mathscr{X}(\Gamma)) & \longrightarrow & T_{\Omega(\rho)}(\mathscr{T}(\Gamma)) & \longrightarrow 0 \\ & \downarrow E_\rho & & \downarrow E_\rho & & \downarrow E_\rho & \\ 0 \longrightarrow & T_i(\mathbf{H}) & \longrightarrow & \mathscr{Z}^1_\rho(\Gamma, \Gamma(\mathbf{H}, \Theta))/\delta_\rho \Gamma(\mathbf{H}, \Theta(-[i]) & \longrightarrow & H^1_\rho(\Gamma, \Gamma(\mathbf{H}, \Theta)) & \longrightarrow 0 \end{array} \tag{7.1.4}$$

(*ii*) *The above isomorphisms are equivariant to the action of* Aut(Γ).

Proof. (*i*) In view of the commutative diagram (2.4.4), one obtains the following commutative diagram:

$$\begin{array}{ccccccccc} 0 \longrightarrow & \mathfrak{sl}(2, \mathbf{R})/\mathfrak{so}(2) & \longrightarrow & \mathscr{Z}^1_\rho(\Gamma, \mathfrak{sl}(2, \mathbf{R}))/\delta_\rho \mathfrak{so}(2) & \longrightarrow & H^1_\rho(\Gamma, \mathfrak{sl}(2, \mathbf{R})) & \longrightarrow 0 \\ & \downarrow & & \downarrow & & \downarrow & \\ 0 \longrightarrow & T_i(\mathbf{H}) & \longrightarrow & \mathscr{Z}^1_\rho(\Gamma, \Gamma(\mathbf{H}, \Theta))/\delta_\rho \Gamma(\mathbf{H}, \Theta(-[i]) & \longrightarrow & H^1_\rho(\Gamma, \Gamma(\mathbf{H}, \Theta)) & \longrightarrow 0. \end{array}$$

Since the first and the last columns are isomorphic ((2.4.3), (7.1.2)), the middle column is also isomorphic. Combining this isomorphism with diagram (6.4.1), we obtain the result.

(*ii*) This is a consequence of (6.5) Assertion.

7.2.

Definition. Almost complex structures on $\mathscr{X}(\Gamma)$ and $\mathscr{T}(\Gamma)$ are defined by the isomorphism E_ρ of (7.1.4).

The integrability of $\mathscr{X}(\Gamma)$ and $\mathscr{T}(\Gamma)$ will be proven in (9.4) and (10.4). Note that the action of $\mathrm{Aut}(\Gamma) \times \mathrm{Ad}(SL(2, \mathbf{R}))$ on the second line of (7.1.4) induces isomorphisms as complex vector spaces so that the action is equivariant with the structure.

7.3.

In the rest of this section, we identify the above complex structure with the complex structure introduced by Weil, Ahlfors and Bers on the Teichmüller space to clarify the relationship between the Fricke moduli space and the Teichmüller space.

The proof for the integrability of the complex structure, which we shall give in the next sections, is independent of this so that one may skip to Section 8.

7.4.

Define

$\mathscr{A}_{\mathbf{H}}(\Theta) :=$ the sheaf on $\mathbf{H}$ of the germs of vector fields $\varphi(z, \bar{z})\, d/dz$, where φ is a complex-valued differentiable function in z.

$\mathscr{A}_{\mathbf{H}}^{0,1}(\Theta) :=$ the sheaf on $\mathbf{H}$ of the germs of $(-1, 1)$ tensor $\varphi(z, \bar{z})\, d\bar{z}/dz$, where φ is a complex-valued differentiable function in z.

An element of $\mathscr{A}_{\mathbf{H}}^{0,1}(\Theta)$ is referred to as a *Beltrami differential.*

The following is an exact sequence due to the Cauchy–Riemann equation:

$$0 \to \Theta \to \mathscr{A}_{\mathbf{H}}(\Theta) \xrightarrow{\bar{\partial}} \mathscr{A}_{\mathbf{H}}^{0,1}(\Theta) \to 0. \tag{7.4.1}$$

The modules of global sections of the sheaves, denoted by $\Gamma(\mathbf{H}, \mathscr{A}_{\mathbf{H}}(\Theta))$ and $\Gamma(\mathbf{H}, \mathscr{A}^{0,1}(\Theta))$, admit the action of the group Γ_ρ by pull-back as before, which will be referred to as an adjoint action from the right. So we obtain the following commutative diagram:

$$\begin{array}{ccccccc}
 & & & & 0 & & 0 & \\
 & & & & \downarrow & & \downarrow & \\
 & & 0 & \rightarrow & \Gamma(\Gamma_\rho\backslash\mathbf{H}, \mathscr{A}_{\mathbf{H}}(\Theta)) & \xrightarrow{\bar{\partial}} & \Gamma(\Gamma_\rho\backslash\mathbf{H}, \mathscr{A}_{\mathbf{H}}^{0,1}(\Theta)) & \\
 & & \downarrow & & \downarrow & & \downarrow & \\
0 \rightarrow & & \Gamma(\mathbf{H}, \Theta) & \rightarrow & \Gamma(\mathbf{H}, \mathscr{A}_{\mathbf{H}}(\Theta)) & \xrightarrow{\bar{\partial}} & \Gamma(\mathbf{H}, \mathscr{A}_{\mathbf{H}}^{0,1}(\Theta)) & \rightarrow 0 \\
 & & \downarrow{\scriptstyle \delta_\rho} & & \downarrow{\scriptstyle \delta_\rho} & & \downarrow{\scriptstyle \delta_\rho} & \\
0 \rightarrow & & \mathscr{Z}_\rho^1(\Gamma, \Gamma(\mathbf{H}, \Theta)) & \rightarrow & \mathscr{Z}_\rho^1(\Gamma, \Gamma(\mathbf{H}, \mathscr{A}_{\mathbf{H}}(\Theta))) & \xrightarrow{\bar{\partial}} & \mathscr{Z}_\rho^1(\Gamma, \Gamma(\mathbf{H}, \mathscr{A}_{\mathbf{H}}^{0,1}(\Theta)). & \\
 & & & & \downarrow & & & \\
 & & & & 0 & & &
\end{array} \tag{7.4.2}$$

Here we mean that $\Gamma_\rho := \rho(\Gamma)$ and

$$\Gamma(\Gamma_\rho\backslash\mathbf{H}, *) := \{v \in \Gamma(\mathbf{H}, *) \colon v = v \cdot \mathrm{ad}(\gamma) \quad \text{for } \forall \gamma \in \Gamma_\rho\}.$$

7.5.

Assertion. *All sequences in the diagram* (7.4.2) *are exact.*

Proof. The following three facts need to be shown.

(1) $\Gamma(\Gamma_\rho\backslash\mathbf{H}, \Theta) = 0$. This module is identified with the set of holomorphic vector fields on the compact Riemann surface $\Gamma_\rho\backslash\mathbf{H}$ of >1. The non-existence of such a vector field is well known (for instance, by the Gauss–Bonnet theorem).

(2) $H^1(\mathbf{H}, \Theta) = 0$. This is a consequence of Dolbeau's Lemma.

(3) $H^1_\rho(\Gamma, \Gamma(\mathbf{H}, \mathscr{A}_{\mathbf{H}}(\Theta))) = 0$. We identify this module (through the Cech cohomology) with $H^1(\Gamma_\rho\backslash\mathbf{H}, \mathscr{A}_{\mathbf{H}}(\Theta))$. Then the vanishing of such a cohomology group with a fine sheaf coefficient is well known. An explicit proof of this fact dependent upon parameters will be given in the next section (equation (8.4.3)).

7.6.

The diagram (7.4.2) induces an isomorphism of cohomology groups as complex vector spaces:

$$H^1_\rho(\Gamma, \Gamma(\mathbf{H}, \Theta)) \simeq \Gamma(\Gamma_\rho\backslash\mathbf{H}, \mathscr{A}^{0,1}_{\mathbf{H}}(\Theta))/\bar{\partial}\Gamma(\Gamma_\rho\backslash\mathbf{H}, \mathscr{A}_{\mathbf{H}}(\Theta)).$$

The module on the right, the Γ_ρ-invariant Beltrami differential, is exactly the one used to define the complex structure on the Teichmüller space by Weil [47], Bers [11].

8. Local Parametrization

The study of infinitesimal structures in Sections 6–7 is not enough preparation to study the integrability of the complex structures and the metrics on the moduli. One needs a description of their parameter dependence as will be given in this paragraph, whose main results are (8.3) Assertion and (8.5) Lemma*.

8.1.

We mean by a local parametrization map (or simply an lp-map) σ, a differentiable map σ,

$$\sigma : \mathscr{U} \longrightarrow R_0(\Gamma, SL(2, \mathbf{R})) = S^1(\tilde{\mathscr{X}}(\Gamma)), \tag{8.1.1}$$

from a domain $\mathscr{U} \subset \mathbf{R}^n$.

The parametrization σ induces a left Γ-action on $\mathcal{U} \times SL(2, \mathbf{R})$ by making $\gamma \in \Gamma$ operate for every $\underline{t} \in \mathcal{U}$, $g \in SL(2, \mathbf{R})$:

$$\gamma \circ (\underline{t}, g) = (\underline{t}, \sigma(\underline{t})(\gamma) \circ g). \tag{8.1.2}$$

Obviously the action is differentiable and fixed-point free. Theorem (4.1) implies that the action is proper and discontinuous.

On the space $\mathcal{U} \times SL(2, \mathbf{R})$ the group $SL(2, \mathbf{R})$ acts from the right by $(\underline{t}, g) \circ h = (\underline{t}, g \circ h)$. Since the action of Γ (8.1.2) commutes with this action of $SL(2, \mathbf{R})$, it induces a proper discontinuous action of Γ on the quotient: $\mathcal{U} \times \mathbf{H} = \mathcal{U} \times SL(2, \boldsymbol{R})/SO(2)$ by putting

$$\gamma \circ (\underline{t}, z) = (\underline{t}, \sigma(\underline{t})(\gamma)(z)). \tag{8.1.3}$$

8.2.

The projection map of $\mathcal{U} \times SL(2, \mathbf{R})$ (resp. $\mathcal{U} \times \mathbf{H}$) to its first factor $\mathcal{U}$ is invariant by the action of $\gamma \in \Gamma$. Hence the action of Γ induces an action on the relative tangent bundle of the map, denoted by $T(\mathcal{U} \times SL(2, \mathbf{R})/\mathcal{U})$ (resp. $T(\mathcal{U} \times \mathbf{H}/\mathcal{U})$), which is the correction of tangent vectors of $\mathcal{U} \times SL(2, \mathbf{R})$ tangent to fibers ($\simeq SL(2, \mathbf{R})$).

We introduce the following modules of sections of the relative tangent bundles as relative versions for $\mathfrak{sl}(2, \mathbf{R})$ and $\Gamma(\mathbf{H}, \Theta)$.

$$\begin{aligned}\Gamma(\mathcal{U} \times SL(2, \mathbf{R}), \mathcal{A}_{\mathcal{U}}(\mathfrak{sl}(2, \mathbf{R}))) &:= \{v\colon v \text{ is a differentiable section of } T(\mathcal{U} \times SL(2, \mathbf{R})/\mathcal{U}), \text{ which is right } SL(2, \mathbf{R}) \text{ invariant.}\} \\ &\simeq \{\varphi\colon \mathcal{U} \to \mathfrak{sl}(2, \mathbf{R})\colon \varphi \text{ is a differentiable map}\},\end{aligned} \tag{8.2.1}$$

$$\begin{aligned}\Gamma(\mathcal{U} \times \mathbf{H}, \mathcal{A}_{\mathcal{U}}(\Theta)) &:= \{v\colon v \text{ is a differentiable section of } T(\mathcal{U} \times \mathbf{H}/\mathcal{U}), \text{ which is a holomorphic vector field in each fiber.}\} \\ &\simeq \{\varphi(\underline{t}, z)\, d/dz\colon \varphi \text{ is differentiable in } (\underline{t}, z) \text{ and holomorphic in } z.\},\end{aligned} \tag{8.2.2}$$

which are naturally left $\Gamma(\mathcal{U}, \mathcal{A}_{\mathcal{U}})$-modules, where $\Gamma(\mathcal{U}, \mathcal{A}_{\mathcal{U}}) :=$ the algebra of real differentiable functions on $\mathcal{U}$.

The action of γ^{-1} on $\mathcal{U} \times \mathbf{H}$ induces actions on the above modules, called right adjoint action, denoted by ad or ad_σ (cf. (8.4.2)). The map $\iota : \mathfrak{sl}(2, \mathbf{R}) \to \Gamma(\mathbf{H}, \Theta)$ (2.4) induces a $\Gamma(\mathcal{U}, \mathcal{A}_{\mathcal{U}})$-homomorphism.

$$\iota : \Gamma(\mathcal{U} \times SL(2, \mathbf{R}), \mathcal{A}_{\mathcal{U}}(\mathfrak{sl}(2, \mathbf{R}))) \to \Gamma(\mathcal{U} \times \mathbf{H}, \mathcal{A}_{\mathcal{U}}(\Theta)), \tag{8.2.3}$$

which is equivariant to the adjoint action of Γ.

The following facts, which are immediate consequences of (7.1) Assertion and its Corollary, will be used later in Sections 9 and 10.

Assertion. *The map* ι (8.2.3) *induces the* $\Gamma(\mathcal{U}, \mathcal{A}_{\mathcal{U}})$*-isomorphisms*:

$$H^1_\sigma(\Gamma, \Gamma(\mathcal{U}\times SL(2,\mathbf{R}), \mathcal{A}_{\mathcal{U}}(\mathfrak{sl}(2,\mathbf{R})))) \simeq H^1_\sigma(\Gamma, \Gamma(\mathcal{U}\times \mathbf{H}, \mathcal{A}_{\mathcal{U}}(\Theta))), \quad (8.2.4)$$

$$\mathscr{Z}^1_\sigma(\Gamma, \Gamma(\mathcal{U}\times SL(2,\mathbf{R}), \mathcal{A}_{\mathcal{U}}(\mathfrak{sl}(2,\mathbf{R}))))/\delta_\sigma\Gamma(\mathcal{U}\times SL(2,\mathbf{R}), \mathcal{A}_{\mathcal{U}}(\mathfrak{so}(2)))$$
$$\simeq \mathscr{Z}^1_\sigma(\Gamma, \Gamma(\mathcal{U}\times \mathbf{H}, \mathcal{A}_{\mathcal{U}}(\Theta)))/\delta_\sigma\Gamma(\mathcal{U}\times \mathbf{H}, \mathcal{A}_{\mathcal{U}}(\Theta(-[i]))), \quad (8.2.5)$$

which are $\Gamma(\mathcal{U}, \mathcal{A}_{\mathcal{U}})$*-modules of rank* $6g-6$ *and* $6g-4$ *respectively.*

8.3. The Differential $d(\gamma^{-1})_*$ of the Action γ^{-1}

We shall denote by $\Gamma(\mathcal{U}, T(\mathcal{U}))$ the set of differentiable vector fields on $\mathcal{U}$, which is a $\Gamma(\mathcal{U}, \mathcal{A}_{\mathcal{U}})$-module of rank $\dim_{\mathbf{R}} \mathcal{U}$.

Assertion. *For a vector field* $v \in \Gamma(\mathcal{U}, T(\mathcal{U}))$ *on* $\mathcal{U}$, *let us denote by* v^* (*resp.* v^{**}) *the vector field on the product space* $\mathcal{U}\times SL(2,\mathbf{R})$ (*resp.* $\mathcal{U}\times\mathbf{H}$) *by a trivial lifting of* v. *Then for* $\gamma\in\Gamma$,

$$d(\gamma^{-1})_*(v^*) = v^* - e(v)(\gamma) \quad (8.3.1)$$

$$d(\gamma^{-1})_*(v^{**}) = v^{**} - E(v)(\gamma). \quad (8.3.2)$$

Here the notations e and E mean the mappings

$$e: \Gamma(\mathcal{U}, T(\mathcal{U})) \to \mathscr{Z}^1_\sigma(\Gamma, \Gamma(\mathcal{U}\times SL(2,\mathbf{R}), \mathcal{A}_{\mathcal{U}}(\mathfrak{sl}(2,\mathbf{R})))), \quad (8.3.3)$$

$$E: \Gamma(\mathcal{U}, T(\mathcal{U})) \to \mathscr{Z}^1_\sigma(\Gamma, \Gamma(\mathcal{U}\times \mathbf{H}, \mathcal{A}_{\mathcal{U}}(\Theta))), \quad (8.3.4)$$

as compositions of the differential $d\sigma_*: T(\mathcal{U}) \to \mathcal{U}\times_{S^1(\tilde{\mathscr{X}}(\Gamma))} T(S^1(\tilde{\mathscr{X}}(\Gamma)))$ *with the mappings e* (6.2.1) *or E* (7.1.2) *respectively. This abuse of the notations for e and E should not cause confusion.*

Definition. We shall call e and E the cocycle maps for σ. (See (3)* and (4)* in the proof for the definition of e and E.)

Proof. First let us redefine the above maps e and E. Let v be a differentiable vector field on $\mathcal{U}$. By definition, the lp-map σ (8.1.1) is differentiable implies that for all $\gamma\in\Gamma$, $\sigma(\underline{t})(\gamma)$ is differentiable in $\underline{t}\in\mathcal{U}$. Hence one may derivate the matrix $\sigma(\underline{t})(\gamma)$ by v at each entry, denoted as $\partial_v\sigma(\underline{t})(\gamma)$. Now e is defined as

(3)* $$e(v)(\gamma) := (\sigma(\underline{t})(\gamma))^{-1}\partial_v\sigma(\underline{t})(\gamma).$$

That $e(v)$ is a cocycle over Γ is proven as (6.2) Assertion.

Composing $e(v)$ with ι of (8.2.3), we define the cocycle map

$$(4)^* \qquad E(v)(\gamma) := \iota \circ e(v)(\gamma).$$

The above defined e and E coincide with the ones in the assertion.

The action (8.1.2) is explictly written as $g_2 = \sigma(\underline{t})(\gamma)g_1$ and $g_1 = \sigma(\underline{t})(\gamma^{-1})g_2 = (\sigma(\underline{t})(\gamma))^{-1}g_2$. Then the lifting v^* of a vector field v on $\mathcal{U}$ is characterized as $\partial_{v^*}\varphi(\underline{t}, g) = \partial_v\varphi(\underline{t}, g)$ for any φ. Let $t = (\underline{t}_1, \ldots, t_n)$ be a coordinate for $\mathcal{U}$. The differential $d(\gamma^{-1})_*$ is calculated as

$$\begin{aligned} d(\gamma^{-1})_*(v^*) &= \sum_{i=1}^{n} (\partial_{v^*}t_i)\frac{\partial}{\partial t_i} + \left\langle \partial_{v^*}((\sigma(\underline{t})(\gamma^{-1})g_2), \frac{d}{dg_1}\right\rangle \\ &= v^* - \left\langle (\sigma(\underline{t})(\gamma)^{-1})(\partial_v\sigma(\underline{t})(\gamma))(\sigma(\underline{t})(\gamma)^{-1})g_2, \frac{d}{dg_1}\right\rangle \\ &= v^* - \left\langle (\sigma(\underline{t})(\gamma)^{-1})(\partial_v\sigma(\underline{t})(\gamma))g_1, \frac{d}{dg_1}\right\rangle \\ &= v^* - (\sigma(\underline{t})(\gamma)^{-1})\partial_v\sigma(\underline{t})(\gamma) \\ &= v^* - e(v)(\gamma). \end{aligned}$$

This is (8.3.1). Here we mean by the notion $\langle A(g), d/dg\rangle$ a vector field on $SL(2, \mathbf{R})$, whose tangent vector at $g \in SL(2, \mathbf{R})$ is given by the matrix $A(g)g^{-1} \in \mathfrak{sl}(2, \mathbf{R})$.

The formula (8.3.2) is reduced to (8.3.1) as follows. Let us denote by $\pi : \mathcal{U} \times SL(2, \mathbf{R}) \to \mathcal{U} \times \mathbf{H}$ the natural projection map. If a vector field w on $\mathcal{U} \times SL(2, \mathbf{R})$ is invariant by the right action of $SO(2)$, then one has a direct image vector field $d\pi_*(w)$ on $\mathcal{U} \times \mathbf{H}$.

For an element $w \in \Gamma(\mathcal{U} \times SL(2, \mathbf{R}), \mathcal{A}_{\mathcal{U}}(\mathfrak{sl}(2, \mathbf{R})))$ the image $d\pi_*(w)$ is equal to $\iota(w)$ (8.2.3). If v is a vector field on $\mathcal{U}$ and v^* and v^{**} are its liftings as in the Assertion, then $d\pi_*(v^*) = v^{**}$. Since the action of γ commutes with the projection π, $d\pi_*$ commutes with $d(\gamma^{-1})_*$. Hence one has $d\pi_*(d(\gamma^{-1})_*(v^*)) = d(\gamma^{-1})_*(v^{**})$. Now we apply $d\pi_*$ to (8.3.1) and obtain (8.3.2).

8.4.

The embeddings of the sheaf Θ into $\mathcal{A}_{\mathbf{H}}(\Theta)$ on $\mathbf{H}$ in (7.4) is relativised as the embedding of $\mathcal{A}_{\mathcal{U}}(\Theta)$ into $\mathcal{A}_{\mathcal{U}\times\mathbf{H}}(\Theta)$ over $\mathcal{U} \times \mathbf{H}$, where

$\mathcal{A}_{\mathcal{U}\times\mathbf{H}}(\Theta) :=$ the sheaf of germs of the form $\varphi(\underline{t}, z)\, d/dz$, where φ is a complex-valued differentiable function in $(\underline{t}, z)$. (8.4.1)

The action γ^{-1} on $\mathcal{U}\times\mathbf{H}$ (8.1.3) induces a Γ-module structure on $\Gamma(\mathcal{U}\times\mathbf{H}, \mathcal{A}_{\mathcal{U}\times\mathbf{H}}(\Theta))$ by $d(\gamma^{-1})_*$ referred to as a right adjoint action, i.e., for $v=\varphi(\underline{t}, z)\, d/dz\in\Gamma(\mathcal{U}\times\mathbf{H}, \mathcal{A}_{\mathcal{U}\times\mathbf{H}}(\Theta))$ and

$$\sigma(\underline{t})(\gamma)=\begin{bmatrix} a(t) & b(t) \\ c(\underline{t}) & d(\underline{t})\end{bmatrix}, \gamma\in\Gamma,$$

$$v\cdot \mathrm{ad}_\sigma(\gamma):=d(\gamma^{-1})_*(v)=\varphi(\gamma(\underline{t}, z))(c(\underline{t})z+d(\underline{t}))^2\frac{d}{dz}. \tag{8.4.2}$$

The following is a key lemma in the whole paper. For a proof of it, we use the result of Weil as quoted in (4.1) Theorem.

Lemma. For any given local parametrization σ, we have

$$H^1_\sigma(\Gamma, \Gamma(\mathcal{U}\times\mathbf{H}, \mathcal{A}_{\mathcal{U}\times\mathbf{H}}(\Theta)))=0. \tag{8.4.3}$$

As a standard proof for the vanishing of the cohomology groups for fine sheaves, we shall use the technique of partitions of unity, where Weil's result is used to construct partitions of unity depending simultaneously on the parameter $\underline{t}\in\mathcal{U}$.

Lemma (Partition of unity). *For any $\rho_0\in R_0(\Gamma, SL(2,\mathbf{R}))$, there exists a neighbourhood $\mathcal{U}$ of ρ_0 in $R_0(\Gamma, SL(2,\mathbf{R}))$ with the following property* (P), *where $\mathcal{U}$ is regarded as a local parameter by the inclusion $\mathcal{U}\subset R_0(\Gamma, SL(2,\mathbf{R}))$.*

(P) *There exists a system $\{\varphi_\gamma\}_{\gamma\in\Gamma}$ of non-negative real-valued differentiable functions defined on $\mathcal{U}\times\mathbf{H}$ such that*

(*i*) *Proper supportness: The projection map* $\mathrm{supp}(\varphi_\gamma)\to\mathcal{U}$ *is proper, where* $\mathrm{supp}(\varphi_\gamma):=$ *closure of* $\{(\underline{t}, w)\in\mathcal{U}\times\mathbf{H}: \varphi_\gamma(\underline{t}, w)\neq 0\}$ *for* $\gamma\in\Gamma$.

(*ii*) *Locally finiteness: Any point* $(\underline{t}, w)\in\mathcal{U}\times\mathbf{H}$ *has a neighbourhood which meets with only a finite number of* $\mathrm{supp}(\varphi_\gamma)$.

(*iii*) *Γ-invariance: $\varphi_{\gamma_2}\circ\gamma_1=\varphi_{\gamma_2\gamma_1}$ for $\forall\gamma_1, \gamma_2\in\Gamma$.*

(*iv*) *Partition of unity: $\sum_{\gamma\in\Gamma}\varphi_\gamma\equiv 1$ on $\mathcal{U}\times\mathbf{H}$.*

Proof of the Lemma. By the Theorem (4.1), there exists a compact set K of $\mathbf{H}$ and a neighbourhood $\mathcal{U}$ of ρ_0 such that $\bigcup_{\gamma\in\Gamma}\rho(\gamma)K=\mathbf{H}$ for all $\rho\in\mathcal{U}$. Let $\mathcal{V}$ be an open subset of $\mathbf{H}$ containing K and $\bar{\mathcal{V}}$ is compact such that

$$\bigcup_{\gamma\in\Gamma}\gamma(\mathcal{U}\times\mathcal{V})=\mathcal{U}\times\mathbf{H}.$$

The properness of the action of Γ on $\mathscr{U}\times\mathbf{H}$ implies the locally finiteness of the system of the open set $\gamma(\mathscr{U}\times\mathscr{V})$ for $\gamma\in\Gamma$.

Let χ be a differentiable function on $\mathbf{H}$ such that (i) $\chi(w)\geq 0$ for $w\in\mathbf{H}$ and $\chi(w)>0$ for $w\in K$, and (ii) $\mathrm{supp}(\chi)\subset\mathscr{V}$. Regarding χ as a function on the product space, we put

$$\chi_\gamma(\underline{t},z):=\chi(\gamma(\underline{t},z))\qquad\text{for }\gamma\in\Gamma.$$

Since $\mathrm{supp}(\chi_\gamma)\subset\gamma^{-1}(\mathscr{U}\times\mathscr{V})$ $(\gamma\in\Gamma)$ are locally finite, one may define $\sum_{\gamma\in\Gamma}\chi_\gamma$ on $\mathscr{U}\times\mathbf{H}$ as a finite-valued differentiable function, which is everywhere positive and invariant by the action of Γ. Put

$$\varphi_\gamma:=\chi_\gamma\Big/\sum_{\delta\in\Gamma}\chi_\delta\qquad\text{for }\gamma\in\Gamma.$$

This is a partition of unity, stated in the Lemma.

Proof of (8.4.3). One may easily check that the Lemma is reduced to particular local parametrization in the above Lemma. Hence, for a cocycle $z\in\mathscr{Z}^1_\sigma(\Gamma,\Gamma(\mathscr{U}\times\mathbf{H},\mathscr{A}_{\mathscr{U}\times\mathbf{H}}(\Theta))$ for such σ, we show an existence of $\xi\in\Gamma(\mathscr{U}\times\mathbf{H},\mathscr{A}_{\mathscr{U}\times\mathbf{H}}(\Theta))$ such that

$$(*)\qquad z(\gamma)=\xi-\xi\cdot\mathrm{ad}_\sigma(\gamma).$$

The following sum is well defined due to Lemma (ii):

$$(**)\qquad \xi:=\sum_{\gamma\in\Gamma}\varphi_\gamma\cdot z(\gamma)\in\Gamma(\mathscr{U}\times\mathbf{H},\mathscr{A}_{\mathscr{U}\times\mathbf{H}}(\Theta)).$$

For this ξ the relation $(*)$ is verified as follows.

$$\begin{aligned}\xi\cdot\mathrm{ad}_\sigma(\gamma)&=\left(\sum_{\delta\in\Gamma}\varphi_\delta\cdot z(\delta)\right)\cdot\mathrm{ad}_\sigma(\gamma)\\&=\sum_{\delta\in\Gamma}(\gamma^*\varphi_\delta)(z(\delta)\cdot\mathrm{ad}_\sigma(\gamma))\\&=\sum_{\delta\in\Gamma}(\varphi_{\delta\gamma})(-z(\gamma)+z(\delta\gamma))\\&=-\left(\sum_{\delta\in\Gamma}\varphi_{\delta\gamma}\right)z(\gamma)+\sum_{\delta\in\Gamma}\varphi_{\delta\gamma}z(\delta\gamma)\\&=-z(\gamma)+\xi.\end{aligned}$$

Remark. The same proof as above shows $H^1_\sigma(\Gamma,\Gamma(\mathscr{U}\times\mathbf{H},\mathscr{M}))=0$ for any $\mathscr{A}_{\mathscr{U}\times\mathbf{H}}$-module sheaf $\mathscr{M}$ on $\mathscr{U}\times\mathbf{H}$.

8.5.

Let an lp-map σ (8.1.1) be given. The vanishing (8.4.3) implies that for any vector field v of $\Gamma(\mathcal{U}, T(\mathcal{U}))$, the cocyle $E(v)$ (8.3.4) is expressed as a coboundary of a vector field $\xi_v \in \Gamma(\mathcal{U} \times \mathbf{H}, \mathcal{A}_{\mathcal{U}\times\mathbf{H}}(\Theta))$ on $\mathcal{U} \times \mathbf{H}$.

$$E(v)(\gamma) = \delta_\sigma(\xi_v)(\gamma) = \xi_v - \xi_v \cdot \mathrm{ad}_\sigma(\gamma) \qquad \text{for } \forall \gamma \in \Gamma. \tag{8.5.1}$$

We shall call ξ_v a *trivialization* of the cocycle $E(v)$, which is unique up to an addition of the element of

$$\Gamma(\Gamma \backslash \mathcal{U} \times \mathbf{H}, \mathcal{A}_{\mathcal{U}\times\mathbf{H}}(\Theta)) := \{\eta \in \Gamma(\mathcal{U} \times \mathbf{H}, \mathcal{A}_{\mathcal{U}\times\mathbf{H}}(\Theta)): \\ \eta = \eta \cdot \mathrm{ad}_\sigma(\gamma) \quad \text{for } \forall \gamma \in \Gamma\}.$$

The correspondence $v \mapsto \xi_v$ can be chosen to be a $\Gamma(\mathcal{U}, \mathcal{A}_{\mathcal{U}})$-homomorphism by a use of free basis of $\Gamma(\mathcal{U}, T(\mathcal{U}))$. Hence we obtain another formulation of (8.4.3), which we shall use later in several variations (cf. (9.3), (10.3), (11.3), (12.3)).

Lemma (Simultaneous trivialization of the cohomology classes). *For a given lp-map* σ (8.1.1), *there exists a* $\Gamma(\mathcal{U}, \mathcal{A}_{\mathcal{U}})$*-homomorphism,*

$$\xi : \Gamma(\mathcal{U}, T(\mathcal{U})) \to \Gamma(\mathcal{U} \times \mathbf{H}, \mathcal{A}_{\mathcal{U}\times\mathbf{H}}(\Theta)), \tag{8.5.2}$$

such that the cocycle map E (8.3.4) *for* σ *is expressed as the coboundary of* ξ,

$$E(v) = \delta_\sigma \xi(v) \qquad \textit{for } \forall v \in \Gamma(\mathcal{U}, T(\mathcal{U})). \tag{8.5.3}$$

The ξ *is unique up to an addition of a* $\Gamma(\mathcal{U}, \mathcal{A}_{\mathcal{U}})$*-homomorphism,*

$$\eta : \Gamma(\mathcal{U}, T(\mathcal{U})) \to \Gamma(\Gamma \backslash \mathcal{U} \times \mathbf{H}, \mathcal{A}_{\mathcal{U}\times\mathbf{H}}(\Theta)). \tag{8.5.4}$$

The map ξ (8.5.2) will be referred to as a *trivialization map* (*t-map*) for the lp-map σ. The formula (8.5.3) is given explicitly as

$$[-1, z]\sigma(t, \gamma)^{-1}\partial_v\sigma(t, \gamma)\begin{bmatrix} z \\ 1 \end{bmatrix}\frac{d}{dz} = \Xi_v(t, z)\frac{d}{dz} - \Xi_v(t, \sigma(t,\gamma)(z))\left(\frac{d\sigma}{dz}\right)^{-1}\frac{d}{dz}, \tag{8.5.3*}$$

where $\xi(v) = \Xi_v(t, z)\, d/dz$.

9. Integrability of the Complex Structure on $\mathcal{T}(\Gamma)$

The integrability of the complex structure on $\mathcal{T}(\Gamma)$ is shown in (9.4) Theorem. For the proof, we study the complex structure on an lp-space $\mathcal{U}$, lifted by $\mathcal{U} \to \mathcal{T}(\Gamma)$, in (9.1)–(9.3).

9.1.

Let us consider a local parametrization $\sigma : \mathcal{U} \to S^1(\mathcal{X}(\Gamma))$ from a domain $\mathcal{U}$ of $\dim_{\mathbf{R}} \mathcal{U} = 6g-6$ such that $\Omega \circ \sigma : \mathcal{U} \to \mathcal{T}(\Gamma)$ is a diffeomorphism onto a domain in $\mathcal{T}(\Gamma)$. For the definition of Ω, see (5.3.1). This means that for $\forall \underline{t} \in \mathcal{U}$ the composition $T_{\underline{t}}(\mathcal{U}) \xrightarrow{d\sigma_*} T_{\sigma(\underline{t})}S^1(\mathcal{X}(\Gamma)) \xrightarrow{d\Omega_*} T_{\Omega\circ\sigma(\underline{t})}\mathcal{T}(\Gamma)$ is an $\mathbf{R}$-linear isomorphism. Hence one obtains a splitting (cf. (6.4.2)),

$$T_{\sigma(\underline{t})}S^1(\mathcal{X}(\Gamma)) = T_{\underline{t}}(\mathcal{U}) \oplus T_{\sigma(\underline{t})}(\sigma(t)\cdot \mathrm{Ad}(SL(2,\mathbf{R}))). \tag{9.1.1}$$

9.2.

Recall the map $E: T_\rho S^1(\mathcal{X}(\Gamma)) \to \mathcal{Z}^1_\rho(\Gamma, \Gamma(\mathbf{H}, \Theta))$, inducing the real isomorphism $T_{\Omega(\rho)}(\mathcal{T}(\Gamma)) \simeq H^1_\rho(\Gamma, \Gamma(\mathbf{H}, \Theta))$ (cf. (7.1.3), (7.1.4)). Combining this fact with (9.1.1), we obtain a splitting/$\mathbf{R}$.

$$\mathcal{Z}^1_{\sigma(\underline{t})}(\Gamma, \Gamma(\mathbf{H}, \Theta)) = E(T_{\underline{t}}(\mathcal{U})) \oplus \delta_{\sigma(\underline{t})}\Gamma(\mathbf{H}, \Theta). \tag{9.2.1}$$

In the same way, the cocycle map E (8.3.4) induces an isomorphism: $\Gamma(\mathcal{U}, T(\mathcal{U})) \simeq H^1(\Gamma, \Gamma(\mathcal{U}\times\mathbf{H}, \mathcal{A}_{\mathcal{U}}(\Theta)))$ as $\Gamma(\mathcal{U}, \mathcal{A}_{\mathcal{U}})$-modules (cf. (8.2.4)), so that we obtain a simultaneous splitting as $\Gamma(\mathcal{U}, \mathcal{A})$-modules.

$$\mathcal{Z}^1_\sigma(\Gamma, \Gamma(\mathcal{U}\times\mathbf{H}, \mathcal{A}_{\mathcal{U}}(\Theta))) = E(\Gamma(\mathcal{U}, T(\mathcal{U})) \oplus \delta_\sigma \Gamma(\mathcal{U}\times\mathbf{H}, \mathcal{A}_{\mathcal{U}}(\Theta)). \tag{9.2.2}$$

We defined an endomorphism $J_{\underline{t}}$: $\mathrm{T}_{\underline{t}}(\mathcal{U}) \to T_{\underline{t}}(\mathcal{U})$ by the relation

$$iE(v) \equiv E(J_{\underline{t}}(v)) \bmod \delta_{\sigma(\underline{t})}\Gamma(\mathbf{H}, \Theta) \qquad \text{for} \quad v \in T_{\underline{t}}(\mathcal{U}). \tag{9.2.3}$$

The tensor $J = (J_{\underline{t}})_{\underline{t}\in\mathcal{U}}$, which is shown to be differentiable by (9.2.2), is an almost complex structure on $\mathcal{U}$ (i.e., $J^2 = -$identity of $T(\mathcal{U})$), which is transferred from that on $\mathcal{T}(\Gamma)$.

As usual, J induces a direct sum decomposition,

$$T(\mathcal{U}) \otimes \mathbf{C} = T^{1,0}(\mathcal{U}) \oplus T^{0,1}(\mathcal{U}), \tag{9.2.4}$$

where $T^{1,0}(\mathcal{U})$ (resp. $T^{0,1}(\mathcal{U})$) is the eigenspace of J belonging to the eigenvalue i (resp. $-i$).

9.3.

The map E (8.3.4) is extended complex linearly to $T(\mathcal{U}) \otimes \mathbf{C}$.

$$E : \Gamma(\mathcal{U}, T(\mathcal{U}) \otimes \mathbf{C}) \to \mathcal{Z}^1_\sigma(\Gamma, \Gamma(\mathcal{U}\times\mathbf{H}, \mathcal{A}_{\mathcal{U}}(\Theta))).$$

Then the subbundle $T^{0,1}(\mathcal{U})$ is characterized as follows. Let $v = x + i\cdot y \in$

$\Gamma(\mathcal{U}, T(\mathcal{U})\otimes\mathbf{C})$ with $x, y\in\Gamma(\mathcal{U}, T(\mathcal{U}))$. Then, using (9.2.2) and (9.2.3)

$$v\in\Gamma(\mathcal{U}, T^{0,1}(\mathcal{U}))\Leftrightarrow J(v)=-i\cdot v\Leftrightarrow x=-J\cdot y \text{ (and } y=J\cdot x)$$
$$\Leftrightarrow E(x+J\cdot y)=0$$
$$\Leftrightarrow E(x+i\cdot y)\equiv 0 \bmod \delta_\sigma\Gamma(\mathcal{U}\times\mathbf{H}, \mathcal{A}_{\mathcal{U}}(\Theta))$$
$$\Leftrightarrow E(v)\in\delta_\sigma\Gamma(\mathcal{U}\times\mathbf{H}, \mathcal{A}_{\mathcal{U}}(\Theta)).$$

Using a trivialization ξ in (8.5.2), this fact is rewritten,

$$\Leftrightarrow\delta_\sigma(\xi(v))\in\delta_\sigma\Gamma(\mathcal{U}\times\mathbf{H}, \mathcal{A}_{\mathcal{U}}(\Theta))$$
$$\Leftrightarrow\xi(v)\in\Gamma(\mathcal{U}\times\mathbf{H}, \mathcal{A}_{\mathcal{U}}(\Theta))\oplus\Gamma(\Gamma\backslash\mathcal{U}\times\mathbf{H}, \mathcal{A}_{\mathcal{U}\times\mathbf{H}}(\Theta)).$$

Here the last term is a direct sum, since $\Gamma(\mathbf{H},\Theta)\cap\Gamma(\Gamma_{\sigma(t)}\backslash\mathbf{H}, \mathcal{A}_{\mathbf{H}}(\Theta))=\Gamma(\Gamma_{\sigma(t)}\backslash\mathbf{H},\Theta)=0$ (cf. (7.5)). Denoting by $h(v)$ the first factor of $\xi(v)$ in this decomposition, we summarize the calculation as follows.

Assertion. *Let a local parametrization* σ *of* (9.1) *be given.*

(*i*) *For a vector field* $v\in\Gamma(\mathcal{U}, T(\mathcal{U})\otimes\mathbf{C})$,

$$v\in\Gamma(\mathcal{U}, T^{0,1}(\mathcal{U}))\Leftrightarrow E(v)\in\delta_\sigma\Gamma(\mathcal{U}\times\mathbf{H}, \mathcal{A}_{\mathcal{U}}(\Theta)).$$

(*ii*) *There exists uniquely a* $\Gamma(\mathcal{U}, \mathcal{A}_{\mathcal{U}}\otimes\mathbf{C})$*-homorphism,*

$$h:\Gamma(\mathcal{U}, T^{0,1}(\mathcal{U})\to\Gamma(\mathcal{U}\times\mathbf{H}, \mathcal{A}_{\mathcal{U}}(\Theta)), \tag{9.3.1}$$

such that the cocycle $E(v)$ *for* $\forall v\in\Gamma(\mathcal{U}, T^{0,1}(\mathcal{U}))$ *is the coboundary of* h:

$$E(v)=\delta_\sigma h(v). \tag{9.3.2}$$

(*iii*) *For any trivialization map* ξ (8.5.2) *for* σ, *the difference* $\xi-h$ *defines a* $\Gamma(\mathcal{U}, \mathcal{A}_{\mathcal{U}})$*-homomorphism*:

$$\xi-h:\Gamma(\mathcal{U}, T^{0,1}(\mathcal{U}))\to\Gamma(\Gamma\backslash\mathcal{U}\times\mathbf{H}, \mathcal{A}_{\mathcal{U}\times\mathbf{H}}(\Theta)), \tag{9.3.3}$$

which can be a 0*-map for a suitable choice of* ξ. *Such* ξ *is unique up to an addition of a* $\Gamma(\mathcal{U}, \mathcal{A}_{\mathcal{U}}\otimes\mathbf{C})$*-homomorphism,*

$$\eta:\Gamma(\mathcal{U}, T(\mathcal{U})\otimes\mathbf{C})\to\Gamma(\Gamma\backslash\mathcal{U}\times\mathbf{H}, \mathcal{A}_{\mathcal{U}\times\mathbf{H}}(\Theta)), \tag{9.3.4}$$

$$\eta\,|\,T^{0,1}(\mathcal{U})=0. \tag{9.3.5}$$

Proof. Let us show that h does not depend on the choice of ξ. The ambiguity η of ξ given in (8.5.4) belongs to the second factor of ξ, so that it does not affect the first factor h. (*iii*) is also obvious by this description.

9.4.

Theorem.

1. *The almost complex structure on* $\mathscr{T}(\Gamma)$ *in* (7.2) *is integrable.*
2. *The left action of* $\mathrm{Aut}(\Gamma)/\Gamma$ *on* $\mathscr{T}(\Gamma)$ *is holomorphic.*

Proof. Statement 2 is included for the sake of completeness. To prove it, it is enough to show that the action of $\mathrm{Aut}(\Gamma)/\Gamma$ ((5.3)(*iii*)) preserves the almost complex structure on the tangent spaces of $\mathscr{T}(\Gamma)$, which is immediate from (7.1) Corollary.

As for Statement 1, the integrability of the complex structure on $\mathscr{T}(\Gamma)$ is reduced to that of J on $\mathscr{U}$ (9.2) for all lp-maps σ of (9.1). Hence we need only show the involutivity of the bundle $T^{0,1}(\mathscr{U})$, i.e., that $\Gamma(\mathscr{U}, T^{0,1}(\mathscr{U}))$ is closed under the bracket product (cf. (9.2)).

Take any $u, v \in \Gamma(\mathscr{U}, T^{0,1}(\mathscr{U}))$. Using (9.3.2) and (8.4.2), there are $h(u)$, $h(v) \in \Gamma(\mathscr{U} \times \mathbf{H}, \mathscr{A}_{\mathscr{U}}(\Theta))$ so that we have expressions for $\gamma \in \Gamma$,

$$(*) \qquad E(u)(\gamma) = h(u) - d(\gamma^{-1})_*(h(u)),$$

$$(**) \qquad E(v)(\gamma) = h(v) - d(\gamma^{-1})_*(h(v)).$$

We want to show that $E([u, v]) \in \delta_\sigma \Gamma(\mathscr{U} \times \mathbf{H}, \mathscr{A}_{\mathscr{U}}(\Theta))$.

Recall the formula (8.3.2) and apply it for u, v and $[u, v]$. Noting the fact that the differential $d(\gamma^{-1})_*$ for $\gamma \in \Gamma$ preserves the bracket product structure, one obtains the relation

$$(***) \qquad [u, v]^{**} - E([u, v])(\gamma) = [u^{**} - E(u)(\gamma), v^{**} - E(v)(\gamma)].$$

Here u^{**}, v^{**}, etc., means the vector fields on $\mathscr{U} \times \mathbf{H}$ trivially lifted from that u, v, etc. on $\mathscr{U}$. Since $[u, v]^{**} = [u^{**}, v^{**}]$, the relation $(***)$ is rewritten as

$$E([u, v])(\gamma) = [u^{**}, E(v)(\gamma)] + [E(u)(\gamma), v^{**}] - [E(u)(\gamma), E(v)(\gamma)].$$

By substituting $(*)$ and $(**)$ in the above right-hand side, we have

$$\begin{aligned} &= [u^{**}, h(v) - d(\gamma^{-1})_*(h(v))] \\ &\quad - [v^{**}, h(u) - d(\gamma^{-1})_*(h(u))] \\ &\quad - [h(u) - d(\gamma^{-1})_*(h(u)), h(v) - d(\gamma^{-1})_*(h(v))] \\ &= \begin{pmatrix} [u^{**}, h(v)] - d(\gamma^{-1})_*[u^{**}, h(v)] \\ [-u^{**} + d(\gamma^{-1})_* u^{**}, h(v) \cdot \mathrm{ad}_\sigma(\gamma)] \end{pmatrix} \end{aligned}$$

$$-\begin{pmatrix} [v**, h(u)]-d(\gamma^{-1})_*[v**, h(u)] \\ [-v**+d(\gamma^{-1})_*v**, h(u)\cdot \mathrm{ad}_\sigma(\gamma)] \end{pmatrix}$$
$$-([h(u), h(v)]-[h(u)\cdot \mathrm{ad}_\sigma(\gamma), h(v)]$$
$$-[h(u), h(v)\cdot \mathrm{ad}_\sigma(\gamma)]$$
$$+[h(u)\cdot \mathrm{ad}_\sigma(\gamma), h(v)\cdot \mathrm{ad}_\sigma(\gamma)])$$

Here we note that $[u**, h(v)], [v**, h(u)] \in \Gamma(\mathscr{U}\times \mathbf{H}, \mathscr{A}_{\mathscr{U}}(\Theta))$. Again apply the formula (8.3.2) for u and v in the second line,

$$= \delta_\sigma([u**, h(v)]-[v**, h(u)]-[h(u), h(v)])(\gamma)$$
$$-[h(u)-h(u)\cdot \mathrm{ad}_\sigma(\gamma), h(v)\cdot \mathrm{ad}_\sigma(\gamma)]$$
$$+[h(v)-h(v)\cdot \mathrm{ad}_\sigma(\gamma), h(u)\cdot \mathrm{ad}_\sigma(\gamma)]$$
$$-(-[h(u)\cdot \mathrm{ad}_\sigma(\gamma), h(v)]-[h(u), h(v)\cdot \mathrm{ad}_\sigma(\gamma)]$$
$$+2[h(u)\cdot \mathrm{ad}_\sigma(\gamma), h(v)\cdot \mathrm{ad}_\sigma(\gamma)])$$
$$= \delta_\sigma([u**, h(v)]-[v**, h(u)]-[h(u), h(v)])(\gamma).$$

Since $[u**, h(v)]-[v**, h(u)]-[h(u), h(v)] \in \Gamma(\mathscr{U}\times \mathbf{H}, \mathscr{A}_{\mathscr{U}}(\Theta))$, the final expression means $E([u, v]) \in \delta_\sigma \Gamma(\mathscr{U}\times \mathbf{H}, \mathscr{A}_{\mathscr{U}}(\Theta))$. This implies $[u, v] \in \Gamma(\mathscr{U}, T^{0,1}(\mathscr{U}))$ due to the equivalence (9.3) Assertion (i). This completes a proof of the Theorem.

The final step in the above proof implies also the formula

$$h([u, v]) = [u**, h(v)]-[v**, h(u)]-[h(u), h(v)]. \tag{9.4.1}$$

9.5.

Due to (9.4) Theorem, we shall regard $\mathscr{T}(\Gamma)$ as a complex manifold of dimension $3g-3$. The domain $\mathscr{U}$ for the map σ of (9.1) may be regarded as a complex chart for $\mathscr{T}(\Gamma)$.

Summarizing the results, we obtain the following.

Theorem *bis*. *Let $\mathscr{U}$ be a domain in $\mathbf{C}^n$. Then an lp-map σ (8.1) on $\mathscr{U}$ induces a holomorphic map $\Omega\circ\sigma : \mathscr{U} \to \mathscr{T}(\Gamma)$ iff there exists a $\Gamma(\mathscr{U}, \mathscr{A}_{\mathscr{U}})$-homomorphism*

$$h : \Gamma(\mathscr{U}, T^{0,1}(\mathscr{U})) \to \Gamma(\mathscr{U}\times \mathbf{H}, A_{\mathscr{U}}(\Theta)),$$

such that

(i) *The cocycle $E(v)$ for $\forall v \in \Gamma(\mathscr{U}, T^{0,1}(\mathscr{U}))$ are expressed as*

$$E(v) = \delta_\sigma(h(v)).$$

I.e., this means explicitly that for every $\gamma \in \Gamma$, *one has*

$$[-1, z]\sigma(\underline{t})(\gamma)^{-1}\partial_v\sigma(\underline{t})(\gamma)\begin{bmatrix} z \\ 1 \end{bmatrix}\frac{d}{dz} = h(v) - h(v)\cdot \mathrm{ad}(\sigma(\underline{t})(\gamma)).$$

(*ii*) *Put*

$$H(u) := u{*}{*} - h(u) \qquad \textit{for } u \in \Gamma(\mathscr{U}, T^{0,1}(\mathscr{U})). \tag{9.5.1}$$

Then H defines a $\Gamma(\mathscr{U}, \mathscr{A}_{\mathscr{U}})$*-Lie-algebra homomorphism*

$$H : \Gamma(\mathscr{U}, T^{0,1}(\mathscr{U})) \to \Gamma(\Gamma\backslash\mathscr{U} \times \mathbf{H}, T(\mathscr{U} \times \mathbf{H})). \tag{9.5.2}$$

Proof. It is enough to show the theorem for the lp-map of the type (9.1). Then the map h is defined in (9.3.1).

(*i*) This is almost the definition of the complex structure on $\mathscr{T}$.

(*ii*) The equality $[u{*}{*} - h(u),\ v{*}{*} - h(v)] = [u, v]{*}{*} - h([u, v])$ is equivalent to (9.5.1).

That $H(u)$ is invariant by the action of Γ follows from (8.3.2), since

$$\begin{aligned} d(\gamma^{-1})_*(H(u)) &= d(\gamma^{-1})_*(u{*}{*}) - d(\gamma^{-1})_*(h(u)) \\ &= u{*}{*} - E(u)(\gamma) - h(u)\cdot \mathrm{ad}_\sigma(\gamma) \\ &= u{*}{*} - \delta_\sigma(h(u))(\gamma) - h(u)\cdot \mathrm{ad}_\sigma(\gamma) \\ &= u{*}{*} - (h(u) - h(u)\cdot \mathrm{ad}_\sigma(\gamma)) - h(u)\cdot \mathrm{ad}_\sigma(\gamma) \\ &= u{*}{*} - h(u) = H(u). \end{aligned}$$

10. Integrability of the Complex Structure on $\mathscr{X}(\Gamma)$

Integrability of the complex structures on $\mathscr{X}(\Gamma)$ and $\tilde{\mathscr{X}}(\Gamma)$ are shown in (10.5) Theorem. For the sake of completeness, we proceed with a brief proof of it, which is a modification of that for $\mathscr{T}$ in Section 9.

10.1.

Let us consider a new local parameterization map,

$$\sigma{*} : \mathscr{V} \longrightarrow S^1(\mathscr{X}(\Gamma)), \tag{10.1.1}$$

where $\mathscr{V}$ is a domain in $\mathbf{R}^{6g-4}$ such that the composition map

$$\Pi \circ \sigma{*} : \mathscr{V} \longrightarrow \mathscr{X}(\Gamma) \tag{10.1.2}$$

is a diffeomorphism onto a domain in $\mathscr{X}(\Gamma)$. For the definition of Π, see (5.3.1). This implies for each $\underline{t}_* \in \mathscr{V}$ the splitting (cf. (6.4.3))

$$T_{\sigma*(\underline{t}*)}S^1(\mathscr{X}(\Gamma)) = T_{\underline{t}*}(\mathscr{V}) \oplus T_{\sigma*(\underline{t}*)}(\sigma*(t*)\cdot \mathrm{ad}(SO(2))). \tag{10.1.3}$$

10.2.

To avoid confusion with the cocycle map E for σ in Section 9, we shall denote by E^* the cocycle map (8.4.3) for $\sigma*$,

$$E^*: \Gamma(\mathscr{V}, T(\mathscr{V})) \to \mathscr{Z}^1_{\sigma*}(\Gamma, \Gamma(\mathscr{V}\times \mathbf{H}, \mathscr{A}_{\mathscr{V}}(\Theta))), \tag{10.2.1}$$

which induces a $\Gamma(\mathscr{V}, \mathscr{A}_{\mathscr{V}})$-isomorphism:

$$\Gamma(\mathscr{V}, T(\mathscr{V})) \simeq \mathscr{Z}^1_{\sigma*}(\Gamma, \Gamma(\mathscr{V}\times \mathbf{H}, \mathscr{A}_{\mathscr{V}}(\Theta)))/\delta_{\sigma*}\Gamma(\mathscr{V}\times \mathbf{H}, \mathscr{A}_{\mathscr{V}}(\Theta(-[i]))),$$

of rank $6g-4$ modules over $\Gamma(\mathscr{V}, \mathscr{A}_{\mathscr{V}})$ (cf. (7.1.4), (8.2.5)).

Hence, the following splitting is induced as $\Gamma(\mathscr{V}, \mathscr{A})$-modules:

$$\begin{aligned}\mathscr{Z}^1_{\sigma*}(\Gamma, \Gamma(\mathscr{V}\times \mathbf{H}, \mathscr{A}_{\mathscr{V}}(\Theta))) \simeq\ & E^*(\Gamma(\mathscr{V}, T(\mathscr{V}))) \\ & \oplus \delta_{\sigma*}\Gamma(\mathscr{V}\times \mathbf{H}, \mathscr{A}_{\mathscr{V}}(\Theta(-[i]))).\end{aligned} \tag{10.2.2}$$

We defined an endomorphism J^*_{t*}: $T_{t*}(\mathscr{V}) \to T_{t*}(\mathscr{V})$ by the relation

$$iE^*(v) \equiv E^*(J^*_{t*}(v)) \quad \mathrm{mod}\ \delta_{\sigma*(t*)}\Gamma(\mathbf{H}, \Theta(-[i])) \tag{10.2.3}$$

for $v \in T_{t*}(\mathscr{V})$.

The tensor $J^* = (J^*_{t*})_{t*\in \mathscr{V}}$, which is shown to be differentiable by (10.2.2), is the most complex structure on $\mathscr{V}$ (i.e., $J^{*2} = -$identity of $T(\mathscr{V})$), which is transferred from that on $\mathscr{X}(\Gamma)$.

As usual, J^* induces a direct sum decomposition,

$$T(\mathscr{V}) \otimes \mathbf{C} = T^{1,0}(\mathscr{V}) \oplus T^{0,1}(\mathscr{V}) \tag{10.2.4}$$

by the eigenspaces for J^* for the eigenvalues i and $-i$.

10.3.

The map E^* (10.2.1) extends complex linearly to $T(\mathscr{V})\otimes \mathbf{C}$. A calculation similar to that in (9.3) characterizes the subbundle $T^{0,1}(\mathscr{V})$: Namely for $v \in \Gamma(\mathscr{V}, T(\mathscr{V})\otimes \mathbf{C})$,

$$v \in \Gamma(\mathscr{V}, T^{0,1}(\mathscr{V})) \Leftrightarrow E^*(v) \in \delta_{\sigma*}\Gamma(\mathscr{V}\times \mathbf{H}, \mathscr{A}_{\mathscr{V}}(\Theta(-[i]))).$$

Using a trivialization ξ^* (8.5.2) for σ^*, this fact is rewritten,

$$\begin{aligned}&\Leftrightarrow \delta_{\sigma^*}(\xi^*(v)) \in \delta_{\sigma^*}\Gamma(\mathscr{U}\times \mathbf{H}, \mathscr{A}_{\mathscr{V}}(\Theta(-[i])))\\ &\Leftrightarrow \xi^*(v) \in \Gamma(\mathscr{V}\times \mathbf{H}, \mathscr{A}_{\mathscr{V}}(\Theta(-[i])))\\ &\qquad\qquad \oplus \Gamma(\Gamma\backslash \mathscr{V}\times \mathbf{H}, \mathscr{A}_{\mathscr{V}\times \mathbf{H}}(\Theta)).\end{aligned}$$

Here the last term is a direct sum as before in Section 9. Let us denote the first factor of ξ^* by h^*. As a summary, we obtain the following Assertion which is parallel to that in (9.3).

Assertion. *Let an lp-map* σ^* *of* (10.1) *be given.*

(*i*) *For* $v \in \Gamma(\mathscr{V}, T(\mathscr{V}) \otimes \mathbf{C})$,

$$v \in \Gamma(\mathscr{V}, T^{0,1}(\mathscr{V})) \Leftrightarrow E^*(v) \in \delta_{\sigma^*}\Gamma(\mathscr{V} \times \mathbf{H}, \mathscr{A}_{\mathscr{V}}(\Theta(-[i]))).$$

(*ii*) *There is a unique* $\Gamma(\mathscr{V}, \mathscr{A}_{\mathscr{V}} \otimes \mathbf{C})$*-homomorphism,*

$$h^*: \Gamma(\mathscr{V}, T^{0,1}(\mathscr{V})) \to \Gamma(\mathscr{V} \times \mathbf{H}, \mathscr{A}_{\mathscr{V}}(\Theta(-[i]))), \tag{10.3.1}$$

such that the cocycle $E^*(v)$ *for* $\forall v \in \Gamma(\mathscr{V}, T^{0,1}(\mathscr{V}))$ *is expressed as*

$$E^*(v) = \delta_{\sigma^*} h^*(v) \qquad \text{for } \forall v \in \Gamma(\mathscr{V}, T^{0,1}(\mathscr{V})). \tag{10.3.2}$$

(*iii*) *For any trivialization map* ξ^* (8.5.2) *for* σ^*, *the difference* $\xi^* - h^*$ *defines a* $\Gamma(\mathscr{V}, \mathscr{A}_{\mathscr{V}})$*-homomorphism,*

$$\xi^* - h^*: \Gamma(\mathscr{V}, T^{0,1}(\mathscr{V})) \to \Gamma(\Gamma \backslash \mathscr{V} \times \mathbf{H}, \mathscr{A}_{\mathscr{V} \times \mathbf{H}}(\Theta)), \tag{10.3.3}$$

which can be a 0*-map for a suitable choice of* ξ^*. *Such a* ξ^* *is unique up to an addition of a* $\Gamma(\mathscr{V}, \mathscr{A}_{\mathscr{V}} \otimes \mathbf{C})$*-homomorphism,*

$$\eta: \Gamma(\mathscr{V}, T(\mathscr{V}) \otimes \mathbf{C}) \to \Gamma(\Gamma \backslash \mathscr{V} \times \mathbf{H}, \mathscr{A}_{\mathscr{V} \times \mathbf{H}}(\Theta)), \tag{10.3.4}$$

with

$$\eta \,|\, T^{0,1}(\mathscr{V}) = 0. \tag{10.3.5}$$

Proof. Omitted.

10.4.

Theorem. 1. *The almost complex structures on* $\mathscr{X}(\Gamma)$ *and on* $\tilde{\mathscr{X}}(\Gamma)$ (7.2) *are integrable.*

2. *The left actions of* $\mathrm{Aut}(\Gamma)/\Gamma$ *on* $\mathscr{X}(\Gamma)$ *and* $\mathrm{Aut}(\Gamma)$ *on* $\tilde{\mathscr{X}}(\Gamma)$ *are holomorphic.*

3. *The projection maps* $\Phi: \mathscr{X}(\Gamma) \to \mathscr{T}(\Gamma)$ *and* $\tilde{\Phi}: \tilde{\mathscr{X}}(\Gamma) \to \mathscr{T}(\Gamma)$ *are holomorphic.*

4. *The fibers* $\tilde{X}_t := \tilde{\Phi}^{-1}(t)$ *and* $X_t := \Phi^{-1}(t)$ *over* $t \in \mathscr{T}(\Gamma)$ *are isomorphic to* $\mathbf{H}$ *and to* $\Gamma_\rho \backslash \mathbf{H}$ *for any* $\rho \in \Omega^{-1}(t)$ *respectively.*

Proof. Statements 2, 3 and 4 are included for the sake of completeness.

2. It is enough to show the action of $\mathrm{Aut}(\Gamma)/\Gamma$ (5.3) (*iii*) preserves the almost complex structure on the tangent space, which follows from (7.1) Corollary.

3. The exact sequence (7.1.4) implies this.

4. Take a point ρ in the fiber $\Omega^{-1}(t)$ (resp. $\tilde{\Omega}^{-1}(t)$).

Then

$$\Omega^{-1}(t) = \rho \cdot \mathrm{Ad}(SL(2, \mathbf{R})) \simeq \rho(\Gamma)\backslash SL(2, \mathbf{R})$$

$$(\text{resp. } \tilde{\Omega}^{-1}(t) = \rho \cdot \mathrm{Ad}(SL(2, \mathbf{R})) \simeq SL(2, \mathbf{R}))$$

and hence $\Phi^{-1}(t) \simeq \Gamma_\rho\backslash SL(2, \mathbf{R})/SO(2)$ (resp. $\tilde{\Phi}^{-1}(t) \simeq SL(2, \mathbf{R})/SO(2)$). Then (7.1.4) implies that the complex structure on the fibers of Φ and $\tilde{\Phi}$ induced from that on $\mathscr{X}$ and $\tilde{\mathscr{X}}$ coincides with the canonical complex structure on $SL(2, \mathbf{R})/SO(2)$.

1. Again (7.1) Corollary (*ii*) implies that the integrability for $\mathscr{X}(\Gamma)$ is equivalent to that for $\tilde{\mathscr{X}}(\Gamma)$.

The integrability of the complex structure on $\tilde{\mathscr{X}}(\Gamma)$ is reduced to that of J^* for all local parametrizations $\sigma*$ of (10.1). Hence we need only to show the involutivity of $T^{0,1}(\mathscr{V})$, i.e., to show that $\Gamma(\mathscr{V}, T^{0,1}(\mathscr{V}))$ is closed under the bracket product.

Take any $u, v \in \Gamma(\mathscr{V}, T^{0,1}(\mathscr{V}))$. Using (10.3.2), there are $h^*(u), h^*(v) \in \Gamma(\mathscr{V} \times \mathbf{H}, \mathscr{A}_{\mathscr{V}}(\Theta(-[i])))$ so that we have expressions for $\gamma \in \Gamma$,

$$(*) \qquad E^*(u)(\gamma) = h^*(u) - d(\gamma^{-1})_*(h^*(u)),$$

$$(**) \qquad E^*(v)(\gamma) = h^*(v) - d(\gamma^{-1})_*(h^*(v)).$$

One only has to show that $E^*([u, v]) \in \delta_{\sigma*}\Gamma(\mathscr{V} \times \mathbf{H}, \mathscr{A}_{\mathscr{V}}(\Theta(-[i])))$.

As in (8.3) Assertion, let us denote by $u**$, $v**$, etc., the vector fields on $\mathscr{V} \times \mathbf{H}$, which are the liftings of that of u, v, etc., on $\mathscr{V}$. By this notation, a parallel calculation to that in the proof of (9.4) Theorem, shows a relation

$$E^*([u, v])(\gamma) = \delta_{\sigma*}([u**, h^*(v)] - [v**, h^*(u)] - [h^*(u), h^*(v)])(\gamma).$$

One checks easily,

$$[u**, h^*(v)] - [v**, h^*(u)] - [h^*(u), h^*(v)] \in \Gamma(\mathscr{U} \times \mathbf{H}, \mathscr{A}_{\mathscr{U}}(\Theta(-[i]))),$$

so that we have shown that

$$E^*([u,v]) \in \delta_{\sigma *}\Gamma(\mathscr{V}\times\mathbf{H}, \mathscr{A}_{\mathscr{V}}(\Theta(-[i]))).$$

Hence $[u, v] \in \Gamma(\mathscr{U}, T^{0,1}(\mathscr{U}))$ due to the equivalence (10.3) Assertion (i). This completes a proof of the Theorem.

10.5.

Due to this Theorem, we shall regard $\mathscr{X}(\Gamma)$ as a complex manifold of dimension $3g-2$ and $\Phi: \mathscr{X}(\Gamma) \to \mathscr{T}(\Gamma)$ as the complex analytic family of Riemann surfaces. The domain $\mathscr{V}$ for the map $\sigma*$ of (10.1) may be regarded as a complex chart for $\mathscr{X}(\Gamma)$.

Summarizing the previous calculations, we obtain the following.

Theorem *bis*. *Let $\mathscr{V}$ be a domain in $\mathbf{C}^n$ and let $\sigma*$ be an lp-map of* (8.1). *Then the induced map $\Pi \circ \sigma* : \mathscr{V} \to \mathscr{X}(\Gamma)$ is holomorphic, iff there exists a $\Gamma(\mathscr{U}, \mathscr{A}_{\mathscr{V}})$-homomorphism*

$$h^* : \Gamma(\mathscr{V}, T^{0,1}(\mathscr{V})) \to \Gamma(\mathscr{V}\times\mathbf{H}, \mathscr{A}_{\mathscr{V}}(\Theta(-[i])))$$

such that

(*i*) $E^*(v) = \delta_{\sigma*}(h^*(v))$ *for* $\forall v \in \Gamma(\mathscr{V}, T^{0,1}(\mathscr{V}))$.

I.e., this means explicitly that for every $\gamma \in \Gamma$, one has

$$[-1, z]\sigma*(\underline{t}*)(\gamma)^{-1}\partial_v\sigma*(\underline{t}*)(\gamma)\begin{bmatrix} z \\ 1 \end{bmatrix}\frac{d}{dz} = h^*(v) - h^*(v)\cdot \mathrm{ad}(\sigma*(\underline{t}*)(\gamma)).$$

(*ii*) *Put,*

$$H^*(u) := u** - h^*(u) \qquad \textit{for } u \in \Gamma(\mathscr{V}, T^{0,1}(\mathscr{U})). \tag{10.5.1}$$

Then H^* defines a $\Gamma(\mathscr{V}, \mathscr{A}_{\mathscr{V}})$-Lie-algebra homomorphism:

$$H^* : \Gamma(\mathscr{V}, T^{0,1}(\mathscr{V})) \to \Gamma(\Gamma\backslash\mathscr{V}\times\mathbf{H}, T(\mathscr{V}\times\mathbf{H})). \tag{10.5.2}$$

Proof. This is shown in a similar manner as the proof for (9.5) Theorem-*bis* so that the proof is omitted.

11. Gauge Transformation I

11.1.

Let us consider an lp-map $\sigma: \mathcal{U} \to S^1(\mathcal{X}(\Gamma))$ of (9.1) and an lp-map $\sigma*: \mathcal{V} \to S^1(\mathcal{X}(\Gamma))$ of (10.1), which are related as follows.

(*i*) *$\mathcal{U}$ is a domain in $\mathbf{C}^{3g-3}$ and $\mathcal{V}$ is a domain in $\mathcal{U} \times \mathbf{C}$. We shall denote by t and $t* = (t, w)$ the coordinates for $\mathcal{U}$ and $\mathcal{V}$ respectively.*

(*ii*) *There exists a differentiable map*

$$Z: \mathcal{V} \to \mathbf{H}, \qquad \text{with } Z(\underline{t}, w) := X(\underline{t}, w) + i \cdot Y(\underline{t}, w), \tag{11.1.1}$$

so that the parametrization maps $\sigma*$ and σ are related as

$$\sigma*(\underline{t}*) = \sigma(\underline{t}) \cdot \mathrm{Ad}(g(Z(\underline{t}*)))(= g(Z(\underline{t}*))^{-1} \sigma(\underline{t}) g(Z(\underline{t}*))). \tag{11.1.2}$$

Here $g(Z)$ is an element of $SL(2, \mathbf{R})$ defined as

$$g(Z) := \begin{bmatrix} 1 & X \\ 0 & 1 \end{bmatrix} \cdot \begin{bmatrix} \sqrt{Y} & 0 \\ 0 & 1/\sqrt{Y} \end{bmatrix} \qquad \text{for } Z = X + i \cdot Y \in \mathbf{H}. \tag{11.1.3}$$

11.2.

Let us assume moreover that the maps $\Omega \circ \sigma: \mathcal{U} \to \mathcal{T}$ and $\Pi \circ \sigma*: \mathcal{V} \to \mathcal{X}$ are locally biholomorphic. I.e., the natural complex structures on $\mathcal{U}$ and $\mathcal{V}$ coincide with the induced ones J and J^*. (For notations Ω, Π, J and J^*, recall (5.3.1), (9.2.3) and (10.2.3).)

The following fact is easy to see.

Let $\varphi: \mathcal{V} \to \mathcal{U}$ be the natural projection map to the first factor. Then one has the following commutative diagram:

$$\begin{array}{ccc} \mathcal{V} & \xrightarrow[\Pi \circ \sigma*]{} & \mathcal{X} \\ \downarrow{\scriptstyle \varphi} & & \downarrow{\scriptstyle \Phi} \\ \mathcal{U} & \xrightarrow[\Omega \circ \sigma]{} & \mathcal{T} \end{array}$$

and for any $t^* \in \mathcal{V}$

$$\begin{array}{ccccc} T_{t*}(\mathcal{V}) & \xrightarrow[d\Pi\sigma**]{\sim} & T_{\Pi \circ \sigma*(t*)}(\mathcal{X}) & \xrightarrow{\sim} & \mathcal{Z}^1_{\sigma*(t*)}(\Gamma, \Gamma(\mathbf{H}, \Theta))/\delta_{\sigma*(t*)}\Gamma(\mathbf{H}, \Theta(-[i])) \\ \downarrow{\scriptstyle d\varphi_*} & & \downarrow{\scriptstyle d\Phi_*} & & \downarrow \\ T_t(\mathcal{U}) & \xrightarrow[d\Omega\sigma_*]{\sim} & T_{\Omega \circ \sigma(t)}(\mathcal{T}) & \xrightarrow{\sim} & H^1_{\sigma(t)}(\Gamma, \Gamma(\mathbf{H}, \Theta)). \end{array}$$

The map $d\varphi_*$ is commutative with the complex structures, i.e.,

$$J_* \circ (d\varphi_*) = (d\varphi_*) \circ J, \tag{11.2.1}$$

so that $d\varphi_*$ preserves the decomposition (9.2.4) and (10.2.4). Particularly $d\varphi_*: T^{0,1}_{t*}(\mathscr{V}) \to T^{0,1}_t(\mathscr{U})$ is surjective. The kernel is a complex 1-dimensional space spanned by $\partial/\partial\bar{w}$.

11.3.

We want to calculate explicitly E^* and its t-map ξ^* by use of the data in (10.1) and in terms of E and ξ in Section 9.

Assertion. *Let lp-maps σ and $\sigma*$ as in (11.1) be given.*

(*i*) *The cocycle map E^* (10.2.1) for $\sigma*$ is given by*

$$E^*(v) = E(d\varphi_*(v)) \cdot \mathrm{ad}(g(Z)) - \delta_{\sigma*}\left(\frac{\partial_v(X+z\cdot Y)}{Y}\frac{d}{dz}\right). \tag{11.3.1}$$

(*ii*) *If ξ is a t-map for σ, then a t-map ξ^* for $\sigma*$ is given by*

$$\xi^*(v) = \xi(d\varphi_*(v)) \cdot \mathrm{ad}(g(Z)) - \frac{\partial_v(X+z\cdot Y)}{Y}\frac{d}{dz}. \tag{11.3.2}$$

Proof. For a vector field $v \in \Gamma(\mathscr{V}, T(\mathscr{V}))$, one calculates that

$$g(Z)^{-1}\partial_v g(Z) = Y^{-1}\begin{bmatrix} \partial_v Y/2 & \partial_v X \\ 0 & -\partial_v Y/2 \end{bmatrix}.$$

Hence composing this map with ι (2.3.2), we obtain

(*) $$\iota(g(Z)^{-1}\partial_v g(Z)) = -Y^{-1}\partial_v(X+z\cdot Y).$$

On the other hand,

(**) $$\begin{aligned} e*(v) := \sigma_*^{-1}\partial_v\sigma* &= g(Z)^{-1}\sigma^{-1}g(Z)\partial_v(g(Z)^{-1}\sigma g(Z)) \\ &= g(Z)^{-1}(\sigma^{-1}\partial_v\sigma)g(Z) + g(Z)^{-1}\partial_v g(Z) \\ &\quad - (g(Z)^{-1}\sigma^{-1}g(Z))(g(Z)^{-1}\partial_v g(Z))(g(Z)^{-1}\sigma g(Z)) \\ &= (\sigma^{-1}\partial_v\sigma)\cdot \mathrm{ad}(g(Z)) + \delta_{\sigma*}(g(Z)^{-1}\partial_v g(Z)) \\ &= e(d\varphi_*(v))\cdot \mathrm{ad}(g(Z)) + \delta_\sigma*(g(Z)^{-1}\partial_v g(Z)). \end{aligned}$$

Hence, combining (*) and (**), one obtains

$$E^*(v)=\iota(e*(v))=\iota(e(d\varphi_*(v)))\cdot \mathrm{ad}(g(Z))+\delta_{\sigma*}\iota(g(Z)^{-1}\partial_v g(Z))$$
$$=E(d\varphi_*(v))\cdot \mathrm{ad}(g(Z))-\delta_{\sigma*}\left(\frac{\partial_v(X+z\cdot Y)}{Y}\frac{d}{dz}\right).$$

If $E(v)=\delta_\sigma\xi(v)$, further substituting this in the above gives

$$E^*(v)=(\delta_\sigma\xi(d\varphi_*(v)))\cdot \mathrm{ad}(g(Z))-\delta_{\sigma*}\left(\frac{\partial_v(X+z\cdot Y)}{Y}\frac{d}{dz}\right)$$
$$=\delta_{\sigma*}\left(\xi(d\varphi_*(v))\cdot \mathrm{ad}(g(Z))-\frac{\partial_v(X+z\cdot Y)}{Y}\frac{d}{dz}\right).$$

These prove (11.3.1) and (11.3.2).

11.4.

From the expression (11.3.2), the holomorphic factors h and h^* for σ and $\sigma*$ (cf. (9.3.1), (10.3.1)) are related as:

Corollary. *The holomorphic factor h^* for $\sigma*$ is given by*

$$h^*(v)=h(d\varphi_*(v))\cdot \mathrm{ad}(g(Z))-\frac{\partial_v(X+z\cdot Y)}{Y}\frac{d}{dz}, \tag{11.4.1}$$

so that the cocycle $E^*(v)$ is expressed as the coboundary of h^*:

$$E^*(v)=\delta_{\sigma*}(h^*(v)) \qquad \text{for } v\in\Gamma(\mathscr{V}, T^{0,1}(\mathscr{V})). \tag{11.4.2}$$

11.5.

The conditions for $\Omega\circ\sigma$ and $\Pi\circ\sigma*$ to be holomorphic are now written down in terms of $Z(t, w)=X(t, w)+iY(t, w)$ (11.1.1).

Assertion. *For $v\in\Gamma(\mathscr{U}, T^{0,1}(\mathscr{U}))$, put*

$$h(v)=: h_v(\underline{t}, z)\frac{d}{dz}, \tag{11.5.1}$$

where $h_v(t, z)$ is a differentiable function on $\mathscr{U}\times\mathbf{H}$, which is holomorphic in the variable z. Then for $v\in\Gamma(\mathscr{V}, T^{0,1}(\mathscr{V}))$,

$$\partial_v Z(t, w)=h_v(\underline{t}, Z(t, w)). \tag{11.5.2}$$

Particularly

$$\frac{\partial}{\partial \bar{w}} Z(\underline{t}, w) = 0. \tag{11.5.3}$$

Proof. One has only to compare Assertions (9.3) and (10.3). An explicit calculation shows that

$$h^*(v)|_{z=i} = \left(\frac{h_v(\underline{t}, X + zY)}{Y} \frac{d}{dz} - \frac{\partial_v(X + zY)}{Y} \frac{d}{dz} \right) \Bigg|_{z=i}.$$

12. Gauge Transformation II

12.1.

Let $\sigma* : \mathscr{V} \to S^1(\mathscr{X}(\Gamma))$ be an lp-map considered in Section 10. We consider a new local lp-map:

$$\sigma^{\vee} : \mathscr{W} \longrightarrow S^1(\mathscr{X}(\Gamma)), \tag{12.1.1}$$

which is related to $\sigma*$ as follows. $\mathscr{W}$ is a domain in $\mathscr{V} \times \mathbf{R}$, with coordinate $\underline{t}^{\vee} = (\underline{t}*, s)$. There exists a differentiable map

$$\theta : \mathscr{W} \longrightarrow S^1, \tag{12.1.2}$$

so that

$$\sigma^{\vee}(\underline{t}^{\vee}) := \sigma*(\underline{t}*) \cdot \mathrm{Ad}(k(\theta(\underline{t}^{\vee}))). \tag{12.1.3}$$

Here

$$k(\theta) := \begin{bmatrix} \cos(\theta) & \sin(\theta) \\ -\sin(\theta) & \cos(\theta) \end{bmatrix}. \tag{12.1.4}$$

Thus one has the following commutative diagram.

$$\begin{array}{ccc} \mathscr{X} & \xrightarrow{\sigma^{\vee}} & S^1\mathscr{X} \\ \downarrow{\scriptstyle \pi} & & \downarrow{\scriptstyle \Pi} \\ \mathscr{V} & \xrightarrow[\Pi \circ \sigma*]{} & \mathscr{X} \end{array}$$

and for any $t^\vee \in \mathcal{W}$

$$\begin{array}{ccccc} T_{t^\vee}(\mathcal{W}) & \xrightarrow[\sim]{d(\sigma\vee)_*} & T(S^1\mathcal{X}) & \overset{E}{\subset} & \mathcal{Z}^1_{\sigma^\vee(t^\vee)}(\Gamma, \Gamma(\mathbf{H}, \Theta)) \\ \Big\downarrow d\pi_* & & \Big\downarrow d\Pi_* & & \Big\downarrow \\ T_{t*}(\mathcal{V}) & \xrightarrow[d(\Pi\sigma*)_*]{\sim} & T(\mathcal{X}) & \xrightarrow{\sim} & \mathcal{Z}^1_{\sigma*(t*)}(\Gamma, \Gamma(\mathbf{H}, \Theta))/\delta_{\sigma*(t*)}\Gamma(\mathbf{H}, \Theta(-[i])), \end{array}$$

where the kernel of $d\pi_*$ is a space of rank 1 spanned by $\partial/\partial s$.

Let us consider the complexification $d\pi_*: T_{t^\vee}(\mathcal{W})\otimes\mathbf{C} \to T_{t*}(\mathcal{V})\otimes\mathbf{C}$. Put

$$\tilde{T}^{1,0}(\mathcal{W}) := d\pi_*^{-1}(T^{1,0}(\mathcal{V})),$$

and

$$\tilde{T}^{0,1}(\mathcal{W}) := d\pi_*^{-1}(T^{0,1}(\mathcal{V})),$$

so that we have the following short exact sequence:

$$0 \to \mathbf{C}\frac{\partial}{\partial s} \to \tilde{T}^{1,0}(\mathcal{W}) \oplus \tilde{T}^{0,1}(\mathcal{W}) \to T(\mathcal{W})\otimes\mathbf{C} \to 0.$$

Here we mean by $\mathbf{C}\,\partial/\partial s$ the complex line bundle over $\mathcal{W}$ spanned by $\partial/\partial s$.

12.2.

Recall the cocycle map (8.3.4) for $\sigma\vee$ and denote it by

$$E^\vee: \Gamma(\mathcal{W}, T(\mathcal{W})) \to \mathcal{Z}^1_{\sigma^\vee}(\Gamma, \Gamma(\mathcal{W}\times\mathbf{H}, \mathcal{A}_{\mathcal{W}}(\Theta))), \tag{12.2.1}$$

whose image is exactly the set of the cocycles of completely integrable vector fields on $\mathbf{H}$ (i.e., $\iota(\mathfrak{sl}(2,\mathbf{R}))$ (cf. (6.2.1)). We calculate $E^\vee$ and its trivialization $\xi^\vee$ in terms of E^*, ξ^* and θ.

12.3.

Assertion. *Let an lp-map $\sigma^\vee$ on $\mathcal{W}$ as in (12.1) be given.*

(*i*) *The cocycle map $E^\vee$ (12.2.1) for $\sigma^\vee$ is given by*

$$E^\vee(v) = E^*(d\pi_*(v))\cdot \mathrm{ad}(k(\theta)) - \delta_{\sigma^\vee}\left((\partial_v\theta)(z^2+1)\frac{d}{dz}\right). \tag{12.3.1}$$

(*ii*) *Let ξ^* be a t-map for σ^*. Then a t-map $\xi^\vee$ for $\sigma^\vee$ is given by*

$$\xi^\vee(v) = \xi^*(d\pi_*(v))\cdot \mathrm{ad}(k(\theta)) - (\partial_v\theta)(z^2+1)\frac{d}{dz}. \tag{12.3.2}$$

Proof. For a vector field $v \in \Gamma(\mathscr{W}, T(\mathscr{W}))$, one calculates that

$$k(\theta)^{-1}\partial_v k(\theta) = \begin{bmatrix} 0 & 1 \\ -1 & 0 \end{bmatrix} \partial_v \theta.$$

Applying ι of (2.3.2) on this,

$$(*) \qquad \iota(k(\theta)^{-1}\partial_v k(\theta)) = -(z^2+1)\partial_v\theta \frac{\partial}{\partial z}.$$

On the other hand,

$$\begin{aligned}(**) \quad e^\vee(v) &:= \sigma^{\vee -1}\partial_v \sigma^\vee = k(\theta)^{-1}\sigma *^{-1} k(\theta)\partial_v(k(\theta)^{-1}\sigma * k(\theta)) \\ &= k(\theta)^{-1}(\sigma *^{-1}\partial_v \sigma *)k(\theta) + k(\theta)^{-1}\partial_v k(\theta) \\ &\quad - (k(\theta)^{-1}\sigma *^{-1} k(\theta))(k(\theta)^{-1}\partial_v k(\theta))(k(\theta)^{-1}\sigma^* k(\theta)) \\ &= (\sigma *^{-1}\partial_v \sigma *)\cdot \mathrm{ad}(k(\theta)) + \delta_{\sigma^\vee}(k(\theta)^{-1}\partial_v k(\theta)). \\ &= e*(d\pi_*(v))\cdot \mathrm{ad}(k(\theta)) + \delta_{\sigma^\vee}(k(\theta)^{-1}\partial_v k(\theta)).\end{aligned}$$

Hence, combining $(*)$ and $(**)$, one obtains

$$\begin{aligned}E^\vee(v) = \iota(e^\vee(v)) &= \iota(e*(d\pi_*(v)))\cdot \mathrm{ad}(k(\theta)) + \delta_{\sigma^\vee}\iota(k(\theta)^{-1}\partial_v k(\theta)) \\ &= E^*(d\pi_*(v))\cdot \mathrm{ad}(k(\theta)) - \delta_{\sigma^\vee}\left((\partial_v\theta)(z^2+1)\frac{d}{dz}\right).\end{aligned}$$

By substituting $E^*(v) = \delta_{\sigma *}\xi^*(v)$ in the above formula, one has

$$E^\vee(v) = \delta_{\sigma^\vee}\left(\xi^*(d\pi_*(v))\cdot \mathrm{ad}(k(\theta)) - (\partial_v\theta)(z^2+1)\frac{d}{dz}\right).$$

These prove (12.3.1) and (12.3.2).

12.4.

From the expression (12.3.2), it is easy to check that, for $v \in \Gamma(\mathscr{W}, T(\mathscr{W}) \otimes \mathbf{C})$,

$$\begin{aligned}v \in \Gamma(\mathscr{W}, \tilde{T}^{0,1}(\mathscr{V})) &\Leftrightarrow E^\vee(v) \in \delta_{\sigma^\vee}\Gamma(\mathscr{W}\times \mathbf{H}, \mathscr{A}_{\mathscr{W}}(\Theta(-[i]))) \\ &\Leftrightarrow \delta_{\sigma^\vee}(\xi^\vee(v)) \in \delta_{\sigma^\vee}\Gamma(\mathscr{U}\times \mathbf{H}, \mathscr{A}_{\mathscr{V}}(\Theta(-[i]))) \\ &\Leftrightarrow \xi^\vee(v) \in \Gamma(\mathscr{V}\times \mathbf{H}, \mathscr{A}_{\mathscr{V}}(\Theta(-[i]))) \\ &\qquad \oplus \Gamma(\Gamma\backslash \mathscr{V}\times \mathbf{H}, \mathscr{A}_{\mathscr{V}\times\mathbf{H}}(\Theta)).\end{aligned}$$

Here the last term is a direct sum as before in Section 9. Let us denote the first factor of $\xi^\vee$ by $h^\vee$. To summarize, we get:

Assertion. *Let an lp-map* $\sigma^\vee$ *of* (12.1) *be given.*

(*i*) *For* $v \in \Gamma(\mathscr{W}, T(\mathscr{W}) \otimes \mathbf{C})$,

$$v \in \Gamma(\mathscr{W}, \tilde{T}^{0,1}(\mathscr{W})) \Leftrightarrow E^\vee(v) \in \delta_{\sigma^\vee}\Gamma(\mathscr{W} \times \mathbf{H}, \mathscr{A}_{\mathscr{W}}(\Theta(-[i]))).$$

(*ii*) *For* $v \in \Gamma(\mathscr{W}, \tilde{T}^{0,1}(\mathscr{W}))$,

$$h^\vee(v) = h^*(d\pi_*(v)) \cdot \operatorname{ad}(k(\theta)) - (\partial_v\theta)(z^2+1)\frac{d}{dz}. \tag{12.4.1}$$

$$E^\vee(v) = \delta_{\sigma^\vee} h^\vee(v). \tag{12.4.2}$$

Proof. Almost obvious and omitted.

13. The Unit Circle Bundle

We realize $S^1\mathscr{X}(\Gamma)$ as a unit circle bundle embedded in $T(\mathscr{X}/\mathscr{T})$. This picture could have been explained immediately after Section 5, if we do not concern ourselves with the complex structures on $\mathscr{X}$ and $\mathscr{T}$.

13.1.

Recall the right adjoint action of $SL(2, \mathbf{R})$ on $S^1(\tilde{\mathscr{X}})$ (4.2):

$$S^1(\tilde{\mathscr{X}}) \times SL(2, \mathbf{R}) \longrightarrow S^1(\tilde{\mathscr{X}}).$$

Taking the orbit space by the action of $SO(2) \subset SL(2,\mathbf{R})$, one defines

$$\tilde{U}\colon S^1(\tilde{\mathscr{X}}) \times \mathbf{H} \longrightarrow \tilde{\mathscr{X}}. \tag{13.1.1}$$

Taking the orbit space by the left action of Γ, one defines

$$U\colon S^1(\mathscr{X}) \times \mathbf{H} \longrightarrow \mathscr{X}. \tag{13.1.2}$$

By construction, U is commutative with the projections to $\mathscr{T}$.

$$\begin{array}{ccc} S^1(\mathscr{X}(\Gamma)) \times \mathbf{H} & \xrightarrow{U} & \mathscr{X}(\Gamma) \\ & {}_{\Omega}\searrow \quad \swarrow_{\Phi} & \\ & \mathscr{T}(\Gamma) & \end{array}$$

This U is the universal uniformization map as mentioned in the introduction (1.4).

Assertion. *The restriction of U on the fiber over a point $t \in \mathcal{T}(\Gamma)$*

$$U_t: S^1(X_t) \times \mathbf{H} \longrightarrow X_t$$

is the universal uniformization in the introduction (1.3). *Namely, for $\rho \in S^1(X_t)$, the restriction map*

$$U(\rho, *): \rho \times \mathbf{H} \longrightarrow X_t \tag{13.1.3}$$

is a uniformization for the Riemann surface X_t such that

$$U(\rho, i) = \Pi(\rho) \tag{13.1.4}$$

and

$$U(\rho \cdot \mathrm{Ad}(g), z) = U(\rho, g(z)), \tag{13.1.5}$$

for $\forall g \in SL(2, \mathbf{R})$ and $\forall z \in \mathbf{H}$.

Proof. Choose a point $\rho \in S^1\mathscr{X}(\Gamma)$ with $t = \Omega(\rho)$. Then

$$X_t := \Phi^{-1}(t) = \rho \cdot \mathrm{Ad}(SL(2, \mathbf{R}))/\mathrm{Ad}(SO(2)) \simeq \rho(\Gamma)\backslash SL(2, \mathbf{R})/SO(2, \mathbf{R}),$$

and

$$S^1(X_t) := \Omega^{-1}(t) = \Omega^{-1} \cdot \Omega(\rho) = \rho \cdot \mathrm{Ad}(SL(2, \mathbf{R})) \simeq \rho(\Gamma)\backslash SL(2, \mathbf{R}).$$

By construction, the restriction U_t of U is given by $\rho(\Gamma)\backslash g \times h/SO(2) \mapsto \rho(\Gamma)\backslash g \cdot h/SO(2)$ for $g, h \in SL(2, \mathbf{R})$, which is the u in (1.3).

Take $g_1, g_2 \in SL(2, \mathbf{R})$ with $z_j = g_j(i) \in \mathbf{H}$ for $j = 1, 2$. The images of z_1 and z_2 by (13.1.3) coincide iff there are elements $\gamma \in \Gamma$ and $k \in SO(2)$ such that $\rho \cdot \mathrm{Ad}(g_1) = \mathrm{Ad}(\gamma) \cdot \rho \cdot \mathrm{Ad}(g_2) \cdot \mathrm{Ad}(k)$. Recalling (4.2)-(ii), the right hand is equal to $\rho \cdot \mathrm{Ad}(\rho(\gamma) \cdot g_2 \cdot k)$. Since the action of $SL(2, \mathbf{R})$ is fixed point free ((5.2) Assertion (i)), we have $g_1 = \rho(\gamma) \cdot g_2 \cdot k$. This is equivalent to saying that $z_1 = \rho(\gamma) z_2$ for $\gamma \in \Gamma$ and hence $\rho(\Gamma)\backslash \mathbf{H} \simeq X_t$.

That $i \in \mathbf{H}$ is a fixed point of $SO(2)$ implies (13.1.4).

If we choose $h \in SL(2, \mathbf{R})$ such that $z = h(i) \in \mathbf{H}$, then both sides of (13.1.5) are represented by the same $\rho \cdot \mathrm{Ad}(g) \cdot \mathrm{Ad}(h)$ due to the associativity of the adjoint action of $SL(2, \mathbf{R})$ on $S^1\mathscr{X}(\Gamma)$.

Note. By substituting

$$g = -\begin{bmatrix} 1 & 0 \\ 0 & 1 \end{bmatrix}$$

in (13.1.5), we see that the map U is factored through a space $PS^1(\mathscr{X}(\Gamma))$, where

$$PS^1(\mathscr{X}(\Gamma)) := S^1(\mathscr{X}(\Gamma)) \Big/ \left\{ \pm \begin{bmatrix} 1 & 0 \\ 0 & 1 \end{bmatrix} \right\}. \tag{13.1.6}$$

13.2.

Since the range of $U(\rho, z)$ for fixed ρ runs in the fiber of $\Phi: \mathscr{X} \to \mathscr{T}$ over $\Omega(t)$, the differential of U induces a map from the tangent space of $\mathbf{H}$ to the relative tangent bundle $T(\mathscr{X}/\mathscr{T})$: a line bundle over $\mathscr{X}(\Gamma)$, of Φ:

$$dU_*: S^1(\mathscr{X}(\Gamma)) \times T(\mathbf{H}) \to T(\mathscr{X}(\Gamma)/\mathscr{T}(\Gamma)). \tag{13.2.1}$$

Take a complex tangent vector $v_0 := d/dz \in T_i(\mathbf{H})$ at the point $z = i$. By fixing the second argument in (13.2.1) to v_0, we define a map

$$V := dU_*(*, v_0): S^1(\mathscr{X}(\Gamma)) \to T(\mathscr{X}(\Gamma)/\mathscr{T}(\Gamma)). \tag{13.2.2}$$

Obvious by definition, the following diagram is commutative:

$$\begin{array}{ccc} S^1(\mathscr{X}(\Gamma)) & \xrightarrow{V} & T(\mathscr{X}(\Gamma)/\mathscr{T}(\Gamma)) \\ & {\scriptstyle \Pi}\searrow \quad \swarrow & \\ & \mathscr{X}(\Gamma). & \end{array} \tag{13.2.3}$$

Assertion. (*i*) *V is equivariant with the action of* $k(\theta) \in SO(2)$:

$$V(\rho \cdot k(\theta)) = e^{2i\theta} V(\rho). \tag{13.2.4}$$

(*ii*) *The map V is equivariant with the action of* $\alpha \in \mathrm{Aut}(\Gamma)/\Gamma$:

$$V(\alpha \cdot \rho) = d\alpha_*(V(\rho)). \tag{13.2.5}$$

Here the action of α *on* $S^1(\mathscr{X}(\Gamma))$ *is given in* (5.3)(*iii*), *and that on the right-hand side,* $d\alpha_*$, *means the differential of the action of* α *on* $\mathscr{X}(\Gamma)$, *which is also given in* (5.3)(*iii*).

Proof. (*i*) Let $k(\theta) \in SO(2)$ be as (12.1.4). Applying the formula (13.1.5), we calculate the action of $k(\theta)$.

$$\begin{aligned} V(\rho \cdot k(\theta)) &:= dU_*(\rho \cdot k(\theta), v_0) = dU_*(\rho, dk(\theta)_*(v_0)) \\ &= dU_*\left(\rho, \left.\frac{dk(\theta)(z)}{dz}\right|_{z=i} \frac{d}{dz}\right) \\ &= dU_*\left(\rho, (\cos(\theta) - i\sin(\theta))^{-2} \frac{d}{dz}\right) = e^{2i\theta} V(\rho). \end{aligned}$$

(*ii*) Without taking the quotient by Γ, one may construct a map

$$\tilde{V} := d\tilde{U}_*(*, v_0): S^1(\tilde{\mathscr{X}}(\Gamma)) \to T(\tilde{\mathscr{X}}(\Gamma)/\mathscr{T}(\Gamma)).$$

Hence we have only to show the Aut(Γ)-equivariance of this map. The tangent vector v_0 is the derivative of the motion $\exp(tv_0)$ $(t \in \mathbf{R})$ at $t=0$ for a $v_0 = \left[\begin{smallmatrix} 0 & 1 \\ 0 & 0 \end{smallmatrix}\right] \in \mathfrak{sl}(2, \mathbf{R})$, and therefore

$$V(\rho) := dU_*(\rho, v_0) = \frac{d}{dt} U(\rho, \exp(tv_0), i)\Big|_{t=0}$$

$$= \frac{d}{dt} U(\rho \cdot \exp(tv_0), i)\Big|_{t=0} \qquad (13.1.5)$$

$$= \frac{d}{dt} \Pi(\rho \cdot \exp(tv_0))\Big|_{t=0} \qquad (13.1.4)$$

Recalling the fact that the action of $\alpha \in \mathrm{Aut}(\Gamma)$ and that of $\mathrm{Aut}(SL(2, \mathbf{R}))$ are commutative (4.2)(*i*), we can now calculate as

$$V(\alpha \cdot \rho) = \frac{d}{dt} \Pi((\alpha \cdot \rho) \cdot \mathrm{Ad}(\exp(tv_0)))\Big|_{t=0}$$

$$= \frac{d}{dt} \Pi(\alpha \cdot (\rho \cdot \mathrm{Ad}(\exp(tv_0))))\Big|_{t=0}$$

Here, the action of α on $\mathscr{X}(\Gamma)$ (5.3)(*iii*) is defined by the relation: $\alpha(\Pi(\rho)) := \Pi(\alpha \cdot \rho)$ for $\rho \in S^1(\mathscr{X}(\Gamma))$. Hence we have

$$= \frac{d}{dt} \alpha(\Pi(\rho \cdot \mathrm{Ad}(\exp(tv_0))))\Big|_{t=0}$$

$$= d\alpha_*\left(\frac{d}{dt} \Pi(\rho \cdot \mathrm{Ad}(\exp(tv_0)))\Big|_{t=0}\right) = d\alpha_*(V(\rho)).$$

Remark. The map V (13.2.2) can be derived from the representation $\imath$ (6.3) of $\mathfrak{sl}(2\,\mathbf{R})$ as follows: Let us fix $v_0 \in \mathfrak{sl}(2, \mathbf{R}) \backslash \mathfrak{so}(2)$ as above. Then the vector field $\imath(v_0)$ defines a map from $\rho \in S^1(\mathscr{X})$ to $\imath(v_0)(\rho) \in T_\rho(S^1(\mathscr{X}))$. In view of (6.4.1),

$$V(\rho) = \imath(v_0 \bmod \mathfrak{so}(2))(\rho) \in T_{\Pi(\rho)}(\mathscr{X}).$$

13.3.

We have the following $\mathrm{Aut}(\Gamma)/\Gamma$-equivariant embedding over $\mathscr{X}(\Gamma)$ (cf. (13.2.3), (13.2.4) and (13.1) Note):

$$V: PS^1(\mathscr{X}(\Gamma)) \subset T(\mathscr{X}(\Gamma)/\mathscr{T}(\Gamma)), \tag{13.3.1}$$

as a unit circle bundle. This defines a Hermitian metric on the complex line bundle $T(\mathscr{X}(\Gamma)/\mathscr{T}(\Gamma))$.

The restriction of the embedding (13.3.1) to a fiber over $t \in \mathscr{T}(\Gamma)$ is described in (2.5) Assertion. From the description, we know that the restriction of the Hermitian metric on $T(\mathscr{X}(\Gamma)/\mathscr{T}(\Gamma))$ to a fiber $X_t := \Phi^{-1}(t)$ over $t \in \mathscr{T}(\Gamma)$ gives the Poincaré metric.

13.4.

Let $\sigma*$ and σ be local parametrizations as given in (11.1) and recall the local immersions with the commutative diagram

$$\begin{array}{ccc} \mathscr{V} & \xrightarrow[\Pi\circ\sigma*]{} & \mathscr{X} \\ \downarrow{\scriptstyle\varphi} & & \downarrow{\scriptstyle\Phi} \\ \mathscr{U} & \xrightarrow[\Omega\circ\sigma]{} & \mathscr{T} \end{array}$$

$$\begin{array}{ccccc} T_{t*}(\mathscr{V}) & \xrightarrow[d(\Pi\sigma*)*]{\sim} & T_{\Pi\circ\sigma*(t*)}(\mathscr{X}) & \xrightarrow{\sim} & \mathscr{Z}^1_{\sigma*(t*)}(\Gamma,\Gamma(\mathbf{H},\Theta))/\delta_{\sigma*(t*)}\Gamma(\mathbf{H},\Theta(-[i])) \\ \downarrow{\scriptstyle d\varphi_*} & & \downarrow{\scriptstyle d\Phi_*} & & \downarrow \\ T_t(\mathscr{U}) & \xrightarrow[d(\Omega\sigma)_*]{\sim} & T_{\Omega\circ\sigma(t)}(\mathscr{T}) & \xrightarrow{\sim} & H^1_{\sigma(t)}(\Gamma,\Gamma(\mathbf{H},\Theta)). \end{array}$$

Then $\Pi^{-1}\Pi(\mathscr{V}) = \Pi(\mathscr{V})\cdot \mathrm{Ad}(SO(2))$ is the unit circle bundle over $\mathscr{V}$.

Assertion. *Let the notations be as above and* (11.1).

(i) *The unit circle bundle over* $\mathscr{V}$ *is parametrized by the vector field*

$$\frac{Y(t,w)}{e^{2i\theta}\dfrac{\partial Z(t,w)}{\partial w}}\frac{d}{dw} \in T_{(t,w)}(\mathscr{V}/\mathscr{U}) \tag{13.4.1}$$

for $k(\theta) \in SO(2)$ *and* $t^* = (t,w) \in \mathscr{V} \subset \mathbf{C}^{3g-3}\times\mathbf{C}$.

(*ii*) *The Hermitian metric for the circle bundle is given by*

$$\frac{\left|\dfrac{\partial X(t, w)}{\partial w}\right|^2}{Y(t, w)^2} dw \otimes d\bar{w}. \tag{13.4.2}$$

Proof. Use (11.3) Assertion.

References

[1] E. Arbarello, C. De Concini, V. Kac, and C. Procesi, Moduli space of curves and representation theory, Preprint (1987).

[2] L. V. Ahlfors, Some remarks on Teichmüller's space of Riemann surfaces, *Ann. of Math.* **74** (1961) 171–191.

[3] L. V. Ahlfors, Curvature properties of Teichmüller space, *J. Analyse Math.* **9** (1961) 161–176.

[4] L. V. Ahlfors, Teichmüller spaces, *Proc. Int. Congr. Math.* Stockholm (1962) 3–9.

[5] L. V. Ahlfors and L. Bers, Riemann's mapping theorem for variable metrics, *Ann. of Math.* **72** (1960) 385–404.

[6] L. V. Ahlfors and A. Weil, A uniqueness theorem for Beltrami equations, *Proc. Amer. Math. Soc.* **13** (1962) 975–978.

[7] A. A. Beilinson and Yu. I. Manin, The Mumford Form and the Polyakov Measure in String Theory, *Commun. Math. Phys.* **107** (1986) 359.

[8] A. A. Beilinson, Yu. I. Manin, and Y. A. Schectman, *Localization of the Virasaoro and Neveu-Schwartz Algebra*, Moscow preprint (1986).

[9] A. A. Belavin and V. G. Knizhnik, Complex Geometry and Theory of Quantum String, *Sov. Phys. JETP* **64** (1986) 214.

[10] A. A. Belavin, A. M. Polyakov, and A. B. Zamolodchikov, Infinite Conformal Symmetry in Two-Dimensional Quantum Field Theory, *Nucl. Phys.* **B241** (1984) 333.

[11] L. Bers, Spaces of Riemann surfaces, *Proc. Int. Congr. Math.* Cambridge (1958) 349–361.

[12] L. Bers, Simultaneous uniformization, *Bull. Amer. Math. Soc.* **66** (1960) 94–97.

[13] L. Bers, *Quasiconformal mappings and Teichmüller's theory*, in *Analytic Functions*, R. Nevanlinna, *et al.*, Princeton University Press, Princeton, New Jersey 1960, pp. 89–119.

[14] E. Date, M. Jimbo, M. Kashiwara, and T. Miwa, Transformation Groups for Soliton Equations, *Proc. of RIMS Symp. on Non-Linear Integrable Systems-Classical Theory and Quantum Theory, Kyoto, Japan*, (M. Jimbo and T. Miwa, eds.), World Science Publ. Co., Singapore 1983.

[15] E. Calabi and E. Vesentini, On compact locally symmetric Kähler manifolds, *Ann. of Math,* **71** (1960) 472–507.

[16] C. J. Earle and J. Eells, A fibre bundle description of Teichmüller theory. *J. Diff. Geom.* **3** (1969) 19–43.

[17] T. Eguchi and H. Ooguri, Conformal and Current Algebras on a General Riemann Surface, *Nucl. Phys.* **B282** (1987) 308.

[18] M. Eichler, Eine Verallgemeinerung der Abelschen Integrale, *Math. Z.* **67** (1957) 267–298.

[19] W. Fenchel and J. Nielsen, Treatise on Fuchsian groups, (unpublished).

[20] R. Fricke and F. Klein, *Vorlesungen über die Theorie der Automorphen Funktionen*, Band I.B.G. Teubner, Leipzig, 1926.
[21] D. Friedan, A New Formulation of String Theory, *Second Nobel Symposium on Elementary Particle Physics*, presented at Marstrand, Sweden (1986), and *Symposium on Geometry and Topology in Field Theory*, Espoo, Finland (1986).
[22] D. Friedan and S. Shenker, The Analytic Geometry of two dimensional Conformal Field Theory, *Nucl. Phys.* **B281** (1987) 509.
[23] A. Grothendieck, Techniques de construction en géometrie analytique, I: Description axiomatique de l'espace de Teichmüller et de ses variantes, *Séminaire Henri Cartan, 13e anée: 1960/61*, Exposés 7-8, Ecole Normale Supérieure.
[24] W. J. Harvey, Spaces of Fuchsian groups and Teichmüller theory, Advances in the theory of Riemann surfaces, *Ann. of Math. Studies* **66** (1971).
[25] W. J. Harvey *et al.*, *Discrete Groups and Automorphic Functions*, Proc. of an instructional Conf. organized by the London Math. Soc. and the Univ. of Cambridge (1977).
[26] N. Kawamoto, Y. Namikawa, A. Tsuchiya, and Y. Yamada, Geometric Realization of Conformal Field Theory On Riemann Surfaces, *Comm. Math. Phy.* **116** (1988) 247–308.
[27] L. Keen, Intrinsic moduli on Riemann surfaces, *Ann. of Math.* **84** (1966) 404–420.
[28] L. Keen, On Fricke moduli, Advances in the theory of Riemann surfaces, *Ann. of Math. Studies* **66** (1971).
[29] K. Kodaira and D. C. Spencer, On deformation of complex analytic structures, *Ann. of Math.* **67** (1958) 328–466.
[30] I. Kra, *Automorphic Forms and Kleinian Groups*, W. A. Benjamin, Inc. (1972).
[31] A. M. Macbeath, Groups of homeomorphisms of a simply connected space, *Ann. of Math.* **79** (1964) 473–488.
[32] A. M. Macbeath and D. Singerman, Spaces of subgroups and Teichmüller space, *Proc. London Math. Soc.* **31** (3) (1975) 211–256.
[33] Y. Matsushima and S. Murakami. On certain cohomology groups attached to Hermitian Symmetric Spaces, *Osaka J. Math.* **2** (1965) 1–35.
[34] S. Murakami, Cohomology of vector-valued forms on symmetric spaces, Lecture at Chicago Univ. (1966).
[35] J. Nielsen, Untersuchungen zur Topologie der geschlossenen zweiseitigen Flächen, *Acta Math.* **50** (1927) 189–358.
[36] T. Oda, K. Saito's Period Map for Holomorphic Functions with Isolated Critical Points, *Advanced Studies in Pure Math.* **10** (1987) 591–648.
[37] H. Poincaré, Sur l'uniformisation des fonctions analytiques, *Acta Math.* **31** (1907) 1–64.
[38] K. Saito, Period mapping associated to a primitive form, *Publ. RIMS, Kyoto Univ.* **19** (1983) 1231–1264.
[39] M. Sato, Soliton equations as dynamical systems on an infinite dimensional Grassmann manifold, *RIMS-Kokyuroku* 439, **30** (1981), Kyoto Univ.
[40] M. Sato, Soliton equations and the infinite dimensional Grassmann manifolds, Lectures delivered at Univ. of Tokyo, Nagoya Univ. (Notes by M. Mulase in Japanese 1981–1982) and Kyoto Univ. (Notes by T. Umeda in Japanese 1984–1985).
[41] A. Selberg, On discontinuous groups in higher-dimensional symmetric spaces, *Contributions to function theory*, Bombay (1960) 945–963.
[42] G. Shimura, Sur les intégrales attachées aux forms automorphes, *J. Math. Soc. Japan* **11** (1959) 291–311.
[43] C. L. Siegel, Some remarks on discontinuous groups, *Ann. of Math.* **46** (1945) 708–718.
[44] C. L. Siegel, Über einige Ungleichungen bei Bewegungsgruppen in der nichteuklidischen Ebene, *Math. Ann.* **133** (1957) 127–138.
[45] C. L. Siegel, *Topics of Complex Function Theory*, Vol. I, II, Wiley-Interscience, New York.

[46] A. J. Tromba and A. E. Fischer, Almost complex principal bundles and complex structure on Teichmüller space, *Crelles J.* **352** (1984) 151–160.
[47] A. Weil, Sur les modules des surfaces de Riemann. *Séminaire Bourbaki*, May, 1958.
[48] A. Weil, On discrete subgroups of Lie groups I, *Ann. of Math.* **72** (1960) 369–384.
[49] A. Weil, On discrete subgroups of Lie groups II, *Ann. of Math.* **75** (1962) 578–602.
[50] A. Weil, Remarks on the cohomology of groups, *Ann. of Math.* **80** (1964) 149–157.

The Hamburger Theorem for the Epstein Zeta Functions

Fumihiro Sato
Mathematisches Institut der Universität
Göttingen
Federal Republic of Germany
and
Department of Mathematics
Rikkyo University, Tokyo
Japan

0. Introduction

Hamburger's theorem, which characterizes the Riemann zeta function by means of its functional equation, has tempted a number of researchers to find an alternative proof and/or a generalization (e.g., [1], [6], [8], [12]). Among others one of the motivations of Hecke's theory of automorphic forms was to supply a better understanding of Hamburger's theorem (cf. [7, Introduction]).

Recently Ehrenpreis and Kawai [2] proved a characterization of hyperfunctions on $\mathbf{R}^n$ whose supports, as well as the supports of their Fourier transforms, are contained in $\mathbf{Z}^n$. As an application of their theorem, they gave a simple proof of Hamburger's theorem. They further obtained a

Algebraic Analysis, Volume II

ISBN 0-12-400466-0

similar theorem for the zeta function of $\mathbf{Q}(\sqrt{-1})$. This result was generalized to an arbitrary algebraic number field by Yoshimoto [16].

The point of their method is as follows. Claiming the existence of functional equations of a certain family of L-functions with Größencharacters, they first derive a formula similar to the Poisson summation formula. The argument is a familiar one in the theory of automorphic forms. Then the uniqueness of the Poisson summation formula (cf. [16, pp. 75–76]) immediately yields the Hamburger theorem for an algebraic number field in question. Here it is appropriate to note that the connection between Hamburger's theorem and the uniqueness of the Poisson summation formula has already been realized by several authors (cf. [1], [8]).

It seems that this method works well within the framework of the theory of prehomogeneous vector spaces, since the functional equation of zeta functions associated with a prehomogeneous vector space is based on the Poisson summation formula (cf. [10], [11]). In the present paper, we apply the method of Ehrenpreis, Kawai and Yoshimoto to prove the Hamburger theorem for the Epstein zeta function, a typical example of zeta functions associated with prehomogeneous vector spaces. In our case Größencharacters should be replaced by harmonic polynomials.

We formulate the Hamburger theorem for the Epstein zeta function in Section 1, and discuss some of its variants. The proof of the Hamburger theorem is given in Section 2. At the end of this paper we shall make a brief remark about a possible generalization of Hamburger's theorem to more general prehomogeneous vector spaces.

This work was done while the author was staying in Göttingen under the support of Sonderforschungsbereich 170 "Geometrie und Analysis," to which he expresses his gratitude.

1. Statement of the Results

As usual we denote by **C**, **R**, **Q** and **Z** the complex number field, the real number field, the rational number field and the ring of rational integers, respectively.

1.1.

Let $V=\mathbf{R}^n$ be the real n-dimensional vector space equipped with the inner product $(x, y)=x_1y_1+\cdots+x_ny_n$ $(x, y \in V)$ and let L be a lattice in V. Set

$V' = V - \{0\}$ and $L' = L \cap V'$. For a non-negative integer d, denote by $H^{n,d}$ the space of all homogeneous harmonic polynomials of degree d in n variables:

$$H^{n,d} = \left\{ P(x) \in \mathbf{C}[x_1, \ldots, x_n];\ \left(x_1 \frac{\partial}{\partial x_1} + \cdots + x_n \frac{\partial}{\partial x_n} \right) P(x) = d \cdot P(x), \right.$$

$$\left. \left(\frac{\partial^2}{\partial x_1^2} + \cdots + \frac{\partial^2}{\partial x_n^2} \right) P(x) = 0 \right\}.$$

For a $P(x) \in H^{n,d}$, we put

$$\zeta_P(L; s) = \sum_{m \in L'} \frac{P(m)}{(m, m)^{s+d/2}} \qquad (s \in \mathbf{C})$$

and

$$\xi_P(L; s) = v(L)^{1/2} \pi^{-s} \Gamma(s + d/2) \zeta_P(L; s),$$

where $v(L)$ is the volume of a fundamental domain of V/L with respect to the standard Euclidean measure $dx = dx_1 \cdots dx_n$. The Dirichlet series $\zeta_P(L; s)$ is absolutely convergent for $\operatorname{Re} s > n/2$. We note that the series $\zeta_P(L; s)$ is identically equal to 0, if d is odd.

Let L^* be the lattice dual to L:

$$L^* = \{ y \in V;\ (x, y) \in \mathbf{Z} \text{ for all } x \in L \}.$$

Then the following theorem is due to Epstein [3] (see also Siegel [13, Chapter 1, Section 5]).

Theorem (Epstein). (*i*) *The Dirichlet series* $\zeta_P(L; s)$ *has an analytic continuation to a meromorphic function of* s *in* $\mathbf{C}$. *Moreover*

$$\zeta_P(L; s) - \pi^{n/2} P(0) / \Gamma(n/2) v(L)(s - n/2)$$

is an entire function of order not greater than 1.

(*ii*) *The following functional equation holds*:

$$e^{d\pi i/2} \xi_P\left(L^*; \frac{n}{2} - s \right) = \xi_P(L; s).$$

Remark. If $d > 0$, then $P(0) = 0$ and hence $\zeta_P(L; s)$ is an entire function.

The aim of this paper is to show that the analytic properties described in the theorem above characterize the Epstein zeta functions $\zeta_P(L; s)$.

1.2.

Let $a: L' \to \mathbf{C}$ and $a^*: (L^*)' \to \mathbf{C}$ be any mappings satisfying the inequalities

$$a(m) < c(m, m)^M \qquad (m \in L') \tag{1.1}$$

and

$$a^*(m^*) < c(m^*, m^*)^M \qquad (m^* \in (L^*)') \tag{1.1}^*$$

for some positive constants c and M.

Now we consider the Dirichlet series

$$\zeta_P(a, L; s) = \sum_{m \in L'} \frac{a(m)P(m)}{(m, m)^{s+d/2}}$$

and

$$\zeta_P(a^*, L^*; s) = \sum_{m^* \in (L^*)'} \frac{a^*(m^*)P(m^*)}{(m^*, m^*)^{s+d/2}}.$$

By (1.1) and (1.1)*, these series are absolutely convergent for $\operatorname{Re} s > n/2 + M$. If $a(m) = 1$ for all $m \in L'$, then $\zeta_P(a, L; s)$ is the original Epstein zeta function. When $d = 0$ and $P(x)$ is identically equal to 1, we write simply $\zeta(a, L; s)$ and $\zeta(a^*, L^*; s)$ for $\zeta_P(a, L; s)$ and $\zeta_P(a^*, L^*; s)$, respectively.

Set

$$\xi_P(a, L; s) = v(L)^{1/2} \pi^{-s} \Gamma(s + d/2) \zeta_P(a, L; s)$$

and

$$\xi_P(a^*, L^*; s) = v(L^*)^{1/2} \pi^{-s} \Gamma(s + d/2) \zeta_P(a^*, L^*; s).$$

Consider the following conditions on these Dirichlet series:

(1) *$\zeta_P(a, L; s)$ and $\zeta_P(a^*, L^*; s)$ are extended to the whole complex plane as meromorphic functions of s.*

(2) *There exists a polynomial $b_P(s)$ of s such that the functions $\zeta_P(a, L; s)$ and $\zeta_P(a^*, L^*; s)$ multiplied by $b_P(s)$ are entire functions of finite order.*

(3) *The following functional equation holds:*

$$e^{d\pi i/2} \xi_P\left(a^*, L^*; \frac{n}{2} - s\right) = \xi_P(a, L; s).$$

Our main result is the following theorem, which will be proved in Section 2.

Theorem 1 (The Hamburger theorem for the Epstein zeta functions). *Assume that*

(*i*) *for any homogeneous harmonic polynomial P, the Dirichlet series* $\zeta_P(a, L; s)$ *and* $\zeta_P(a^*, L^*; s)$ *satisfy the conditions* (1), (2) *and* (3),

and

(*ii*) *if* $d = \deg P$ *is sufficiently large, then* $\zeta_P(a, L; s)$ *and* $\zeta_P(a^*, L^*; s)$ *are entire functions.*

Then a and a^* *are constant functions on* L' *and* $(L^*)'$, *respectively, and we have*

$$\begin{aligned} a(m) &= \pi^{-n/2}\Gamma(n/2)v(L)r(a) \\ &= \pi^{-n/2}\Gamma(n/2)v(L^*)r^*(a^*) \\ &= a^*(m^*) \end{aligned}$$

for all $m \in L'$ *and* $m^* \in (L^*)'$, *where* $r(a)$ (*resp.* $r^*(a^*)$) *stands for the residue of* $\zeta(a, L; s)$ (*resp.* $\zeta(a^*, L^*; s)$) *at* $s = n/2$. *In particular* $\zeta_P(a, L; s)$ (*resp.* $\zeta_P(a^*, L^*; s)$) *coincides with the original Epstein zeta function* $\zeta_P(L; s)$ (*resp.* $\zeta_P(L^*; s)$) *up to a constant factor.*

In the rest of this section, assuming Theorem 1, we shall derive some of its consequences.

Theorem 2. *The following three assertions are equivalent*:

(*i*) *the Dirichlet series* $\zeta_P(a, L; s)$ *and* $\zeta_P(a^*, L^*; s)$ *satisfy the conditions* (1), (2) *and* (3) *for any harmonic polynomial P of odd degree. Moreover they are entire functions, if* $d = \deg P$ *is sufficiently large.*

(*ii*) *a and* a^* *are even functions, namely,* $a(m) = a(-m)$ $(m \in L')$ *and* $a^*(m^*) = a^*(-m^*)$ $(m^* \in (L^*)')$.

(*iii*) $\zeta_P(a, L; s) = \zeta_P(a^*, L^*; s) = 0$ *for any harmonic polynomial P of odd degree.*

Proof. It is obvious that the second (resp. third) assertion implies the third (resp. first). Let us prove that the first assertion implies the second. Let b (resp. b^*) be the mapping of L' (resp. $(L^*)'$) into $\mathbf{C}$ defined by

$$b(m) = -b(-m) = \frac{a(m) - a(-m)}{2} \qquad (m \in L')$$

(resp. $b^*(m^*) = -b^*(-m^*) = \{a^*(m^*) - a^*(-m^*)\}/2$ $(m^* \in (L^*)')$). Then

$$\zeta_P(b, L; s) = \begin{cases} \zeta_P(a, L; s) & \text{if } d \text{ is odd,} \\ 0 & \text{if } d \text{ is even,} \end{cases}$$

and

$$\zeta_P(b^*, L^*; s) = \begin{cases} \zeta_P(a^*, L^*; s) & \text{if } d \text{ is odd,} \\ 0 & \text{if } d \text{ is even.} \end{cases}$$

Hence $\zeta_P(b, L; s)$ and $\zeta_P(b^*, L^*; s)$ satisfy all the assumptions in Theorem 1; consequently $b(m)$ (resp. $b^*(m^*)$) is independent of $m \in L'$ (resp. $m^* \in (L^*)'$). In particular we have $b(m) = b(-m) = -b(m)$. Therefore $b(m) = 0$ for any $m \in L'$. Similarly we have $b^*(m^*) = 0$ for any $m^* \in (L^*)'$. This implies the second assertion.

Theorem 2 reveals the meaning of the conditions (1), (2) and (3) for harmonic polynomials of odd degree. With regard to the conditions for harmonic polynomials of even degree, we are able to obtain the following modification of Theorem 1.

Theorem 3. *Assume that*

(*i*) *for any harmonic polynomial P of even degree, the Dirichlet series* $\zeta_P(a, L; s)$ *and* $\zeta_P(a^*, L^*; s)$ *satisfy the conditions* (1), (2) *and* (3),

and

(*ii*) *if* $d = \deg P$ *is sufficiently large, then* $\zeta_P(a, L; s)$ *and* $\zeta_P(a^*, L^*; s)$ *are entire functions.*

Then we have

$$\frac{a(m) + a(-m)}{2} = \pi^{-n/2}\Gamma(n/2)v(L)r(a)$$
$$= \pi^{-n/2}\Gamma(n/2)v(L^*)r^*(a^*)$$
$$= \frac{a^*(m^*) + a^*(-m^*)}{2}$$

for any $m \in L'$ *and any* $m^* \in (L^*)'$, *where* $r(a)$ *and* $r^*(a^*)$ *are the same as in Theorem* 1. *In particular* $\zeta_P(a, L; s)$ (*resp.* $\zeta_P(a^*, L^*; s)$) *coincides with* $\zeta_P(L; s)$ (*resp.* $\zeta_P(L^*; s)$) *up to a constant factor, if* $d = \deg P$ *is even.*

Proof. Let b (resp. b^*) be the mapping of L' (resp. $(L^*)'$) into $\mathbf{C}$ defined by

$$b(m) = b(-m) = \frac{a(m) + a(-m)}{2} \qquad (m \in L')$$

(resp. $b^*(m^*) = b^*(-m^*) = \{a^*(m^*) + a^*(-m^*)\}/2 \quad (m^* \in (L^*)')$). It is obvious that

$$\zeta_P(b, L; s) = \begin{cases} \zeta_P(a, L; s) & \text{if } d \text{ is even,} \\ 0 & \text{if } d \text{ is odd,} \end{cases}$$

and

$$\zeta_P(b^*, L^*; s) = \begin{cases} \zeta_P(a^*, L^*; s) & \text{if } d \text{ is even,} \\ 0 & \text{if } d \text{ is odd.} \end{cases}$$

Hence $\zeta_P(b, L; s)$ and $\zeta_P(b^*, L^*; s)$ satisfy all the assumptions in Theorem 1; consequently we obtain

$$\begin{aligned} b(m) = b^*(m^*) &= \pi^{-n/2}\Gamma(n/2)v(L)r(a) \\ &= \pi^{-n/2}\Gamma(n/2)v(L^*)r^*(a^*) \end{aligned}$$

for all $m \in L'$ and all $m^* \in (L^*)'$. This implies the theorem.

Remark. If $n=1$, then homogeneous harmonic polynomials are necessarily $P(x)=1$ or x up to constant multiples. Therefore for $n=1$ Theorems 2 and 3 take the following form:

Theorem. *Let $\{a_n\}_{n=1}^{\infty}$ and $\{a_n^*\}_{n=1}^{\infty}$ be sequences of complex numbers of polynomial growth. Assume that the Dirichlet series*

$$\zeta_a(s) = \sum_{n=1}^{\infty} a_n n^{-s} \qquad \textit{and} \qquad \zeta_{a^*}(s) = \sum_{n=1}^{\infty} a_n^* n^{-s}$$

have meromorphic continuations to the whole complex plane and the functions $b(s)\zeta_a(s)$ and $b(s)\zeta_{a^*}(s)$ are entire functions of finite order for some polynomial $b(s)$ of s.

(i) *If they satisfy the functional equation*

$$\pi^{-s/2}\Gamma(s/2)\zeta_a(s) = \pi^{-(1-s)/2}\Gamma((1-s)/2)\zeta_{a^*}(1-s),$$

then $\zeta_a(s) = \zeta_{a^*}(s) =$ a constant multiple of $\zeta(s)$.

(ii) *If they satisfy the functional equation*

$$\pi^{-s/2}\Gamma((s+1)/2)\zeta_a(s) = \pi^{-(1-s)/2}\Gamma((2-s)/2)\zeta_{a^*}(1-s),$$

then $\zeta_a(s) = \zeta_{a^*}(s) = 0$.

The first part of this theorem is nothing but the classical Hamburger theorem for the Riemann zeta function. The second part is also a known fact (cf. [1], [7]).

1.3.

As is seen in Theorem 4 below, it follows from Theorem 1 that a mapping $a: L' \to \mathbf{C}$ subject to the condition (1.1) is determined uniquely by the family of Dirichlet series $\zeta_P(a, L; s)$.

Theorem 4. *Let a and b be mappings of L' into $\mathbf{C}$ satisfying* (1.1). *If $\zeta_P(a, L; s) = \zeta_P(b, L; s)$ for any harmonic polynomial $P(x)$, then $a = b$.*

Proof. Let $c: L' \to \mathbf{C}$ be the mapping defined by $c(m) = a(m) - b(m)$ ($m \in L'$). We also define a mapping $c^*: (L^*)' \to \mathbf{C}$ by $c^*(m^*) = 0$ ($m^* \in (L^*)'$). Then $\zeta_P(c, L; s) = \zeta_P(c^*, L^*; s) = 0$ for any harmonic polynomial $P(x)$. Since the pair of $\zeta_P(c, L; s)$ and $\zeta_P(c^*, L^*; s)$ satisfies all the assumptions in Theorem 1, we obtain $c(m) = c^*(m^*) = 0$ for any $m \in L'$. Therefore $a = b$.

1.4.

Let $V_{\mathbf{Q}}$ be a $\mathbf{Q}$-structure of V. Namely $V_{\mathbf{Q}}$ is an n-dimensional $\mathbf{Q}$-vector subspace in V such that $V_{\mathbf{Q}} \otimes_{\mathbf{Q}} \mathbf{R} = V$. A complex valued function ϕ on $V_{\mathbf{Q}}$ is called a *Schwartz-Bruhat function* if there exist two lattices L_1 and L_2 in $V_{\mathbf{Q}}$ such that the support of ϕ is contained in L_1 and $\phi(x) = \phi(y)$ whenever $x - y \in L_2$. We denote by $\mathscr{S}(V_{\mathbf{Q}})$ the totality of Schwartz-Bruhat functions on $V_{\mathbf{Q}}$.

We fix a lattice L in $V_{\mathbf{Q}}$ such that $L \otimes_{\mathbf{Z}} \mathbf{Q} = V_{\mathbf{Q}}$. Let L^* be the lattice dual to L and put $V^*_{\mathbf{Q}} = L^* \otimes_{\mathbf{Z}} \mathbf{Q}$, which we consider as another $\mathbf{Q}$-structure of V.

Now we define the Fourier transform $\hat{\phi}$ of $\phi \in \mathscr{S}(V_{\mathbf{Q}})$ by setting

$$\hat{\phi}(x^*) = [L : L_\phi]^{-1} \sum_{x \in V_{\mathbf{Q}}/L_\phi} \phi(x)\, e^{2\pi i(x, x^*)} \qquad (x^* \in V^*_{\mathbf{Q}}),$$

where L_ϕ is a lattice in $V_{\mathbf{Q}}$ such that $L \supset L_\phi$, $x^* \in L^*_\phi$ and $\phi(x) = \phi(x')$ if $x - x' \in L_\phi$. It is clear that $\hat{\phi}(x^*)$ is independent of the choice of L_ϕ and $\hat{\phi}$ defines a Schwartz-Bruhat function on $V^*_{\mathbf{Q}}$. The Fourier transformation $\phi \to \hat{\phi}$ induces a linear isomorphism of $\mathscr{S}(V_{\mathbf{Q}})$ onto $\mathscr{S}(V^*_{\mathbf{Q}})$.

We put

$$\zeta_P(\phi; s) = \sum_{m \in V'_{\mathbf{Q}}} \frac{\phi(m) P(m)}{(m, m)^{s + d/2}} \qquad (\phi \in \mathscr{S}(V_{\mathbf{Q}}),\ P \in H^{n,d})$$

and

$$\xi_P(\phi; s) = \pi^{-s} \Gamma(s + d/2) \zeta_P(\phi; s),$$

where $V'_{\mathbf{Q}}$ denotes $V_{\mathbf{Q}} \cap V'$. Note that the Dirichlet series include Stark's L-function of a positive definite quadratic form as a special case (cf. [14]).

Theorem 5. (*i*) *$\zeta_P(\phi; s)$ is absolutely convergent for* $\operatorname{Re} s > n/2$ *and has an analytic continuation to a meromorphic function of s in* $\mathbf{C}$.

(*ii*) *The function* $\zeta_P(\phi; s)-\pi^{n/2}\hat{\phi}(0)P(0)/\Gamma(n/2)v(L)(s-n/2)$ *is an entire function of order not greater than* 1.

(*iii*) *For any* $\phi\in\mathscr{S}(V_{\mathbf{Q}})$, *the functions* $\zeta_P(\phi; s)$ *and* $\zeta_P(\hat{\phi}; s)$ *satisfy the functional equation*

$$\xi_P(\hat{\phi}, n/2-s)=e^{d\pi i/2}v(L)\xi_P(\phi; s).$$

For the proof of Theorem 5, see Remark 2 following the proof of Lemma 2.2.

Theorem 6. *Let* ϕ *and* ψ *be in* $\mathscr{S}(V_{\mathbf{Q}})$ *and* $\mathscr{S}(V_{\mathbf{Q}}^*)$, *respectively. If the functional equation*

$$\xi_P\left(\psi; \frac{n}{2}-s\right)=e^{d\pi i/2}v(L)\xi_P(\phi; s)$$

holds for any harmonic polynomial P, then $\psi=\hat{\phi}$.

Proof. By Theorem 5, we have $\zeta_P(\hat{\phi}; s)=\zeta_P(\psi; s)$ for any harmonic polynomial P. Take a lattice L_0 in $V_{\mathbf{Q}}^*$ such that L_0 contains the supports of both ψ and $\hat{\phi}$. We denote the restriction of $\hat{\phi}$ and ψ to L_0' by a and b, respectively. Then

$$\zeta_P(a, L_0; s)=\zeta_P(\hat{\phi}; s)=\zeta_P(\psi; s)=\zeta_P(b, L_0; s).$$

By Theorem 4 we have $a=b$ and hence $\psi=\hat{\phi}$.

2. Proof of Theorem 1

2.1.

We denote by $SO(n)$ the special orthogonal group acting on $V=\mathbf{R}^n$. We identify $SO(n-1)$ with the isotropy subgroup of $SO(n)$ at $v_0={}^t(0,\dots,0,1)\in V$. Namely we consider $SO(n-1)$ as a subgroup of $SO(n)$ via the mapping

$$SO(n-1)\ni h\mapsto\begin{pmatrix} h & 0\\ 0 & 1\end{pmatrix}\in SO(n).$$

For any $x\in V'=V-\{0\}$, we denote by k_x an element in $SO(n)$ such that

$$x=(x, x)^{1/2}k_x v_0. \tag{2.1}$$

Now we shall recall some facts on representations of $SO(n)$ of class 1 with respect to $SO(n-1)$. A representation ρ of $SO(n)$ is called *of class* 1 *with respect to* $SO(n-1)$, if its representation space $W=W_\rho$ contains a non-zero vector fixed under $SO(n-1)$. We define a Hermitian inner product on the space $H^{n,d}$ of harmonic polynomials by setting

$$\langle P, Q\rangle = \int_{S^{n-1}} P(\xi)\overline{Q(\xi)}\omega(\xi) \qquad (P, Q\in H^{n,d}),$$

where $S^{n-1}=\{\xi\in V;\ (\xi,\xi)=1\}$ and $\omega(\xi)$ is the $SO(n)$-invariant measure on S^{n-1} so normalized that the total volume of S^{n-1} is equal to 1. Then the representation ρ_d of $SO(n)$ on $H^{n,d}$ defined by

$$(\rho_d(k)P)(x)=P(k^{-1}x)$$

is an irreducible unitary representation of class 1 with respect to $SO(n-1)$.

Let ρ be an irreducible unitary representation of $SO(n)$ of class 1 with respect to $SO(n-1)$. It is known that ρ is equivalent to ρ_d for some non-negative integer d. Let w_0 be a unit vector in the representation space W_ρ of ρ fixed under $SO(n-1)$. For any $w\in W_\rho$, set

$$P_w(x)=d(\rho)^{1/2}(x,x)^{d/2}\langle\rho(k_x^{-1})w, w_0\rangle,$$

where $d(\rho)=\dim W_\rho$. Then $P_w(x)$ is in $H^{n,d}$ and the mapping $w\mapsto P_w$ induces an isometry of W_ρ onto $H^{n,d}$, which is an intertwining operator between ρ and ρ_d.

For the theory of representations of $SO(n)$, we refer to [15, Chapter IX].

2.2.

Let a be a mapping of L' into $\mathbf{C}$ satisfying (1.1) and ρ an irreducible unitary representation of $SO(n)$. Denote by $\mathcal{S}(V)$ the space of rapidly decreasing functions on V. For an $f\in\mathcal{S}(V)$, consider the integral

$$Z_\rho(f, a, L; s)=\int_0^\infty t^{2s}d^\times t\int_{SO(n)}\rho(k)\sum_{m\in L'}a(m)f(tkm)\,dk,$$

where dk is the normalized Haar measure on $SO(n)$ and $d^\times t=dt/t$. Then the integral $Z_\rho(f, a, L; s)$ is absolutely convergent for $\operatorname{Re} s>n/2+M$ and represents a holomorphic function of s with values in $\operatorname{End}(W_\rho)=\operatorname{Hom}_{\mathbf{C}}(W_\rho, W_\rho)$.

For any $Q\in H^{n,d}$ and $f\in\mathcal{S}(V)$, put

$$\Phi(Q,f;s)=\int_{V'}(x,x)^{s-(d+n)/2}Q(x)f(x)\,dx.$$

The integral $\Phi(Q,f;s)$ converges absolutely for $\operatorname{Re} s>n/2$ and has an analytic continuation to a meromorphic function of s in $\mathbf{C}$. In fact, using the formula

$$\Phi(Q, f; s) = \frac{2\pi^{n/2}}{\Gamma(n/2)} \int_0^\infty t^{2s}\, d^\times t \int_{S^{n-1}} Q(\xi) f(t\xi)\omega(\xi),$$

we can easily see that $\Gamma(2s)^{-1}\Phi(Q, f; s)$ is an entire function of s.

Lemma 2.1. *(i) The integral $Z_\rho(f, a, L; s)$ is identically equal to 0 unless ρ is of class 1 with respect to $SO(n-1)$.*

(ii) Assume that ρ is of class 1 with respect to $SO(n-1)$. When Re $s > n/2 + M$, *the following identity holds for any $w, w^* \in W_\rho$:*

$$\langle Z_\rho(f, a, L; s)w, w^*\rangle = 2^{-1} d(\rho)^{-1} \pi^{-n/2}\Gamma(n/2)\zeta_{P_w}(a, L; s)\Phi(\overline{P_{w^*}}, f; s),$$

where $\overline{}$ denotes the complex conjugate.

Proof. It is easy to check the following integral formula:

$$\int_0^\infty d^\times t \int_{SO(n)} f(t, k)\, dk = \frac{\Gamma(n/2)}{2\pi^{n/2}} \int_{V'} \frac{dx}{(x, x)^{n/2}} \int_{SO(n-1)} f((x, x)^{1/2}, k_x h)\, dh$$

($f \in L^1(\mathbf{R}_+^\times \times SO(n))$), where dh is the normalized Haar measure on $SO(n-1)$. Using this formula, we obtain

$$Z_\rho(f, a, L; s) = \frac{\Gamma(n/2)}{2\pi^{n/2}} \sum_{m \in L'} \frac{a(m)}{(m, m)^s} \times \int_{V'} (x, x)^{s-n/2} f(x)\, dx \int_{SO(n-1)} \rho(k_x h k_m^{-1})\, dh. \quad (2.2)$$

If ρ is not of class 1 with respect to $SO(n-1)$, by the orthogonality relation of matrix elements, we have $\int_{SO(n-1)} \rho(h)\, dh = 0$. This implies the first assertion. For any $w \in W_\rho$, we have

$$\int_{SO(n-1)} \rho(h)\, dh \cdot w = \langle w, w_0\rangle w_0.$$

Hence

$$\left\langle \int_{SO(n-1)} \rho(k_x h k_m^{-1})\, dh \cdot w, w^* \right\rangle = \langle \rho(k_m^{-1})w, w_0\rangle \overline{\langle \rho(k_x^{-1})w^*, w_0\rangle}$$

$$= d(\rho)^{-1}(m, m)^{-d/2} P_w(m)(x, x)^{-d/2}\overline{P_{w^*}(x)}$$

(for the definition of k_x, see (2.1)).

Therefore, from (2.2) we have

$$\langle Z_\rho(f, a, L; s)w, w^*\rangle = 2^{-1} d(\rho)^{-1}\pi^{-n/2}\Gamma(n/2)\zeta_{P_w}(a, L; s)\Phi(\overline{P_{w^*}}, f; s).$$

Remark. We shall briefly explain the relation between our integral representation of (generalized) Epstein zeta functions in Lemma 2.1(*ii*) and that used in [3] and [13, Chapter 1, Section 5]. Let P be a harmonic polynomial in $H^{n,d}$ and ρ be a representation of $SO(n)$ equivalent to ρ_d. Put $f_0(x) = e^{-\pi(x,x)}$ and $f(x) = P(x) f_0(x)$. It is obvious that $f(x)$ is in $\mathcal{S}(V)$. Consider the integral

$$\mathrm{Tr}(Z_\rho(f, L; s)) = \int_0^\infty t^{2s}\, d^\times t \int_{SO(n)} \chi_\rho(k) \sum_{m \in L'} f(tkm)\, dk,$$

where Tr stands for the trace of $Z_\rho(f, L; s)$ as an element in $\mathrm{End}(W_\rho)$ and $\chi_\rho(k) = \mathrm{Tr}\,\rho(k)$ is the character of ρ. Then it is easy to see that

$$2d(\rho)\,\mathrm{Tr}(Z_\rho(f, L; s)) = \int_0^\infty t^{s+d/2} \sum_{m \in L'} P(m)\, e^{-\pi t(m,m)}\, d^\times t.$$

The right-hand side of the identity coincides with the integral representation employed by Epstein [3] and Siegel [13].

We define the Fourier transform $\hat{f}$ of $f \in \mathcal{S}(V)$ by

$$\hat{f}(y) = \int_V f(x)\, e^{2\pi i (x,y)}\, dx.$$

Lemma 2.2 (Local functional equation). *The following identity holds for any* $Q \in H^{n,d}$ *and any* $f \in \mathcal{S}(V)$:

$$\Phi(Q, \hat{f}; s) = e^{d\pi i/2} \pi^{-2s+n/2} \left\{ \Gamma\left(s + \frac{d}{2}\right) \Big/ \Gamma\left(\frac{d+n}{2} - s\right) \right\} \cdot \Phi\left(Q, f; \frac{n}{2} - s\right).$$

Proof. If $d = 0$ and Q is a constant function, then the functional equation is well known (cf. [4, Chapter II, Section 3.3]). So we may assume that $d \geq 1$. Let $Q(\partial_x)$ be the partial differential operator on V defined by

$$Q(\partial_x) = Q\left(\frac{\partial}{\partial x_1}, \ldots, \frac{\partial}{\partial x_n}\right).$$

Then it is easy to see that

$$Q(x^*) \hat{f}(x^*) = (2\pi)^{-d} e^{d\pi i/2} (Q(\partial_x) f)\hat{}\,(x^*). \tag{2.3}$$

Moreover we have

$$Q(\partial_x)(x, x)^s = 2^d \{\Gamma(s+1)/\Gamma(s-d+1)\} Q(x)(x, x)^{s-d}. \tag{2.4}$$

In fact the space $H^{n,d}$ is spanned by the harmonic polynomials of the form

$$Q_\eta(x) = (\eta_1 x_1 + \cdots + \eta_n x_n)^d,$$

where η ($\in \mathbf{C}^n$) is subject to the condition $(\eta, \eta)=0$, if $d \geq 2$. Hence it suffices to show the identity (2.4) for $Q = Q_\eta$. In this case we can prove (2.4) by a straightforward calculation. It follows from (2.3) and the functional equation for $d=0$ that

$$\Phi(Q, \hat{f}; s) = (2\pi)^{-d} e^{d\pi i/2} \pi^{-2s+d+(n/2)} \left\{ \Gamma\left(s - \frac{d}{2}\right) \Big/ \left(\frac{d+n}{2} - s\right) \right\}$$
$$\times \int_{V'} (x, x)^{d/2-s} (Q(\partial_x) f)(x)\, dx.$$

Integrating by parts and using (2.4), we obtain

$$\int_{V'} (x, x)^{d/2-s} (Q(\partial_x) f)(x)\, dx$$
$$= (-2)^d \left\{ \Gamma\left(\frac{d}{2} + 1 - s\right) \Big/ \Gamma\left(-\frac{d}{2} + 1 - s\right) \right\} \Phi\left(Q, f; \frac{n}{2} - s\right).$$

Hence from the formula $\Gamma(z)\Gamma(1-z) = \pi/\sin(\pi z)$ we have the identity in the lemma.

Remark 1. The functional equation in Lemma 2.2 is not really new. In [9, Section 5] Rallis and Schiffmann proved Lemma 2.2 in a form including the case of indefinite quadratic forms.

Remark 2. Let the notation be as in Section 1.4. For a $\phi \in \mathcal{S}(V_\mathbf{Q})$ and $f \in \mathcal{S}(V)$, the following variant of the Poisson summation formula holds:

$$\sum_{m \in V_\mathbf{Q}} \phi(m) \hat{f}(m) = v(L^*) \sum_{m^* \in V_\mathbf{Q}^*} \hat{\phi}(m^*) f(m^*).$$

Moreover, as a special case of Lemma 2.1(*ii*), we obtain an integral representation of $\zeta_P(\phi; s)$:

$$\langle Z_\rho(f, \phi; s) w, w^* \rangle = \frac{\Gamma(n/2)}{2d(\rho)\pi^{n/2}} \zeta_{P_w}(\phi, s) \Phi(\overline{P_{w^*}}, f; s),$$

where

$$Z_\rho(f, \phi; s) = \int_0^\infty t^{2s}\, d^\times t \int_{SO(n)} \rho(k) \sum_{m \in V'_\mathbf{Q}} \phi(m) f(tkm)\, dk.$$

Therefore, using the local functional equation in Lemma 2.2, we can give a proof of Theorem 5 by quite the same argument as in the proof of [10, Theorem 2].

Lemma 2.3. *For any $Q \in H^{n,d}$, any $f \in \mathcal{S}(V)$ and any positive number ν, there exists a constant c such that the estimate*

$$|\Phi(Q, f; \sigma + it)| \leq c \cdot |t|^{-\nu}$$

holds on $\{\sigma + it \in \mathbf{C}; \sigma_1 \leq \sigma \leq \sigma_2, |t| \geq \varepsilon\}$, where $\sigma_1, \sigma_2 \in \mathbf{R}$ and $\varepsilon > 0$.

Proof. Since $Q(x)f(x)$ is in $\mathcal{S}(V)$, it suffices to prove the lemma for $Q(x) = 1$. In this case we write $\Phi(f; s)$ for $\Phi(Q, f; s)$. Put $b(s) = s(s + n/2 - 1)$ and $b_m(s) = b(s)b(s-1) \cdots b(s-m+1)$. Then it is easy to check the identity

$$\Delta^m (x, x)^s = 2^{2m} b_m(s)(x, x)^{s-m},$$

where Δ is the Euclidean Laplacian. Hence, by integrating by parts, we get

$$\Phi(f; s) = 2^{-2m} b_m\left(s - \frac{n}{2} + m\right)^{-1} \Phi(\Delta^m f; s+m). \tag{2.5}$$

Therefore, for $m > n/2 - \sigma_1$ and $t \neq 0$,

$$|\Phi(f, \sigma + it)| \leq 2^{-2m} \left| b_m\left(\sigma + it - \frac{n}{2} + m\right)\right|^{-1}$$
$$\times \left\{ \int_{V'} (x, x)^{\sigma_2 - n/2 + m} |\Delta^m f(x)| \, dx + \int_{V'} (x, x)^{\sigma_1 - n/2 + m} |\Delta^m f(x)| \, dx \right\}.$$

The lemma follows immediately from this inequality.

2.3. Proof of Theorem 1

Let ρ be an irreducible unitary representation of $SO(n)$ equivalent to ρ_d. By the assumption (3), Lemma 2.1(*ii*) and Lemma 2.2, we have

$$\langle Z_\rho(f, a, L; s) w, w^* \rangle = v(L^*) \left\langle Z_\rho\left(\hat{f}, a^*, L^*; \frac{n}{2} - s\right) w, w^* \right\rangle$$

for any $w, w^* \in W_\rho$. The assumptions (*i*) and (*ii*) in the theorem imply that there exists a monic polynomial $B(s)$ of s such that $B(s)\zeta_P(a, L; s)$ and $B(s)\zeta_P(a^*, L^*; s)$ are entire functions of finite order for any harmonic polynomial P. Let

$$B(s) = \prod_{i=1}^{r} (s - \alpha_i)^{m_i} \qquad (\alpha_i \neq \alpha_j, \text{ if } i \neq j)$$

and

$$\sum_{j=1}^{m_i} c_{ij}(w)(s-\alpha_i)^{-j}$$

be the principal part of the Laurent expansion of $\zeta_{P_w}(a, L; s)$ at $s=\alpha_i$:

$$\zeta_{P_w}(a, L; s) - \sum_{j=1}^{m_i} c_{ij}(w)(s-\alpha_i)^{-j} = \text{a holomorphic function of } s \text{ in a neighbourhood of } s=\alpha_i.$$

In the following we assume that f is in $C_0^\infty(V')$. Then $\Phi(P, f; s)$ is an entire function of s and hence $B(s)\langle Z_\rho(f, a, L; s)w, w^*\rangle$ is an entire function of s for any $f \in C_0^\infty(V')$. From Lemma 2.3 and the Phragmén-Lindelöf theorem, it follows that, for any $\nu > 0$ and any $k > n/2 + M$, there exists a positive constant $c = c_{\nu,k}$ satisfying

$$|B(s)\langle Z_\rho(f, a, L; s)w, w^*\rangle| < c \cdot (1+|\text{Im } s|)^{-\nu}$$

in the vertical strip $\{s \in \mathbf{C};\ n/2 - k \le \text{Re } s \le k\}$.

Therefore putting $c_{\rho,n} = \Gamma(n/2)/2d(\rho)\pi^{n/2}$, we have

$$\begin{aligned}\frac{1}{\pi i}\int_{k-i\infty}^{k+i\infty} &\langle Z_\rho(f, a, L; s)w, w^*\rangle t^{-2s}\, ds \\ &= \frac{v(L^*)}{\pi i}\int_{n/2-k-i\infty}^{n/2-k+i\infty} \left\langle Z_\rho\left(\hat{f}, a^*, L^*; \frac{n}{2}-s\right)w, w^*\right\rangle t^{-2s}\, ds \\ &\quad + c_{\rho,n}\sum_{i=1}^{r}\sum_{j=1}^{m_i}\frac{c_{ij}(w)}{(j-1)!}\left(\frac{d^{j-1}}{ds^{j-1}}\Phi(\overline{P_{w^*}}, f; s)t^{-2s}\Big|_{s=\alpha_i}\right).\end{aligned}$$

Hence by the inversion theorem for the Mellin transformation,

$$\begin{aligned}\int_{SO(n)} &\langle \rho(k)w, w^*\rangle \sum_{m\in L'} a(m)f(tkm)\, dk \\ &= v(L^*)t^{-n}\int_{SO(n)}\langle\rho(k)w, w^*\rangle \sum_{m^*\in(L^*)'} a^*(m^*)\hat{f}(t^{-1}km^*)\, dk \\ &\quad + c_{\rho,n}\sum_{i=1}^{r}\sum_{j=1}^{m_i}\frac{c_{ij}(w)}{(j-1)!}\left(\frac{d^{j-1}}{ds^{j-1}}\Phi(\overline{P_{w^*}}, f; s)t^{-2s}\Big|_{s=\alpha_i}\right).\end{aligned}$$

Now we fix an orthonormal basis $\{P_1^{(d)},\dots,P_{d(\rho_d)}^{(d)}\}$ of $H^{n,d}$ and put

$$R(t,k)=\frac{\Gamma(n/2)}{2\pi^{n/2}}\sum_{i=1}^{r}\sum_{j=1}^{m_i}\sum_{d=0}^{e}\sum_{\mu,\nu=1}^{d(\rho_d)}\overline{\rho_d(k)_{\mu,\nu}}$$
$$\times\frac{c_{ij}(P_\mu^{(d)})}{(j-1)!}\left(\frac{d^{j-1}}{ds^{j-1}}\Phi(\overline{P_\nu^{(d)}},f;s)t^{-2s}\bigg|_{s=\alpha_i}\right),$$

where $\rho_d(k)_{\nu,\mu}=\langle\rho_d(k)P_\nu^{(d)},P_\mu^{(d)}\rangle$ and e is the largest non-negative integer such that $\zeta_P(a,L;s)$ is not an entire function for some $P\in H^{n,e}$.

From the orthogonality relation of matrix elements of irreducible representations of $SO(n)$, we have

$$\int_{SO(n)}\rho_d(k)_{\mu,\nu}\left\{\sum_{m\in L'}a(m)f(tkm)\right.$$
$$\left.-t^{-n}v(L^*)\sum_{m^*\in(L^*)'}a^*(m^*)\hat{f}(t^{-1}km^*)-R(t,k)\right\}dk=0$$

for all d, μ, ν. By the proof of Lemma 2.1(i), the identity also holds, if we replace $\rho_d(k)_{\mu,\nu}$ by any matrix element of irreducible representations of $SO(n)$ not necessarily of class 1. From the Peter–Weyl theorem we see that

$$\sum_{m\in L'}a(m)f(tkm)-t^{-n}v(L^*)\sum_{m^*\in(L^*)'}a^*(m^*)\hat{f}(t^{-1}km^*)=R(t,k)$$

as a function of k on $SO(n)$. In particular, putting $t=1$ and $k=1$, we have

$$\sum_{m\in L'}a(m)f(m)-v(L^*)\sum_{m^*\in(L^*)'}a^*(m^*)\hat{f}(m^*)$$
$$=\frac{\Gamma(n/2)}{2\pi^{n/2}}\sum_{i=1}^{r}\sum_{j=1}^{m_i}\sum_{d=0}^{e}\sum_{\mu=1}^{d(\rho_d)}\frac{c_{ij}(P_\mu^{(d)})}{(j-1)!}$$
$$\times\int_{V'}(x,x)^{\alpha_i-(d+n)/2}(\log(x,x))^{j-1}\overline{P_\mu^{(d)}(x)}f(x)\,dx$$

for all $f\in C_0^\infty(V')$. Fix an $m_0\in L'$ and let $U_0=\{x\in V;(x-m_0,x-m_0)^{1/2}<\varepsilon\}$ be a sufficiently small neighbourhood of m_0. For an $m\in L'$ and an $f\in C_0^\infty(U_0)$, put $f_m(x)=f(x-m+m_0)$. Then $\hat{f}_m(y)=e^{2\pi i(y,m-m_0)}\hat{f}(y)$. Hence

$$a(m)f(m_0)-v(L^*)\sum_{m^*\in(L^*)'}a^*(m^*)\hat{f}(m^*)$$
$$=\int_{V'}K(x+m)f(x)\,dx\qquad(f\in C_0^\infty(U_0)),$$

where

$$K(x)=\frac{\Gamma(n/2)}{2\pi^{n/2}}\sum_{i=1}^{r}\sum_{j=1}^{m_i}\sum_{d=0}^{e}\sum_{\mu=1}^{d(\rho_d)}\frac{c_{ij}(P_\mu^{(d)})}{(j-1)!}$$
$$\times(x-m_0,x-m_0)^{\alpha_i-(d+n)/2}(\log(x-m_0,x-m_0))^{j-1}\,\overline{P_\mu^{(d)}(x-m_0)}.$$

Since the second term of the left-hand side of the identity is independent of m, we obtain

$$\{a(m)-a(m_0)\}f(m_0)=\int_{V'}\{K(x+m)-K(x+m_0)\}f(x)\,dx$$

($f\in C_0^\infty(U_0)$). In other words, the identity

$$\{a(m)-a(m_0)\}\delta(x-m_0)=K(x+m)-K(x+m_0)$$

between distributions holds on a neighbourhood of $x=m_0$. Since $K(x+m)-K(x+m_0)$ is real analytic on U_0, the constant $a(m)-a(m_0)$ must be equal to 0, namely a is a constant function on L'. By the same argument we can see that a^* is also a constant function on $(L^*)'$. Now other assertions in Theorem 1 are obvious.

Remark. It is a distinctive feature of the proof of functional equations from the viewpoint of the theory of prehomogeneous vector spaces that the integral representation of zeta functions involves an arbitrary test function $f\in\mathcal{S}(V)$ (cf. [10], [11]). If we restrict our attention only to proving functional equations, it is often sufficient to consider a particular test function f, such as $P(x)e^{-\pi(x,x)}$ in the remark following Lemma 2.1. However as is shown in the proof above of Theorem 1, the flexibility of the choice of a test function is a great advantage of the theory of prehomogeneous vector spaces. It is also noteworthy that Lemma 2.3 can be proved for more general prehomogeneous vector spaces by using b-functions.

2.4.

Let (G,ρ,V) be a prehomogeneous vector space defined over $\mathbf{Q}$. Following [10], we assume that G is reductive and the singular set S is an irreducible hypersurface. Let $P(x)$ be an irreducible polynomial in $\mathbf{Q}[V]$ defining S and let χ be the rational character of G corresponding to P. Denote by G^+ the identity component of the real Lie group $G_{\mathbf{R}}$ and put $G_1^+=\{g\in G^+;\chi(g)=1\}$. In order to obtain a generalization of the Hamburger theorem to (G,ρ,V), first we must construct a theory of zeta functions associated with (G,ρ,V) whose coefficients are periods of automorphic forms on G_1^+ with respect to an arithmetic subgroup of G_1^+. For this purpose harmonic analysis of homogeneous spaces of G_1^+ contained in $\{x \in V_{\mathbf{R}};\ P(x) = \pm 1\}$ will be indispensable. The most serious difficulty will arise when we try to deal with Eisenstein series on G_1^+.

The result of Yoshimoto [16] is regarded as the Hamburger theorem for the prehomogeneous vector space $R_{K/\mathbf{Q}}(GL(1), M(1))$ obtained from the standard 1-dimensional representation of $GL(1)$ by restricting the field of definition from an algebraic number field K to $\mathbf{Q}$. Then automorphic forms on G_1^+ are nothing but Größencharacters of K.

In our case where $G = GL(1) \times SO(n)$ and $G_1^+ = SO(n)$, automorphic forms on G_1^+ are given by matrix elements of irreducible representations of $SO(n)$ and the necessary harmonic analysis was the theory of spherical functions of $SO(n)/SO(n-1)$.

References

[1] S. Bochner and K. Chandrasekharan, On Riemann's functional equation, *Ann. of Math.* **63** (1956) 336–360.

[2] L. Ehrenpreis and T. Kawai, Poisson's summation formula and Hamburger's theorem, *Publ. Res. Inst. Math. Sci.* **18** (1982) 413–426.

[3] P. Epstein, Zur Theorie allgemeiner Zetafunktionen, I, II, *Math. Ann.* **56** (1903) 615–644; **63** (1907) 205–216.

[4] I. M. Gelfand and G. E. Shilov, *Generalized functions* Vol. 1, Academic Press, New York, 1964.

[5] H. Hamburger, Über die Riemannsche Funktionalgleichung der ζ-Funktion, *Math. Z.* **10** (1921) 240–254.

[6] E. Hecke, Über die Lösungen der Riemannschen Funktionalgleichung, *Math. Z.* **16** (1923) 301–307.

[7] E. Hecke, *Dirichlet series, modular functions and quadratic forms*, Vandenhoeck & Ruprecht, 1983.

[8] J. P. Kahane and S. Mandelbrojt, Sur l'equation fonctionelle de Riemann et la formule sommatoire de Poisson, *Ann. Sci. Ecole Norm. Sup.* **75** (1958) 57–80.

[9] S. Rallis and G. Schiffmann, Distributions invariantes par le groupe orthogonal, *Analyse harmonique des groupes de Lie, Seminaire Nancy-Strasbourg 1973–75*, Lect. Notes in Math. No. 497, Springer Verlag, Berlin-Heidelberg-New York, 1975, 494–642.

[10] M. Sato and T. Shintani, On zeta functions associated with prehomogeneous vector spaces, *Ann. of Math.* **100** (1974) 131–170.

[11] F. Sato, Zeta functions in several variables associated with prehomogeneous vector spaces I: Functional equations, *Tôhoku Math. J.* **34** (1982) 437–483.

[12] C. L. Siegel, Bemerkungen zu einem Satz von Hamburger über die Funktionalgleichungen der Riemannschen Zeta-funktion, *Math. Ann.* **86** (1922) 276–279.

[13] C. L. Siegel, *Lectures on Advanced Analytic Number Theory*, Lecture notes, Tata Institute of Fundamental Research, Bombay, 1961.

[14] H. M. Stark, L-functions and character sums for quadratic forms (I), *Acta Arith.* **16** (1968) 35–50.
[15] N. Vilenkin, *Special Functions and the Theory of Group Representations*, AMS translations of monographs 22, 1968.
[16] A. Yoshimoto, On a generalization of Hamburger's theorem, *Nagoya Math. J.* **98** (1985) 67–76.

Microfunctions for Boundary Value Problems

Pierre Schapira

Centre Scientifique et Polytechnique
Département de Mathématiques
Université Paris-Nord, Villetaneuse, France

1. The Functor μhom (cf. [K–S 2])

Let X be a C^∞-manifold, $\pi: T^*X \to X$ its cotangent bundle. We denote by $D^b(X)$ the derived category of the category of complexes of sheaves of abelian groups on X, with bounded cohomology. If F belongs to $D^b(X)$ and $H^j(F)=0$ for $j<m$ (resp. $>m$, resp. $\neq m$) we say F is concentrated in degree $\geq$m, (resp. $\leq m$, resp. m). If F is concentrated in degree 0 we identify F and $H^0(F)$. Recall that if $F \in D^b(X)$, its micro-support $SS(F)$ is a closed conic involutive subset of T^*X.

If M is a submanifold of X, we denote by $T^*_M X$ the conormal bundle to M in X.

We denote by ω_X the orientation sheaf on X. If $f: Y \to X$ is a continuous map, we set $\omega_{Y/X} = \omega_Y \otimes f^{-1}\omega_X$.

Around 1969, M. Sato introduced the functor $\mu_M: D^b(X) \to D^b(T^*_M X)$, of "microlocalization along M" (cf. [S–K–K]). Using the diagonal pro-

Algebraic Analysis, Volume II

ISBN 0-12-400466-0

cedure it is possible to slightly generalize this functor as follows. Let q_1 and q_2 denote the first and second projection on $X \times X$, and let G and F belong to $D^b(X)$. We set

$$\mu\text{hom}(G, F) = \mu_\Delta R\,\mathcal{H}om(q_2^{-1}G, q_1^!F), \tag{1.1}$$

where Δ is the diagonal of $X \times X$. Identifying X with Δ, and T^*X with $T^*_\Delta(X \times X)$, we obtain a bifunctor $\mu\text{hom}: D^b(X)^0 \times D^b(X) \to D^b(T^*X)$. Note that

$$\mu\text{hom}(\mathbf{Z}_M, *) \simeq \mu_M(*), \tag{1.2}$$

$$R\pi_{X*}\mu\text{hom}(*, *) \simeq R\,\mathcal{H}om(*, *), \tag{1.3}$$

$$\text{supp}(\mu\text{hom}(G, F)) \subset \text{SS}(G) \cap \text{SS}(F). \tag{1.4}$$

We shall need the following result, which is easily deduced from [K–S 2].

Theorem 1.1. *Let U be an open subset of T^*X, h a real C^2-function on U. Assume* $\text{SS}(F) \subset \{h \geq 0\}$, $\text{SS}(G) \subset \{h \leq 0\}$ *and let u be a section of* $H^j(\mu\text{hom}(G, F))$ *on U. Then* $\text{supp}(u)$ *is a union of positive half-integral curves of H_h.*

2. Wave Front Sets at the Boundary (cf. [S 4])

Let M be a real analytic manifold of dimension n, X a complexification of M, A a locally closed subset of M.

Definition 2.1. We set

$$\mathcal{C}_{A|X} = \mu\text{hom}(\mathbf{Z}_A, \mathcal{O}_X) \otimes \omega_{M/X}[n].$$

In particular, $\mathcal{C}_{M|X} = \mu_M(\mathcal{O}_X) \otimes \omega_{M/X}[n]$ is the sheaf (concentrated in degree 0) of Sato's microfunctions. One denotes it simply by $\mathcal{C}_M$.

We have, by (1.3),

$$R\pi_{X*}\mathcal{C}_{A|X} \simeq \Gamma_A(\mathcal{B}_M),$$

where $\mathcal{B}_M$ denotes the sheaf of Sato's hyperfunctions on M. If A is a closed half-space with real analytic boundary, the sheaf $\mathcal{C}_{A|X}$ has been defined by a different method by Kataoka [Ka].

Since the flabby dimension of $\mathcal{O}_X$ is n, the complex $\mathcal{C}_{A|X}$ is concentrated in degree ≤ 0.

Notice that by (1.4),

$$\operatorname{supp}(\mathscr{C}_{A|X}) \subset SS(\mathbf{Z}_A). \tag{2.1}$$

Assume now A is closed in M and set $\Omega = M\backslash A$. The exact sequence $0 \to \mathbf{Z}_\Omega \to \mathbf{Z}_M \to \mathbf{Z}_A \to 0$ yields the distinguished triangle

$$\mathscr{C}_{A|X} \to \mathscr{C}_M \to \mathscr{C}_{\Omega|X} \xrightarrow[+1]{} . \tag{2.2}$$

Note that $\mathscr{C}_{A|X} \simeq \mathscr{C}_{\Omega|X}[-1]$ outside of T^*_MX. Therefore, $\mathscr{C}_{\Omega|X}$ is not concentrated in degree zero in general. On the other hand, $H^0(\mathscr{C}_{\Omega|X})$ is supported by T^*_MX.

Example 2.2. Assume that, locally on M, A is analytically diffeomorphic to a closed convex subset of $\mathbf{R}^n$. Then $\mathscr{C}_{A|X}$ is concentrated in degree 0, and the morphism $\mathscr{C}_{A|X} \to \mathscr{C}_M$ is injective on T^*_MX. In particular, if $\Omega = M - A$, $(\mathscr{C}_{\Omega|X})|_{T^*_MX}$ is concentrated in degree 0 (for the proof, cf. [S 4]). In fact, using the proof of [loc. cit.], it is not difficult to get the same result when A is the union of two closed half-spaces which intersect transversally.

Conjecture 2.3. Let F be a sheaf on M. Denote by j the embedding $M \hookrightarrow X$. Then $(\mu\mathrm{hom}(j_*F, \mathscr{O}_X)[n])|_{T^*_MX}$ is concentrated in degree 0.

We consider now the case where Ω is open in M. Let j denote the embedding $\Omega \hookrightarrow X$. There are natural morphisms

$$\alpha : \pi_X^{-1}\Gamma_\Omega(B_M) \to H^0(\mathscr{C}_{\Omega|X}) \tag{2.3}$$

and

$$\beta : \mathscr{C}_M \to H^0(\mathscr{C}_{\Omega|X}). \tag{2.4}$$

Recall that

$$\Gamma_\Omega(\mathscr{B}_M) \simeq R\Gamma_\Omega(\mathscr{O}_X) \otimes \omega_{M/X}[n] \simeq j_* j^{-1}\mathscr{B}_M. \tag{2.5}$$

Definition 2.4. Let u be a section of $\Gamma_\Omega(\mathscr{B}_M)$ (resp. of $\mathscr{C}_M$). We set

$$SS_\Omega(u) = \operatorname{supp}(\alpha(u)) \qquad (\text{resp. } \operatorname{supp}(\beta(u))).$$

By its construction, $SS_\Omega(u)$ is a closed conic subset of T^*_MX. Moreover if $\Omega = M$, $SS_M(u)$ is nothing but the analytic wavefront set of u.

We still denote by j the embedding $\Omega \times_M T_M^*X \hookrightarrow T_M^*X$. By considering the natural morphisms

$$\mathscr{C}_M \to \mathscr{C}_{\Omega|X} \to j_*j^{-1}\mathscr{C}_M \simeq j_*j^{-1}\mathscr{C}_{\Omega|X}, \tag{2.6}$$

we obtain that if u is a section of $\mathscr{C}_M$,

$$\mathrm{SS}_M(u) \supset \mathrm{SS}_\Omega(u) \supset \overline{\mathrm{SS}_\Omega(u|_\Omega)} \tag{2.7}$$

(of course, $\mathrm{SS}_\Omega(u|_\Omega)$ denotes the analytic wavefront set of u above Ω). In general the inclusions (2.7) are strict; cf. [S 4] for an example in dimension one.

One can extend Definition 2.4 to solutions of differential systems. Let $\mathscr{M}$ be a left coherent module over the ring $\mathscr{D}_X$ of finite-order differential operators on X, (cf. [S-K-K] or else [Bj] or [S 3] for an introduction to the theory of $\mathscr{D}_X$-modules). We still denote by α and β the morphisms

$$\alpha : \pi_X^{-1}\mathscr{E}xt^j_{\mathscr{D}_X}(\mathscr{M}, \Gamma_\Omega(\mathscr{B}_M)) \to H^j(R\mathscr{H}om_{\mathscr{D}_X}(\mathscr{M}, \mathscr{C}_{\Omega|X})) \tag{2.8}$$

$$\beta : \mathscr{E}xt^j_{\mathscr{D}_X}(\mathscr{M}, \mathscr{C}_M) \to H^j(R\mathscr{H}om_{\mathscr{D}_X}(\mathscr{M}, \mathscr{C}_{\Omega|X})). \tag{2.9}$$

(We write $\mathscr{M}$ and $\mathscr{D}_X$ instead of $\pi^{-1}\mathscr{M}$ and $\pi^{-1}\mathscr{D}_X$ in (2.8) and (2.9).)

Note that the complex $R\mathscr{H}om_{\mathscr{D}_X}(\mathscr{M}, \mathscr{C}_M)$ being no more concentrated in degree 0 in general, the sheaves $H^j(R\mathscr{H}om_{\mathscr{D}_X}(\mathscr{M}, \mathscr{C}_{\Omega|X}))$ have no reason to be supported by T_M^*X for $j \geq 0$.

Definition 2.5. Let u be a section of $\mathscr{E}xt^j_{\mathscr{D}_X}(\mathscr{M}, \Gamma_\Omega(\mathscr{B}_M))$ (resp. $\mathscr{E}xt^j_{\mathscr{D}_X}(\mathscr{M}, \mathscr{C}_M)$). We set

$$\mathrm{SS}_\Omega^{\mathscr{M},j}(u) = \mathrm{supp}(\alpha(u)) \qquad (\text{resp. } \mathrm{supp}(\beta(u))).$$

The set $\mathrm{SS}_\Omega^{\mathscr{M},j}(u)$ is a closed conic subset of T^*X contained in $\mathrm{SS}(\mathbf{Z}_\Omega) \cap \mathrm{char}(\mathscr{M})$, where $\mathrm{char}(\mathscr{M})$ is the characteristic variety of $\mathscr{M}$.

Remark 2.6. Assume $\mathscr{M} = \mathscr{D}_X/\mathscr{J}$, and $\mathscr{C}_{\Omega|X}|_{T_M^*X}$ is concentrated in degree 0. Then the natural morphism

$$H^0(R\mathscr{H}om_{\mathscr{D}_X}(\mathscr{M}, \mathscr{C}_{\Omega|X})) \to H^0(\mathscr{C}_{\Omega|X})$$

is injective on T_M^*X. Thus if u is a section of $\Gamma_\Omega(\mathscr{B}_M)$, solution of the system of equations $\mathscr{J}u = 0$, we obtain

$$\mathrm{SS}_\Omega^{\mathscr{M},0}(u) \cap T_M^*X = \mathrm{SS}_\Omega(u). \tag{2.10}$$

3. Boundary Values (cf. [S 4])

Let X be a C^0-manifold, M a closed submanifold of codimension n, N a closed submanifold of M of codimension d in M, Ω an open subset of M. We shall always assume

$$\bar{\Omega} \supset N, \tag{3.1}$$

and

$$R\mathcal{H}om_{\mathbf{Z}_M}(\mathbf{Z}_{\bar{\Omega}}, \mathbf{Z}_M) \simeq \mathbf{Z}_{\Omega}. \tag{3.2}$$

Note that (3.2) is satisfied if, locally on M, Ω is homeomorphic to a convex subset of an affine space.

The morphism $\mathbf{Z}_{\Omega} \to \mathbf{Z}_N$ defines by duality (i.e., by applying $R\mathcal{H}om_{\mathbf{Z}_M}(\cdot, \mathbf{Z}_M)$) the morphism

$$b : \mathbf{Z}_N \to \mathbf{Z}_{\Omega} \otimes \omega_{N/M}[d]. \tag{3.3}$$

We call b the "*boundary value morphism.*" Let $F \in D^b(X)$. Applying the functor $R\mathcal{H}om(\cdot, F)$ to b we get the morphism

$$R\Gamma_{\Omega}(F) \to R\Gamma_N(F) \otimes \omega_{N/M}[d]. \tag{3.4}$$

More generally, applying $\mu\mathrm{hom}(\cdot, F)$ to b we get

$$\mu\mathrm{hom}(\mathbf{Z}_{\Omega}, F) \to \mu_N(F) \otimes \omega_{N/M}[d]. \tag{3.5}$$

We shall still denote by b the morphisms obtained in (3.4) and (3.5).

Now we come back to the situation where M is a real analytic manifold, X a complexification of M. We assume N is real analytic and denote by Y its complexification in X. Let $\mathcal{M}$ be a left coherent $\mathcal{D}_X$-module. We shall always assume

$$Y \text{ is non-characteristic for } \mathcal{M} \tag{3.6}$$

(cf. [S 4] for a definition of boundary values in case $\mathcal{M}$ is only regular along Y).

We denote as usual by $\mathcal{M}_Y$ the induced system on Y.

Since Y is non-characteristic for $\mathcal{M}$, the division theorem gives

$$R\mathcal{H}om_{\mathcal{D}_X}(\mathcal{M}, \Gamma_N(\mathcal{B}_M)) \otimes \omega_{N/M}[d] \simeq R\mathcal{H}om_{\mathcal{D}_Y}(\mathcal{M}_Y, \mathcal{B}_N). \tag{3.7}$$

Using (3.4) with $F = R\mathcal{H}om_{\mathcal{D}_X}(\mathcal{M}, \mathcal{O}_X)$ and using (3.7) we get the morphism

$$R\mathcal{H}om_{\mathcal{D}_X}(\mathcal{M}, \Gamma_\Omega(\mathcal{B}_M))|_N \to R\mathcal{H}om_{\mathcal{D}_Y}(\mathcal{M}_Y, \mathcal{B}_N). \tag{3.8}$$

Note that this construction of the boundary values extends that of [S 1] and [Ko] (or also [Ô]).

It is also possible to define a "microlocal boundary value morphism."

Let $\mathcal{E}_X$ denote the ring of finite-order microdifferential operators. Let U be an open subset of T^*X, $\mathcal{M}$ a coherent $\mathcal{E}_X$-module defined on U, such that Y is non-characteristic for $\mathcal{M}$.

Denote by ρ and $\bar{\omega}$ the natural maps

$$T^*Y \underset{\rho}{\leftarrow} Y \underset{X}{\times} T^*X \underset{\bar{\omega}}{\rightarrow} T^*X, \tag{3.9}$$

and set $\rho_U = \rho|_{Y \underset{X}{\times} U}$, and

$$\mathcal{M}_Y^U = \rho_{U*}(\mathcal{E}_{Y\to X} \otimes_{\mathcal{E}_X} \mathcal{M}), \tag{3.10}$$

where $\mathcal{E}_{Y\to X}$ is the $(\rho^{-1}\mathcal{E}_Y, \bar{\omega}^{-1}\mathcal{E}_X)$-bimodule of [S–K–K], (cf. [Bj], [S 1]). Then the division theorem (3.7) extends as an isomorphism on $\rho\bar{\omega}^{-1}(U)$ (cf. [K–K], [K–S 1]):

$$\rho_{U*}\bar{\omega}^{-1}R\mathcal{H}om_{\mathcal{E}_X}(\mathcal{M}, \mathcal{C}_{N|X}) \otimes \omega_{N/M}[d] \simeq R\mathcal{H}om_{\mathcal{E}_Y}(\mathcal{M}_Y^U, \mathcal{C}_N). \tag{3.11}$$

Now assume again $\mathcal{M}$ is a coherent $\mathcal{D}_X$-module (it would be possible, but a little more difficult, to develop a similar theory for $\mathcal{E}_X$-modules), and use (3.5) with $F = R\mathcal{H}om_{\mathcal{D}_X}(\mathcal{M}, \mathcal{O}_X)$, and (3.11). We get the morphism

$$\rho_{U*}\bar{\omega}^{-1}R\mathcal{H}om_{\mathcal{D}_X}(\mathcal{M}, \mathcal{C}_{\Omega|X}) \to R\mathcal{H}om_{\mathcal{E}_Y}(\mathcal{M}_Y^U, \mathcal{C}_N). \tag{3.12}$$

In particular, we get the commutative diagram on T^*Y:

$$\begin{array}{ccc}
\rho_*\bar{\omega}^{-1}R\mathcal{H}om_{\mathcal{D}_X}(\mathcal{M}, \mathcal{C}_{\Omega|X}) & \longrightarrow & \rho_*\bar{\omega}^{-1}R\mathcal{H}om_{\mathcal{D}_X}(\mathcal{M}, \mathcal{C}_{N|X}) \otimes \omega_{N/M}[d] \\
 & & \downarrow \wr \\
\uparrow & & R\mathcal{H}om_{\mathcal{D}_Y}(\mathcal{M}_Y, \mathcal{C}_N) \\
 & & \uparrow \\
\pi_Y^{-1}R\mathcal{H}om_{\mathcal{D}_X}(\mathcal{M}, \Gamma_\Omega(\mathcal{B}_M))|_N & \longrightarrow & \pi_Y^{-1}R\mathcal{H}om_{\mathcal{D}_Y}(\mathcal{M}_Y, \mathcal{B}_N).
\end{array} \tag{3.13}$$

Proposition 3.1. *Let* $u \in \mathcal{E}xt^j_{\mathcal{D}_X}(\mathcal{M}, \Gamma_\Omega(\mathcal{B}_M))|_N$. *Then*

$$\mathrm{SS}_N^{\mathcal{M}_Y, j}(b(u)) \subset \rho\bar{\omega}^{-1}\mathrm{SS}_\Omega^{\mathcal{M}, j}(u).$$

Proof. Since ρ is finite on char$(\mathcal{M})$, the functor $H^j(\cdot)$ commutes with ρ_* in (3.13) and the result follows.

Remark 3.2. Assume $\mathscr{C}_{\Omega|X}$ is concentrated in degree 0 on $T^*_M X$, and assume moreover $\mathrm{SS}(\mathbf{Z}_\Omega) \cap \mathrm{char}(\mathcal{M}) \subset T^*_M X$, and $\mathcal{M} = \mathscr{D}_X/\mathscr{J}$. Then it follows from Remark 2.6 that if $u \in \mathscr{H}om_{\mathscr{D}_X}(\mathcal{M}, \Gamma_\Omega(\mathscr{B}_M))$, we have

$$\mathrm{SS}_N(b(u)) \subset \rho\bar{\omega}^{-1}(\mathrm{SS}_\Omega(u)). \tag{3.14}$$

Proposition 3.3. *In the situation of Proposition* 3.1, *assume* $j=0$, $d=1$ *(and* Ω *is a half-space). Then*

$$\mathrm{SS}_N^{\mathcal{M}_Y,0}(b(u)) = \rho\bar{\omega}^{-1}\mathrm{SS}_\Omega^{\mathcal{M},0}(u).$$

Proof. By considering diagram (3.13) one sees it is enough to prove the injectivity of the morphism

$$H^0(\bar{\omega}^{-1}(R\mathscr{H}om_{\mathscr{D}_X}(\mathcal{M}, \mathscr{C}_{\Omega|X}))) \to H^0(\bar{\omega}^{-1}(R\mathscr{H}om_{\mathscr{D}_X}(\mathcal{M}, \mathscr{C}_{N|X}[1]))) \tag{3.15}$$

(we may forget $\omega_{N/M}$ since Ω defines a relative orientation of N in M).

On $\mathrm{SS}(\mathbf{Z}_\Omega)\setminus T^*_M X$ this morphism is in fact an isomorphism since $\mathscr{C}_{\Omega|X} \simeq \mathscr{C}_{N|X}[+1]$ on this set.

Set $H = R\mathscr{H}om_{\mathscr{D}_X}(\mathcal{M}, \mathscr{C}_{N|X})[1]$ and $Z = \mathrm{SS}(\mathbf{Z}_\Omega)$. By the division theorem (3.11) we know that H is concentrated in degree $\geqslant 0$. Hence the morphism $H^0(R\Gamma_Z(H)) \to H^0(H)$ is injective. Since the morphism (3.15) factors through $H^0(R\Gamma_Z(H))$, it is enough to prove that for each $p \in T^*_M X$,

$$H^0_Z(\mathscr{C}_{N|X})_p = 0 \tag{3.16}$$

$$(\mathscr{C}_{\Omega|X})_p \to (H^1_Z(\mathscr{C}_{N|X}))_p \text{ is injective.} \tag{3.17}$$

(Recall that $\mathscr{C}_{\Omega|X}$ is concentrated in degree 0 on $T^*_M X$.)

By a quantized complex contact transformation ([K–S 2, Ch. 11]), we are led to the following problem. Let $(z_1, \ldots, z_n) = (z_1, z')$ be the holomorphic coordinates on $\mathbf{C}^n$, with $z = x + \sqrt{-1}y$. Set

$$\Omega_0 = \left\{ z \in \mathbf{C}^n;\ y_n > \sum_{j=1}^{n-1} y_j^2 \right\},$$

$$\Omega_1 = \left\{ z \in \mathbf{C}^n;\ y_n > \sum_{j=2}^{n-1} y_j^2 \right\},$$

$$\Omega_1^{\pm} = \Omega_0 \cup (\Omega_1 \cap \{z \in \mathbf{C}^n;\ y_1 \lesssim 0\}),$$

and define the sheaves

$$\mathscr{C}_{i*} = H^1_{\mathbf{C}^n \setminus \Omega_{i*}}(\mathcal{O}_{\mathbf{C}^n}), \qquad i=0, * = \varnothing, \text{ or } i=1, *=\varnothing, +, -.$$

In other words, $\mathscr{C}_{i*}$ is the sheaf of boundary values of holomorphic functions on $\partial\Omega_{i*}$. Then (3.16) is equivalent to

$$(H^0_{\partial\Omega_1^-}(\mathscr{C}_1))_0 = 0, \tag{3.18}$$

and (3.17) is equivalent to

$$(\mathscr{C}_0/\mathscr{C}_{1^-})_0 \to (H^1_{\partial\Omega_1^-}(\mathscr{C}_1))_0 \text{ is injective.} \tag{3.19}$$

If f is holomorphic in Ω_1 in a neighborhood of 0, and extends holomorphically through $\partial\Omega_1$ for $y_1 < 0$, then f is holomorphic in a neighborhood of 0, in view of the local Bochner's tube theorem. This proves (3.18).

If f is holomorphic in Ω_0 in a neighborhood of 0, one can decompose f as $f = f^+ + f^-$, $f^\pm \in \mathcal{O}(\Omega_1^\pm)$, and f^+ will define a section $\tilde{f}^+$ of

$$(H^1_{\partial\Omega_1^-}(\mathscr{C}_1))_0 = \lim_{0 \in U} \Gamma(U \setminus \partial\Omega_1^-; \mathscr{C}_1)/\Gamma(U; \mathscr{C}_1).$$

Therefore $\tilde{f}^+ = 0$ implies that f^+ belongs to $(\mathscr{C}_1)_0$, and the proof is complete.

4. Application to Diffraction

Let $(z) = (z_1, z')$ be the coordinates on $\mathbf{C}^n$, $(z; \zeta)$ the associated coordinates on $T^*\mathbf{C}^n$, with $z = x + iy$, $\zeta = \xi + i\eta$. Let X be an open subset of $\mathbf{C}^n$, $M = X \cap \mathbf{R}^n$, $\Omega = M \cap \{x; x_1 > 0\}$, and let U be an open subset of T^*X. Let $f(z, \zeta)$ be a holomorphic function on U, homogeneous of degree two with respect to ζ. Assume

$$f(z, \zeta) = \zeta_1^2 - g(z, \zeta') \text{ where } g(x, i\eta') \text{ is real.} \tag{4.1}$$

We set

$$V_0 = \{(z, \zeta); \operatorname{Im} f(z, \zeta) = 0\}. \tag{4.2}$$

If $p \in V_0$, we denote by b_p^+ and b_p^- the positive and negative half-bicharacteristic curve associated to $\operatorname{Im} f$ passing through p, and we set $b_p = b_p^- \cup b_p^+$.

We shall make one of the following hypotheses:

$$\text{(a)} \quad \frac{\partial g}{\partial x_1}(p) < 0 \qquad \text{or} \qquad \text{(b)} \quad \frac{\partial g}{\partial x_1} \equiv 0. \tag{4.3}$$

Let $\mathcal{M}$ be a coherent $\mathcal{D}_X$-module with $\mathrm{char}(\mathcal{M}) \cap U \subset V_0$.

Theorem 4.1. *Let* $p \in \partial\Omega \underset{M}{\times} T^*_M X \cap U$, *and let* u *be a section of* $\mathcal{H}om_{\mathcal{D}_X}(\mathcal{M}, H^0(\mathcal{C}_{\Omega|X}))|_{b_p}$. *Assume* $p \in \mathrm{supp}(u)$. *Then* b_p^- *or* b_p^+ *is contained in* $\mathrm{supp}(u)$, *in a neighborhood of* p.

Sketch of proof. If $\zeta_1 \neq 0$ *at* p, the hypothesis (4.3) is not necessary. In fact $SS(\mathbf{Z}_\Omega) = \{(z, \zeta);\ x_1\xi_1 = 0,\ \xi_1 \leq 0,\ x_1 \geq 0,\ y' = \xi' = 0\}$ and this set is contained in $\{\mathrm{Im} f \cdot \mathrm{sgn}(\eta_1) \leq 0\}$. Hence the result follows from Theorem 1.1 applied with $G = \mathbf{Z}_\Omega$, $F = R\mathcal{H}om_{\mathcal{D}_X}(\mathcal{M}, \mathcal{O}_X)$.

Now assume $\zeta_1 = 0$ at p. Let $*$ denote $+$, $-$, 0 or $\varnothing$ and define

$$\begin{aligned}\Omega_* = \{z \in X;\ x_1 > 0,\ y' = 0,\ &y_1 \geq 0\ (* = +),\\ &\text{or } y_1 \leq 0\ (* = -), \text{ or } y_1 \in \mathbf{R}\ (* = 0),\\ &\text{or } y_1 = 0\ (* = \varnothing)\}.\end{aligned}$$

We have

$$\begin{aligned}SS(\mathbf{Z}_{\Omega_*}) = \{(z; \zeta);\ y' = 0,\ \xi' = 0,\ x_1\xi_1 = 0,\ x_1 \geq 0,\ \xi_1 \leq 0,\ y_1\eta_1 = 0\\ \text{and } y_1 \geq 0,\ \eta_1 \leq 0\ (* = +) \text{ or } y_1 \leq 0,\ \eta_1 \geq 0\ (* = -)\\ \text{or } y_1 \in \mathbf{R},\ \eta_1 = 0\ (* = 0)\}.\end{aligned}$$

Set $F = R\mathcal{H}om_{\mathcal{D}_X}(\mathcal{M}, \mathcal{O}_X)$ and $K_* = \mu\mathrm{hom}(\mathbf{Z}_{\Omega_*}, F)[n]$. We get a distinguished triangle,

$$K_0[-1] \to K \to K_+ \oplus K_- \xrightarrow[+1]{} . \tag{4.4}$$

Moreover, $SS(K_-) \subset \{\mathrm{Im} f \geq 0\}$ and $SS(K_+) \subset \{\mathrm{Im} f \leq 0\}$ on U, and one proves that the complexes appearing in (4.4) are all concentrated in degree ≥ 0 on b_p. If u is a section of $H^0(K)|_{b_p}$ with a sufficiently small compact support, then u belongs to $H^{-1}(K_0)$ in view of (4.4) and Theorem 1.1. If one assumes (4.3b), one concludes that $u = 0$ by applying Theorem 1.1 to $H^{-1}(K_0)$.

Now assume (4.3a). Set $\tilde{M} = \mathbf{C} \times \mathbf{R}^{n-1}$, denote by $\tilde{j}$ the embedding $\Omega_0 \hookrightarrow \tilde{M}$, and define $\mathcal{C}^h_{\Omega_0|X} = \mu\mathrm{hom}(\mathbf{Z}_{\Omega_0}, \mathcal{O}_X)[n-1]$ and $\mathcal{C}^h_{\tilde{M}|X} = \mu_{\tilde{M}}(\mathcal{O}_X)[n-1]$. Then one proves (using [S 2] for example) the injectivity of the morphisms $H^0(\mathcal{C}^h_{\Omega_0|X})|_{T^*_M X} \to \tilde{j}_*\tilde{j}^{-1}H^0(\mathcal{C}^h_{\tilde{M}|X})$ and $\tilde{j}_*\tilde{j}^{-1}H^0(\mathcal{C}^h_{\tilde{M}|X})|_{T^*_M X} \to j_*j^{-1}\mathcal{C}_M$. One concludes again that $u = 0$ by applying the interior propagation theorem.

Comments. (a) The arguments developed here do not make use of the "second microlocalization" (of [K 1]), as in [S 2] or [Ka].

(b) A wavefront set at the boundary of convex open subsets of $\mathbf{R}^n$, which is surely closely related to Definition 1.4, has been defined in [Li].

(c) It appears that the sheaf $\hat{\mathscr{C}}_{N/M^+}$ of "mild microfunctions" of Kataoka [Ka] is nothing but the sheaf $H^0(R\rho_!\bar{\omega}^{-1}\mathscr{C}_{\Omega|X})$ (cf. [S-Z]).

(d) The results of Sections 1 and 2 have been announced in [S 4], with other results and remarks, as well as bibliographical comments.

(e) In view of Proposition 2.1, if in Theorem 3.1 one assumes $\zeta_1 \neq 0$ at p, then one regains a result of [S 2]. If one assumes (3.3a) then one regains Kataoka's diffraction theorem (cf. [Ka], [La], [Le], [Sj]), but we were not able to recover the general version of Lebeau (loc. cit.).

(f) In a forthcoming paper, we will apply these methods to non-smooth boundary.

References

[Bj] J. E. Bjork, *Rings of Differential Operators*, North-Holland, 1979.

[Kl] M. Kashiwara, Exposé à Nice, 1972.

[K-K] M. Kashiwara and T. Kawai, On boundary value problems for elliptic systems of linear differential equations. *Proc. Japan Acad.* **48** (1971) 712–715; **49** (1972) 164–168.

[K-S 1] M. Kashiwara and P. Schapira, Micro-hyperbolic systems. *Acta Mathematica* **142** (1979) 1–55.

[K-S 2] M. Kashiwara and P. Schapira, Microlocal study of sheaves. *Asterisque* **128** (1985).

[Ka] K. Kataoka, Microlocal theory of boundary value problems I, II. *J. Fac. Sci. Univ. Tokyo Sect. A* **27** (1980) 355–399; **28**, 31–56 (1981).

[Ko] H. Komatsu, Boundary values for solutions of elliptic equations. *Proc. Int. Conf. Func. Anal. Rel. Topics*, Univ. Tokyo, 1970, pp. 107–121.

[La] P. Laubin, Asymptotic solutions of hyperbolic boundary value problems with diffraction. *Adv. in Microlocal Analysis.* Reidel Publ. Co. 168, 1985, pp. 165–202.

[Le] G. Lebeau, Deuxième microlocalisation sur les sous-variétés isotropes. *Ann. Inst. Fourier, Grenoble* **35** (2) (1985) 145–216.

[Li] J. L. Lieutenant, Microlocalization at the boundary of a convex set. *J. Fac. Sci. Univ. Tokyo Sect.* 1*A* (1986) 83–130.

[Ô] T. Ôaku, Higher codimensional boundary value problem. *Proc. Conf. Alg. Analys. Oct. 86.* Sûrikaireki-Kenkusko-Kôkynrokn.

[S-K-K] M. Sato, M. Kashiwara and T. Kawai, *Hyperfunctions and pseudo-differential equations.* Lecture Notes in Math. 287, Springer-Verlag, 1973, pp. 265–529.

[S 1] P. Schapira, Problème de Dirichlet et solutions hyperfonctions des équations elliptiques. *Bull. U.M.I.* **4** (1969) 367–372.

[S 2] P. Schapira, Propagation at the boundary of analytic singularities. *Nato Adv. Study Inst. on Boundary Values Problems.* Reidel Publ. Co., C65, 1981, 185–212.

[S 3] P. Schapira, *Microdifferential systems in the complex domain.* Grundlehren der Math. 269, Springer-Verlag, 1985.

[S 4] P. Schapira, Front d'onde analytique au bord II. Sem. E.D.P. Ecole Polyt. Exp. **13** (1986).

[S–Z] P. Schapira and G. Zampieri, to appear.

[Sj] J. Sjostrand, Propagation of analytic singularities for second order Dirichlet problem. I, II. *Comm. PDE* **5** (1) (1980) 41–94; **5** (2) (1980) 187–207.

Regularization of the Product of Complex Powers of Polynomials and Its Application

Jiro Sekiguchi
Department of Mathematics
University of Electro-Communications
Tokyo, Japan

Introduction

The meromorphic property of complex powers of polynomials established by Bernstein–Gel'fand [BG] is a powerful tool in the construction of hyperfunctions or distributions invariant under group actions. For example, one may recall the construction of fundamental solutions of partial differential equations with constant coefficients, those of relative invariant hyperfunctions on prehomogeneous vector spaces and so on. The original proof of the result of Bernstein–Gel'fand is based on Hironaka's desingularization theorem. Later, a simpler proof was given by I. N. Bernstein [B], showing the existence of b-functions (= Bernstein–Sato polynomials) of polynomials on $\mathbf{C}^n$. It is stressed here that it is M. Sato who showed the existence of b-functions of relative invariant polynomials of prehomogeneous vector spaces and conjectured the existence of b-functions of arbitrary polynomials

ISBN 0-12-400466-0

in the early 1960s. An extension to the case of holomorphic functions was shown by J-E. Björk [Bj] and M. Kashiwara [K].

Let G/H be a semisimple symmetric space. Then to construct Poisson kernels on G/H is equivalent to do hyperfunctions on G which are invariant under left H-action and "relative invariant" under right action of an appropriate parabolic subgroup of G. The latter is reduced to the problem of a regularization of distributions which are defined by the product of complex powers of analytic functions on G. So it is possible to apply the result of Bernstein–Gel'fand mentioned above. Such distributions play an important role in the study of representations of G because they generate principal series representations for G/H (cf. [O2]).

We now explain the contents shortly. In Section 1, we shall formulate a generalization (cf. Theorem 1) of the result of I. N. Bernstein and S. I. Gel'fand [BG] to the case of $\mathbf{C}$-valued polynomials under an assumption (Theorem 1, condition (A)) concerning the continuity of complex powers of $\mathbf{C}$-valued polynomials. This assumption seems less trivial to check for given polynomials. Section 2 is devoted to an application of Theorem 1. Let $\tilde{G}$ be the universal covering group of $SU(p+q, p)$ and let $\tilde{H}$ be a connected closed subgroup of $\tilde{G}$ whose Lie algebra is $\mathfrak{so}(p+q, p)$. Introduce real analytic functions $\tilde{f}_j(g)$ $(1 \le j \le 2p+q-1)$ on $\tilde{G}$ which are left $\tilde{H}$-invariant. As an application of Theorem 1, it will be shown in Theorem 9 that if $l_1, \ldots, l_{p-1} \in \mathbf{Z}$, $m_1, \ldots, m_q \in \mathbf{N}$, then

$$\zeta_{\lambda,\alpha,\beta}(g) = \tilde{f}_p(g)^{\alpha} \overline{\tilde{f}_p(g)}^{\beta} \prod_{j=1}^{p-1} |\tilde{f}_j(g)|^{\lambda_j} \tilde{f}_j(g)^{l_j} \prod_{k=1}^{q} \tilde{f}_{p+k}(g)^{m_k}$$

defines distributions on $\tilde{G}$ depending on complex parameters α, β, $\lambda_1, \ldots, \lambda_{p-1}$ meromorphically. The key to its proof is to check the condition (A) of Theorem 1. In this paper, we do not explain the role of $\zeta_{\lambda,\alpha,\beta}(g)$ in the study of analysis on semisimple symmetric spaces. The reader who is interested in this subject should consult [OS1, 2]. In Section 3, we restrict our attention to the case $p=1$, that is, $\tilde{G} = SU(q+1, 1)^{\sim}$ and study spherical functions $\phi_{\alpha,\beta}(g)$ on $\tilde{H}\backslash\tilde{G}$ which are obtained from $\zeta_{\lambda,\alpha,\beta}(g)$ by integrating over a certain closed subgroup of $\tilde{G}$ (cf. Section 3, (1)). In particular, we shall prove Theorem 12 which states that $\phi_{\alpha,\beta}$ is expressed in terms of Gaussian hypergeometric functions. In Remark 14, a connnection between the study of $\phi_{\alpha,\beta}(g)$ and c-functions for semisimple symmetric spaces is explained.

1. Meromorphic Continuation of Distributions Defined by the Product of Complex Powers of Polynomials

Let $P(x_1, \ldots, x_n)$ be an $\mathbf{R}$-valued polynomial. Then the function $|P(x)|^\lambda$ is continuous if $\operatorname{Re}\lambda > 0$, and as a function of λ, it is extended to a $\mathscr{D}'(\mathbf{R}^n)$-valued meromorphic function on $\mathbf{C}$. This is a result of I. N. Bernstein and S. I. Gel'fand [BG]. An extension of the meromorphic property of $|P(x)|^\lambda$ to the case of $\mathbf{C}$-valued polynomials is formulated in the following manner.

Theorem 1. *Let $f_1(x), \ldots, f_l(x)$ be $\mathbf{R}$-valued polynomials and let $g_1(x), \ldots, g_m(x)$ be $\mathbf{C}$-valued polynomials of $x = (x_1, \ldots, x_n)$. Let Ω be a connected component of the set $\{x = \mathbf{R}^n : f_i(x) \neq 0\ (1 \le i \le l),\ g_j(x) \neq 0\ (1 \le j \le m)\}$ and fix $y \in \Omega$. Suppose that $f_i(y) = 1$ $(1 \le i \le l)$, $g_j(y) = 1$ $(1 \le j \le m)$. Assume the condition:*

(A) *Take $\lambda_i \in \mathbf{C}$ $(1 \le i \le l)$, $\mu_j, \nu_j \in \mathbf{C}$ $(1 \le j \le m)$ such that $\operatorname{Re}\lambda_i > 0$, $\operatorname{Re}(\mu_j + \nu_j) > 0$. Then each branch of the function $\prod_{i=1}^{l} |f_i(x)|^{\lambda_i} \prod_{j=1}^{m} g_j(x)^{\mu_j} \overline{g_j(x)}^{\,\nu_j}$ on Ω is single-valued.*

Let $\phi'_{\lambda,\mu,\nu}(x)$ be the branch of the function $\prod_{i=1}^{l} |f_i(x)|^{\lambda_i} \prod_{j=1}^{m} g_j(x)^{\mu_j} \overline{g_j(x)}^{\,\nu_j}$ such that $\phi'_{\lambda,\mu,\nu}(y) = 1$. Define a function $\phi_{\lambda,\mu,\nu}(x)$ on $\mathbf{R}^n$ so that $\phi_{\lambda,\mu,\nu}(x) = \phi'_{\lambda,\mu,\nu}(x)$ if $x \in \Omega$ and $\phi_{\lambda,\mu,\nu}(x) = 0$ otherwise. Then the following hold.

(*i*) *For a non-negative integer p, define $S(p) = \{(\lambda, \mu, \nu) \in \mathbf{C}^{l+2m};\ \operatorname{Re}\lambda_i > p\ (1 \le i \le l),\ \operatorname{Re}\mu_j > p,\ \operatorname{Re}\nu_j > p\ (1 \le j \le m)\}$. Then $\phi_{\lambda,\mu,\nu}(x)$ is of class $\mathscr{C}^p$ if $(\lambda, \mu, \nu) \in S(p)$. Moreover, as a function of (λ, μ, ν), $\phi_{\lambda,\mu,\nu}(x)$ is holomorphic on the open subset $S(0)$ of $\mathbf{C}^{l+2m}$.*

(*ii*) *$\phi_{\lambda,\mu,\nu}(x)$ is extended to a $\mathscr{D}'(\mathbf{R}^n)$-valued meromorphic function on the whole (λ, μ, ν)-space $\mathbf{C}^{l+2m}$.*

This theorem can be shown by an argument similar to that in Bernstein–Gel'fand [BG], where Hironaka's desingularization theorem plays a central role. This also follows from the existence of b-functions of multi-parameters which is stated as follows.

Theorem 2 (Kashiwara-Kawai [KK], Sabbah [S]). *Let $h_i(x)$ $(1 \le i \le l)$ be polynomials of $x = (x_1, \ldots, x_n)$ and define a multi-valued analytic function $\phi_\lambda(x) = \prod_{i=1}^{l} h_i(x)^{\lambda_i}$ on an open subset $X = \{x \in \mathbf{C}^n : h_i(x) \neq 0\ (1 \le i \le (l)\}$ of $\mathbf{C}^n$. Then for each i $(1 \le i \le l)$, there exist a polynomial $P_i(\lambda)$ of λ whose coefficients are differential operators of the variable x and a non-zero polynomial $b_i(\lambda) \in \mathbf{C}[\lambda]$ such that $P_i(\lambda)(h_i(x)\phi_\lambda(x)) = b_i(\lambda)\phi_\lambda(x)$ holds on X.*

Remark 3. Theorem 2 is a generalization to the multi-parameter case of Bernstein's result on the existence of b-functions. Theorem 2 as well as its references was communicated to the author by M. Kashiwara.

Under the situation of Theorem 1, it seems difficult to decide whether the condition (A) holds for the given polynomials $f_1(x), \ldots, f_l(x)$, $g_1(x), \ldots, g_m(x)$ or not. In the rest of this section, we give some examples which agree with the condition (A).

Example 1. $x_+^\lambda(x+\sqrt{-1}y)^\mu$.
First note that $(x+\sqrt{-1}y)^\mu$ defines a continuous function on $\{(x, y) \in \mathbf{R}^2; x \ge 0\}$ if $\operatorname{Re}\mu > 0$. So Theorem 1 is applicable and we conclude that $x_+^\lambda(x+\sqrt{-1}y)^\mu$ defines distributions on $\mathbf{R}^2$ meromorphically depending on the parameters λ, μ.

Example 2. Let $P(x)$, $Q(x)$ be polynomials of $x = (x_1, \ldots, x_n)$. Assume that $P(x) \ge 0$ for all x. Let $\phi_{\lambda,\mu}(x)$ be a branch of $(P(x)+\sqrt{-1}Q(x))^\lambda(P(x)-\sqrt{-1}Q(x))^\mu$. Then it is easy to see that $\phi_{\lambda,\mu}(x)$ is a single-valued continuous function if $\operatorname{Re}(\lambda+\mu) > 0$. So, by Theorem 1, $\phi_{\lambda,\mu}(x)$ is extended to a $\mathscr{D}'(\mathbf{R}^n)$-valued distribution on the whole (λ, μ)-space.

Example 3. Put $f(x, y, t) = 1+2z^2+(\sqrt{-1}t-|z|^2)^2$ $(z = x+\sqrt{-1}y)$. If $\operatorname{Re}(\lambda+\mu) > 0$, each branch of the function $f(x, y, t)^\lambda\overline{f(x, y, t)}^\mu$ defines a single-valued continuous function on $\mathbf{R}^3$. This is shown in the following way. First note that

$$f(x, y, t) = \tfrac{1}{2}(1+|z|^2-\sqrt{-1}t)^2(1+h(x, y, t)),$$

where

$$h(x, y, t) = \frac{(1-|z|^2+\sqrt{-1}t)^2+4z^2}{(1+|z|^2-\sqrt{-1}t)^2}.$$

A simple calculation shows the inequality $|h(x, y, t)| \le 1$. Therefore $\operatorname{Re}(1+h(x, y, t)) \ge 0$ for all $(x, y, t) \in \mathbf{R}^3$. So each branch of $(1+h(x, y, t))^{1/2}$

is single-valued and

$$f(x, y, t)^{1/2} = \frac{1}{\sqrt{2}}(1+|z|^2-\sqrt{-1}t)(1+h(x, y, t))^{1/2}.$$

Then the claim follows. Applying Theorem 1, we find that $f(x, y, t)^{\lambda}\overline{f(x, y, t)}^{\mu}$ defines a family of distributions on $\mathbf{R}^3$ meromorphically depending on (λ, μ). This example is a special case of Theorem 9 in Section 2. Moreover, it is an interesting problem to regularize the divergent integral $\int f(x, y, z)^{\lambda}\overline{f(x, y, z)}^{\mu}\, dx\, dy\, dz$ (cf. Remark 13(*iii*)).

2. An Application of Theorem 1

Take positive integers p, q and put $n = 2p+q$. Define

$$G = \{g \in SL(n, \mathbf{C});\ {}^t\bar{g}S_{pq}g = S_{pq}\},$$

where

$$S_{pq} = \begin{pmatrix} & & J_p \\ & I_q & \\ J_p & & \end{pmatrix} \quad \text{with} \quad J_p = \begin{pmatrix} & & 1 \\ & \cdot^{\cdot^{\cdot}} & \\ 1 & & \end{pmatrix} \text{ a } p \times p \text{ matrix.}$$

By definition, $G \simeq SU(p+q, p)$. Take mutually commuting involutions θ, σ of G defined by $\theta(g) = {}^t\bar{g}^{-1}$, $\sigma(g) = \bar{g}$ for any $g \in G$. Using θ, σ, define $K = \{g \in G;\ \theta(g) = g\}$ and $H = \{g \in G;\ \sigma(g) = g\}$. Then, clearly we have $K \simeq S(U(p+q) \times U(p))$, $H \simeq SO(p+q, p)$ and in particular, K is a maximal compact subgroup of G. Let $\mathfrak{g}$ be the Lie algebra of G.

Introduce some notation:

$\mathfrak{a}_p = \{X \in \mathfrak{g};\ X \text{ is a real diagonal matrix}\}$

$\quad = \{\mathrm{diag}(t_1, \ldots, t_p, 0, \ldots, 0, -t_p, \ldots, -t_1);\ t_1, \ldots, t_p \in \mathbf{R}\}$

$\mathfrak{n} = \{X \in \mathfrak{g};\ X \text{ is upper triangular and nilpotent}\}$

$\bar{\mathfrak{n}} = \{X \in \mathfrak{g};\ X \text{ is lower triangular and nilpotent}\}$

$\mathfrak{k} =$ Lie algebra of K

$\mathfrak{h} =$ Lie algebra of H

$\mathfrak{m} = Z_{\mathfrak{k}}(\mathfrak{a}_p)$

$M = Z_K(\mathfrak{a}_p)$

$A_p = \exp \mathfrak{a}_p$, $N = \exp \mathfrak{n}$, $\bar{N} = \exp \bar{\mathfrak{n}}$

H_0: The identity component of H.

For a matrix $g=(g_{ij})_{1\le i,j\le n}$, let $D_k(g)$ denote the determinant of the matrix $(g_{ij})_{1\le i,j\le k}$. Using these, define functions $f_k(g)=D_k(\sigma(g)^{-1}g)$ for any $g\in G$. The next lemma follows from the definition.

Lemma 4. *(i) $f_k(hg)=f_k(g)$ for any $h\in H$ $(k=1,\ldots,n-1)$.*
(ii) $\overline{f_k(g)}=f_{n-k}(g)$ $(k=1,\ldots,p)$.

Let $\tilde{G}$ be the universal covering group of G and let $\pi:\tilde{G}\to G$ be a natural projection. Let $\tilde{K}$, $\tilde{H}$ be the analytic subgroups of $\tilde{G}$ corresponding to $\mathfrak{k}$, $\mathfrak{h}$, respectively. Put $\tilde{A}_p=\exp\mathfrak{a}_p$, $\tilde{N}=\exp\mathfrak{n}$, $\bar{N}^{\sim}=\exp\bar{\mathfrak{n}}$ in $\tilde{G}$ and $\tilde{M}=Z_{\tilde{K}}(\mathfrak{a}_p)$.

A simple calculation shows that

$$\tilde{M}\simeq\{(\phi,\alpha_1,\ldots,\alpha_p,m);\ \phi\in\mathbf{R},\ \alpha_i\in\mathbf{C},\ |\alpha_i|=1\ (1\le i\le p),$$

$$m \text{ a unitary matrix of degree } q,\ \alpha_1\cdots\alpha_p=e^{-n\sqrt{-1}\phi}=\det m\},$$

$$\mathfrak{m}\simeq\{(\phi,\theta_1,\ldots,\theta_p,X);\ \phi,\theta_1,\ldots,\theta_p\in\mathbf{R},$$

$$X \text{ a skew Hermitian matrix of degree } q,$$

$$\sqrt{-1}(\theta_1+\cdots+\theta_p)=-\sqrt{-1}n\phi=-\mathrm{tr}\,X\},$$

and that under this identification

$$\pi(\phi,\alpha_1,\ldots,\alpha_p,m)=e^{\sqrt{-1}\phi}\begin{pmatrix}\alpha_1 &&&&&\\ &\ddots&&&&\\ &&\alpha_p&&&\\ &&&m&&\\ &&&&\bar{\alpha}_p&&\\ &&&&&\ddots&\\ &&&&&&\bar{\alpha}_1\end{pmatrix}$$

for any $(\phi,\alpha_1,\ldots,\alpha_p,m)\in\tilde{M}$. Since $\tilde{M}$ is reductive, there are many one-dimensional representations of $\tilde{M}$. In particular, define representations χ_j $(1\le j\le p)$ of $\tilde{M}$ by

$$\chi_j(\phi,\alpha_1,\ldots,\alpha_p,m)=e^{2j\sqrt{-1}\phi}(\alpha_1\cdots\alpha_j)^2$$

for any $(\phi, \alpha_1, \ldots, \alpha_p, m) \in \tilde{M}$. Their differentials are denoted by $\delta\chi_j$ $(1 \leq j \leq p)$. Then

$$\delta\chi_j(\phi, \theta_1, \ldots, \theta_p, X) = \sqrt{-1}(\theta_1 + \cdots + \theta_j + 2j\phi)$$

for any $(\phi, \theta_1, \ldots, \theta_p, X) \in \mathfrak{m}$.

Lemma 5. *Assume that* $1 \leq j \leq p-1$. *Then* $\mu\delta\chi_j$ *is lifted to a character of* $\tilde{M}$ *if and only if* $\mu \in \mathbf{Z}$. *On the other hand,* $\chi_p(\exp Y)^\mu = \exp \mu\delta\chi_p(Y)$ $(Y \in \mathfrak{m})$ *defines a character of* $\tilde{M}$ *for any* $\mu \in \mathbf{C}$.

This lemma follows from the concrete form of $\tilde{M}$ given before.

For simplicity, put

$$a(t) = \exp(\operatorname{diag}(t_1, \ldots, t_p, 0, \ldots, 0, -t_p, \ldots, -t_1)) \in \tilde{A}_p$$

for any $t_1, \ldots, t_p \in \mathbf{R}$ and $\tilde{f}_j = f_j \circ \pi$ $(1 \leq j \leq n-1)$. The next lemma follows from the definition.

Lemma 6. *If* $1 \leq j \leq p$, *then*

$$\tilde{f}_j(gma(t)n) = e^{2t_1 + \cdots + 2t_j}\chi_j(m)\tilde{f}_j(g)$$

for any $g \in \tilde{G}$, $m \in \tilde{M}$, $a(t) \in \tilde{A}_p$, $n \in \tilde{N}$.

Lemma 7. (*i*) *The set* $\Omega = \tilde{H}\tilde{M}\tilde{A}_p\tilde{N}$ *is connected and open dense in* $\tilde{G}$.
(*ii*) $\Omega = \{g \in \tilde{G}; \tilde{f}_j(g) \neq 0 \ (1 \leq j \leq p)\}$.

Proof. Since $\tilde{H}$, $\tilde{M}$, $\tilde{A}_p$, $\tilde{N}$ are all connected, so is Ω. That Ω is open dense in $\tilde{G}$ is shown by direct calculation or follows from [M]. (*ii*) is a special case of [OS2].

The next proposition is the key to the subsequent theorem.

Proposition 8. *Take* $\lambda \in \mathbf{C}$ *with* $\operatorname{Re} \lambda > 0$.
(*i*) *Suppose that* $1 \leq j < p$. *Then* $\tilde{f}_j(g)^\lambda$ *is single-valued if and only if* λ *is an integer.*
(*ii*) *Each branch of* $\tilde{f}_p(g)^\lambda$ *defines a single-valued continuous function on* $\tilde{G}$.

Proof. (*i*) The "only if" part is clear. To prove the "if" part, consider the case $j=1$. For any $z \in \mathbf{C}$,

$$\bar{n}_1(z) = \pi^{-1} \begin{pmatrix} 1 & & & & \\ z & 1 & & & \\ & & \ddots & & \\ & & & 1 & \\ & & & -\bar{z} & 1 \end{pmatrix}$$

is contained in $\bar{N}^{\sim}$ and a simple computation shows that $\tilde{f}_j(\bar{n}_1(z)) = 1 + z^2$. If λ is not an integer, the function $(1+z^2)^\lambda$ is not single-valued and therefore $\tilde{f}_j(g)^\lambda$ is not single-valued. So the result follows when $j=1$. The other cases are shown in a similar way.

(*ii*) Define a function $\Phi_\lambda(g)$ on $\tilde{G}$ as follows. If $g = hma(t)n$ with $h \in \tilde{H}$, $m = \exp(Y)$, $Y \in m$, $a(t) \in \tilde{A}_p$, $n \in \tilde{N}$, then $\Phi_\lambda(g) = \exp\{\lambda\delta\chi_p(Y) + 2\lambda(t_1 + \cdots + t_p)\}$ and if $g \notin \Omega$, then $\Phi_\lambda(g) = 0$. Since $\tilde{M} \simeq M_1 \times \mathbf{R}$ for a connected compact Lie group M_1, $\Phi_\lambda(g)$ is well-defined. We first show that $\Phi_\lambda(g)$ is continuous on $\tilde{G} - S$, where $S = \{g \in \tilde{G}; \prod_{j=1}^{p-1} \tilde{f}_j(g) = 0\}$. Since Ω is connected, it easily follows from the definition that $\Phi_\lambda(g)$ is continuous on Ω. On the other hand, since $|\Phi_\lambda(g)| = |\tilde{f}_p(g)|^{\mathrm{Re}\,\lambda}$ if $g \in \Omega$ and since $\tilde{G} - (\Omega \cup S)$ $(\subset \{\tilde{f}_p(g) = 0\})$ is the boundary of Ω in $\tilde{G} - S$, we find that $\Phi_\lambda(g)$ is continuous on $\tilde{G} - S$. We next show that $\Phi_\lambda(g)$ is actually continuous on $\tilde{G}$. Now fix j $(1 \leq j < p)$. By the concrete form of $\tilde{f}_j(\bar{n})$, we find that $\mathrm{codim}_{\bar{N}^{\sim}}\{\bar{n} \in \bar{N}^{\sim}; \tilde{f}_j(\bar{n}) = 0\} \geq 2$. Put $\tilde{P} = \tilde{M}\tilde{A}_p\tilde{N}$ for simplicity. Then for any $\bar{n} \in \bar{N}^{\sim}$, $p \in \tilde{P}$, $\tilde{f}_j(\bar{n}p) = 0$ if and only if $\tilde{f}_j(\bar{n}) = 0$. So we find that $\mathrm{codim}_{\bar{N}^{\sim}\tilde{P}}\, S_j \cap \bar{N}^{\sim}\tilde{P} \geq 2$, where $S_j = \{g \in \tilde{G}; \tilde{f}_j(g) = 0\}$. On the other hand, the complement of $\bar{N}^{\sim}\tilde{P}$ in $\tilde{G}$ also has codimension ≥ 2 in $\tilde{G}$. So $\mathrm{codim}_{\tilde{G}}\, S_j \geq 2$. Since $S = \bigcup_{j=1}^{p-1} S_j$, we find that $\mathrm{codim}_{\tilde{G}}\, S \geq 2$. Now take $g \in S$ with $\tilde{f}_p(g) \neq 0$. Let U be a simply connected neighbourhood of g such that $\tilde{f}_p$ is invertible on it. Noting that $\Phi_\lambda(g)$ is locally a branch of the multi-valued function $\tilde{f}_p(g)^\lambda$, take a single-valued branch $\Phi'_\lambda(g)$ of $\tilde{f}_p(g)^\lambda$ on U coinciding with $\Phi_\lambda(g)$ on $U - S$. Since $\mathrm{codim}_{\tilde{G}}\, S \geq 2$, $U - S$ is connected. Hence $\Phi_\lambda(g)$ and $\Phi'_\lambda(g)$ must coincide on U. Then $\Phi_\lambda(g)$ defines a single-valued continuous function on $\tilde{G}$. By definition, $\Phi_\lambda(g)$ is the branch of $\tilde{f}_p(g)^\lambda$ with $\Phi_\lambda(e) = 1$. Therefore, each branch of $\tilde{f}_p(g)^\lambda$ is a constant multiple of $\Phi_\lambda(g)$ and the result follows. Q.E.D.

For $\alpha, \beta \in \mathbf{C}$ with $\mathrm{Re}\,(\alpha + \beta) > 0$, let $\eta_{\alpha,\beta}(g)$ be the branch of $\tilde{f}_p(g)^\alpha \overline{\tilde{f}_p(g)}^\beta$ such that $\eta_{\alpha,\beta}(e) = 1$. Then it follows from Proposition 8 that $\eta_{\alpha,\beta}(g)$ is a single-valued continuous function on $\tilde{G}$. Take $l_1, \ldots, l_{p-1} \in \mathbf{Z}$, $m_1, \ldots, m_q \in \mathbf{N}$ and fix them. For any $\lambda = (\lambda_1, \ldots, \lambda_{p-1}) \in \mathbf{C}^{p-1}$, $\alpha, \beta \in \mathbf{C}$, define a function $\Phi_{\lambda,\alpha,\beta}(g)$ on $\tilde{G}$ with parameters $(\lambda, \alpha, \beta) \in \mathbf{C}^{p-1} \times \mathbf{C}^2$ as follows. If $g \in \Omega$, then

$$\zeta_{\lambda,\alpha,\beta}(g) = \eta_{\alpha,\beta}(g) \prod_{j=1}^{p-1} |\tilde{f}_j(g)|^{\lambda_j} \tilde{f}_j(g)^{l_j} \prod_{k=1}^{q} \tilde{f}_{p+k}(g)^{m_k},$$

and $\zeta_{\lambda,\alpha,\beta}(g) = 0$ otherwise.

Theorem 9. *Fix $l_1, \ldots, l_{p-1} \in \mathbf{Z}$ and $m_1, \ldots, m_q \in \mathbf{N}$. Then the function $\zeta_{\lambda,\alpha,\beta}(g)$ on $\tilde{G}$ with parameters $(\lambda, \alpha, \beta) \in \mathbf{C}^{p-1} \times \mathbf{C}^2$ satisfies the following properties.*

(*i*) *If* $\operatorname{Re}(\lambda_i + l_i) > 0$ $(1 \le i \le p)$ *and* $\operatorname{Re}(\alpha + \beta) > 0$, *then $\zeta_{\lambda,\alpha,\beta}(g)$ is a single-valued continuous function on $\tilde{G}$.*

(*ii*) *As a function of (λ, α, β), $\zeta_{\lambda,\alpha,\beta}(g)$ is extended to a $\mathscr{D}'(\tilde{G})$-valued meromorphic function on the whole (λ, α, β)-plane.*

Proof. The idea of the proof is the same as in [OS 2] extending that in [OS 1].

(*i*) follows from Proposition 8.

(*ii*) Since $\prod_{i=1}^{q} \tilde{f}_{p+i}(g)^{m_i}$ is real analytic and since the product of a distribution and a real analytic function is a distribution, to prove the theorem, it suffices to show the case where $m_1 = \cdots = m_q = 0$. So assume the condition $m_1 = \cdots = m_q = 0$. As in the proof of Proposition 8, put $\tilde{P} = \tilde{M}\tilde{A}_p\tilde{N}$. Since $g\bar{N}^{\sim}\tilde{P}$ is an open subset of $\tilde{G}$ for any $g \in \tilde{G}$, it also suffices to show that the restriction of $\zeta_{\lambda,\alpha,\beta}$ to $g\bar{N}^{\sim}\tilde{P}$ can be extended to a $\mathscr{D}'(g\bar{N}^{\sim}\tilde{P})$-valued meromorphic function of (λ, α, β). Let $\bar{N}^{\sim} \times \tilde{P} \to g\bar{N}^{\sim}\tilde{P}$ be a natural product map and consider the pull-back $\zeta^*_{\lambda,\alpha,\beta}$ of $\zeta_{\lambda,\alpha,\beta}$ for this map. Then it follows from the definition that $\zeta^*_{\lambda,\alpha,\beta}(\bar{n}, man) = \zeta_{\lambda,\alpha,\beta}(g\bar{n})\zeta_{\lambda,\alpha,\beta}(ma)$ $(\bar{n} \in \bar{N}^{\sim}$, $m \in \tilde{M}$, $a \in \tilde{A}_p$, $n \in \tilde{N})$. By definition, $\zeta_{\lambda,\alpha,\beta}(ma)$ is invertible and is holomorphically extended to the (λ, α, β)-plane. On the other hand, take a base $X_1, \ldots, X_d$ of $\bar{\mathfrak{n}}$ $(d = \dim \bar{\mathfrak{n}})$ and put $\zeta'_{\lambda,\alpha,\beta}(x_1, \ldots, x_d) = \zeta_{\lambda,\alpha,\beta}(g\bar{n}(x))$, where $\bar{n}(x) = \exp(x_1X_1 + \cdots + x_dX_d)$. Since $\tilde{f}_j(g\bar{n}(x))$ $(1 \le j \le p)$ are complex-valued polynomials, Theorem 1 combined with Proposition 8 implies that $\zeta'_{\lambda,\alpha,\beta}(x)$ can be extended to a $\mathscr{D}'(\mathbf{R}^d)$-valued meromorphic function of (λ, α, β). Hence $\zeta_{\lambda,\alpha,\beta}$ is also extended to a $\mathscr{D}'(g\bar{N}^{\sim}\tilde{P})$-valued meromorphic function of (λ, α, β) and the theorem follows. Q.E.D.

3. Properties of Functions Related with the Distribution $\eta_{\alpha,\beta}$

In this section, we restrict our attention to the case $p=1$, namely, the case $\tilde{G}=$the universal covering group of $SU(q+1,1)$. We have already defined the distribution $\eta_{\alpha,\beta}(g)$ in the previous section. Put

$$\phi_{\alpha,\beta}(g)=\int_{K_s}\eta_{\alpha,\beta}(gk)\,dk, \tag{1}$$

where $K_s=(\tilde{K},\tilde{K})$ and dk is the Haar measure on it normalized by $\int_{K_s}dk=1$. The purpose of this section is to study this function in detail. As a consequence, we shall compute the value of "a regularization in a certain sense of the divergent integral

$$I(\alpha,\beta)=\int_{\bar{N}^{\sim}}\eta_{\alpha,\beta}(\bar{n})\,d\bar{n}\text{''} \tag{2}$$

(cf. Theorem 12). T. Oshima [O 1] introduced c-functions for semisimple symmetric spaces, generalizing Harish-Chandra's c-function for Riemannian symmetric spaces. In the case of the universal covering space of $SU(p+1,q+1)/SO(p+1,q+1)$, the c-function is a regularization of the divergent integral $\int_{\bar{N}^{\sim}}\zeta_{\lambda,\alpha,\beta}(\bar{n})\,d\bar{n}$ in some sense, and the above integral $I(\alpha,\beta)$ is its special case. (See Remark 14 in this section.)

Let E_{ij} be a matrix of degree $q+2$ whose (i',j') entry is 1 if $(i',j')=(i,j)$ and 0 otherwise. Then $\mathfrak{g}_c$ (=the complexification of $\mathfrak{g}$) is spanned by E_{ij} $(i\neq j)$, $E_{ii}-E_{jj}$. Let $\mathcal{U}(\mathfrak{g})$ be the universal enveloping algebra of $\mathfrak{g}_c$ and let ω be its Casimir element. Then, by definition, we have

$$2(q+2)\omega=\sum_{i<j}\{(E_{ii}-E_{jj})^2+2(q+2)E_{ji}E_{ij}\}+(q+2)\sum(q+1-2j)E_{jj}.$$

In the sequel, we identify elements of $\mathfrak{g}_c$ with left invariant vector fields on $\tilde{G}$ and also identify elements of $\mathcal{U}(\mathfrak{g})$ with left invariant differential operators on $\tilde{G}$. By direct calculation, we have

Lemma 10.

$$\omega\eta_{\alpha,\beta}=\left\{(\alpha+\beta)(\alpha+\beta+q+1)+\frac{q}{q+2}(\alpha-\beta)^2\right\}\eta_{\alpha,\beta}.$$

We now start to study the function $\phi_{\alpha,\beta}$. In the sequel, $i=\sqrt{-1}$ unless otherwise stated.

Proposition 11. (*i*) $\phi_{\alpha,\beta}(g)$ *is* $(\tilde{H}, K_s)$*-invariant.*
(*ii*) $\omega\phi_{\alpha,\beta}=\{(\alpha+\beta)(\alpha+\beta+q+1)+[q/(q+2)](\alpha-\beta)^2\}\phi_{\alpha,\beta}$.
(*iii*) $\phi_{\alpha,\beta}$ *is real analytic.*
(*iv*) *Put* $S_{\alpha,\beta}(t)=\phi_{\alpha,\beta}(\exp tY)$, *where* $Y=\frac{1}{2}(E_{11}-E_{q+2,q+2})$.

Then $S_{\alpha,\beta}(-t)=S_{\alpha,\beta}(t)$ *and if* $\mathrm{Re}\,(\alpha+\beta)>0$, *the following integral formula holds*:

$$S_{\alpha,\beta}(t)=\frac{1}{a_q}\int_{\mathbf{R}\times\mathbf{C}^q}\frac{\{e^t+2(z,z)+e^{-t}(|z|^2-ix)^2\}^\alpha}{(1+|z|^2-ix)^{2\alpha+q+1}}$$
$$\times\frac{\{e^t+2\overline{(z,z)}+e^{-t}(|z|^2+ix)^2\}^\beta}{(1+|z|^2+ix)^{2\beta+q+1}}\,dx\,dz\,d\bar{z}, \tag{3}$$

where, for any $z=(z_1,\ldots,z_q)\in\mathbf{C}^q$, $(z,z)=z_1^2+\cdots+z_q^2$, $|z|^2=|z_1|^2+\cdots+|z_q|^2$, *and* $dz\,d\bar{z}=dz_1\,d\bar{z}_1\cdots dz_q\,d\bar{z}_q$. *Moreover,*

$$a_q=\int_{\mathbf{R}\times\mathbf{C}^q}\{(1+|z|^2)^2+x^2\}^{-q-1}\,dx\,dz\,d\bar{z}=\frac{(-\pi i)^q\pi}{2^q\Gamma(q+1)}.$$

Proof. (*i*) is clear from the definition. Since ω is contained in the center of $\mathcal{U}(\mathfrak{g})$ and since $\phi_{\alpha,\beta}$ is obtained by integrating $\eta_{\alpha,\beta}$ over K_s, (*ii*) follows from Lemma 10. Noting that $\phi_{\alpha,\beta}$ is relative invariant by the right $\tilde{K}$-action, we find that (*ii*) implies (*iii*). Using a standard technique of changing the integration over K_s by that over $\bar{N}^{\sim}$, we obtain the integral formula in (*iv*). It is easy to check that the integral on the right side is actually convergent if $\mathrm{Re}\,(\alpha+\beta)>0$.

Q.E.D.

Theorem 12.

(*i*) $S_{\alpha,\beta}(t)=2^{-\alpha-\beta}$

$$\times\frac{\Gamma\left(\frac{q+2}{2}\right)\Gamma\left(\alpha+\beta+\frac{q+2}{2}\right)}{\Gamma\left(\alpha+\frac{q+2}{2}\right)\Gamma\left(\beta+\frac{q+2}{2}\right)}(\mathrm{ch}\,t)^{\alpha+\beta}F\left(-\alpha,-\beta,\frac{q+1}{2};(\mathrm{th}\,t)^2\right).$$

(*ii*) *If* $\mathrm{Re}\,(\alpha+\beta)>-(q+1)/2$, *then*

$$\lim_{t\to+\infty}e^{-(\alpha+\beta)t}S_{\alpha,\beta}(t)=2^{-2\alpha-2\beta}\frac{\Gamma(q+1)\Gamma(2\alpha+2\beta+q+1)}{\Gamma(2\alpha+q+1)\Gamma(2\beta+q+1)}. \tag{4}$$

On the other hand, if $\mathrm{Re}\,(\alpha+\beta)<-(q+1)/2$, *then*

$$\lim_{t\to+\infty} e^{(\alpha+\beta+q+1)t}S_{\alpha,\beta}(t)$$
$$=\frac{2}{\sqrt{\pi}}\frac{\sin\pi\alpha\sin\pi\beta}{\sin\pi\left(\alpha+\beta+\dfrac{q-1}{2}\right)}$$
$$\times\frac{\Gamma(q+1)\Gamma(\alpha+1)\Gamma(\beta+1)\Gamma\left(\alpha+\beta+\dfrac{q+2}{2}\right)}{\Gamma\left(\alpha+\dfrac{q+2}{2}\right)\Gamma\left(\beta+\dfrac{q+2}{2}\right)\Gamma\left(\alpha+\beta+\dfrac{q+3}{2}\right)}. \tag{5}$$

Proof. First of all, put $r=(q+1)/2$ for simplicity. Modifying the technique of calculating the radial component of the Casimir operator (cf. [W, p. 277]), we can obtain the concrete form of $(\omega\phi_{\alpha,\beta})(\exp tY)$. Then Proposition 11, (*ii*) leads to

$$\left\{\frac{d^2}{dt^2}+\left(q\frac{\mathrm{ch}\,t}{\mathrm{sh}\,t}+\frac{\mathrm{sh}\,t}{\mathrm{ch}\,t}\right)\frac{d}{dt}+\frac{(\alpha-\beta)^2}{(\mathrm{ch}\,t)^2}-(\alpha+\beta)(\alpha+\beta+q+1)\right\}S_{\alpha,\beta}(t)=0.$$

Put $x=(\mathrm{ch}\,t)^2$ and $T_{\alpha,\beta}(x)=S_{\alpha,\beta}(t)$. Then $T_{\alpha,\beta}(x)$ satisfies the differential equation

$$\left\{x(x-1)\frac{d^2}{dx^2}+((r+1)x-1)\frac{d}{dx}+\frac{(\alpha-\beta)^2}{4x}-\tfrac{1}{4}(\alpha+\beta)(\alpha+\beta+2r)\right\}u(x)=0.$$

Each solution of this differential equation which is real analytic in a neighbourhood of $t=0$, that is, in a neighbourhood of $x=1$, equals $x^{(\alpha+\beta)/2}F(-\alpha,-\beta,r;(x-1)/x)$ up to a constant factor. Therefore we find that

$$S_{\alpha,\beta}(t)=a(\alpha,\beta)(\mathrm{ch}\,t)^{\alpha+\beta}F(-\alpha,-\beta,r;(\mathrm{th}\,t)^2) \tag{6}$$

for some constant $a(\alpha,\beta)$ depending on α,β meromorphically. In virtue of the integral representation (3) of $S_{\alpha,\beta}(t)$ and the central limit theorem, we have

$$\lim_{t\to+\infty} e^{-(\alpha+\beta)t}S_{\alpha,\beta}(t)$$
$$=\frac{1}{a_q}\int_{\mathbf{R}\times\mathbf{C}^q}(1+|z|^2-ix)^{-2\alpha-2r}(1+|z|^2+ix)^{-2\beta-2r}\,dx\,dz\,d\bar{z}$$
$$=\frac{1}{a_q}\int_{\mathbf{C}^q}(1+|z|^2)^{-2\alpha-2\beta-2r+1}\,dz\,d\bar{z}\int_{-\infty}^{\infty}(1-ix)^{-2\alpha-2r}(1+ix)^{-2\beta-2r}\,dx \tag{7}$$

under the condition $\operatorname{Re}(\alpha+\beta)>0$. A simple computation shows that

$$\int_{\mathbf{C}^q} (1+|z|^2)^{-2\alpha-2\beta-2r+1}\, dz\, d\bar{z} = (-2\pi i)^q \frac{\Gamma(2\alpha+2\beta+2r)}{\Gamma(2\alpha+2\beta+4r-1)}$$

if $\operatorname{Re}(\alpha+\beta)>0$. On the other hand, it follows from [E, p. 12, formula (30)] that

$$\int_{-\infty}^{\infty} (1-ix)^{-2\alpha-2r}(1+ix)^{-2\beta-2r}\, dx$$

$$= 2\int_0^{\pi/2} \cos(2(\alpha-\beta)\theta)(\cos\theta)^{2(\alpha+\beta+2r-1)}\, d\theta \qquad (x=\tan\theta)$$

$$= \frac{2^{-2\alpha-2\beta-4r+4}\pi\Gamma(2\alpha+2\beta+4r-1)}{\Gamma(2\alpha+2r)\Gamma(2\beta+2r)}$$

if $\operatorname{Re}(\alpha+\beta)>-2r+\frac{1}{2}$. Therefore we obtain (4) under the condition $\operatorname{Re}(\alpha+\beta)>0$. Since both sides of (4) depend on α, β meromorphically, (4) holds if $\operatorname{Re}(\alpha+\beta)>-r$.

Now we recall a connection formula for Gaussian hypergeometric functions (cf. [E, p. 107, formula (33)]):

$$\begin{aligned} F(a, b, c; x) = {} & \frac{\Gamma(c)\Gamma(a+b-c)}{\Gamma(a)\Gamma(b)} \\ & \times |1-x|^{c-a-b} F(c-a, c-b, c-a-b+1; 1-x) \\ & + \frac{\Gamma(c)\Gamma(c-a-b)}{\Gamma(c-a)\Gamma(c-b)} F(a, b, a+b-c+1; 1-x) \end{aligned}$$

$(0<x<1)$. Substituting $a=-\alpha$, $b=-\beta$, $c=r$, $x=(\operatorname{th} t)^2$ in this formula and multiplying both sides by $a(\alpha,\beta)(\operatorname{ch} t)^{\alpha+\beta}$, we obtain

$$\begin{aligned} S_{\alpha,\beta}(t) = {} & a(\alpha,\beta)\frac{\Gamma(r)\Gamma(\alpha+\beta+r)}{\Gamma(\alpha+r)\Gamma(\beta+r)}(\operatorname{ch} t)^{\alpha+\beta} \\ & \times F(-\alpha, -\beta, -\alpha-\beta-r+1; (\operatorname{ch} t)^{-2}) \\ & + a(\alpha,\beta)\frac{\Gamma(r)\Gamma(-\alpha-\beta-r)}{\Gamma(-\alpha)\Gamma(-\beta)}(\operatorname{ch} t)^{-\alpha-\beta-2r} \\ & \times F(\alpha+r, \beta+r, \alpha+\beta+r+1; (\operatorname{ch} t)^{-2}). \end{aligned} \qquad (8)$$

If $\mathrm{Re}(\alpha+\beta)>-r$, this formula implies

$$\lim_{t\to+\infty} e^{-(\alpha+\beta)t}S_{\alpha,\beta}(t)=2^{-\alpha-\beta}a(\alpha,\beta)\frac{\Gamma(r)\Gamma(\alpha+\beta+r)}{\Gamma(\alpha+r)\Gamma(\beta+r)}.$$

Comparing this with (4), we have

$$a(\alpha,\beta)=2^{-\alpha-\beta}\frac{\Gamma(r+\frac{1}{2})\Gamma(\alpha+\beta+r+\frac{1}{2})}{\Gamma(\alpha+r+\frac{1}{2})\Gamma(\beta+r+\frac{1}{2})}.$$

Then (i) follows. Finally, it follows from (8) that if $\mathrm{Re}(\alpha+\beta)<-r$, then

$$\lim_{t\to+\infty} e^{(\alpha+\beta+2r)t}S_{\alpha,\beta}(t)=2^{\alpha+\beta+2r}a(\alpha,\beta)\frac{\Gamma(r)\Gamma(-\alpha-\beta-r)}{\Gamma(-\alpha)\Gamma(-\beta)}.$$

So we obtain (5). Q.E.D.

Remark 13. (i) The idea of the determination of Theorem 12 (ii) is based on that in [Se 1].

(ii) The integral in (7) in the case where $2(\alpha-\beta)\in\mathbf{Z}$ is treated in H. Schlichtkrull [Sc].

(iii) By changing variables of the integrand in (3), we have

$$e^{(\alpha+\beta+q+1)t}a_q S_{\alpha,\beta}(t)$$
$$=\int_{\mathbf{R}\times\mathbf{C}^q}\frac{\{1+2(z,z)+(|z|^2-ix)^2\}^{\alpha}}{\{1+e^{-t}(|z|^2-ix)\}^{2\alpha+q+1}}\frac{\{1+2\overline{(z,z)}+(|z|^2+ix)^2\}^{\beta}}{\{1+e^{-t}(|z|^2+ix)\}^{2\beta+q+1}}\,dx\,dz\,d\bar{z}.$$

If the integration and the limit process were commutative, the divergent integral

$$\frac{1}{a_q}\int_{\mathbf{R}\times\mathbf{C}^q}\{1+2(z,z)+(|z|^2-ix)^2\}^{\alpha}\{1+2\overline{(z,z)}+(|z|^2+ix)^2\}^{\beta}\,dx\,dz\,d\bar{z} \quad (9)$$

should coincide with the limit $\lim_{t\to+\infty} e^{(\alpha+\beta+q+1)t}S_{\alpha,\beta}(t)$ in the case $\mathrm{Re}(\alpha+\beta)<-(q+1)/2$. For this reason, the right-hand side of (5) is regarded as the value of a regularization of the divergent integral (9) in a certain sense. It should be noted that if $\mathrm{Re}(\alpha+\beta)>0$, then the integrand of (9) satisfies the condition (A) of Theorem 1.

Remark 14. Let G/H be a general semisimple symmetric space, where G is a connected semisimple Lie group. T. Oshima [O 1] introduced c-functions for G/H, generalizing Harish-Chandra's c-function for Riemannian symmetric spaces. In the case where G is linear, he explained four statements (i)–(iv) in [O, p. 365] which in fact give a method of the explicit calculation of the c-function. But to treat the case where G is not linear, we need a modification of the part (iv) of [O 1, p. 365] as follows.

> (B) "If G/H is split rank one, *K-relative invariant* joint eigenfunctions on G/H of invariant differential operators are expressed by Gaussian hypergeometric functions. (Here K is a closed subgroup of G such that $\mathrm{Ad}_G(K)$ is a maximal compact subgroup of $\mathrm{Ad}_G(G)$.)"

We return to our case, namely, $\tilde{G} = SU(q+1, 1)^{\tilde{}}$, $\tilde{H}$ = a connected closed subgroup of $\tilde{G}$ locally isomorphic to $SO_0(q+1, 1)$. Then $\psi_{\alpha,\beta}(g) = \phi_{\alpha,\beta}(g^{-1})$ is regarded as a $\tilde{K}$*-relative invariant* eigenfunction on $\tilde{G}/\tilde{H}$ of invariant differential operators. Theorem 12 states that $\psi_{\alpha,\beta}(g)$ is actually expressed by a Gaussian hypergeometric function and that its asymptotics are determined by using a connection formula of Gaussian hypergeometric functions. In particular, the statement (B) holds for the symmetric space $\tilde{G}/\tilde{H}$.

A proof of the statement (B) for other cases will be published elsewhere.

References

[B] I. N. Bernstein, The analytic continuation of generalized functions with respect to a parameter, *Functional Anal. Appl.* **6** (1972) 26–40.

[BG] I. N. Bernstein and S. I. Gel'fand, The meromorphic property of the function P^λ, *Functional Anal. Appl.* **3** (1969) 84–86.

[Bj] J-E. Björk, The global homological dimension of some algebras of differential operators, *Invent. Math.* **17** (1972) 67–78.

[E] A. Erdélyi *et al.*, *Higher Transcendental Functions*, I. McGraw-Hill Book Company, Inc. 1981.

[K] M. Kashiwara, B-functions and holonomic systems, *Invent. Math.* **38** (1976) 33–54.

[KK] M. Kashiwara and T. Kawai, On holonomic systems for $\prod_{l=1}^{N} (f_l + \sqrt{-1}0)^{\lambda_l}$, *Publ. RIMS, Kyoto Univ.* **15** (1979) 551–575.

[M] T. Matsuki, The orbits of affine symmetric spaces under the action of minimal parabolic subgroups, *J. Math. Soc. Japan* **31** (1979) 331–357.

[O 1] T. Oshima, Fourier analysis of semisimple symmetric spaces, in "*Non-Commutative Harmonic Analysis and Lie Groups IV*", Lecture Notes in Math. 880, Springer-Verlag, 1981, pp. 357–369.

[O 2] T. Oshima, A realization of semisimple symmetric spaces and construction of boundary value maps, preprint.

[OS 1] T. Oshima and J. Sekiguchi, Eigenspaces of invariant differential operators on an affine symmetric space, *Invent. Math.* **57** (1980) 1–81.

[OS 2] T. Oshima and J. Sekiguchi. In preparation.

[S] C. Sabbah, Proximité évanescente, II. Equations fonctionnelles pour plusieurs fonctions analytiques, to appear.

[Sc] H. Schlichtkrull, One-dimensional K-types in finite dimensional representations of semisimple Lie groups: A generalization of Helgason's theorem, *Math. Scand.* **54** (1984) 279–294.

[Se 1] J. Sekiguchi, Eigenfunctions of the Laplace–Beltrami operator on a hyperboloid, *Nagoya Math. J.* **79** (1980) 151–185.

[Se 2] J. Sekiguchi, Fundamental groups of semisimple symmetric spaces, to appear in *Advanced Studies in Pure Math.*

[W] G. Warner, *Harmonic Analysis on Semi-Simple Lie Groups*, II, Springer-Verlag, Berlin-Heidelberg-New York, 1972.

On the Local Solvability of Fuchsian Type Partial Differential Equations

Hidetoshi Tahara
Department of Mathematics
Sophia University
Tokyo, Japan

Fuchsian type partial differential equations have been studied by many authors in various problems. As to the local solvability of them, we can survey it as follows.

(1) (Analogue of Cauchy–Kowalewski Theorem). In analytic function spaces, the unique solvability is proved by Hasegawa [7], Baouendi-Goulaouic [2], Tahara [12].

(2) (Hyperbolic case). The local solvability (or the well-posedness of the Cauchy problem) is proved by Alinhac [1], Hanges [6], Tahara [13] in C^∞ function spaces, by Uryu [16], Tahara [14], Itoh-Uryu [8] in Gevrey function spaces, by Tahara [12] in hyperfunction spaces, and by Bove–Lewis–Parenti [3, 4], Bove–Lewis–Parenti–Tahara [5] in distribution spaces.

(3) (Non-hyperbolic case). The local solvability is proved by Ōaku [11] in hyperfunction spaces for elliptic equations, and by Parenti-Tahara [10] in distribution spaces for second-order equations of various types.

ISBN 0-12-400466-0

In this paper, the author will present a class of higher-order Fuchsian type partial differential equations, which contains both hyperbolic cases and elliptic cases and for which the local solvability is valid in distribution spaces.

1. Main Result

Let us consider the following type of operators:

$$P = t^m P_m + t^{m-1} P_{m-1} + \cdots + P_0, \tag{1.1}$$

where $(t, x) = (t, x_1, \ldots, x_n) \in \mathbf{R}_t \times \mathbf{R}^n_x$, $m \in \mathbf{N}(=\{1, 2, \ldots\})$, $P_j = P_j(t, x, \partial_t, \partial_x)$ is a linear partial differential operator of order j with C^∞ coefficients near the origin ($0 \le j \le m$, $\partial_t = \partial/\partial t$ and $\partial_x = (\partial/\partial x_1, \ldots, \partial/\partial x_n)$), and P_m is assumed to be non-characteristic in the direction dt (that is, the coefficient of ∂_t^m in $P_m = P_m(t, x, \partial_t, \partial_x)$ does not vanish at the origin).

Then, P is a typical model of partial differential operators of Fuchsian type with respect to t. The indicial polynomial $C(\rho, x)$ is defined by

$$C(\rho, x) = t^{-\rho} P(t^\rho)|_{t=0} \qquad (\rho \in \mathbf{C}) \tag{1.2}$$

(note that $C(\rho, x)$ is a polynomial of degree m in ρ whose coefficients are functions in x), and the characteristic exponents $\rho_1(x), \ldots, \rho_m(x)$ are defined by the roots of the equation $C(\rho, x) = 0$ in ρ.

The following is one of the most fundamental problems: determine the condition under which $Pu = f$ is locally solvable in distribution spaces. In this paper, we say that $Pu = f$ is *locally solvable in* $\mathscr{D}'_0$ if the map

$$P : \mathscr{D}'_0 \to \mathscr{D}'_0 \tag{1.3}$$

is surjective, where $\mathscr{D}'_0$ denotes the stalk of the sheaf of distribution in (t, x) at the origin.

The main result of this paper, which gives a sufficient condition for the local solvability in $\mathscr{D}'_0$, is as follows. Denote by $\mathring{p}(t, x, \tau, \xi)$ the principal symbol of P_m, that is,

$$\mathring{p}(t, x, \tau, \xi) = \sigma_m(P_m)(t, x, \tau, \xi), \tag{1.4}$$

where τ and ξ are the dual variables of t and x, respectively. We put the following conditions on $\mathring{p}(t, x, \tau, \xi)$.

(A-1) The coefficients of $\mathring{p}(t, x, \tau, \xi)$ (as a polynomial in (τ, ξ)) are real-valued.

(A-2) For any $\xi \in \mathbf{R}^n \backslash \{0\}$, the polynomial

$$\mathbf{C} \ni \lambda \to \mathring{p}(0, 0, \lambda, \xi)$$

has only simple roots.

Then, we have

Theorem 1. *Let P be as in* (1.1). *Assume that* (A-1), (A-2) *and* $\rho_1(0), \ldots, \rho_m(0) \notin \{-1, -2, \ldots\}$ *hold. Then,* $Pu = f$ *is locally solvable in* $\mathscr{D}'_0$, *that is, the map* (1.3) *is surjective.*

Remark 1. By the argument in Section 2, we can also see that the map

$$P : \mathscr{D}'(\bar{U}) \to \mathscr{D}'(\bar{U})$$

is surjective for a small neighbourhood U of $(0, 0)$ in $\mathbf{R} \times \mathbf{R}^n$, where $\bar{U}$ means the closure of U and $\mathscr{D}'(\bar{U})$ denotes the set of all sections of the sheaf of distributions on $\bar{U}$.

2. Proof of Theorem 1

For an open neighbourhood U of $(0, 0)$ in $\mathbf{R} \times \mathbf{R}^n$, we write $U(+) = U \cap \{t > 0\}$ and $U(-) = U \cap \{t < 0\}$. Put $\mathscr{D}'_{\{t=0\}}$, $\mathscr{D}'(+)$, $\mathscr{D}'(-)$, $\mathscr{D}'_{\text{ext}}(+)$ and $\mathscr{D}'_{\text{ext}}(-)$ as follows:

$$\mathscr{D}'_{\{t=0\}} = \{u \in \mathscr{D}'_0;\ \text{supp}(u) \subset \{t = 0\}\},$$

$$\mathscr{D}'(\pm) = \varinjlim_{U \ni (0,0)} \mathscr{D}'(U(\pm)), \tag{2.1}$$

$$\mathscr{D}'_{\text{ext}}(\pm) = \{v \in \mathscr{D}'(\pm);\ \text{there exists a } u \in \mathscr{D}'_0 \text{ such that } u = v \text{ on } \{\pm t > 0\}\}.$$

Note that $\mathscr{D}'_{\text{ext}}(\pm)$ is the set of all distributions $v \in \mathscr{D}'(\pm)$ which is extendable to a full neighbourhood of $t = 0$ as a distribution.

Then, Theorem 1 is an easy consequence of the following propositions.

Proposition 1. *Let P be as in* (1.1). *Assume that* $\rho_1(0), \ldots, \rho_m(0) \notin \{-1, -2, \ldots\}$ *holds. Then, the map*

$$P : \mathscr{D}'_{\{t=0\}} \to \mathscr{D}'_{\{t=0\}} \tag{2.2}$$

is bijective.

Proposition 2. *Let P be as in* (1.1). *Assume that* (A-1) *and* (A-2) *hold. Then, the map*

$$P:\mathscr{D}'_{\mathrm{ext}}(\pm)\to\mathscr{D}'_{\mathrm{ext}}(\pm) \tag{2.3}$$

is surjective.

In fact, if we notice that the sequence

$$0\to\mathscr{D}'_{\{t=0\}}\to\mathscr{D}'_0\to\mathscr{D}'_{\mathrm{ext}}(+)\oplus\mathscr{D}'_{\mathrm{ext}}(-)\to 0$$

is exact, by applying Propositions 1 and 2 to the commutative diagram

$$\begin{array}{ccccccccc} 0\to & \mathscr{D}'_{\{t=0\}} & \to & \mathscr{D}'_0 & \to & \mathscr{D}'_{\mathrm{ext}}(+)\oplus\mathscr{D}'_{\mathrm{ext}}(-) & \to 0 \\ & \downarrow P & & \downarrow P & & \downarrow P & \\ 0\to & \mathscr{D}'_{\{t=0\}} & \to & \mathscr{D}'_0 & \to & \mathscr{D}'_{\mathrm{ext}}(+)\oplus\mathscr{D}'_{\mathrm{ext}}(-) & \to 0 \end{array} \tag{2.4}$$

we can easily obtain Theorem 1.

Hence, to prove Theorem 1 it is sufficient to show Propositions 1 and 2.

Proof of Proposition 1. Let u and f be of the form

$$u=\sum_{i=0}^{N}\delta^{(i)}(t)\otimes\varphi_i(x),\qquad f=\sum_{i=0}^{N}\delta^{(i)}(t)\otimes\psi_i(x), \tag{2.5}$$

where $N\in\mathbf{Z}_+(=\{0,1,2,\dots\})$ and $\varphi_i(x)$, $\psi_i(x)$ are germs of distributions in x at the origin in $\mathbf{R}^n$. Then, by an easy calculation we can see that $Pu=f$ is equivalent to

$$\begin{cases} C(-N-1,x)\varphi_N=\psi_N, \\ \quad C(-N,x)\varphi_{N-1}=\psi_{N-1}+L_N^{(N-1)}(x,\partial_x)\varphi_N, \\ \qquad\qquad\vdots \\ \qquad C(-1,x)\varphi_0=\psi_0+L_N^{(0)}(x,\partial_x)\varphi_N+\cdots+L_1^{(0)}(x,\partial_x)\varphi_1, \end{cases} \tag{2.6}$$

where $L_j^{(k)}(x,\partial_x)$ are linear differential operators depending only on P and N. Since $\rho_1(0),\dots,\rho_m(0)\notin\{-1,-2,\dots\}$ is assumed, we have $C(\rho,0)\neq 0$ for $\rho\in\{-1,-2,\dots\}$ and therefore we can uniquely determine $\{\varphi_i\}_{i=0}^N$ from the given $\{\psi_i\}_{i=0}^N$ so that (2.6) is satisfied. This proves Proposition 1, because any $u,f\in\mathscr{D}'_{\{t=0\}}$ are expressed in the form (2.5). Q.E.D.

For $p\in\mathbf{Z}_+$, $s\in\mathbf{R}$ and an open neighbourhood U of $(0,0)$ in $\mathbf{R}\times\mathbf{R}^n$, we define $W_{-s}^{-p}(U(\pm))$ by

$$W_{-s}^{-p}(U(\pm))=\{u\in\mathscr{D}'(U(\pm));\ t^su\in H^{-p}(U(\pm))\},$$

where $H^{-p}(U(\pm))$ means the usual Sobolev space on $U(\pm)$. Then, we have

$$\mathscr{D}'_{\text{ext}}(\pm) = \varinjlim_{U \ni (0,0)} \left(\bigcup_{p=0}^{\infty} \bigcup_{s>0} W_{-s}^{-p}(U(\pm)) \right).$$

Therefore, Proposition 2 is obtained from the following theorem, which is a particular result of our class here.

Theorem 2. *Let P be as in* (1.1). *Assume that* (A-1) *and* (A-2) *hold. Then, there are $s_k > 0$ ($k \in \mathbf{Z}_+$) and an open neighbourhood U_0 of $(0,0)$ in $\mathbf{R} \times \mathbf{R}^n$ such that for any $U \subset U_0$, $k \in \mathbf{Z}_+$, $s > s_k$ and for any $f \in W_{-s+m-1}^{-k-m+1}(U(\pm))$ there exists a solution $u \in W_{-s}^{-k}(U(\pm))$ of $Pu = f$ on $U(\pm)$.*

Proof. Let P_{-s} be the differential operator defined by

$$P_{-s}u = t^s P(t^{-s}u) \qquad (s \in \mathbf{R}), \tag{2.7}$$

and let $(P_{-s})^*$ be the formal adjoint operator of P_{-s}. Then, by Proposition 3 in Section 3 we can see the following; there are $\delta_k > 0$ ($k \in \mathbf{Z}_+$), $s_k > 0$ ($k \in \mathbf{Z}_+$) and an open neighbourhood U_0 of $(0,0)$ in $\mathbf{R} \times \mathbf{R}^n$ such that for any $k \in \mathbf{Z}_+$ the estimate

$$\|(P_{-s})^*\varphi\|_{H^k}^2 \geq \delta_k s \|t^{m-1}\varphi\|_{H^{k+m-1}}^2 \tag{2.8}$$

holds for any $\varphi \in C_0^\infty(U_0(\pm))$ and $s > s_k$, where H^p ($p \in \mathbf{Z}_+$) means the usual Sobolev space.

Let $U \subset U_0$, $k \in \mathbf{Z}_+$, $s > s_k$ and let $f \in W_{-s+m-1}^{-k-m+1}(U(\pm))$, that is, $t^{s-m+1}f \in H^{-k-m+1}(U(\pm))$. Put $Z = \{(P_{-s})^*\varphi;\ \varphi \in C_0^\infty(U(\pm))\}$ and define a linear functional T on Z by

$$T((P_{-s})^*\varphi) = \langle t^s f, \varphi \rangle. \tag{2.9}$$

Then, by (2.8) we can see that (2.9) is well defined and

$$\begin{aligned}
|T((P_{-s})^*\varphi)| &= |\langle t^{s-m+1}f, t^{m-1}\varphi\rangle| \\
&\leq \|t^{s-m+1}f\|_{H^{-k-m+1}(U(\pm))} \times \|t^{m-1}\varphi\|_{H^{k+m-1}(U(\pm))} \\
&\leq \frac{1}{\sqrt{\delta_k s}} \|t^{s-m+1}f\|_{H^{-k-m+1}(U(\pm))} \times \|(P_{-s})^*\varphi\|_{H^k(U(\pm))}.
\end{aligned} \tag{2.10}$$

Let $\mathring{H}^k(U(\pm))$ be the closure of $C_0^\infty(U(\pm))$ in $H^k(U(\pm))$. Then, (2.10) implies that T can be extended to a continuous linear functional on $\mathring{H}^k(U(\pm))$. Hence, we can find $v \in H^{-k}(U(\pm))$ such that

$$T((P_{-s})^*\varphi) = \langle v, (P_{-s})^*\varphi \rangle \tag{2.11}$$

for any $\varphi \in C_0^\infty(U(\pm))$, because $H^{-k}(U(\pm))$ is the dual space of $\mathring{H}^k(U(\pm))$.

Thus, by (2.9) and (2.11) we have $\langle t^s f, \varphi\rangle = \langle v, (P_{-s})^* \varphi\rangle$ for any $\varphi \in C_0^\infty(U(\pm))$, that is, $P_{-s}v = t^s f$ on $U(\pm)$. Since $t^{-s}P_{-s}v = P(t^{-s}v)$, by putting $u = t^{-s}v$ we can obtain a solution $u \in W_{-s}^{-k}(U(\pm))$ desired in Theorem 2.

Q.E.D.

3. Proof of (2.8)

The purpose of this section is to establish the following proposition:

Proposition 3. *Let P be as in* (1.1). *Assume that* (A-1) *and* (A-2) *hold. Then, there are* $\delta_k > 0$ ($k \in \mathbf{Z}_+$), $s_k > 0$ ($k \in \mathbf{Z}_+$) *and an open neighbourhood* U_0 *of* $(0, 0)$ *in* $\mathbf{R} \times \mathbf{R}^n$ *such that for any* $k \in \mathbf{Z}_+$ *the estimate*

$$\|(P_{-s})^* \varphi\|_{H^k}^2 \geq \delta_k s \|t^{m-1}\varphi\|_{H^{k+m-1}}^2 \tag{3.1}$$

holds for any $\varphi \in C_0^\infty(U_0(\pm))$ *and* $s > s_k$.

First, let us present some preparatory lemmas.

Lemma 1. *Let* $a > \frac{1}{2}$. *Then, for any* $\varphi \in C_0^\infty((0, T) \times \mathbf{R}^n)$ $(T > 0)$, *we have*

$$\|(t\partial_t + a)\varphi\|_{L^2} \geq \frac{2a-1}{2}\|\varphi\|_{L^2}, \frac{2a-1}{4a-1}\|t\partial_t\varphi\|_{L^2}. \tag{3.2}$$

Proof. Put $f = (t\partial_t + a)\varphi$. Then, we have

$$\varphi(t, x) = t^{-a}\int_0^t \tau^{a-1} f(\tau, x)\, d\tau. \tag{3.3}$$

Put $\lambda = \log t$, $\mu = \log \tau$, $\Phi(\lambda, x) = \sqrt{t}\varphi(t, x)$ and $F(\mu, x) = \sqrt{\tau} f(\tau, x)$. Then, (3.3) is equivalent to

$$\Phi(\lambda, x) = \int_{-\infty}^{\lambda} e^{-(a-1/2)(\lambda-\mu)} F(\mu, x)\, d\mu. \tag{3.4}$$

By applying Hausdorff–Young's inequality to (3.4), we obtain $\|\Phi\| \leq (2/(2a-1))\|F\|$. This proves (3.2), since $\|\Phi\| = \|\varphi\|$, $\|F\| = \|f\| = \|(t\partial_t + a)\varphi\|$ and $\|t\partial_t\varphi\| \leq \|(t\partial_t + a)\varphi\| + a\|\varphi\|$ hold. Q.E.D.

By $S^p([0, T])$, $p \in \mathbf{R}$, we denote the set of all functions $a(t, x, \xi) \in C^\infty([0, T] \times \mathbf{R}^n \times \mathbf{R}^n)$ satisfying

$$\sup_{[0,T]\times\mathbf{R}^n\times\mathbf{R}^n} \{(1+|\xi|)^{-p+|\beta|}|\partial_t^i \partial_x^\alpha \partial_\xi^\beta a(t, x, \xi)|\} < +\infty$$

for any $i \in \mathbf{Z}_+$ and $\alpha, \beta \in \mathbf{Z}_+^n$. By $\mathcal{S}^p([0, T])$, we denote the class of pseudo-differential operators $a(t, x, D_x)$ corresponding to $a(t, x, \xi) \in S^p([0, T])$. By $S^p([0, T], k \times k)$ (resp. $\mathcal{S}^p([0, T], k \times k)$), we denote the set of all $k \times k$ matrices with components in $S^p([0, T])$ (resp. $\mathcal{S}^p([0, T])$).

For $\lambda(t, x, \xi) \in S^1([0, T])$, we put the following conditions (H) or (E).

(H) $\operatorname{Im}(\lambda(t, x, \xi)) = 0$ holds on $[0, T] \times \mathbf{R}^n \times \{|\xi| \geq 1\}$.

(E) There is a $c > 0$ such that $|\operatorname{Im}(\lambda(t, x, \xi))| \geq c|\xi|$ holds on $[0, T] \times \mathbf{R}^n \times \{|\xi| \geq 1\}$.

Then, we have

Lemma 2. *Let $\lambda(t, x, \xi) \in S^1([0, T])$ and let us consider*

$$J_s = -t\partial_t - s + \sqrt{-1}\, t\lambda(t, x, D_x) \qquad (s \in \mathbf{R})$$

on $[0, T] \times \mathbf{R}^n$. Then, if $\lambda(t, x, \xi)$ satisfies (H) *or* (E), *there are $\delta > 0$, $s_0 > 0$ and $T_0 > 0$ such that the estimate*

$$\|J_s \varphi\|_{L^2}^2 \geq \delta s \|\varphi\|_{L^2}^2$$

holds for any $\varphi \in C_0^\infty((0, T_0) \times \mathbf{R}^n)$ and $s > s_0$.

Lemma 3. *Let $k \in \mathbf{N}$, $A(t, x, D_x) \in \mathcal{S}^1([0, T], k \times k)$, $B(t, x, D_x) \in \mathcal{S}^0([0, T], k \times k)$ and let us consider*

$$L_s = -t\partial_t - s + \sqrt{-1}\, tA(t, x, D_x) + B(t, x, D_x) \qquad (s \in \mathbf{R})$$

on $[0, T] \times \mathbf{R}^n$. Assume that there are $D(t, x, D_x) \in \mathcal{S}^1([0, T], k \times k)$ and $N(t, x, D_x)$, $M(t, x, D_x) \in \mathcal{S}^0([0, T], k \times k)$ which satisfy the following conditions: (i) $D(t, x, \xi)$ is a diagonal matrix with components satisfying (H) *or* (E), *(ii) $N(t, x, D_x)A(t, x, D_x) - D(t, x, D_x)N(t, x, D_x) \in \mathcal{S}^0([0, T], k \times k)$, and (iii) $M(t, x, D_x)N(t, x, D_x) - I_k \in \mathcal{S}^{-1}([0, T], k \times k)$. Then, there are $\delta > 0$, $s_0 > 0$ and $T_0 > 0$ such that the estimate*

$$\|L_s \varphi\|_{L^2}^2 \geq \delta s \|\varphi\|_{L^2}^2$$

holds for any $\varphi \in C_0^\infty((0, T_0) \times \mathbf{R}^n)^k$ and $s > s_0$.

Lemma 2 is essentially contained in Uryu [15, Proposition 3.1]. Lemma 3 is verified by applying Lemma 2 to the diagonalizable system L_s (modulo $\mathcal{S}^0([0, T], k \times k)$).

Now, let us return to the situation in Proposition 3. Let P be the operator in (1.1) and let P_{-s} be as in (2.7). Then, without loss of generality we may assume that $(P_{-s})^*$ is expressed in the form

$$(P_{-s})^* = (-t\partial_t - s)^m + \sum_{\substack{j+|\alpha| \le m \\ j<m}} a_{j,\alpha}(t, x)(-t\partial_t - s)^j(-t\partial_x)^\alpha, \tag{3.5}$$

that the coefficients $a_{j,\alpha}(t, x)$ $(j+|\alpha| \le m$ and $j<m)$ belong to $C^\infty([-T, T] \times \mathbf{R}^n)$ and are constant outside a small neighbourhood of $(0, 0)$ in $[-T, T] \times \mathbf{R}^n$, and that $\mathring{p}(t, x, \tau, \xi)$ in (1.4) is given by

$$\mathring{p}(t, x, \tau, \xi) = \tau^m + \sum_{\substack{j+|\alpha|=m \\ j<m}} a_{j,\alpha}(t, x)\tau^j \xi^\alpha.$$

Moreover, from (A-1) and (A-2) we may assume the following: (a-1) $a_{j,\alpha}(t, x)$ $(j+|\alpha|=m$ and $j<m)$ are real-valued, and (a-2) for any $(t, x, \xi) \in [-T, T] \times \mathbf{R}^n \times (\mathbf{R}^n \setminus \{0\})$ the equation $\mathring{p}(t, x, \lambda, \xi)=0$ (in $\lambda \in \mathbf{C}$) has only simple roots.

Let Γ be a proper open convex cone in $\mathbf{R}^n \setminus \{0\}$, and let $\lambda_i(t, x, \xi)$ $(i=1, \ldots, m)$ be the roots of $\mathring{p}(t, x, \lambda, \xi)=0$ (in $\lambda \in \mathbf{C}$) for $(t, x, \xi) \in [-T, T] \times \mathbf{R}^n \times \Gamma$. Then, by the above setting we have the following conditions: $(b-1)$ $\lambda_i(t, x, \xi) \in C^\infty([-T, T] \times \mathbf{R}^n \times \Gamma)$ $(i=1, \ldots, m)$ are well-defined, (b-2) there is a $c>0$ such that

$$|\lambda_i(t, x, \xi) - \lambda_j(t, x, \xi)| \ge c|\xi|$$

for any $1 \le i \ne j \le m$ and $(t, x, \xi) \in [-T, T] \times \mathbf{R}^n \times \Gamma$, and (b-3) each $\lambda_i(t, x, \xi)$ satisfies (H) or (E) on $[-T, T] \times \mathbf{R}^n \times \Gamma$. (Note the following fact: when $n \ne 2$, we may take $\Gamma = \mathbf{R}^n \setminus \{0\}$; but when $n=2$, we cannot take $\Gamma = \mathbf{R}^n \setminus \{0\}$, since in general $\lambda_i(t, x, \xi)$ $(i=1, \ldots, m)$ are multi-valued functions on $[-T, T] \times \mathbf{R}^n \times (\mathbf{R}^n \setminus \{0\})$.)

Anyhow, in this setting we have

Proposition 4. *Let $(P_{-s})^*$ be as in (3.5). Assume the conditions posed above. Then there are $\delta>0$, $s_0>0$ and $T_0>0$ such that the estimate*

$$\|(P_{-s})^*\varphi\|_{L^2}^2 \ge \delta s \sum_{j+|\alpha| \le m-1} \|(t\partial_t + s)^j (t\partial_x)^\alpha \varphi\|_{L^2}^2 \tag{3.6}$$

holds for any $\varphi \in C_0^\infty((0, T_0) \times \mathbf{R}^n)$ and $s>s_0$.

Proof. Let $\varphi \in C_0^\infty((0, T) \times \mathbf{R}^n)$, let Λ be the pseudo-differential operator with symbol $(1+|\xi|^2)^{1/2}$, and put

$$u_j^{(i)} = (-t\partial_t - s)^{i-1}(\sqrt{-1}\,t\Lambda)^{j-i}\varphi$$

for $1 \le i \le j \le m$. Put $k = m(m+1)/2$ and define k-vectors U, F by $U = {}^t(u_1^{(1)}, u_2^{(1)}, u_2^{(2)}, \ldots, u_m^{(1)}, \ldots, u_m^{(m)})$, $F = {}^t(0, \ldots, 0, (P_{-s})^*\varphi)$. Then, by a calculation we can get

$$(-t\partial_t - s + \sqrt{-1}\,tA(t, x, D_x) + B(t, x, D_x))U = F \tag{3.7}$$

for some $A(t, x, D_x) \in \mathcal{S}^1([0, T], k \times k)$ and $B(t, x, D_x) \in \mathcal{S}^0([0, T], k \times k)$ such that $A(t, x, D_x)$ has the form

$$A(t, x, D_x) = \begin{bmatrix} 0 & \vdots & 0 \\ \cdots & \cdots & \cdots \\ 0 & \vdots & A_0(t, x, D_x) \end{bmatrix}$$

for some $A_0(t, x, D_x) \in \mathcal{S}^1([0, T], m \times m)$, that

$$\det(\lambda I_m - A_0(t, x, \xi)) = \mathring{p}(t, x, \lambda, \xi)$$

holds for any $\lambda \in \mathbf{C}$ and $(t, x, \xi) \in [0, T] \times \mathbf{R}^n \times \mathbf{R}^n$, and therefore the eigenvalues $\lambda_i(t, x, \xi)$ $(i = 1, \ldots, m)$ of $A_0(t, x, \xi)$ satisfy (b-1)–(b-3).

When $\lambda_i(t, x, \xi)$ $(i = 1, \ldots, m)$ are well-defined as single-valued functions on $[0, T] \times \mathbf{R}^n \times (\mathbf{R}^n \backslash \{0\})$ and when $n \ne 3$, by the standard method we can construct $D(t, x, D_x) \in \mathcal{S}^1([0, T], k \times k)$ and $N(t, x, D_x)$, $M(t, x, D_x) \in \mathcal{S}^0([0, T], k \times k)$ so that the conditions (i)–(iii) in Lemma 3 are satisfied. Therefore, by applying Lemma 3 to (3.7) we obtain the following: there are $\delta > 0$, $s_0 > 0$ and $T_0 > 0$ such that if $\operatorname{supp}(U) \subset (0, T_0) \times \mathbf{R}^n$ we have

$$\|F\|_{L^2}^2 \ge \delta s \|U\|_{L^2}^2 \tag{3.8}$$

for any $s > s_0$. This implies Proposition 4.

To prove the general case, we need a microlocalization. Let Γ be a proper open convex cone in $\mathbf{R}^n \backslash \{0\}$. Since (b-1)–(b-3) holds on $[0, T] \times \mathbf{R}^n \times \Gamma$, we can construct $d(t, x, \xi)$, $n(t, x, \xi)$, $m(t, x, \xi) \in C^\infty([0, T] \times \mathbf{R}^n \times \Gamma, k \times k)$ so that the following conditions (0)–(iii) are satisfied on $[0, T] \times \mathbf{R}^n \times \Gamma$: (0) $d(t, x, \xi)$, $n(t, x, \xi)$ and $m(t, x, \xi)$ are positively homogeneous of degree 1, 0 and 0 in ξ, (i) $d(t, x, \xi)$ is a diagonal matrix with components satisfying (H) or (E), (ii) $n(t, x, \xi)A(t, x, \xi) = d(t, x, \xi)n(t, x, \xi)$, and (iii) $m(t, x, \xi)n(t, x, \xi) = I_k$. Hence, by combining Lemma 3 with the method of microlocalization in Mizohata [9], we can prove (3.8) also in the general case.

Q.E.D.

Corollary. *Under the situation in Proposition* 4 *we have the following: there are* $\delta_k > 0$ $(k \in \mathbf{Z}_+)$, $s_k > 0$ $(k \in \mathbf{Z}_+)$ *and* $T_0 > 0$ *such that for any* $k \in \mathbf{Z}_+$ *the estimate*

$$\sum_{i+|\beta| \le k} \|(t\partial_t)^i (t\partial_x)^\beta (P_{-s})^* \varphi\|_{L^2}^2 \ge \delta_k s \sum_{j+|\alpha| \le k+m-1} \|(t\partial_t)^j (t\partial_x)^\alpha \varphi\|_{L^2}^2 \tag{3.9}$$

holds for any $\varphi \in C_0^\infty((0, T_0) \times \mathbf{R}^n)$ *and* $s > s_k$.

Proof. When $k=0$, (3.9) is clear from (3.6) and Lemma 1. Since T_0 in Proposition 4 depends only on $\mathring{p}(t, x, \tau, \xi)$, we can also prove (3.9) for $k \geq 1$ in the same way. Q.E.D.

Proof of Proposition 3. Note that for any $p \in \mathbf{Z}_+$ there are $c_1 > 0$ and $c_2 > 0$ such that

$$c_1 \| t^p \psi \|_{H^p}^2 \leq \sum_{j+|\alpha| \leq p} \| (t\partial_t)^j (t\partial_x)^\alpha \psi \|_{L^2}^2 \leq c_2 \| t^p \psi \|_{H^p}^2 \tag{3.10}$$

holds for any $\psi \in C_0^\infty((0, T_0) \times \mathbf{R}^n)$. In fact, the first inequality is clear and the second inequality is verified by the following: if $i + |\alpha| = p$, by Lemma 1 we have for some $c > 0$

$$\begin{aligned} \| \partial_t^i \partial_x^\alpha (t^p \psi) \|_{L^2}^2 &= \| (t\partial_t + 1) \cdots (t\partial_t + i)(t\partial_x)^\alpha \psi \|_{L^2}^2 \\ &\geq c \sum_{j \leq i} \| (t\partial_t)^j (t\partial_x)^\alpha \psi \|_{L^2}^2 . \end{aligned}$$

Hence, by (3.9) and (3.10) we obtain

$$\| t^k (P_{-s})^* \psi \|_{H^k}^2 \geq \mu_k s \| t^{k+m-1} \psi \|_{H^{k+m-1}}^2$$

for some $\mu_k > 0$. Since $t^k (P_{-s})^* \psi = (P_{-s+k})^* (t^k \psi)$, by putting $\varphi = t^k \psi$ we obtain

$$\| (P_{-s+k})^* \varphi \|_{H^k}^2 \geq \mu_k s \| t^{m-1} \varphi \|_{H^{k+m-1}}^2 .$$

This proves Proposition 3. Q.E.D.

4. Generalization

Let us give here a generalization. Let P be of the form

$$P = (t\partial_t)^m + \sum_{\substack{j+|\alpha| \leq m \\ j<m}} a_{j,\alpha}(t, x)(t\partial_t)^j (t^k \partial_x)^\alpha, \tag{4.1}$$

where $k \in \mathbf{N}$, $(t^k \partial_x) = (t^k \partial/\partial x_1, \ldots, t^k \partial/\partial x_n)$, $(t^k \partial_x)^\alpha = t^{k|\alpha|} \partial_x^\alpha$, and $a_{j,\alpha}(t, x)$ ($j + |\alpha| \leq m$ and $j < m$) are C^∞ functions near the origin. Define the reduced principal symbol $\mathring{p}(t, x, \tau, \xi)$ of P by

$$\mathring{p}(t, x, \tau, \xi) = \tau^m + \sum_{\substack{j+|\alpha| = m \\ j<m}} a_{j,\alpha}(t, x) \tau^j \xi^\alpha, \tag{4.2}$$

and define $\rho_1(x), \ldots, \rho_m(x)$ as before. Then, we have

Theorem 3. *Let P be as in* (4.1). *Assume that $\mathring{p}(t, x, \tau, \xi)$ in* (4.2) *satisfies* (A-1) *and* (A-2), *and that $\rho_1(0), \ldots, \rho_m(0) \notin \{-1, -2, \ldots\}$ holds. Then $Pu = f$ is locally solvable in $\mathscr{D}_0'$.*

Proof. Since Proposition 1 is clear, it is sufficient to prove Proposition 2. Let R be the transformation of P by the change of variables: $t^k \to t$ and $x \to x$. Then, R has the form (1.1) and every discussion concerning Proposition 2 in Sections 2 and 3 is valid for R (note that the coefficients of R have the form $a(t^{1/k}, x)$). Therefore, we can conclude that the map

$$R : \mathscr{D}'_{\text{ext}}(\pm) \to \mathscr{D}'_{\text{ext}}(\pm)$$

is surjective. This means that Proposition 2 is valid also for P in (4.1), since $\mathscr{D}'_{\text{ext}}(\pm)$ is preserved under the change of variables: $t^{1/k} \to t$ and $x \to x$ (by Lemma 4 given below). Q.E.D.

Lemma 4. *Let $\lambda > 0$. Then $\mathscr{D}'_{\text{ext}}(+)$ is preserved under the change of variables $t \to t^\lambda$ and $x \to x$. In other words, if $u(t, x) \in \mathscr{D}'_{\text{ext}}(+)$ holds, then we have $u(t^\lambda, x) \in \mathscr{D}'_{\text{ext}}(+)$ as a distribution in (t, x) (not in (t^λ, x)).*

Proof. Let $\lambda > 0$, $u(t, x) \in \mathscr{D}'(\mathbf{R}_+ \times \mathbf{R}^n)$ (where $\mathbf{R}_+ = \{t \in \mathbf{R};\ t > 0\}$) and put $w(t, x) = u(t^\lambda, x) \in \mathscr{D}'(\mathbf{R}_+ \times \mathbf{R}^n)$. Assume that there is a $U(t, x) \in \mathscr{D}'(\mathbf{R} \times \mathbf{R}^n)$ such that $U = u$ on $\mathbf{R}_+ \times \mathbf{R}^n$. Then, we have

$$\langle w, \varphi \rangle = \langle U, \gamma[\varphi] \rangle, \qquad \varphi \in C_0^\infty(\mathbf{R}_+ \times \mathbf{R}^n), \tag{4.3}$$

where $\gamma[\varphi](t, x) = (1/\lambda) t^{(1-\lambda)/\lambda} \varphi(t^{1/\lambda}, x) (\in C_0^\infty(\mathbf{R}_+ \times \mathbf{R}^n))$.

Let Z be the subspace of the linear topological space $\mathscr{D}(\mathbf{R} \times \mathbf{R}^n)$ which coincides with $C_0^\infty(\mathbf{R}_+ \times \mathbf{R}^n)$ as a set. Then, we can see that $\varphi_j \to 0$ in Z (as $j \to \infty$) implies $\gamma[\varphi_j] \to 0$ in Z (as $j \to \infty$). Since U is continuous on Z, by (4.3) we can conclude that w is also continuous on Z. Hence, by the Hahn–Banach theorem we can extend w as a continuous linear functional on $\mathscr{D}(\mathbf{R} \times \mathbf{R}^n)$. This means that there is a $W \in \mathscr{D}'(\mathbf{R} \times \mathbf{R}^n)$ such that $W = w$ on $\mathbf{R}_+ \times \mathbf{R}^n$. Q.E.D.

Remark 2. In the case $k \in \mathbf{Q}$ and $k > 0$, we can also see the following: Theorem 3 is valid for P in (4.1) with $k \in \mathbf{Q}$ and $k > 0$, if the coefficients $a_{j,\alpha}(t, x)$ satisfy $a_{j,\alpha}(t, x) \in C^0(\mathbf{R} \times \mathbf{R}^n)$ and $t^{k|\alpha|} a_{j,\alpha}(t, x) \in C^\infty(\mathbf{R} \times \mathbf{R}^n)$.

Example. Let P be of the form

$$P = (t\partial_t)^2 \pm t^h \sum_{i,j=1}^n a_{ij}(t, x) \partial_{x_i} \partial_{x_j} + a(t, x)(t\partial_t) + t^l \sum_{i=1}^n b_i(t, x) \partial_{x_i} + c(t, x),$$

where $h, l \in \mathbf{N}$, $l \geq h/2$, $a_{ij}(t, x), a(t, x), b_i(t, x), c(t, x) \in C^\infty(\mathbf{R} \times \mathbf{R}^n)$, $a_{ij}(t, x)$ are real-valued, and $\sum_{i,j=1}^n a_{ij}(0, 0)\xi_i \xi_j > 0$ (for $\xi \in \mathbf{R}^n \backslash \{0\}$). Then, by putting

$k=h/2$ and $C(\rho, x)=\rho^2+a(0, x)\rho+c(0, x)$ we can apply our result: Theorem 3 in the case $h/2\in\mathbf{N}$ and Remark 2 in the case $h/2\in\mathbf{N}+(1/2)$. (This operator is discussed in Parenti-Tahara [10].)

References

[1] S. Alinhac, Systèmes hyperboliques singuliers. *Asterisque* **19** (1974) 3-24.

[2] M. S. Baouendi and C. Goulaouic, Cauchy problems with characteristic initial hypersurface. *Comm. Pure Appl. Math.* **26** (1973) 455-475.

[3] A. Bove, J. E. Lewis and C. Parenti, Cauchy problem for Fuchsian hyperbolic operators. *Hokkaido Math. J.* **14** (1985) 175-248.

[4] A. Bove, J. E. Lewis and C. Parenti, Structure properties of solutions of some Fuchsian hyperbolic equations. *Math. Ann.* **273** (1986) 553-571.

[5] A. Bove, J. E. Lewis, C. Parenti and H. Tahara, Cauchy problem for Fuchsian hyperbolic operators, II. *J. Fac. Sci. Univ., Tokyo Sect. IA Math.* **34** (1987) 127-157.

[6] N. Hanges, Parametrices and local solvability for a class of singular hyperbolic operators. *Comm. in PDE* **3** (1978) 105-152.

[7] Y. Hasegawa, On the initial-value problems with data on a characteristic hypersurface. *J. Math. Kyoto Univ.* **13** (1973) 579-593.

[8] S. Itoh and H. Uryu, Conditions for well-posedness in Gevrey classes of the Cauchy problems for Fuchsian hyperbolic operators, II.

[9] S. Mizohata, Note sur le traitement par les opérateurs d'intégrale singulière de problème de Cauchy. *J. Math. Soc. Japan* **11** (1959) 234-240.

[10] C. Parenti and H. Tahara, Examples of locally solvable second order Fuchsian operators.

[11] T. Ōaku, A short communication in the meeting "Hyperbolic Equations of Fuchs Type and of Gevrey Classes" held on July 8-11, 1986 at the University of Tokyo (Japan).

[12] H. Tahara, Fuchsian type equations and Fuchsian hyperbolic equations. *Japan. J. Math. New Ser.* **5** (1979) 245-347.

[13] H. Tahara, Singular hyperbolic systems, III. On the Cauchy problem for Fuchsian hyperbolic partial differential equations. *J. Fac. Sci. Univ. Tokyo Sect. IA Math.* **27** (1980) 465-507.

[14] H. Tahara, Singular hyperbolic systems, VI. Asymptotic analysis for Fuchsian hyperbolic equations in Gevrey classes. *J. Math. Soc. Japan* **39** (1987) 551-580.

[15] H. Uryu, Uniqueness for the characteristic Cauchy problem and its application. *Tokyo J. Math.* **5** (1982) 117-136.

[16] H. Uryu, Conditions for well-posedness in Gevrey classes of the Cauchy problems for Fuchsian hyperbolic operators. *Publ. RIMS, Kyoto Univ.* **21** (1985) 355-383.

$\bar{\partial}_b$-Cohomology and the Bochner–Martinelli Kernel

Shinichi Tajima
Faculty of General Education
Niigata University
Niigata, Japan

Let Ω be a domain in $\mathbf{C}^2$ with boundary $b\Omega$, and let j be the natural inclusion map $\Omega \to \mathbf{C}^2$. Let us denote by $\mathcal{O}$ the sheaf on $\mathbf{C}^2$ of germs of holomorphic functions. Setting $F := \mathbf{C}^2 - \Omega$, we have the following isomorphisms for the sheaves of local cohomology with supports in F:

$$\mathscr{H}_F^1(\mathcal{O}) \cong (j_* j^{-1}\mathcal{O})/\mathcal{O},$$

$$\mathscr{H}_F^2(\mathcal{O}) \cong R^1 j_* j^{-1}\mathcal{O}.$$

We first prove the following theorem.

Theorem 1. *In the above situation, we have the following*:

$$\operatorname{supp} \mathscr{H}_F^1(\mathcal{O}) \cup \operatorname{supp} \mathscr{H}_F^S(\mathcal{O}) = b\Omega.$$

Remark. This result is related to the notion of micro-support (see Kashiwara–Schapira [4]).

Algebraic Analysis, Volume II

ISBN 0-12-400466-0

Proof. Let P be a boundary point, and let (z_1, z_2) be complex coordinates. For the sake of simplicity we assume that $P=(0,0)$. Let $U_1=\{(z_1, z_2);\ z_1\neq 0\}$, $U_2=\{(z_1, z_2);\ z_2\neq 0\}$. We set

$$B=\frac{\bar{z}_1\, d\bar{z}_2-\bar{z}_2\, d\bar{z}_1}{(z_1\bar{z}_1+z_2\bar{z}_2)^2},$$

$$b_1=\frac{\bar{z}_2}{z_1(z_1\bar{z}_1+z_2\bar{z}_2)},$$

$$b_2=\frac{-\bar{z}_1}{z_2(z_1\bar{z}_1+z_2\bar{z}_2)}.$$

Calculation gives

$$\bar{\partial} b_1=B \qquad \text{on } U_1, \qquad \bar{\partial} b_2=B \qquad \text{on } U_2,$$

and

$$b_1-b_2=\frac{1}{z_1z_2} \qquad \text{on } U_1\cap U_2.$$

Hence the Bochner–Martinelli form B is a closed $(0,1)$ form on $\mathbf{C}^2-P$, which corresponds to the Cech cohomology class $(1/z_1z_2)\in H^1(\mathbf{C}^2-P, \mathcal{O})\cong H^2_P(\mathcal{O})$. Restricting B to Ω, we see that

$$B\in(R^1j_*j^{-1}\mathcal{O})_p\cong\mathscr{H}^2_F(\mathcal{O})_P.$$

Assume

$$\text{(A)} \quad \mathscr{H}^1_F(\mathcal{O})_P=\mathscr{H}^2_F(\mathcal{O})_P=0.$$

By the assumption (A) there exist a neighbourhood U of P and a hyperfunction α defined on $\Omega\cap U$, which satisfies $\bar{\partial}\alpha=B$ on $\Omega\cap U$. Setting $h_1:=z_1\alpha-z_1b_1$, $h_2:=z_2\alpha-z_2b_2$, we have

$$\bar{\partial}h_1=\bar{\partial}h_2=0 \qquad \text{on } \Omega\cap U.$$

By the assumption (A) it follows that h_1 and h_2 extend uniquely as holomorphic functions, also denoted by h_1 and h_2, to some neighbourhood V of P.

We set

$$g(z_1, z_2)=\begin{cases}\dfrac{h_1}{z_1}+b_1 & \text{for } (z_1, z_2)\in U_1\cap V,\\[2ex] \dfrac{h_2}{z_2}+b_2 & \text{for } (z_1, z_2)\in U_2\cap V.\end{cases}$$

Lemma. *$g(z_1, z_2)$ is a well-defined function on $V-P$.*

Proof. By definition we get

$$\bar{\partial}\left\{\left(\frac{h_1}{z_1}+b_1\right)-\left(\frac{h_2}{z_2}+b_2\right)\right\}=0 \qquad \text{on } U_1\cap U_2\cap V.$$

Since $(h_1/z_1+b_1)-(h_2/z_2+b_2)=0$ on the open set $\Omega\cap U_1\cap U_2\cap V$, $(h_1/z_1+b_1)-(h_2/z_2+b_2)$ vanishes identically on $U_1\cap U_2\cap V$. (Q.E.D.

We set

$$h(z_1, z_2)=z_1 g(z_1, z_2)-\frac{\bar{z}_2}{z_1\bar{z}_1+z_2\bar{z}_2}.$$

It follows that $\bar{\partial}h=0$ on $V-P$, so by the assumption (A), h is holomorphic in V. But by definition, we have $h(0, z_2)=-1/z_2$. Hence h is not bounded in any neighbourhood of P. This is a contradiction. Q.E.D.

We next apply Theorem 1 to the study of the tangential Cauchy-Riemann equations.

Let $\Omega=\{\rho<0\}$ be a domain in $\mathbf{C}^2$ with real analytic boundary N. Here ρ is a real valued real analytic function such that grad $\rho|_N\neq 0$. Let Y be a complexification of N, $S_N^* Y$ be the spherical conormal bundle. We regard $N_\pm=\{(P, \pm\text{grad}\,\rho(P);\, P\in N\}$ as subsets of $S_N^* Y$. Let $\bar{\partial}_b$ be the tangential Cauchy-Riemann operator induced on N. Since the real locus of the characteristic variety of the tangential Cauchy-Riemann equation $\bar{\partial}_b$ can be identified with $S_N^*\mathbf{C}^2=N_+\cup N_-$ it follows that

$$\text{supp}(\text{Ker}(\bar{\partial}_b:\mathscr{C}_N\to\mathscr{C}_N))\cup\text{supp}(\text{Coker}(\bar{\partial}_b:\mathscr{C}_N\to\mathscr{C}_N))\subseteqq N_+\cup N_-,$$

where $\mathscr{C}_N$ is the sheaf on $S_N^* Y$ of microfunctions.

Theorem 2. *In the above situation, we have*

$$\text{supp}(\text{Ker}\,\bar{\partial}_b)|_{N_+}\cup\text{supp}(\text{Coker}\,\bar{\partial}_b)|_{N_+}=N_+.$$

Proof. Recall that the following isomorphisms hold ([3], [7]):

$$\mathscr{H}_F^1(\mathscr{O})\cong\text{Ker}(\bar{\partial}_b:\mathscr{C}_N\to\mathscr{C}_N)|_{N_+}$$

$$\mathscr{H}_F^2(\mathscr{O})\cong\text{Coker}(\bar{\partial}_b:\mathscr{C}_N\to\mathscr{C}_N)|_{N_+}.$$

By Theorem 1 we have

$$\text{supp}\,\mathscr{H}_F^1(\mathscr{O})\cup\text{supp}\,\mathscr{H}_F^2(\mathscr{O})=b\Omega=N.$$

This implies the result. Q.E.D.

If one uses the notion of micro-support ([4]), one can prove the following theorem in the same way.

Theorem 3. *Let Ω be a domain in $\mathbf{C}^n$ with real analytic boundary N, and let Y be a complexification of N. Let $\bar{\partial}_b$ be the tangential Cauchy-Riemann system. Then we have*

$$\operatorname{supp}(R\mathcal{H}om_{\mathcal{D}_Y}(\bar{\partial}_b, \mathcal{C}_N)|_{N_+}) = N_+,$$

where $\mathcal{D}_Y$ is the sheaf of holomorphic differential operators on Y.

References

[1] A. Andreotti, G. Fredericks, and M Nacinovich, On the absence of Poincaré lemma in tangential Cauchy-Riemann complexes. *Annali Scuola Norm. Sup. Pisa*, Ser IV **8** No. 3 (1981) 365-404.

[2] G. M. Henkin and J. Leiterer, *Theory of Functions on Complex Manifolds.* Birkhaüser, Boston, 1984.

[3] M. Kashiwara and T. Kawai, On the boundary value problem for elliptic system of linear differential equations, I. *Proc. Japan Acad.* **48** (1972) 712-715.

[4] M. Kashiwara and P. Schapira, Microlocal Study of Sheaves. *Astérisque* **128** (1985).

[5] N. Øvrelid, Integral representation formulas and L^p-estimates for the $\bar{\partial}$-equation. *Math. Scand.* **29** (1971) 137-160.

[6] M. Sato, T. Kawai, and M. Kashiwara, Microfunctions and pseudo-differential equations. *Lecture Notes in Math.*, **287** (1973) 265-529.

[7] S. Tajima, Analyse microlocale sur les variétés de Cauchy-Riemann et problème du prolongement des solutions holomorphes des équations aux dérivées partielles. *Publ. RIMS, Kyoto Univ.* **18** (1982) 911-945.

[8] S. Tajima, CR-microfunctions and the Henkin-Ramirez reproducing kernel. *Proc. Japan Acad.* **61** No. 5 (1985) 137-139.

Issues of Multi-Dimensional Integrable Systems

K. Takasaki
Research Institute of Mathematical Sciences
Kyoto University
Kyoto, Japan

0. Introduction

At present it is a very hard task to give an exact definition to the notion of integrability or integrable system. What we have in hand now is rather an enormous list of examples accumulated for many years of intensive studies from both the physical and mathematical sides (cf. Jimbo, Miwa [1] and references cited therein). Most of this list is occupied by the so-called soliton equations, which are so named because of their origin in soliton phenomena. Mathematically, soliton equations describe nonlinear waves propagating in one-dimensional space like a canal. Even the KP (Kadomtsev–Petviashvili) equation should be considered as such (Sato and Sato [2]) though physically it was introduced as a two-space-dimensional generalisation of the KdV (Korteweg–de Vries) equation. A natural question coming then, would be whether there are multi-dimensional analogues of soliton equations. Frankly speaking, naive attempts at such a generalisation have almost all failed up to now; they could at best just (re)produce a new type of soliton equations.

ISBN 0-12-400466-0

A few examples of what can be really called multi-dimensional integrable systems have been discovered from a somewhat distinct point of view, that is, twistor theory. The equations of motion of self-dual connections (Yang-Mills fields) [3], self-dual metrics (gravitational fields) [4] and their extensions to dimensions greater than four (e.g., hyper-Kähler versions) [5, 6] provide such examples. Of course this assertion would be meaningless unless the notion of integrability is made more definite, though regrettably an ultimate definition of integrability with mathematical rigor still lies far beyond our scope. A series of my work since 1983 [7] has aimed at the analysis of in what sense the nonlinear systems mentioned above are integrable, and how that can be understood from a more general principle. This article presents an intermediate summary of this research, including current interest and some outlook towards the future.

1. Self-Dual Connections, Linear System, and Twistors

From here through Section 5 we examine self-dual connections from various aspects. The integrability of the equations of motion of self-dual connections becomes most manifest in a complexified setting. Let $\mathbf{x} = (y, z, \bar{y}, \bar{z})$ be a set of complex coordinates in $\mathbf{C}^4$ with complex metric $ds^2 = dy\, d\bar{y} + dz\, d\bar{z}$ and $\nabla = d + A$, $A = A_y\, dy + A_z\, dz + A_{\bar{y}}\, d\bar{y} + A_{\bar{z}}\, d\bar{z}$, a connection with a complex structure group, say, $GL(r, \mathbf{C})$. Its coordinate components are $\nabla_u = \partial_u + A_u$, $u \in \mathbf{x}$, the A_u's being the so-called gauge potentials that play the role of unknown functions. The curvature form $F = dA + A \wedge A$ has six components $F_{uv} = [\partial_u + A_u, \partial_v + A_v]$, $u, v \in \mathbf{x}$. The equations of motion of self-dual connections are

$$F_{yz} = 0, \qquad F_{\bar{y}\bar{z}} = 0, \qquad F_{y\bar{y}} + F_{z\bar{z}} = 0.$$

With the aid of an auxiliary parameter λ (called a *spectral parameter* after the terminology of inverse scattering theory) these equations are written more compactly as

$$[-\lambda \nabla_y + \nabla_{\bar{z}}, \lambda \nabla_z + \nabla_{\bar{y}}] = 0.$$

This nonlinear system, the so-called zero-curvature representation, is the integrability condition in the sense of Frobenius of the following linear system (Belavin, Zakharov [8], Pohlmeyer [9], Chau, Prasad, Sinha [10]):

$$(-\lambda \nabla_y + \nabla_{\bar{z}}) W = 0, \qquad (\lambda \nabla_z + \nabla_{\bar{y}}) W = 0,$$

where $W = W(\mathbf{x}, \lambda)$ is an unknown function taking values in $GL(r, \mathbf{C})$. Another way of stating this fact is by using the Pfaffian system of (cf. Gindikin [11])

$$dW + AW = 0,$$

$$dy + \lambda\, d\bar{z} = 0, \qquad dy + \lambda\, d\bar{z} = 0, \qquad d\lambda = 0,$$

in $GL(r, \mathbf{C}) \times \mathbf{C}^4 \times \mathbf{P}^1$, $(W, \mathbf{x}, \lambda)$ being coordinates in this manifold. Its integrability condition can be written

$$(dA + A \wedge A) \wedge (dy + \lambda\, d\bar{z}) \wedge (dz - \lambda\, d\bar{y}) \wedge d\lambda = 0,$$

giving another equivalent expression of the previous zero-curvature representation. The last three equations of the above Pfaffian system form a Pfaffian system integrable in itself, whose first integrals $y + \lambda\bar{z}$, $z - \lambda\bar{y}$, λ may be thought of as taking values in the twistor space $\mathbf{P}^3$. (As we shall see later, twistor theory of self-dual metrics is nothing other than an extension of this picture into curved manifolds.) With this correspondence, self-dual connections can be described in terms of vector bundles on the twistor space. This story has now become very familiar to us (cf. Atiyah [12] and references therein).

2. Riemann–Hilbert Transformations

Riemann–Hilbert transformations (RHT) had been studied for the case of soliton equations, but it was Ueno and Nakamura [13] who first introduced them to the case of self-dual connections. We shall not repeat the construction here, and refer details to their original paper, but the reason why such transformations exist may be understood without difficulty from the following heuristic argument.

Given a self-dual connection ∇, suppose there are two solutions $W = W(\mathbf{x}, \lambda)$, $\hat{W} = \hat{W}(\mathbf{x}, \lambda)$ of the linear system which are holomorphic functions of $(\mathbf{x}, \lambda)$ but with different domains of definition with respect to λ, D and $\hat{D}$, which are discs covering $\mathbf{P}^1$ and centered at respectively $\lambda = \infty(D)$ and $\lambda = 0(\hat{D})$. Then evidently the product $g = W^{-1}\hat{W}$ satisfies

$$(-\lambda\partial_y + \partial_{\bar{z}})g = 0, \qquad (\lambda\partial_z + \partial_{\bar{y}})g = 0,$$

which means that g is a function (with values in $GL(r, \mathbf{C})$) of $(y + \lambda\bar{z}, z - \lambda\bar{y}, \lambda)$, λ running over the anulus $D \cap \hat{D}$. Thus naturally arise the three variables of the twistor picture. In fact this g can be used as a transition function to construct a vector bundle on the twistor space, and this is exactly the correspondence of self-dual connections and vector bundles mentioned above. The inverse correspondence is achieved by factorising a given $GL(r, \mathbf{C})$-valued holomorphic function g of $(y+\lambda\bar{z}, z-\lambda\bar{y}, \lambda)$, $\lambda \in D \cap \hat{G}$, into the product of two functions as

$$g = g(y+\lambda\bar{z}, z-\lambda\bar{y}, \lambda) = W(\mathbf{x}, \lambda)^{-1}\hat{W}(\mathbf{x}, \lambda),$$

where W and $\hat{W}$ are required to have the same analyticity properties as above. It is indeed not difficult to see that the W and $\hat{W}$ thus obtained from g do satisfy the linear system of self-dual connections for an appropriate choice of A. Strictly speaking, there are some cases where the above argument breaks down because of some topological reasons, but in a generic case this gives a complete description of self-dual connections.

In the above setting, RHT's are nothing other than the left and right multiplication of the above g with $GL(r, \mathbf{C})$-valued functions of the same analytical properties as

$$g \to g_L g g_R^{-1}, \qquad g_{L(R)} = g_{L(R)}(y+\lambda\bar{z}, z-\lambda\bar{y}, \lambda).$$

This action on g by (g_L, g_R) causes a transformation of the triple $(A, W, \hat{W})$ (to be more precise, its gauge equivalence class), thus there are in fact *two* types of RHT's, *left* and *right* (cf. Wu [14]). Note that they mutually commute, that is, the result of the successive action of g_L and g_R does not depend on the order. What Ueno and Nakamura considered were only one of them, say left RHT's. A simplification gained by considering one-sided RHT's is that they can be described in terms of just two components, (A, W) (left case) or $(A, \hat{W})$ (right case), of the triple $(A, W, \hat{W})$. To be more precise, under the action of, say, a left RHT g_L the result of transformation of A and W is independent of $\hat{W}$, so that one can recognise g_L simply as acting on the pair (A, W); just the same is true for g_R and $(A, \hat{W})$. For details and related topics, cf. Takasaki [7, 1985].

3. Cauchy Problem

The following observation is also a variation of the factorisation problem mentioned above, but before my work [7] no one has noticed such a point of view.

We first note that the $GL(r, \mathbf{C})$-valued function g is uniquely determined by the "Cauchy data" $g^{(0)}(y, z, \lambda) = g(\bar{y} = \bar{z} = 0)$. The linear system of g

with the initial condition $g(\bar{y}=\bar{z}=0)=g^{(0)}$ can be solved as

$$g=\exp(\bar{z}\lambda\partial_y-\bar{y}\lambda\partial_z)g^{(0)}.$$

Here evidently $(\bar{y}, \bar{z})$ play the role of "time" variables, and (y, z) that of "space" variables, though this has nothing to do with the original physical interpretation of these variables.

What about the W? To make clear this point, one has to "fix the gauge" as $W(\lambda=\infty)=1$. Then from the linear system of W the gauge potentials are fixed as

$$A_y=0, \qquad A_z=0, \qquad A_{\bar{y}}=-\partial_z W_1, \qquad A_{\bar{z}}=\partial_y W_1,$$

where W_1 denotes the next-to-leading coefficient of the Laurent expansion of W around $\lambda=\infty$, $W=1+\sum_{n\geq 1} W_n\lambda^{-n}$, $W_n=W_n(\mathbf{x})$. Taking back these expressions of the gauge potentials, one finds that the linear system for W becomes now a nonlinear system for the W_n's, which read

$$-\partial_y W_n+\partial_{\bar{z}} W_{n+1}+(\partial_y W_1)W_n=0,$$

$$\partial_z W_n+\partial_{\bar{y}} W_{n+1}-(\partial_z W_1)W_n=0.$$

With a simple argument one can check that each solution (for example, holomorphic) of this system is uniquely determined by the "Cauchy data" $W_n^{(0)}=W_n(\bar{y}=\bar{z}=0)$ and, besides, the Cauchy data can be given arbitrarily, thus providing a set of good "functional coordinates" in the solution space of the above nonlinear system.

Recovering W from $W^{(0)}$ is not as simple as in the case of g, but can be reduced to a factorisation problem of the same nature as before. Eliminating g from the previous relations, one can indeed deduce the relation

$$\exp(\bar{z}\lambda\partial_y-\bar{y}\lambda\partial_z)W^{(0)-1}=W^{-1}V,$$

where $V=\hat{W}\cdot\exp(\bar{z}\lambda\partial_y-\bar{y}\lambda\partial_z)\hat{W}(\bar{y}=\bar{z}=0)^{-1}$. The explicit form of V is irrelevant; the point is that $V=V(\mathbf{x},\lambda)$ is holomorphic with respect to λ in $\hat{D}$ (at least for sufficiently small values of $\bar{y}$ and $\bar{z}$). Thus we encounter another factorisation problem, which describes the "time evolution" from $W^{(0)}$ to W.

The point of view of the Cauchy problem, though it appears fairly artificial at first glance, is one of the indispensable elements of the method developed in [7]. Without this, any comparison with soliton equations will remain at a superficial level. We shall see some other aspects of this in the following.

4. Grassmann Manifold

The goal set up in [7, 1983–85] was to give a framework for reformulating self-dual connections à la Sato [2]. An infinite-dimensional Grassmann manifold plays a basic role in Sato's approach, as a parameter (or moduli) space of solutions to soliton equations. As a result, both an infinite hierarchy of time evolutions and a variety of transformations of solutions (Bäcklund transformations etc.) can be understood on an equal footing, as part of the large general linear group $GL(\infty)$ acting on the Grassmann manifold. This, of course, clearly explains why soliton equations are related to infinite-dimensional Lie algebras (cf. Jimbo and Miwa [1]).

A conclusion presented in [7, 1983–85] is that almost the same structure exists in the case of self-dual connections, so that this case, too, may be judged as "integrable." The essence is as follows.

We construct an infinite matrix $\xi=(\xi_{ij})_{i\in\mathbf{Z},j\in\mathbf{N}^c}$ of size $\mathbf{Z}\times\mathbf{N}^c$, $\mathbf{Z}=\{0,\pm 1,\pm 2,\dots\}$, $\mathbf{N}^c=\{-1,-2,\dots\}$, from the Laurent coefficients of W as

$$\xi_{ij}=\sum_{k<0} W^*_{i-k}W_{k-j},$$

where the W^*_n's stand for the Laurent coefficients of W^{-1} in the sense of formal series, $W^{-1}=1+\sum_{n\geq 1} W^*_n\lambda^{-n}$. The entries of ξ then satisfy the algebraic relations

$$\xi_{ij}=\delta_{ij}\ \text{(Kronecker's delta)} \qquad \text{for } i<0, j<0,$$

$$\xi_{i+1,j}=\xi_{i,j-1}+\xi_{i,-1}\xi_{0j} \qquad \text{for } i\in\mathbf{Z}, j<0,$$

which conversely characterise a matrix ξ that can be obtained from some W as above.

Such a matrix ξ of maximal rank, in general, can be thought of as representing a vector subspace V of a fixed larger vector space $\mathbf{V}$ formed by column vectors of size $\mathbf{Z}$; the V being spanned by the columns of ξ. Actually the correspondence between ξ and V is by no means one-to-one; ξ and ξh, where h is an invertible (in an appropriate sense) $\mathbf{N}^c\times\mathbf{N}^c$ matrix, should be considered equivalent to the effect that they correspond to the same V. One can thus freely replace a ξ-matrix by an equivalent one. An extreme choice, allowed as far as the $\mathbf{N}^c\times\mathbf{N}^c$ part $\xi_{(-)}=(\xi_{ij})_{i,j<0}$ of ξ is invertible, is to retake ξ to be such that $\xi_{ij}=\delta_{ij}$ for $i,j<0$, by multiplying $\xi_{(-)}^{-1}$ from the right. The ξ-matrix constructed above from W is exactly of that nature.

In terms of the above ξ, the action of RHT's and the solution of the Cauchy problem discussed in the preceding sections can be represented as follows:

$$g_{\mathrm{L}} \circ \xi = g_L(y+\Lambda\bar{z}, z-\Lambda\bar{y}, \Lambda)\xi \cdot \text{(invertible matrix)},$$

$$\xi = \exp(\bar{z}\Lambda\partial_y - \bar{y}\Lambda\partial_z)\xi^{(0)} \cdot \text{(invertible matrix)},$$

where $g_{\mathrm{L}} \circ \xi$ and $\xi^{(0)}$ denote the ξ-matrices corresponding to, respectively, $g_{\mathrm{L}} \circ W$ (= the action of a left RHT g_{L} on W) and $W^{(0)}$ (= the Cauchy data of W on $\bar{y}=\bar{z}=0$); Λ stands for the shift matrix $(\delta_{i+1,j})_{i,j\in\mathbf{Z}}$ that acts on the entries of, say ξ, as $\Lambda\xi = (\xi_{i+1,j})_{i\in\mathbf{Z},j<0}$. The above relations are not of symbolic nature, but provide a practical solution process [7, 1983–84].

A description of bi-sided RHT's along a similar line is argued in [7, 1985], in which case one has to enlarge the ξ-matrix so as to incorporate both W and $\hat{W}$ in its structure.

The method to Grassmann manifold presented here has also been applied to some other equations by Suzuki [15], Harnad, Jacques [16], Nagatomo [17], and Nakamura [18].

5. Enlarged Groups and Lie Algebras

The content of this section belongs to a research just beginning [19].

As we have seen above, the method of Grassmann manifold provides a framework to unify two seemingly distinct objects, that is, RHT's (= the action of matrix functions $g_{\mathrm{L}}(y+\Lambda\bar{z}, z-\Lambda\bar{y}, \Lambda)$ on ξ, or more preferably that of $g_{\mathrm{L}}(y, z, \Lambda)$ on $\xi^{(0)}$) and time evolution (= the action of operators such as $\exp(\bar{z}\Lambda\partial_y - \bar{y}\Lambda\partial_z)$ on $\xi^{(0)}$). In view of the work of Sato and Sato [2], a natural question would be what kind of groups or Lie algebras lie behind this picture.

To make the situation simpler, we here focus on the infinitesimal version of the above objects and formulate them in an abstract setting.

Let R be a differential ring on which ∂_y and ∂_z act as derivations; for example, $R=\mathbf{C}[[y,z]]$ if one wishes to deal with formal power series solutions. Then the Lie algebra of infinitesimal one-sided (say, left) RHT's can be identified with the formal loop algebra $\mathfrak{gl}(r, R((\lambda^{-1}))) = \mathfrak{gl}(r, \mathbf{C}) \otimes R((\lambda^{-1}))$, where $R((\lambda^{-1})) = \{\sum a_n\lambda^n;\ a_n \in R\ (n\in\mathbf{Z}),\ a_n = 0\ \ (n \gg 0)\}$. This should be supplemented by another Lie algebra including the infinitesimal generators $\lambda\partial_y$ and $\lambda\partial_z$ of time evolutions in $(\bar{y}, \bar{z})$; a candidate would be $R[\lambda^{-1}]\partial_y + R[\lambda^{-1}]\partial_z$. Thus as a Lie algebra relevant to $GL(r, \mathbf{C})$-self-dual connections, one may take the following:

$$\mathfrak{g}_{\mathrm{SDYM}} = \mathfrak{gl}(r, R((\lambda^{-1}))) + R[\lambda^{-1}]\partial_y + R[\lambda^{-1}]\partial_z.$$

This is not a simple Lie algebra in the sense that the first component on the right side forms a nontrivial ideal.

The second and third components include a mutually commutative set of derivations $\lambda^n\partial_y$, $\lambda^n\partial_z$, $n = 1, 2, \ldots$, which generate an infinite hierarchy of simultaneous time evolutions parallel to the case of soliton equations. A similar idea was also presented by Nakamura [18].

Since the Lie algebra $\mathfrak{g}_{\mathrm{SDYM}}$ incorporates these time evolutions in a natural manner, one may expect to develop by use of this an orbit method of the Kostant–Kirillov type. In fact, it appears hard to give a variational formulation, but the argument of Flaschka *et al.* [20] on an abstract Poisson bracket structure described for a class of soliton equations can be extended to the present setting.

Another, more challenging issue would be whether there is a natural extension $\hat{\mathfrak{g}}_{\mathrm{SDYM}}$ of $\mathfrak{g}_{\mathrm{SDYM}}$ with the exact sequence $0 \to \mathfrak{o} \to \hat{\mathfrak{g}}_{\mathrm{SDYM}} \to \mathfrak{g}_{\mathrm{SDYM}} \to 0$, $\mathfrak{o}$ being an abelian ideal (but possibly not central). Such a Lie algebra, if it exists, provides an interesting multi-dimensional analogue of Kac-Moody Lie algebras with central charge.

6. Issue of Multi-Dimensional Spectral Parameters

We now discuss a somewhat delicate subject related to the problem mentioned at the beginning of this article.

The nonlinear system of self-dual connections is by no means an isolated example. What we have seen up to here can be readily extended to a general setting that also covers, for example, the hyper-Kähler version of self-dual connections in $4k$, $k = 1, 2, \ldots$, dimensions included in the table of Ward [5]. A common feature of these nonlinear systems is the presence of an associated linear system with a spectral parameter, λ, which reproduces the relevant nonlinear system as its integrability conditions in the sense of Frobenius. In fact, as we shall see later, this is still insufficient to ensure "integrability" in the same sense as self-dual connections are integrable, but our primary subject in the following is the case where a nonlinear system has an associated linear system with *more than one* spectral parameter.

Several examples of such nonlinear systems with multi-dimensional spectral parameters are already known, all related to gauge theory, which are: (*i*) eight-dimensional gauge fields introduced by Witten [21] and Isenberg *et al.* [22], and discussed from the point of view of Grassmann manifold by Suzuki [15]; (*ii*) constraint equations of super-Yang–Mills fields in four dimensions studied in the context of integrability by Volovich [23], Devchand [24], Chau, Ge, Popowicz [25], Harnad and Jacques [16] on the basis of Witten's twistorial interpretation [21]; (*iii*) super-Yang–Mills fields in ten dimensions by Harnad and Schnider [26]; (*iv*) examples presented by Ward [5], etc. In fact, all the above papers on case (*ii*) are based on a linear system with just one spectral parameter, which is derived through a certain process from another linear system with two spectral parameters that Witten's argument originally suggests (cf. Devchand [27]); this however still inherits a serious problem as we shall see later.

The problem is, *whether the equations above are really integrable.* The answer to this question may depend on what one expects as the contents of integrability. If one requires as a basic ingredient of integrability the presence of a large (e.g., transitive) transformation group or Lie algebra acting on the solution space, there are several evidences suggesting that the above examples are *non-integrable*; at present there is no hope to construct such a group or Lie algebra.

A heuristic argument to support this observation is as follows. According to Ward [5] and Isenberg *et al.* [22], solving these equations is equivalent to finding certain (family of) trivial vector bundles on a complex manifold Σ; spectral parameters are coordinates of Σ. For example, Σ is $\mathbf{P}^1 \times \mathbf{P}^1$ in cases (*i*) and (*ii*), a nonsingular quadratic in $\mathbf{P}^9$ in case (*iii*), and case (*iv*) includes examples with $\Sigma = \mathbf{P}^m$. Such a vector bundle is described by a finite covering $\{U_i\}$ of Σ and a set of patching functions $\{g_{ij};\ U_i \cap U_j \neq \varnothing\}$ with the cocyclic conditions $g_{ij}g_{jk} = g_{ik}$ on nonempty intersections $U_i \cap U_j \cap U_k$. The case of $\dim \Sigma = 1$ (e.g., $\Sigma = \mathbf{P}^1$) is exceptional; one can take the covering in such a way that three distinct patches never intersect, so that there are in fact no cocycle conditions to be imposed on the g_{ij}'s. The set of such g_{ij}'s acquires a structure of a group of componentwise matrix multiplication as $\{g_{ij}\}, \{h_{ij}\} \to \{g_{ij}h_{ij}\}$. This is exactly the origin of RHT's (cf. Section 2). If $\dim \Sigma > 1$, there is no such natural group structure of patching functions retaining cocycle conditions, so probably no analogue of RHT's.

One might, however, still expect a loophole in the direction indicated in [23–25], that is, by means of a linear system with just one spectral parameter derived by putting some relations among multi-dimensional spectral parameters. Regrettably, this leads to no substantial improvement of the situation. The reduced linear system in, for example, case (*ii*) becomes

$$(\nabla_1^i + \lambda \nabla_2^i) W = 0, \qquad (\bar{\nabla}_{1i} + \lambda^2 \bar{\nabla}_{2i}) W = 0,$$

$$(\nabla_1 + \lambda \nabla_2 + \lambda^2 \nabla_3 + \lambda^3 \nabla_4) W = 0,$$

where ∇_1^i, etc., are components of a super-connection; their precise form is irrelevant here. What is crucial is the absence of the λ-term in the second set of equations. Because of this, it turns out that RHT's do not act on this linear system, since the action of a RHT in general produces a new λ-term therein. This difficulty has been also noticed by several people [28].

Of course these observations never reduce the importance of the nonlinear systems mentioned above, their "integrability" however thus being considerably problematical. They should be understood, rather, as defining a subset of the solution space of some other nonlinear systems which *are* integrable.

7. Self-Dual Metrics, or Deformation of Integrable *G*-Structures

We now turn to another, less familiar class of nonlinear systems, which are related to metrics or G-structures. A typical one, in a position to be compared with self-dual connections, is provided by self-dual metrics in four dimensions. Also in this case, a complexified setting is much more convenient than the ordinary Riemannian geometry, in order to see aspects of integrability.

We start from a complex metric ds^2 (= nondegenerate holomorphic bilinear form on the holomorphic tangent bundle) of a four-dimensional complex manifold X; actually we focus on local geometry. In a local coordinate patch, one can take a set of linearly independent 1-forms e_1, e_2, $\bar{e}_1$, $\bar{e}_2$ (the bar does not mean complex conjugation) with which ds^2 can be written

$$ds^2 = e_1 \bar{e}_1 + e_2 \bar{e}_2 .$$

One may recognise this as generalising the representation $ds^2 = dy\,d\bar{y} + dz\,d\bar{z}$ encountered in Section 1. Self-duality is defined, originally, by a set of linear equations among the components of the Riemann curvature R (cf. Eguchi *et al.* [29]) (compare this with the case of self-duality of Yang–Mills connections), a rather cumbersome object. Fortunately there is another way to handle this notion just in terms of the 1-forms $e_1, \ldots, \bar{e}_2$ without referring to R; it is due to the fact (cf. Boyer [30], Gindikin [31]) that ds^2 is self-dual iff after an appropriate linear transformation of $e_1, \ldots, \bar{e}_2$, the following exterior differential equations are satisfied:

$$d(e_1 \wedge e_2) = 0, \qquad d(\bar{e}_1 \wedge \bar{e}_2) = 0, \qquad d(e_1 \wedge \bar{e}_1 + e_2 \wedge \bar{e}_2) = 0.$$

The use of a spectral parameter λ, again, brings a remarkable simplification. The above equations can be gathered up to become the following:

$$\hat{d}((e_1 + \lambda \bar{e}_2) \wedge (e_2 - \lambda \bar{e}_1)) \wedge \hat{d}\lambda = 0,$$

where $\hat{d}$ denotes the total differentiation on $X \times \mathbf{P}^1$, $\hat{d} = d + d_{\mathbf{P}^1}$, introduced so as to distinguish it from $d = d_X$ on X. One may therefore write the above equation as

$$d((e_1 + \lambda \bar{e}_2) \wedge (e_2 - \lambda \bar{e}_1)) = 0,$$

λ then being considered a parameter. Anyway, as Darboux's theorem ensures, this is nothing other than the integrability condition for the existence, at each point $(\mathbf{x}_0, \lambda_0) \in X \times \mathbf{P}^1$, of a pair of functions ("canonical variables") u_1 and u_2 defined in a neighborhood of $(\mathbf{x}_0, \lambda_0)$ for which

$$(e_1 + \lambda \bar{e}_2) \wedge (e_2 - \lambda \bar{e}_1) = du_1 \wedge du_2.$$

These functions (u_1, u_2) play the same role as a solution W of the linear system of self-dual connections (cf. [7, 1986]).

An analogue of the notion of RHT's was already pointed out by Boyer and Plebanski [32]; the main part of [7, 1986] is devoted to an analysis based on their results, leading to a close analogy with the results on self-dual connections in [7, 1984–85]. In the above setting, the existence of such a transformation group can be seen as follows. Suppose there are two pairs (u_1, u_2) and $(\hat{u}_1, \hat{u}_2)$ of "canonical variables" with different analytical properties almost parallel to W and $\hat{W}$ in Section 2, that is, $\lambda^{-1} u_A$ and $\hat{u}_A$ are holomorphic in, respectively, D and $\hat{D}$ with regard to λ. Since $du_1 \wedge du_2 = d\hat{u}_1 \wedge d\hat{u}_2$, there should be a two-dimensional parametric canonical transformation $f=(f_1(y,z,\lambda),f_2(y,z,\lambda))$, $df_1\wedge df_2=dy\wedge dz$, $\lambda\in D\cap\hat{D}$, such that

$$u_A=f_A(\hat{u}_1,\hat{u}_2,\lambda),\qquad A=1,2.$$

This canonical transformation f provides a counterpart of the transition function g in Section 2; in an appropriate setting, indeed, a complete description of (local) self-dual metrics can be given along this line. Evidently such f's form a (pseudo)group by composition of maps, and give rise to an analogue of the notion of RHT's. The role of matrix groups such as $GL(r,\mathbf{C})$ in the case of conventional integrable systems, is thus played by a (pseudo)group of (local) diffeomorphisms here.

One can recognise the present situation from a geometrical point of view (cf. Goldschmidt [33]), as a sort of deformation of integrable G-structures; this leads to various possibilities of generalisation. Details and further progress in this direction will be presented in a forthcoming series of papers [34]. In the context of Riemannian or Kählerian geometry, basically the same structure is usually understood as deformation of complex structures, whose parameter space is exactly the corresponding twistor space (cf. Hitchin *et al.* [6]). In our setting the twistor space can be derived as follows (cf. Gindikin [11, 31]). Let us consider the Pfaffian system on $X\times\mathbf{P}^1$:

$$e_1+\lambda\bar{e}_2=0,\qquad e_2-\lambda\bar{e}_1=0,\qquad \hat{d}\lambda=0.$$

This Pfaffian system is integrable in the sense of Frobenius if the previous equations to $e_1,\ldots,\bar{e}_2$ are satisfied, the triple (u_1,u_2,λ) then giving a set of first integrals. The set of maximal integral manifolds (leaves of the corresponding complex foliation) is nothing other than the twistor space. The canonical transformation f above (in fact, one has to "twist it" [7, 1986]) plays the role of transition function gluing together coordinate patches to form the twistor space.

References

[1] M. Jimbo and T. Miwa, *Publ. RIMS* **19** (1983) 943–1101.

[2] M. Sato and Y. Sato, In *Nonlinear PDEs in Applied Science*, Proc. U.S.-Japan seminar, Tokyo 1982, North Holland/Kinokuniya, 1982, pp. 259–271.

[3] M. F. Atiyah and R. S. Ward, *Commun. Math. Phys.* **58** (1977) 117–124.

[4] R. Penrose, *Gen. Rel. Grav.* **7** (1976) 31–52.

[5] R. S. Ward, *Nucl. Phys.* **B236** (1984) 381–396.

[6] N. J. Hitchin, A. Kahlhede, U. Lindström, and M. Roček, *Commun. Math. Phys.* **108** (1987) 535–589.

[7] K. Takasaki, *Proc. Japan Acad.* **59**, Ser. A (1983) 308–311; *Commun. Math. Phys.* **94** (1984), 35–59; *Saitama Math. J.* **3** (1985) 11–40; *Publ. RIMS, Kyoto Univ.* **22** (1986) 949–990.
[8] A. A. Belavin and V. G. Zakharov, *Phys. Lett.* **65A** (1978) 53–57.
[9] K. Pohlmeyer, *Commun. Math. Phys.* **72** (1980) 37–47.
[10] L. L. Chau, M. Prasad, and K. Sinha, *Phys. Rev.* **D24** (1981) 1578–1580.
[11] S. G. Gindikin, *Funct. Ana. Appl.* **18** (1985) 278–298.
[12] M. F. Atiyah, *Geometry of Yang–Mills Fields*, Scola Normale Superoire, Pisa, 1979.
[13] K. Ueno and Y. Nakamura, *Phys. Lett.* **B109** (1982) 273–278; *Publ. RIMS* **19** (1983) 519–547.
[14] Y. S. Wu, *Commun. Math. Phys.* **90** (1983) 461–472.
[15] N. Suzuki, *Proc. Japan Acad.* **60**, Ser. A (1984) 141–144.
[16] J. Harnad and M. Jacques, *J. Math. Phys.* **27** (1986) 2394–2400.
[17] K. Nagatomo, An Approach to the Stationary Axially Symmetric Vacuum Einstein Equations, preprint, 1987.
[18] Y. Nakamura, Riemann–Hilbert Transformations for a Toeplitz Matrix Equation: Some Ideas and Applications to Linear Prediction Problem, preprint 1987; Transformation Group Acting on a Self-Dual Yang–Mills Hierarchy, preprint, 1987.
[19] K. Takasaki, In preparation.
[20] H. Flaschka, A. C. Newell, and T. Ratiu, *Physica* **9D** (1983) 300–323.
[21] E. Witten, *Phys. Lett.* **77B** (1978) 394–498.
[22] J. Isenberg, P. B. Yasskin, and P. S. Green, *Phys. Lett.* **78B** (1979) 462–464.
[23] I. V. Volovich, *Lett. Math. Phys.* **7** (1983) 517–521.
[24] C. Devchand, *Nucl. Phys.* **B238** (1984) 1269–1272.
[25] L. L. Chau, M. L. Ge, and Z. Popowicz, *Phys. Rev. Lett.* **52** (1984) 1940–1943.
[26] J. Harnad and S. Schnider, *Commun. Math. Phys.* **106** (1986) 183–199.
[27] C. Devchand, In *Field Theory, Quantum Theory and Supergravity*, Lect. Notes Phys. No. 246, Springer, 1986, pp. 190–205.
[28] J. Harnad, Private communication.
[29] T. Eguchi, P. B. Gilkey, and A. J. Hanson, *Phys. Rep.* **66** (1980) 213–293.
[30] C. P. Boyer, In *Nonlinear Phenomena*, Lect. Notes. Phys. No. 189, Springer, 1983, pp. 25–46.
[31] S. G. Gindikin, *Soviet J. Nucl. Phys.* **36** (1982) 537–548.
[32] C. P. Boyer and J. F. Plebanski, *J. Math. Phys.* **26** (1985) 229–234.
[33] H. Goldschmidt, *Bull. Amer. Math. Soc.* **84** (1978) 531–546.
[34] K. Takasaki, In preparation.

Second Microlocalization and Conical Refraction (II)

Nobuyuki Tose
Department of Mathematics
Faculty of Science
University of Tokyo
Tokyo, Japan

1. Introduction

We study microdifferential equations with simple characteristics in the sense of second microlocalization. Explicitly, let M be a real analytic manifold with a complexification X and let P be a microdifferential operator defined in a neighborhood of $\rho_0 \in T^*_M X \backslash M$ satisfying the following conditions (1.1)-(1.3).

$$p = \sigma(P) \text{ is real-valued on } T^*_M X \backslash M. \tag{1.1}$$

$$\Sigma = \{\rho \in T^*_M X \backslash M;\, p(\rho) = 0,\, dp(\rho) = 0\} \text{ is a regular involutory submanifold of codimension } d(\geq 2) \text{ through } \rho_0 \text{ in } T^*_M X \backslash M. \tag{1.2}$$

$$\text{Hess } (p)(\rho) \text{ has rank } d \text{ with positivity } d'(>0) \text{ if } \rho \in \Sigma. \tag{1.3}$$

This class of equations was studied in the C^∞ case by R. B. Melrose and G. A. Uhlmann [M-U] and V. Ja. Ivrii [I] and in the C^ω case by P. Laubin [Lb_1], [Lb_2]. See also E. Bernardi [Be] and N. Dencker [D].

Algebraic Analysis, Volume II

ISBN 0-12-400466-0

In $[T_3]$, the author studied the class of equations above in the case $d'=1$, utilizing the theory of 2-microdifferential operators due to Y. Laurent [Lr], and gave the 2-microlocal canonical form for the equations considered under the assumption on the lower terms:

P has regular singularities along $\Sigma^{\mathbf{C}}$ (a complexification of Σ in T^*X) in the sense of M. Kashiwara and T. Oshima [K–O]. (1.4)

Moreover, he gave in $[T_4]$ another proof for the theorem about propagation of 2-microlocal singularities in $[T_3]$, where the assumption (1.4) was neglected. More precisely, he employed the theory of Microlocal Study of Sheaves due to M. Kashiwara and P. Schapira $[K\text{–}S_2]$ to study the systems of microdifferential equations with characteristic varieties satisfying the conditions analogous to (1.1)–(1.3).

In this paper, we will give the 2-microlocal canonical form for the equation $Pu=0$ without assumption (1.4). In Section 2, we prepare some notation and some prerequisites. We announce the main theorem in Section 3, and its proof will appear in Section 4.

The author would like to express his gratitude to Prof. Y. Laurent for the discussion they had in Paris in November, 1986. He is also grateful to Prof. H. Komatsu and Prof. P. Schapira for guidance and encouragement.

2. Preliminary

2.1. 2-Microdifferential Operators

We review the theory of 2-microdifferential operators due to Y. Laurent [Lr].

Let X be an open subset in $\mathbf{C}^{n+d}$ and let T^*X be its cotangent bundle. We take a coordinate of X as (w, z) with $w \in \mathbf{C}^n$ and $z \in \mathbf{C}^d$. Then $\rho = (w, z; \theta\, dw + \zeta\, dz)$ denotes a point of T^*X with $\theta \in \mathbf{C}^n$ and $\zeta \in \mathbf{C}^d$. T^*X is endowed with the sheaf $\mathscr{E}_X$ of microdifferential operators constructed by M. Sato *et al.* [SKK]. See also P. Schapira [Sc] for details about $\mathscr{E}_X$.

Hereafter in Section 2.1, Λ is the regular involutory submanifold in $\mathring{T}^*X(\simeq T^*X\backslash X)$:

$$\Lambda = \{(w, z; \theta, \zeta);\ \zeta = 0\}. \tag{2.1}$$

We identify Λ with a submanifold of $\Lambda \times \Lambda$ through

$$T^*X \simeq T^*_X(X \times X) \to T^*(X \times X).$$

By definition, $\tilde{\Lambda}$ is the union of bicharacteristic leaves of $\Lambda \times \Lambda$ issued from Λ. We take a coordinate of $T^*_\Lambda\tilde{\Lambda}$ as $(w, z; \theta, z^*)$ with $(w, z; \theta) \in \Lambda$ and $z^* \in \mathbf{C}^d$.

$T^*_\Lambda\tilde{\Lambda}$ is endowed with the sheaf $\mathscr{E}^{2,\infty}_\Lambda$ of 2-microdifferential operators of infinite order constructed in Y. Laurent [Lr].

Definition 2.1. For an open subset U of $T^*_\Lambda\tilde{\Lambda}$, a formal sum $\sum_{(i,j)\in \mathbf{Z}^2} P_{ij}(w, z, \theta, z^*)$ belongs to $\mathscr{E}^{2,\infty}_\Lambda(U)$ if and only if the following conditions (2.2) and (2.3) are satisfied.

P_{ij} is holomorphic on U and homogeneous of order j with respect to (θ, z^*) and of order i with respect to z^*. (2.2)

For any compact subset K of U, there exists a positive number C_K, and for any positive ε and a compact subset K, we can take a positive $C_{\varepsilon,K}$ such that (2.3)

$$\sup_K |P_{i,i+k}| \le \begin{cases} C_{\varepsilon,K}\varepsilon^{i+k}/i!k! & (i, k \ge 0) \\ C_{\varepsilon,K}^{-k}\varepsilon^{i}(-k)!/i! & (i \ge 0, k < 0) \\ C_{\varepsilon,K}\varepsilon^{k}C_K^{-i}(-i)!/k! & (i < 0, k \ge 0) \\ C_K^{-i-k}(-k)!(-i)! & (i, k < 0). \end{cases}$$

Y. Laurent [Lr] also defined the sheaf $\mathscr{E}^{2,(r,1)}_\Lambda$ of 2-microdifferential operators of type $(r, 1)$.

Definition 2.2. Let U be an open subset of $T^*_\Lambda\tilde{\Lambda}$ and let P be an element of $\mathscr{E}^{2,\infty}_\Lambda(U)$. For $r(>1) \in \mathbf{Q} \cup \{\infty\}$ and $(i_0, j_0) \in \mathbf{Z}^2$, $P = \sum_{ij} P_{ij}(w, z, D_z, D_w) \in \mathscr{E}^{2,(r,1)}_\Lambda[i_0, j_0](U)$ if and only if

$$P_{ij} \equiv 0 \quad \text{when } (1/r)\cdot i + (j-i) > (1/r)\cdot i_0 + (j_0 - i_0) \text{ or } j > j_0. \tag{2.4}$$

We put

$$\mathscr{E}^{2,(r,1)}_\Lambda = \bigcup_{i,j} \mathscr{E}^{2,(r,1)}_\Lambda[i, j]. \tag{2.5}$$

For a section P of $\mathscr{E}^{2,(r,1)}_\Lambda$, the principal symbol of P of type $(r, 1)$ is defined by

$$\sigma^{(r,1)}_\Lambda(P) = P_{i_0 j_0}, \tag{2.6}$$

where P is not a section of $\mathscr{E}^{2,(r,1)}_\Lambda[i, j]$ that is strictly smaller than $\mathscr{E}^{2,(r,1)}_\Lambda[i_0, j_0]$.

See Y. Laurent [Lr] for details about 2-microdifferential operators.

2.2. Bisymplectic Structure of $T_\Lambda^*\tilde{\Lambda}$

Y. Laurent introduced in [Lr] the transformation theory for 2-microdifferential operators. The class of transformations is called *quantized bicanonical transformations* and is wider than the class of quantized contact transformations.

Let X be a complex manifold and Λ be a regular involutory submanifold in $\mathring{T}^*X(=T^*X\backslash X)$. The regular involutory submanifold in (2.1) is denoted by Λ_0 in Section 2.2. We identify Λ with a submanifold of $\Lambda\times\Lambda$ through $T^*X \simeq T_X^*(X\times X)\to T^*(X\times X)$. Then $\tilde{\Lambda}$ is the union of all bicharacteristic leaves of $\Lambda\times\Lambda$ issued from Λ.

$T_\Lambda^*\tilde{\Lambda}$ is equipped with the canonical 1-form $\omega_\Lambda = p^*\omega_X$ and the canonical 2-form $\Omega_\Lambda = d\omega$. Here

$$p: T_\Lambda^*\tilde{\Lambda}\to\Lambda\to T^*X$$

and ω_X is the canonical 1-form of T^*X. In case $\Lambda=\Lambda_0$, ω_Λ is expressed by coordinates as

$$\omega_\Lambda = \sum_{j=1}^{n} \theta_j\, dw_j. \tag{2.7}$$

The canonical 2-form Ω_Λ induces a map

$$TT_\Lambda^*\tilde{\Lambda}\to T^*T_\Lambda^*\tilde{\Lambda}.$$

We put its kernel as $T_{\text{rel}}T_\Lambda^*\tilde{\Lambda}$. We dualize the exact sequence

$$0\to T_{\text{rel}}T_\Lambda^*\tilde{\Lambda}\to TT_\Lambda^*\tilde{\Lambda} \tag{2.8}$$

and obtain the exact sequence

$$0\leftarrow T_{\text{rel}}^*T_\Lambda^*\tilde{\Lambda}\leftarrow T^*T_\Lambda^*\tilde{\Lambda}. \tag{2.9}$$

We can take a section ω_Λ^r of $T_{\text{rel}}^*T_\Lambda^*\tilde{\Lambda}$ canonically, which is called the *relative canonical* 1*-form.* We also define the *canonical* 2*-form* $\Omega_\Lambda = d\omega_\Lambda$. Refer to Section 2.9.4 of [Lr] for the construction of ω_Λ^r. See also Section 2.2. of [T_3]. In the case $\Lambda=\Lambda_0$, ω_Λ is written by coordinates as

$$\omega_\Lambda^r = \sum_{j=1}^{d} z_j^*\, dz_j. \tag{2.10}$$

The relative 2-form Ω_Λ induces an isomorphism

$$H_\Lambda^r\colon T_{\text{rel}}^*T_\Lambda^*\tilde{\Lambda}\to T_{\text{rel}}T_\Lambda^*\tilde{\Lambda}. \tag{2.11}$$

For a function f defined in an open subset U of $T^*_\Lambda\tilde{\Lambda}$, we put

$$H^r_f = H^r_\Lambda(\overline{df}). \tag{2.12}$$

Here $\overline{df}$ is the image of df by $T^*T^*_\Lambda\tilde{\Lambda} \to T^*_{\text{rel}}T^*_\Lambda\tilde{\Lambda}$. H^r_f is a vector field on U and is called the *relative Hamiltonian vector field of f*. In case $\Lambda = \Lambda_0$, H^r_f is expressed by coordinates as

$$H^r_f = \sum_{j=1}^{d}\left(\frac{\partial f}{\partial z_j^*}\cdot\frac{\partial}{\partial z_j} - \frac{\partial f}{\partial z_j}\cdot\frac{\partial}{\partial z_j^*}\right).$$

Hereafter in Section 2.2. we restrict ourselves to the case $\Lambda = \Lambda_0$.

Suppose that the map φ: $U \to U'$ between open subsets in $\mathring{T}^*_\Lambda\tilde{\Lambda}$ preserves the two $\mathbf{C}^\times$ actions

$$(z, w; \theta; z^*) \to (z, w; \lambda\theta; \lambda z^*) \tag{2.13}$$

and

$$(z, w; \theta; z^*) \to (z, w; \theta; \lambda z^*) \qquad (\lambda \in \mathbf{C}^\times), \tag{2.14}$$

and satisfies

$$\varphi^*\Omega_\Lambda = \Omega_\Lambda. \tag{2.15}$$

Then the morphism

$$\varphi^*\colon T^*_{\text{rel}}T^*_\Lambda\tilde{\Lambda}|_U \to T^*_{\text{rel}}T^*_\Lambda\tilde{\Lambda}|_{U'} \tag{2.16}$$

is induced. Moreover we assume

$$\varphi^*\Omega^r_\Lambda = \Omega^r_\Lambda. \tag{2.17}$$

Then φ is called a *homogeneous bicanonical transformation.*

Associated with φ, we can construct ring isomorphisms

$$\Phi : \mathscr{E}^{2,(r,1)}_\Lambda|_U \to \varphi^{-1}(\mathscr{E}^{2,(r,1)}_\Lambda|_{U'}) \tag{2.18}$$

and

$$\Phi : \mathscr{E}^{2,\infty}_\Lambda|_U \to \varphi^{-1}(\mathscr{E}^{2,\infty}_\Lambda|_{U'}). \tag{2.19}$$

Φ is called a *quantized bicanonical transformaton* associated with φ and satisfies the following properties.

$$\Phi(\mathscr{E}^{2,(r,1)}_\Lambda[i, j]) \subset \mathscr{E}^{2,(r,1)}_\Lambda[i, j]. \tag{2.20}$$

$$\sigma^{(r,1)}_\Lambda(\Phi(P)) = \sigma^{(r,1)}_\Lambda(P)\circ\varphi^{-1} \qquad \text{for any } P \in \mathscr{E}^{2,(r,1)}_\Lambda. \tag{2.21}$$

See [Lr] for details about quantized bicanonical transformations.

2.3. 2-Microfunctions

Let M be a real analytic manifold with a complexification X. Let Σ be a regular involutory submanifold in $\mathring{T}^*_M X (= T^*_M X \backslash M)$ with a complexification Λ in T^*X.

As a model for Σ, we set

$$\Sigma_0 = \{(t, x; \sqrt{-1}(\tau\, dt + \xi\, dx)) \in \mathring{T}^*_{M_0} X_0;\ \xi = 0\}. \tag{2.2.2}$$

Here M_0 is an open subset in $\mathbf{R}^n_t \times \mathbf{R}^d_x$ and X_0 is a complex neighborhood of M_0 in $\mathbf{C}^n_w \times \mathbf{C}^d_z$. Then $(x, t; \sqrt{-1}(\tau\, dt + \xi\, dx))$ denotes a point of $T^*_{M_0} X_0$ with $\tau \in \mathbf{R}^n$ and $\xi \in \mathbf{R}^d$.

By definition, $\tilde{\Sigma}$ is the union of all bicharacteristic leaves of Λ issued from Σ. On $\tilde{\Sigma}$, there exists the sheaf $\mathscr{C}_{\tilde{\Sigma}}$ of microfunctions along $\tilde{\Sigma}$. $\mathscr{C}_{\tilde{\Sigma}}$ is the image of the sheaf of microfunctions with holomorphic parameters by a suitable quantized contact transformation. $\tilde{\Sigma}$ is foliated by the canonical foliation of Λ and for any section u of $\mathscr{C}_{\tilde{\Sigma}}$, u has the unique continuation property along the leaves.

$T^*_{\Sigma}\tilde{\Sigma}$ is endowed with the sheaf $\mathscr{C}^2_{\Sigma}$ of 2-microfunctions along Σ, which was constructed by M. Kashiwara about 10 years ago in Nice. The sheaf $\mathscr{C}^2_{\Sigma}$ plays a powerful role to study properties of microfunctions defined on Σ. Precisely, we have exact sequences

$$0 \to \mathscr{C}_{\tilde{\Sigma}}|_{\Sigma} \to \mathscr{B}^2_{\Sigma} \to \pi_*(\mathscr{C}^2_{\Sigma}|_{T^*_{\Sigma}\tilde{\Sigma}\backslash\Sigma}) \to 0 \tag{2.23}$$

and

$$0 \to \mathscr{C}_M|_{\Sigma} \to \mathscr{B}^2_{\Sigma}. \tag{2.24}$$

Here $\mathscr{B}^2_{\Sigma} = \mathscr{C}^2_{\Sigma}|_{\Sigma}$ and π: $T^*_{\Sigma}\tilde{\Sigma}\backslash\Sigma \to \Sigma$.

Moreover there exists the canonical spectral map

$$\mathrm{Sp}^2_{\Sigma}: \pi^{-1}\mathscr{B}^2_{\Sigma} \to \mathscr{C}^2_{\Sigma}. \tag{2.25}$$

We set for $u \in \mathscr{C}_M|_{\Sigma}$

$$\mathrm{SS}^2_{\Sigma}(u) = \mathrm{supp}(\mathrm{Sp}^2_{\Sigma}(u)), \tag{2.26}$$

which is called *2-singular spectrum* of u along Σ.

For details about 2-microfunctions, see M. Kashiwara and Y. Laurent [K-L].

$T^*_{\Lambda}\tilde{\Lambda}$ is a natural complexification of $T^*_{\Sigma}\tilde{\Sigma}$. Thus $T^*_{\Sigma}\tilde{\Sigma}$ is equipped with the real bisymplectic structure. For a function f defined on an open subset U in $T^*_{\Sigma}\tilde{\Sigma}$, we set its relative Hamiltonian vector field as H^r_f. In case $\Sigma = \Sigma_0$,

H_f^r is expressed by coordinates as

$$H_f^r = \sum_{j=1}^{d} \left(\frac{\partial f}{\partial x_j^*} \cdot \frac{\partial}{\partial x_j} - \frac{\partial f}{\partial x_j} \cdot \frac{\partial}{\partial x_j^*} \right). \tag{2.27}$$

Here we take a coordinate of $T^*_\Sigma \tilde{\Sigma}$ as $(t, x; \sqrt{-1}\tau\, dt; \sqrt{-1}x^*\, dx)$ with $(t, x; \sqrt{-1}\tau\, dt) \in \Sigma$ and $x^* \in \mathbf{R}^d$.

3. Statement of the Main Theorem

Let M be a real analytic manifold with a complexification X. We consider a microdifferential equation $Pu = 0$ defined in a neighborhood of $\rho_0 \in T^*_M X \backslash M$ that satisfies the conditions (1.1), (1.2) and (1.3). Let

$$\Sigma = \{\rho \in T^*_M X \backslash M;\ p(\rho) = 0,\ dp(\rho) = 0\} \tag{3.1}$$

and Λ be a complexification of Σ in T^*X. $\tilde{\Sigma}$ is the union of all bicharacteristics of Λ emanated from Σ. Then we have an isomorphism

$$H_\Sigma : T^*_\Sigma \tilde{\Sigma} \xrightarrow{\sim} T_\Sigma(T^*_M X) \tag{3.2}$$

through the Hamiltonian isomorphism $H : T^*T^*X \xrightarrow{\sim} TT^*X$. Take a point $\rho \in \Sigma$ and put for $\tau \in T^*_\Sigma \tilde{\Sigma}|_\rho$

$$p_\Sigma(\rho, \tau) = \langle \mathrm{Hess}(p)(\rho) \cdot H_\Sigma(\tau), H_\Sigma(\tau) \rangle. \tag{3.3}$$

Then p_Σ is a well-defined function on $T^*_\Sigma \tilde{\Sigma}$.

We give

Theorem 3.1. *Let u be a microfunction solution to $Pu = 0$. Then*

$$\mathrm{SS}^2_\Sigma(u) \backslash \Sigma \subset \{\gamma \in T^*_\Sigma \tilde{\Sigma} \backslash \Sigma;\ p_\Sigma(\gamma) = 0\}. \tag{3.4}$$

Moreover $\mathrm{SS}^2_\Sigma(u) \backslash \Sigma$ is invariant under $H^r_{p_\Sigma}$.

The proof of Theorem 3.1 will appear in Section 4.

By Theorem 3.1 above, we can deduce a theorem of Holmgren type below. Let

$$\Gamma_{\rho_0} = \{\tau \in T^*_\Sigma \tilde{\Sigma} \backslash \Sigma|_{\rho_0};\ p_\Sigma(\rho_0, \tau) = 0,\ \tau \neq 0\},$$

and set

$$\tilde{\Gamma} = \pi_\Sigma(\{\exp(sH^r_{p_\Sigma})(\rho_0, \tau);\ \tau \in \Gamma,\ s \geq 0\}).$$

Here $\exp(s\Theta)(q)$ denotes the flow of a vector field Θ starting from q and $\pi_\Sigma : T^*_\Sigma \tilde{\Sigma} \backslash \Sigma \to \Sigma$. Then we have

Theorem 3.2. *There exists a neighborhood Ω of ρ_0 in $T^*_M X$ such that for any microfunction solution u to $Pu=0$, the condition*

$$\Omega \cap \operatorname{supp}(u) \cap (\tilde{\Gamma} \backslash \{\rho_0\}) = \varnothing \tag{3.5}$$

implies $\rho_0 \notin \operatorname{supp}(u)$.

In the case $d'=1$, Γ_{ρ_0} consists of two conic subset Γ_+ and Γ_- in $(T^*_\Sigma \tilde{\Sigma} \backslash \Sigma)|_{\rho_0}$. We set

$$\tilde{\Gamma}_+ = \pi_\Sigma(\{\exp(sH^r_{p_\Sigma})(\rho_0, \tau);\ \tau \in \Gamma_+,\ s \geq 0\}). \tag{3.6}$$

Then we have

Theorem 3.3. *Assume $d'=1$. Then there exists a neighborhood Ω of ρ_0 in $T^*_M X$ such that for any microfunction solution u to $Pu=0$, the condition*

$$\Omega \cap \operatorname{supp}(u) \cap (\tilde{\Gamma}_+ \backslash \{\rho_0\}) = \varnothing \tag{3.7}$$

implies $\rho_0 \notin \operatorname{supp}(u)$.

Theorem 3.2 and Theorem 3.3 are obtained by Theorem 3.1 if we consult the exact sequences (2.23) and (2.24) and the unique continuation property along the leaves for $\mathscr{C}_{\tilde{\Sigma}}$.

Remark 3.4. The result above in Theorem 3.1 is first proved in [T_3] by finding the 2-microlocal canonical form for $Pu=0$, where the condition (1.4) on the lower terms is required. Later, Theorem 3.1 is regained in [T_4] with no assumption on the lower terms by another method using Microlocal Analysis of Sheaves due to M. Kashiwara and P. Schapira [K–S_2]. Moreover the result in [T_4] relies only on the geometry of characteristic varieties and was given for systems without any conditions on the multiplicities of equations. We emphasize here in this note that the result in Theorem 3.1 is obtained in the same way as in [T_3] without any assumption on the lower terms.

Remark 3.5. Independently, P. Laubin [Lb_3] obtained the same result about propagation of 2-microlocal singularities as given in [T_3] and this note. He employed the theory of J. Sjöstrand [Sj] and G. Lebeau [Le].

4. Proof of the Main Theorem

4.1.

First of all, we prove the following theorem about the canonical form for 2-microdifferential equations with simple characteristics in the sense of 2-microlocalization.

Let X_0 be an open subset in $\mathbf{C}^n_z$ and T^*X_0 denote its cotangent bundle. Then $(z, \zeta dz)$ takes for a point of T^*X_0 with $\zeta \in \mathbf{C}^n$. We set

$$\Lambda = \{(z, \zeta\, dz) \in T^*X_0;\, \zeta_1 = \cdots = \zeta_d = 0\} \tag{4.1}$$

and take a coordinate of $T^*_\Lambda \tilde{\Lambda}$ as $(z, \zeta''\, dz''; z'^*\, dz')$ with $\zeta'' = (\zeta_{d+1}, \ldots, \zeta_n)$, $z'' = (z_{d+1}, \ldots, z_n)$, $z'^* = (z_1^*, \ldots, z_d^*)$ and $z' = (z_1, \ldots, z_d)$. Here we give

Theorem 4.1. *Let P be a section of $\mathscr{E}_\Lambda^{2,(r,1)}[1, 1]$ defined in a neighborhood of $\rho_0 = (0, dz_n; dz_d)$ with $\sigma = r/(r-1) \in N$ and $\alpha \geq 2$. We assume*

$$\sigma_\Lambda^{(r,1)}(P) = z_1^* \tag{4.2}$$

and that

$$S(P) = \{(j, i) \in \mathbf{Z}^2;\, P_{ij} = 0\} \subset \{(j, i);\, i \geq (1-\sigma) + j,\, i \geq \sigma(j-1) + 1,\, j \leq 1\}. \tag{4.3}$$

Then we can find an invertible $R \in \mathscr{E}_\Lambda^{2,\infty}$ defined in a neighborhood of ρ_0 satisfying

$$P(z, D)R(z, D) = R(z, D)D_1. \tag{4.4}$$

Using the preparation theorem for $\mathscr{E}_\Lambda^{2,(r,1)}$ (see Theorem 2.7.4 of [Lr].), we can write P in the form

$$P = Q(D_1 + B(z, D')) \tag{4.5}$$

with $D' = (D_2, \ldots, D_n)$. Here

$$Q \text{ is invertible in } \mathscr{E}_\Lambda^{2,(r,1)}. \tag{4.6}$$

Moreover if we set $\tilde{P} = D_1 + B(z, D') \in \mathscr{E}_\Lambda^{2,(r,1)}[1, 1]$, $\tilde{P}$ satisfies the conditions (4.2) and (4.3). Thus we may assume from the beginning that $B = P - D_1$ is independent of D_1.

In the same way as Chapter 2 in [SKK], we define formally

$$R(z, D') = \sum_{l \geq 0} R^{(l)}(x, D')$$

by finding $\{R^{(l)}\}$ recursively as follows.

$$R^{(0)} = 1. \tag{4.7}$$

$$\partial/\partial z_1 \cdot R^{(l)}(z, D') = B(z, D') R^{(l-1)}(z, D') \qquad (l \geq 1). \tag{4.8}$$

$$R^{(l)}(z, D')|_{z_1=0} = 0 \qquad (l \geq 1). \tag{4.9}$$

$R^{(l)}$ is given explicitly by

$$R^{(l)} = \int_0^{z_1} B(s_l, \hat{z}, D') \int_0^{s_l} B(s_{l-1}, \hat{z}, D') \cdots \int_0^{s_2} B(s_1, \hat{z}, D')\, ds_1 \cdots ds_l \tag{4.10}$$

with $\hat{z} = (z_2, \ldots, z_n)$. Since the coefficients of $B(z, D')$ are holomorphic, $R^{(l)}$ is given by

$$R^{(l)} = \int \cdots \int_{V_l} B(s_l, \hat{z}, D') B(s_{l-1}, \hat{z}, D') \cdots B(s_1, \hat{z}, D')\, ds_1 \cdots ds_l.$$

Here V_l is a real l-dimensional simplex whose volume is $|z_1|^l / l\,!$. We remark that

$$S(R^{(l)}) = \{(j, i);\ R_{ij} = 0\} \subset \{j \leq l,\ i \geq j - (\sigma - 1)l,\ i \geq (\sigma + 1)j - (\sigma - 1)l\} \tag{4.11}$$

and that

$$R^{(l)} \in \mathscr{E}_\Lambda^{2,(r+1,1)}[2l, l]. \tag{4.12}$$

We put for $l \in \mathbf{N}$

$$\mathscr{E}_\Lambda^2\langle l\rangle = \{P \in \mathscr{E}_\Lambda^{2,(r+1,1)}[2l, l];\ P_{ij} = 0 \text{ if } i < j - (\sigma - 1)l\}. \tag{4.13}$$

Then we have for $l, l' \in \mathbf{N}$

$$\mathscr{E}_\Lambda^2\langle l\rangle \mathscr{E}_\Lambda^2\langle l'\rangle \subset \mathscr{E}_\Lambda^2\langle l + l'\rangle \tag{4.14}$$

and

$$B \in \mathscr{E}_\Lambda^2\langle 1\rangle. \tag{4.15}$$

To prove the convergence for $\sum_l R^{(l)}$ in $\mathscr{E}_\Lambda^{2,\infty}$, we define *Formal Norm* for sections of $\mathscr{E}_\Lambda^2\langle l\rangle$.

Definition 4.2. Let U be an open subset of $T_\Lambda^*\tilde{\Lambda}$ and let $E(z, D) = \sum_{ij} E_{ij}(z, D)$ be a section of $\mathscr{E}_\Lambda^2\langle l\rangle$ on U. For a compact subset K of U, we define *Formal Norm* for E on K by

$$N_K^{(l)}(E, s, t) = \sum_{i',j',\alpha,\beta} \frac{2(2(n+d))^{j'}(-j')!}{(-j'+|\alpha|)!(-j'+|\beta|)!} \sup_K |E^{\alpha,\beta}_{2l+i',l+j'}|$$
$$\times s^{-2i'+|\alpha_1|+|\beta_1|} t^{-2(j'-i')+|\alpha_2|+|\beta_2|}$$
$$= \sum_{i,k,\alpha,\beta} \frac{2(2(n+d))^{i+k-l}(l-i-k)!}{(l-i-k+|\alpha|)!(l-i-k-|\beta|)!} \sup_K |E^{\alpha,\beta}_{i,i+k}|$$
$$\times s^{4l-2i+|\alpha_1|+|\beta_1|} t^{-2l-2k+|\alpha_2|+|\beta_2|} \tag{4.16}$$

where $P^{\alpha\beta}_{ij} = (\partial/\partial z'^*)^{\alpha_1}(\partial/\partial \zeta'')^{\alpha_2}(\partial/\partial z)^{\beta} P_{ij}$ with $\alpha = (\alpha_1, \alpha_2) \in \mathbf{Z}^d_+ \times \mathbf{Z}^{n-d}_+$ and $\beta \in \mathbf{Z}^n_+$.

Remark 4.2. For $P_1 \in \mathscr{E}^2_\Lambda\langle l\rangle$ and $P_2 \in \mathscr{E}^2_\Lambda\langle l'\rangle$, we have

$$N_K^{(l+l')}(P_1P_2, s, t) \ll N^{(l)}(P_1, s, t) \cdot N^{(l')}(P_2, s, t). \tag{4.17}$$

We can prove the formula (4.17) above by modifying Theorem 2.4.9 of Y. Laurent [Lr].

We give a proposition which plays a crucial role in the following.

Proposition 4.3. *For $P = \sum_{ij} P_{ij} \in \mathscr{E}^2_\Lambda\langle l\rangle$ and $Q = \sum_{ij} Q_{ij} \in \mathscr{E}^2_\Lambda\langle l'\rangle$ satisfying the condition*

$$P_{i,l+j'} = 0 \text{ (when } -j' \geq p) \text{ and } Q_{i,l'+j'} = 0 \text{ (when } -j' \geq q), \tag{4.18}$$

we have

$$N_K^{(l+l')}(PQ, s, t) \ll \frac{C'}{p+q} \cdot N_K^{(l)}(P, s, t) \cdot N_K^{(l')}(Q, s, t). \tag{4.19}$$

Here C' is a positive constant independent of l, l', p and q.

The proof of Proposition 4.3 can be given essentially in the same way as Theorem 2.4.9 of [Lr].

Now we go back to the proof of Theorem 4.1.

We put

$$B_2 = \sum_i B_{i1}(z, D') \qquad \text{and} \qquad B_1 = B - B_2 = \sum_{j \leq 0} B_{ij}.$$

Moreover we set

$$R = \sum R_{i,i+k} \qquad \text{and} \qquad R^{(l)} = \sum R^{(l)}_{i,i+k}.$$

Then we have the estimate on a compact neighborhood K of ρ_0:

$$\sup_K |R^{(l)}_{i,i+k}| \leq \frac{(l-i-k)!}{2(2(n+d))^{i+k-l}} \cdot \left(\frac{t^2}{s^4}\right)^l \cdot s^{2i} t^{2k} \cdot N^{(l)}_K(R^{(l)}, s, t)$$

$$\leq \left(\frac{s^2}{2(n+d)}\right)^i \cdot \left(\frac{t^2}{2(n+d)}\right)^k \sum_l \left\{\frac{t^2}{s^4}(2(n+d))\right\}^l$$

$$\cdot (l-i-k)! \cdot N^{(l)}_k(R, s, t). \tag{4.20}$$

To estimate $N^{(l)}_K(R, s, t)$, we give

Lemma 4.4. *Let* $Q = \prod_{l=1}^m B_1^{g_l} \cdot B_2^{h_l}$ *and* $g = g_1 + \cdots + g_m$ *and* $h = h_1 + \cdots + h_m$. *Then there exists a constant* $C > 0$ *satisfying*

$$N^{(h+g)}_K(Q, s, t) \ll \frac{C^g}{g!} \cdot N^{(1)}_K(B_1, s, t)^g \cdot N^{(1)}_K(B_2, s, t)^h. \tag{4.21}$$

Here C *is independent of* $(g_1, \ldots, g_m)$ *and* $(h_1, \ldots, h_m)$.

Lemma 4.4 above can be verified in the same way as Lemma 2.4.13 of [Lr] using Proposition 4.3.

By Lemma 4.4, we have

$$N^{(l)}_K(R^{(l)}, s, t) \ll \sum_{a+b=l} \left\{\frac{(a+b)!}{a!b!}\right\} \cdot \frac{C^a}{a!} \cdot N^{(1)}_K(B_1)^a \cdot N^{(1)}_K(B_2)^b \cdot \frac{A^l}{l!} \tag{4.22}$$

with $A = \sup_K |z_1|$. By (4.20) and (4.22), we have

$$\sup_K |R_{i,i+k}|$$

$$\leq \left(\frac{s^2}{2(n+d)}\right)^i \left(\frac{t^2}{2(n+d)}\right)^k \sum_{a+b=l} \left\{\frac{(a+b)!}{a!b!}\right\}$$

$$\cdot \frac{C^a}{a!} \cdot N^{(1)}_K(B_1)^a \cdot N^{(1)}_K(B_2)^b \cdot A^l G^l \frac{(l-i-k)!}{l!}$$

$$\leq \left(\frac{s^2}{2(n+d)}\right)^i \left(\frac{t^2}{2(n+d)}\right)^k \sum_{a,b\geq 0} 2^{a+b}$$

$$\cdot \frac{C^a}{a!} \cdot N^{(1)}_K(B_1)^a N^{(1)}_K(B_2)^b (AG)^{a+b} 3^{a+b} C_{ik}$$

$$\leq \left(\frac{s^2}{2(n+d)}\right)^i \left(\frac{t^2}{2(n+d)}\right)^k C_{ik} \sum_{a,b\geq 0} \frac{\{6ACG \cdot N^{(1)}_K(B_1)\}^a}{a!} \cdot \{6AG \cdot N^{(1)}_K(B_2)\}^b$$

$$\le\left(\frac{s^2}{2(n+d)}\right)^i\left(\frac{t^2}{2(n+d)}\right)^k\cdot C_{ik}$$

$$\cdot\exp\{6ACG\cdot N_K^{(1)}(B_1)\}\cdot\sum_{b\ge 0}\{6AG\cdot N_K^{(1)}(B_2)\}^b. \tag{4.23}$$

Here $G=t^2/s^4(2(n+d))$ and $C_{ik}=1/i!k!\ (i,k\ge 0)$, $(-k)!/i!\ (i\ge 0, k<0)$, $(-i)!/k!\ (i<0, k\ge 0)$ and $(-i)!(-k)!\ (i,k<0)$.

Here we give a lemma about the estimate for $N_K^{(1)}(B_1,s,t)$ and $N_K^{(1)}(B_2,s,t)$.

Lemma 4.5. (*i*) *There exists a constant $C>0$ such that for any $\alpha>0$ we can find a positive number C_α satisfying*

$$|N_K^{(1)}(B_1,s,t)|\le C_\alpha\qquad\left(\alpha|s|\le|t|\le\frac{|s|}{C}\le\frac{1}{C^2}\right). \tag{4.24}$$

(*ii*) *There exist constants C_1 and C_2 such that*

$$|N_K^{(1)}(B_2,s,t)|\le C_2\qquad\text{when}\qquad 0\le|t|\le\frac{|s|}{C_1}\le\frac{1}{C_1^2}. \tag{4.25}$$

Since Lemma 4.5 above can be easily proved, we omit its proof.

By Lemma 4.5, we can find a positive constant C such that

$$|6ACG\cdot N_K^{(1)}(B_1)|<+\infty\qquad\text{when } 0<|t|\le\frac{|s|}{C}\le\frac{1}{C^2}. \tag{4.26}$$

Moreover there exist positive numbers C_1 and C_2 satisfying

$$|6AG\cdot N_K^{(1)}(B_2)|\le C_2A\qquad\text{when } 0<|t|\le\frac{|s|^2}{C_1}\le\frac{1}{C_1^3}. \tag{4.27}$$

Thus if we take K small enough, we can take a positive number C satisfying

$$\varphi(s,t)=\left|\exp\{6ACG\cdot N_K^{(1)}(B_1)\}\cdot\sum_{b\ge 0}\{6AG\cdot N_K^{(1)}(B_2)\}^b\right|<+\infty \tag{4.28}$$

when $0<|t|\le|s|^2/C^2\le 1/C^3$. After all, we have the estimate

$$\sup_K|R_{i,i+k}|\le\left(\frac{|s|^2}{2(n+d)}\right)^i\left(\frac{|t|^2}{2(n+d)}\right)^k C_{ik}\varphi(s,t) \tag{4.29}$$

when $0<|t|\le|s|^2/C_1\le 1/C_1^3$.

By (4.28) and (4.29) we can conclude that $R \in \mathscr{E}_\Lambda^{2,\infty}$ in a neighborhood of ρ_0. We prove that $R(z, D')$ is invertible. But we can verify it by applying the same argument of Theorem 5.2.1 in Chapter 2 of [SKK].

4.2. [Proof of Theorem 3.1]

By finding a suitable real quantized contact transformation, the problem is reduced to studying the microdifferential equation

$$P_0 u = \left\{ D_1^2 - \sum_{i,j=2}^{d} A^{ij}(z, D) D_i D_j + (\text{lower order}) \right\} u = 0 \qquad (4.30)$$

defined in a neighborhood of $\rho_0 = (0, \sqrt{-1}\, dx_n) \in \sqrt{-1}\, T^*\mathbf{R}^n$. Here we assume that

$$\{A^{ij}\} \text{ are of order } 0 \qquad (4.31)$$

and that

$$(\sigma(A^{ij}))_{2 \le i,j \le d} \text{ is definite in a neighborhood of } \rho_0. \qquad (4.32)$$

In the case above, we find

$$\Sigma = \{(x, \sqrt{-1}\xi\, dx) \in \sqrt{-1}\, \mathring{T}^*\mathbf{R}^n;\ \xi_1 = \cdots = \xi_d = 0\}. \qquad (4.33)$$

Here we take a coordinate of $\sqrt{-1}\, T^*\mathbf{R}^n$ as $(x, \sqrt{-1}\xi\, dx)$ with $x \in \mathbf{R}^n$ and $\xi \in \mathbf{R}^n$ and that of $T^*\mathbf{C}^n$ as $(z, \zeta\, dz)$ with $z \in \mathbf{C}^n$ and $\zeta \in \mathbf{C}^n$. We set

$$\Lambda = \{(z, \zeta\, dz) \in T^*\mathbf{C}^n;\ \zeta_1 = \cdots = \zeta_d = 0\} \qquad (4.34)$$

as a complexification of Σ. Then $(x; \sqrt{-1}\xi''\, dx''; \sqrt{-1}x'^*\, dx')$ (resp. $(z; \zeta''\, dz''; z'^*\, dz')$) denotes a point of $T^*_\Sigma \tilde{\Sigma}$ (resp. $T^*_\Lambda \tilde{\Lambda}$) with $\xi'' = (\xi_{d+1}, \ldots, \xi_n)$ and $x'^* = (x_1^*, \ldots, x_d^*)$ (resp. $\zeta'' = (\zeta_{d+1}, \ldots, \zeta_n)$ and $z'^* = (z_1^*, \ldots, z_d^*)$).

We regard P_0 as a 2-microdifferential operator defined in a neighborhood of $\tau_0 \in \pi_\Sigma^{-1}(\rho_0)$. Here $\pi_\Sigma : T^*_\Sigma \tilde{\Sigma} \backslash \Sigma \to \Sigma$.

We may assume $z_d^* \neq 0$ at τ_0. Then 2-microlocally it is sufficient to consider the equation

$$P_1 u = D_d^{-1} P_0 u = 0. \qquad (4.35)$$

Here we remark that

$$P_1 \in \mathscr{E}_\Lambda^{2,(2,1)}[1, 1] \qquad (4.36)$$

and that

$$S(P_1) = \{(j, i);\ (P_1)_{ij} = 0\} \subset \{i \ge j - 1,\ i \ge 2j - 1,\ j \le 1\}. \qquad (4.37)$$

In the same way as in Section 4 of [T_3], we can construct a quantized bicanonical transformation by which the equation $P_1 u = 0$ can be transformed into the 2-microdifferential equation $P_2 u = 0$ defined in a neighborhood of $\tau_1 = (0, \sqrt{-1}\, dx_n; \sqrt{-1}\, dx_d)$ with

$$\sigma_\Lambda^{(2,1)}(P_2) = z_1^*. \tag{4.38}$$

Then we have by (2.20)

$$P_2 \in \mathscr{E}_\Lambda^{2,(2,1)}[1, 1] \tag{4.39}$$

and

$$S(P_2) = \{(j, i); (P_2)_{ij} = 0\} \subset \{i \geq j-1, i \geq 2j-1, j \leq 1\}. \tag{4.40}$$

The conditions (4.38), (4.39) and (4.40) assure us to apply Theorem 4.1. Thus we can find an invertible $R(z, D) \in \mathscr{E}_\Lambda^{2,\infty}$ in a neighborhood of τ_1 satisfying

$$P_2(z, D) R(z, D) = R(z, D) D_1. \tag{4.41}$$

By (4.41), we can easily prove the propagation of 2-microlocal singularities.

References

[Be] E. Bernardi, Propagation of singularities for hyperbolic operators with multiple involutive characteristics, to appear in *Osaka J. of Math.*

[D] N. Dencker, On the propagation of polarization in conical refraction, preprint.

[I] V. Ja. Ivrii, Wave fronts of the systems of crystal optics, *Soviet Math. Dokl.* **18** (1977) 139–141.

[K-L] M. Kashiwara and Y. Laurent, Théorèmes d'annulation et deuxième microlocalisation, Prépublication d'Orsay (1983).

[K-O] M. Kashiwara and T. Oshima, Systems of differential equations with regular singularities and their boundary value problems, *Annals of Math.* **106** (1975) 145–200.

[K-S_1] M. Kashiwara and P. Schapira, Microhyperbolic systems, *Acata Math.* **142** (1979) 1–55.

[K-S_2] M. Kashiwara and P. Schapira, Microlocal Study of Sheaves, *Astérisque* **128** (1985).

[Lb_1] P. Laubin, Thesis presented to Univ. Liège (1983).

[Lb_2] P. Laubin, Réfraction conique et propagation des singularités analytique, *J. Math. pure et appl.* **63** (1984) 149–169.

[Lb_3] P. Laubin, Propagation of the second analytic wave front set in conical refraction, preprint.

[Le] G. Lebeau, Deuxième microlocalisation à croissance, Séminaire Goulaouic–Meyer–Schwarz **XV**, 1982–1983.

[Lr] Y. Laurent, *Théorie de la deuxième microlocalisation dans le domaine complexe: opérateurs 2-microdifferentiels*, Progress in Math. 53, Birkhauser (1985).

[M-U] R. B. Melrose and G. A. Uhlmann, Microlocal structure of conical refraction, *Duke Math. J.* **46** (1979) 571-582.

[SKK] M. Sato, T. Kawai, and M. Kashiwara, *Microfunctions and Pseudodifferential Equations*, Lecture Notes in Math. 287, Springer, 1973.

[Sc] P. Schapira, *Microdifferential Systems in the Complex Domain*, Grundlehren der Math. 269, Springer, 1985.

[Sj] J. Sjöstrand, Singularités analytiques microlocales, *Astérisque* **96** (1982).

[T_1] N. Tose, On a class of microdifferential equations with involutory double characteristics, *J. Fac. Sci., Univ. of Tokyo* **33** (1986) 619-634.

[T_2] N. Tose, The 2-microlocal canonical form for a class of microdifferential equations and propagation of singularities, *Publ. of RIMS, Kyoto Univ.* **23** (1987) 101-116.

[T_3] N. Tose, 2nd Microlocalisation and Conical Refraction, *Ann. Inst. Fourier* **37-2** (1987) 239-260.

[T_4] N. Tose, On a class of 2-microhyperbolic systems, *J. Math. pure et appl.*, **67** (1988) 23-37.

Pseudodifferential Operators in Hyperfunction Theory

Keisuke Uchikoshi
Department of Mathematics
National Defense Academy
Yokosuka, Japan

In this article, we briefly explain about the most general theory of pseudodifferential operators in hyperfunction theory. It turns out that such a theory has some formal similarity to the $S^m_{\rho,\delta}$-theory, which is standard in distribution theory. Symbol formulae for adjoint operators and composite operators require careful treatment. We give such formulae under reasonable assumptions. We also give some examples of our theory important in microlocal analysis of partial differential equations. The details will be published elsewhere.

0 Introduction

Let $X = \mathbf{C}^n$ and let $M = \mathbf{R}^n$. We denote the variables of X or of M by the same letter x, if confusion is not likely. We denote by ξ the dual variables of x. We identify M with the diagonal set in $M \times M$. Therefore we have an

Algebraic Analysis, Volume II

ISBN 0-12-400466-0

isomorphism

$$\sqrt{-1}S^*M \ni (x, \xi) \overset{\sim}{\mapsto} (x, x, \xi, -\xi) \in \sqrt{-1}S^*_M(M \times M).$$

The sheaf $\mathscr{L} = \mathscr{L}_M$ of microlocal operators is defined on $\sqrt{-1}S^*M \cong \sqrt{-1}S^*_M(M \times M)$ by

$$\mathscr{L}_M = \mathscr{H}^0_{\sqrt{-1}S^*_M(M\times M)}(\mathscr{C}_{M\times M} \otimes \mathscr{V}_M),$$

where $\mathscr{C}_{M\times M}$ denotes the sheaf of microfunctions on $M \times M$, and $\mathscr{V}_M$ that of densities on M with analytic coefficients (see M. Kashiwara and T. Kawai [3] and M. Sato, T. Kawai, and M. Kashiwara [5]). To help the understanding of the beginners, we explain the meaning of the above definition. For instance, let $P(x, D) = \sum_{|\alpha| \le m} a_\alpha(x) D^\alpha$, $D = \partial/\partial x$, be a partial differential operator with analytic coefficients on M. The kernel function $L(x, x')$ of $P(x, D)$ can be written as

$$L(x, x') = \sum_{|\alpha| \le m} a_\alpha(x) \delta^{(\alpha)}(x - x').$$

Here (x, x') denotes the variables of $M \times M$. As a microfunction, the support of $L(x, x')$ is contained in $\sqrt{-1}S^*_M(M \times M)$, and thus $L(x, x')\, dx'$ is a section of $\mathscr{L}_M$. A section of $\mathscr{L}_M$ is called a microlocal operator on M, and this notion is a generalization of the notion of partial differential operators from the above point of view. A microlocal operator acts on microfunctions on M, and it has the microlocal property, i.e., under this action, the support of each microfunction does not increase. An operator possessing the microlocal property should be called a pseudo-differential operator, and, roughly speaking, one may understand that the notion of microlocal operators is the most general one of such operators. Several subclasses of microlocal operators have been investigated in detail, using symbol functions (see Section 1). Our aim is to extend such an investigation for general microlocal operators. In the next section we first show that we can give a simple characterization of microlocal operators from a symbol theoretical point of view.

1. Symbol Functions of Microlocal Operators

Let $\mathring{x}^* = (0; 0, \ldots, 0, \sqrt{-1}) \in \sqrt{-1}S^*M$, and let $A = u(x, x')\, dx' \in \mathscr{L}_{\mathring{x}^*}$. We define the (total) symbol $\sigma(A)$ of A by

$$\sigma(A) = \int e^{-(x-x')\xi} u(x, x')\, dx'. \tag{1}$$

We do not give the precise meaning of this integral, because it requires some preliminaries. In fact we consider the defining function of $u(x, x')$, and take the path of integration certainly in a complex domain. One may understand that (1) is defined in a natural way in the sense of microfunction. To give the characterization of such symbol functions, we prepare two definitions.

Definition 1. Let $r > 0$ be small, and let $\lambda(t)$ be continuous for $0 < t < r$. We say that $\lambda(t)$ is a *scaling function* if

$$\lim_{t \to +0} \frac{\lambda(t)}{t} = 0$$

and

$$t < t' \Rightarrow \frac{\lambda(t)}{t} < \frac{\lambda(t')}{t'}.$$

It is easy to see that a scaling function is positive and monotonically increasing. The simplest examples are $\lambda(t) = mt^i$, $m > 0$, $i > 1$, and $\lambda(t) = mt/|\log t|^i$, $m > 0$, $i > 0$. If $t = 0$, we always define $\lambda(0) = 0$.

Definition 2. (*i*) Let $r > 0$ be small. We define $V_r \subset T^*X$ by

$$\begin{aligned} V_r = \{(x, \xi) \in T^*X;\ &|x| < r,\ r \operatorname{Im} \xi_n > 1, \\ &r \operatorname{Im} \xi_n > |\operatorname{Im} \xi_j|,\ 1 \le j \le n-1, \\ &r \operatorname{Im} \xi_n > |\operatorname{Re} \xi_j|,\ 1 \le j \le n\}. \end{aligned}$$

(*ii*) Let $\lambda(t)$ and $\mu(t)$ be two scaling functions. We define $\mathcal{S}_{\lambda,\mu} = (\mathcal{S}_{\lambda,\mu})_{\mathring{x}^*}$ to be the space of all holomorphic functions $a(x, \xi)$ defined on V_r for some $r > 0$, such that for any $\varepsilon > 0$ there exists some $C_\varepsilon > 0$ satisfying

$$|a(x, \xi)| \le C_\varepsilon \exp\{(\lambda(|\operatorname{Im} x|) + \mu(|\operatorname{Re} \xi|/\operatorname{Im} \xi_n) + \varepsilon) \operatorname{Im} \xi_n\} \tag{2}$$

on V_r. We define $\mathcal{S} = \mathcal{S}_{\mathring{x}^*}$ by $\mathcal{S} = \bigcup_{\lambda,\mu} \mathcal{S}_{\lambda,\mu}$.

(*iii*) We denote by $\mathcal{N} = \mathcal{N}_{\mathring{x}^*}$ the space of all holomorphic functions $a(x, \xi)$ defined on V_r for some $r > 0$, such that there exists some $C > 0$ satisfying

$$|a(x, \xi)| \le C \exp(-\operatorname{Im} \xi_n / C)$$

on V_r.

Our first result is the following

Theorem 3. *The mapping* (1) *gives the following isomorphism*:

$$\mathscr{L}_{\mathring{x}^*} \ni A \mapsto \sigma(A) \in \mathscr{S}_{\mathring{x}^*}/\mathscr{N}_{\mathring{x}^*}.$$

Examples. (*i*) $e^{x\cdot\xi} \notin \mathscr{S}_{\mathring{x}^*}$. In fact, we only have

$$|e^{x\cdot\xi}| \le \exp\{(|\mathrm{Im}\, x| + |\mathrm{Re}\, \xi|/\mathrm{Im}\, \xi_n)\, \mathrm{Im}\, \xi_n\}$$

on a conical neighborhood of $\mathring{x}^*$, and it does not satisfy the above requirement. Neither do the symbol functions of (analytic) Fourier integral operators satisfy the above requirement in general, by the same reason.

(*ii*) We have

$$\exp(\sqrt{-1}\xi_j^2/\xi_n) \in \mathscr{S}_{\mathring{x}^*}, \qquad 1 \le j \le n-1,$$

and

$$\exp(\sqrt{-1}x_j^2\xi_n) \in \mathscr{S}_{\mathring{x}^*}, \qquad 1 \le j \le n.$$

(*iii*) Let $n=2$, and thus let $\mathring{x}^* = (0, 0, 0, \sqrt{-1}) \in \sqrt{-1}S^*\mathbf{R}^2$. We define $K(x, \xi)$ by

$$K(x, \xi) = \sqrt{2}\exp\left\{\frac{\sqrt{-1}}{2}x_1^2\xi_2 + \frac{\sqrt{-1}}{2}\xi_1^2\xi_2^{-1} - x_1\xi_1\right\}.$$

A direct calculation shows that $K(x, \xi) \in \mathscr{S}_{\mathring{x}^*}$. It has an important meaning, and we denote the corresponding operator by $K(x, D)$ (see Section 3).

(*iv*) Other examples will be given in Section 3.

Remark. We denote by $\mathscr{S}_1 = (\mathscr{S}_1)_{\mathring{x}^*}$ the space of all holomorphic functions $a(x, \xi) \in \mathscr{S}_{\mathring{x}^*}$ which satisfy (2) with $\lambda(t) = \mu(t) = 0$, and $\mathscr{S}_2 = (\mathscr{S}_2)_{\mathring{x}^*}$ the space of all $a(x, \xi) \in (\mathscr{S}_1)_{\mathring{x}^*}$ which have asymptotic expansions

$$a(x, \xi) \sim \sum_{j=-\infty}^{\infty} a_j(x, \xi),$$

where each $a_j(x, \xi)$ is homogeneous in ξ of degree j (see [5] for the precise definition). In [3] and [5] it is proved that $\mathscr{E}^\infty_{\mathring{x}^*} \cong (\mathscr{S}_2)_{\mathring{x}^*}/\mathscr{N}_{\mathring{x}^*}$ and in [1] that $\mathscr{E}^{\mathbf{R}}_{\mathring{x}^*} \cong (\mathscr{S}_1)_{\mathring{x}^*}/\mathscr{N}_{\mathring{x}^*}$. Here $\mathscr{E}^\infty$ denotes the sheaf of microdifferential operators of infinite orders, and $\mathscr{E}^{\mathbf{R}}$ that of holomorphic microlocal operators. They are subsheaves of the sheaf of microlocal operators, and we have the following inclusions:

$$\begin{array}{ccccc} \mathscr{E}^{\infty}_{\mathring{X}^*} & \subset & \mathscr{E}^{\mathbf{R}}_{\mathring{X}^*} & \subset & \mathscr{L}_{\mathring{X}^*} \\ \wr\!\parallel & & \wr\!\parallel & & \wr\!\parallel \\ \mathscr{S}_2/\mathscr{N} & \subset & \mathscr{S}_1/\mathscr{N} & \subset & \mathscr{S}/\mathscr{N}. \end{array}$$

To help the understanding of the problem which occurs in the next section, we compare the above symbol spaces with those of distribution theory. Note that a symbol function in hyperfunction theory is defined on (a complex neighborhood of) $\sqrt{-1}S^*M$, and a symbol function in distribution theory is defined on S^*M. And there exist several differences of notation between these two theories. In this article we always use the convention of hyperfunction theory, and we think that even a symbol function in distribution theory is defined on $\sqrt{-1}S^*M$. Note also that symbol formulae which appear in the next section look a little different from those of distribution theory, by this reason.

Now we denote by $S^m_{\rho,\delta}(\mathbf{R}^n)$, $m \in \mathbf{Z}$, $0 \leq \rho$, $\delta \leq 1$, the symbol classes (on $\sqrt{-1}S^*\mathbf{R}^n$) defined by [2]. Let us denote by $S^m_{\mathrm{cl}}(\mathbf{R}^n)$, $m \in \mathbf{Z}$, the spaces of all $a(x,\xi) \in S^m_{1,0}$, which have homogeneous asymptotic expansions: $a(x,\xi) \sim \sum a_j(x,\xi)$. It is easy to see that if $a(x,\xi) \in \mathscr{S}_1$ (resp. $\mathscr{S}_2$) is of finite order, then we have $a(x,\xi) \in S^m_{1,0}$ (resp. S^m_{cl}) for some m. In this sense there is some formal similarity between $\mathscr{S}_1$ (resp. $\mathscr{S}_2$) and $\bigcup_m S^m_{1,0}$ (resp. $\bigcup_m S^m_{\mathrm{cl}}$). As for our symbol space $\mathscr{S}$, it turns out that, roughly speaking, there exists formal similarity between $\mathscr{S}$ and $\bigcup_{m,\rho,\delta} S^m_{\rho,\delta}$. It causes a problem, and we explain this fact in the next section.

2. Symbol Formulae for Adjoint Operators and for Composite Operators

As announced in Section 1, we first estimate the derivatives of $a(x,\xi) \in \mathscr{S}$. For the sake of simplicity, we only consider the most typical case.

Proposition 4. *Let $m \in \mathbf{Z}$ and assume that $a(x,\xi) \in \mathscr{S}_{\mathring{X}^*}$ satisfies*

$$|a(x,\xi)| \leq C|\xi|^m \cdot \exp\{(\lambda(|\mathrm{Im}\, x|) + \mu(|\mathrm{Re}\, \xi|/\mathrm{Im}\, \xi_n))\mathrm{Im}\, \xi\}$$

on V_r. Here V_r is as in Definition 2, and $\lambda(t) = Ct^i$, $\mu(t) = Ct^j$, $C > 0$, $i, j > 1$.

Then we have

$$a(x, \xi)|_{\sqrt{-1}S^*M} \in S^m_{\rho,\delta}(\mathbf{R}^n)$$

with $\rho = 1 - 1/j$ *and* $\delta = 1/i$.

Proof. Let $(x, \xi) \in \sqrt{-1}S^*\mathbf{R}^n \cap V_{r/4}$. To estimate $\partial_x^\alpha \partial_\xi^\beta a(x, \xi)$ for such (x, ξ), we first restrict ourselves to the case

$$\operatorname{Im} \xi_n > \frac{|\alpha|+1}{r^i}, \qquad \operatorname{Im} \xi_n > \frac{|\beta|+1}{r^j}. \tag{3}$$

Under this assumption, we assume that $(\tilde{x}, \tilde{\xi}) \in \mathbf{C}^n \times \mathbf{C}^n$ satisfies

$$|\tilde{x}_k| \leq \frac{((|\alpha|+1)/\operatorname{Im} \xi_n)^{1/i}}{2n}, \qquad 1 \leq k \leq n,$$

$$|\tilde{\xi}_k| \leq \frac{((|\beta|+1)/\operatorname{Im} \xi_n)^{1/j} \operatorname{Im} \xi_n}{4n}, \qquad 1 \leq k \leq n.$$

From (3) it follows that $|\tilde{x}_k| \leq r/2n$, and $|\tilde{\xi}_k| \leq r \operatorname{Im} \xi_n / 4n$, $1 \leq k \leq n$, and thus $(x + \tilde{x}, \xi + \tilde{\xi}) \in V_r$. We therefore obtain

$$|a(x, \xi)| \leq \left| \frac{a!\beta!}{(2\pi\sqrt{-1})^{\partial n}} \iint \frac{a(x+\tilde{x}, \xi+\tilde{\xi})}{\prod_{k=1}^n \tilde{x}_k^{\alpha_k+1} \tilde{\xi}_k^{\beta_k+1}} d\tilde{x}\, d\tilde{\xi} \right|,$$

where the integral is calculated for

$$|\tilde{x}_k| = \frac{((|\alpha|+1)/\operatorname{Im} \xi_n)^{1/i}}{2n}, \qquad 1 \leq k \leq n,$$

$$|\tilde{\xi}_k| = \frac{((|\beta|+1)/\operatorname{Im} \xi_n)^{1/j} \operatorname{Im} \xi_n}{4n}, \qquad 1 \leq k \leq n.$$

A direct calculation shows

$$|a(x, \xi)| \leq C'^{|\alpha|+|\beta|+1} \alpha!^{1+\delta} \beta!^{\rho} |\xi|^{m+\delta|\alpha|-\rho|\beta|}$$

with the above ρ and δ, if $(x, \xi) \in \sqrt{-1}S^*\mathbf{R}^n \cap V_{r/4}$ satisfies (3). If (3) does not hold, then the statement is empty, and we have proved the proposition. Q.E.D.

If one considers more general scaling functions, then one obtains formal similarity between $\mathscr{S}$ and $S^m_{\nu,\rho,\delta}(\mathbf{R}^n)$, where $m \in \mathbf{Z}$, $0 \leq \rho$, $\delta \leq 1$, and $\nu(\xi)$ is a "basic weight function," defined by H. Kumanogo [4]. However, if one does not go into such details, one may roughly understand that our space $\mathscr{S}$ has a similar property as the general space $S^m_{\rho,\delta}$.

We next want to derive the symbol formula for adjoint operators. Let $A \in \mathscr{L}_{\mathring{x}^*}$ and let A^* be the adjoint operator defined at $\mathring{x}'^* = (0; 0, 0, \ldots, 0, -\sqrt{-1})$. Let $a(x, \xi) = \sigma(A)$ be the (total) symbol of A defined in Section 1. We want to prove the following formula:

$$\sigma(A^*)(x, -\xi) \sim \sum_{\alpha} \frac{(-1)^{|\alpha|}}{\alpha!} \partial_x^\alpha \partial_\xi^\beta a(x, \xi). \tag{4}$$

We remind the reader of the result in distribution theory: If $a(x, \xi) \in S^m_{\rho,\delta}(\mathbf{R}^n)$, then we have

$$\left| \frac{1}{\alpha!} \partial_x^\alpha \partial_\xi^\alpha a(x, \xi) \right| \le C_\alpha |\xi|^{m+(\delta-\rho)|\alpha|}.$$

If $\rho > \delta$, the asymptotic expansion (4) formally decreases in $|\xi|$ as $|\alpha| \to \infty$. However, if $\rho < \delta$ to the contrary, it increases as $|\alpha| \to \infty$, and thus we cannot expect to obtain such a formula. And therefore we need to restrict ourselves to the case $\rho > \delta$ (or $\rho \ge \delta$), and in such a case we can in fact prove (4) in the strict sense.

In our hyperfunction theory, an analogous situation occurs, and as is easily understood from the above explanations, we need to assume some conditions in terms of scaling functions. Let $C_0, C_1 > 0$ be two given constants. We say that two scaling functions $\lambda(t)$ and $\mu(t)$ satisfy *Condition* C_0, C_1 if

$$0 < t < \frac{1}{C_0} \Rightarrow \lambda^{-1}(t)\mu^{-1}(t) > C_1 t.$$

This condition corresponds to the above condition "$\rho > \delta$." Now we can state the following

Theorem 5. *Let $C_0, C_1 > 0$ be large enough, and let $\lambda(t)$, $\mu(t)$ satisfy Condition C_0, C_1. If $\sigma(A) = a(x, \xi) \in \mathscr{S}_{\lambda,\mu}$, then we have* (4). *More precisely, the kernel function $u(x, x')$ of A^* is given by*

$$u(x, x') = \frac{1}{(2\pi\sqrt{-1})^n} \sum_{j=0}^{\infty} \int_{\Delta(j)} e^{(x-x')\xi} b_j(x, \xi)\, d\xi, \tag{5}$$

where

$$b_j(x, -\xi) = \sum_{|\alpha|=j} \frac{(-1)^{|\alpha|}}{\alpha!} \partial_x^\alpha \partial_\xi^\alpha a(x, \xi)$$

and

$$\Delta(j) = \{\xi \in \sqrt{-1}\mathbf{R}^n;\ \mathrm{Im}\ \xi_n < -C_0 |\mathrm{Im}\ \xi_k|,\ 1 \le k \le n-1,\ \mathrm{Im}\ \xi_n < -C_0(j+1)\}.$$

Remark. We can prove that the power series (5) converges in an appropriate complex domain, and it defines a microlocal operator, which coincides with A^* at $\mathring{x}'^*$.

As for the symbol formula for composite operators, the situation is similar, and we can prove the following

Theorem 6. *Let $A_i \in \mathcal{L}_{\mathring{x}^*}$, $i=1,2$, and let $\sigma(A_i)=a_i(x,\xi)\in\mathcal{S}_{\lambda_i,\mu_i}$, $i=1,2$. If $\mu_1(t)$ and $\lambda_2(t)$ satisfy Condition C_0, C_1 with C_0 and C_1 large enough, we have*

$$\sigma(A_1A_2)\sim\sum_\alpha\frac{1}{\alpha!}\partial_\xi^\alpha a_1(x,\xi)\partial_x^\alpha a_2(x,\xi). \tag{6}$$

(*The precise meaning of the asymptotic expansion* (6) *is the same as that of Theorem* 5.)

3. Examples

In this section we give two important examples of our theory. We first consider the Lewy-Mizohata type operators $P_\pm(x,D)=D_1\pm\sqrt{-1}x_1D_2$ at $\mathring{x}^*=(0,0,0,\sqrt{-1})\in\sqrt{-1}S^*\mathbf{R}^2$. As is well known, $P_-(x,D)$ has a right inverse at $\mathring{x}^*$. We calculate it heuristically from a symbol-theoretical point of view. We assume that $E(x,D)\in\mathcal{L}_{\mathring{x}^*}$ satisfies $P_-(x,D)E(x,D)=\mathrm{Id}$ at $\mathring{x}^*$. Denoting the symbol of $E(x,D)$ also by $E(x,\xi)$, we should have

$$(\xi_1-\sqrt{-1}x_1\xi_2)E(x,\xi)+\partial_{x_1}E(x,\xi)=1. \tag{7}$$

The general solution of (7) can be written as

$$E(x,\xi)=\int_{\mathring{x}_1}^{x_1}\exp\varphi(x_1,t,\xi)\,dt,$$

where $\varphi(x_1,t,\xi)=(\sqrt{-1}/2)x_1^2\xi_2-x_1\xi_1-(\sqrt{-1}/2)t^2\xi_2+t\xi_1$, and $\mathring{x}_1$ is some fixed point. We define $\mathring{x}_1$ by $\mathring{x}_1=\sqrt{-1}c\xi_1/\xi_2$, where $c>0$ is arbitrary, but we choose $c=\sqrt{2}-1$. If t lies on the line segment from $\mathring{x}_1$ to x_1, a direct calculation shows

$$\operatorname{Re}\varphi(x_1,t,\xi)\le C(|\operatorname{Im}x|^2+(|\operatorname{Re}\xi|/\operatorname{Im}\xi_2)^2)\operatorname{Im}\xi_2$$

with some $C>0$, and thus we have $E(x,\xi)\in\mathcal{S}_{\mathring{x}^*}$, $P_-(x,D)E(x,D)=\mathrm{Id}$. It may be of interest to calculate the symbol of $E(x, D)P_-(x, D)$. It is given by

$$E(x, \xi)(\xi_1 - \sqrt{-1}x_1\xi_2) + \partial_{\xi_1} E(x, \xi)(-\sqrt{-1}\xi_2)$$
$$= \int_{\mathring{x}_1}^{x_1} (\xi_1 - \sqrt{-1}t\xi_2) \exp \varphi(x_1, t, \xi)\, dt + \sqrt{-1}\xi_2 \frac{\partial \mathring{x}_1}{\partial \xi_1} \exp \varphi(x_1, \mathring{x}_1, \xi)$$
$$= \int_{\mathring{x}_1}^{x_1} \frac{\partial}{\partial t} \exp \varphi(x_1, t, \xi)\, dt - (\sqrt{2}-1) \exp \varphi(x_1, \mathring{x}_1, \xi)$$
$$= \exp \varphi(x_1, x_1, \xi) - \sqrt{2} \exp \varphi(x_1, \mathring{x}_1, \xi)$$
$$= 1 - K(x, \xi).$$

Here $K(x, \xi)$ is as in the example in Section 1. Thus we have $E(x, D)P_-(x, D) = \mathrm{Id} - K(x, D)$. As for $P_+(x, D)$, we define $F(x, \xi)$ by

$$F(x, \xi) = \int_{\mathring{\xi}_1}^{\xi_1} \exp \psi(x_1, \xi, \tau)\, d\tau$$

where $\psi(x_1, \xi, \tau) = (\sqrt{-1}/2)\xi_1^2\xi_2^{-1} - x_1\xi_1 - (\sqrt{-1}/2)\tau^2\xi_2^{-1} + x_1\tau$ and $\mathring{\xi}_1 = \sqrt{-1}(\sqrt{2}-1)x_1\xi_2$. Then we can prove that $F(x, \xi) \in \mathscr{S}_{\mathring{x}^*}$ and the corresponding operator $F(x, D)$ satisfies

$$P_+(x, D)F(x, D) = \mathrm{Id} - K(x, D), \qquad F(x, D)P_+(x, D) = \mathrm{Id}.$$

We next consider a Grusin type operator $P(x, D) = D_1^2 + x_1^2 D_2^2 + A(x, D)$ at $\mathring{x}^* = (0, 0, 0, \sqrt{-1}) \in \sqrt{-1}S^*\mathbf{R}^2$. Here $A(x, D)$ is a microdifferential operator of order 1. It is well known that if the principal symbol $A_1(x, \xi)$ of $A(x, D)$ satisfies

$$A_1(x, \xi)\xi_2^{-1} \notin \{-(2j+1)\sqrt{-1};\, j = 0, 1, 2, \ldots\},$$

then $P(x, D)$ has the right and left inverse at $\mathring{x}^*$. However, the author has not yet been able to calculate it under the above general condition, and here we assume the following stronger condition:

$$\mathrm{Im}(A_1(x, \xi)\xi_2^{-1}) > -1. \tag{8}$$

Under the assumption (8), we can calculate the right and left parametrix $G(x, D)$ of $P(x, D)$. Since the calculation is rather long, we only give a sketch of it. We want to solve the following equation for $G_j(x, \xi)$, $j = 0, 1, 2, \ldots$, inductively:

$$\sum_{|\alpha|+i=j} \frac{1}{\alpha!} \partial_\xi^\alpha G_i \partial_x^\alpha (\xi_1^2 + x_1^2\xi_2^2 + A(x, \xi)) = \delta_{j0}.$$

Here $A(x, \xi)$ denotes the total symbol of $A(x, D)$. From technical reason, we add meaningless terms, and consider

$$\begin{aligned}\sum_{|\alpha|+i=j} \frac{1}{\alpha!} \partial_\xi^\alpha G_i \partial_x^\alpha (\xi_1^2 + x_1^2 \xi_2^2 + A(x, \xi)) \\ + (2\sqrt{-1}\xi_1 \xi_2 \partial_{\xi_1} + 2x_1 \xi_2^2 \partial_{\xi_1} + \sqrt{-1}\xi_2)(G_j - G_{j-1}) \\ + 2\xi_2^2 \partial_{\xi_1}^2 (G_{j-1} - G_{j-2}) = \delta_{j0}. \qquad (9)_j\end{aligned}$$

We define $G_{-1} = G_{-2} = 0$, and define $G_0, G_1, G_2, \ldots$, inductively by solving $(9)_j$. We can prove that there exist $G_j(x, \xi)$, $j = 0, 1, 2, \ldots$, satisfying $(9)_j$, such that

$$|G_j(x, \xi)| \leq CR^j \exp\{C(|\text{Im } x|^2 + (|\text{Re } \xi|/\text{Im } \xi_2)^2)\text{Im } \xi_2\}$$

on

$$\{(x, \xi) \in \mathbf{C}^n \times \mathbf{C}^n;\ C|x| < 1,$$
$$\text{Im } \xi_2 > C|\text{Im } \xi_1|,\ \text{Im } \xi_2 > C|\text{Re } \xi_k|,\ k = 1, 2,\ \text{Im } \xi_2 > C(j+1)\}.$$

Here $C > 0$ is some constant, and $0 < R < 1$. The total symbol $G(x, \xi)$ of $G(x, D)$ is given by $G(x, \xi) \sim \sum_{j=0}^\infty G_j(x, \xi)$. Precisely speaking, the kernel function $L(x, x')$ of $G(x, D)$ is given by

$$L(x, x') = \frac{1}{(2\pi\sqrt{-1})^n} \int_{\Delta'(j)} e^{(x-x')\xi} G_j(x, \xi)\, d\xi$$

where

$$\Delta'(j) = \{\xi \in \sqrt{-1}\mathbf{R}^2;\ \text{Im } \xi_2 > C|\text{Im } \xi_1|,\ \text{Im } \xi_2 > C(j+1)\}.$$

From $(9)_j$ it is easy to see that we have $G(x, D)P(x, D) = \text{Id}$. Considering the adjoint operators, we can also prove that $P(x, D)G(x, D) = \text{Id}$.

References

[1] T. Aoki, Symbols and formal symbols of pseudodifferential operators, *Advanced Studies in Pure Mathematics* **4** (1984) 181–208.

[2] L. Hörmander, Pseudo-differential operators and hypoelliptic equations, *Proc. Symposium on Singular Integrals*, Amer. Math. Soc. **10** (1967) 138–183.

[3] M. Kashiwara and T. Kawai, Pseudo-differential operators in the theory of hyperfunctions, *Proc. Japan Acad.* **46** (1970) 1130–1134.

[4] H. Kumanogo, Pseudo-differential operators and the uniqueness of the Cauchy problem, *Comm. Pure and Appl. Math.* **22** (1969) 73–129.

[5] M. Sato, T. Kawai, and M. Kashiwara, *Microfunctions and Pseudo-differential Equations*, Lect. Notes in Math. 287, Springer, 1973, pp. 265–529.

Some Observations on Geometric Representations of the Superconformal Algebras and a Super Analogue of the Mumford Sheaves

Kimio Ueno
Department of Mathematics
Yokohama City University
Yokohama, Japan

and

Hirofumi Yamada
Department of Mathematics
College of Science
University of the Ryukyus
Okinawa, Japan

Introduction

In this paper we make an attempt to "superize" the results of Arbarello, De Concini, Kac and Procesi [6]. We will discuss the super analogue of the Mumford sheaves.

Recently many physicists and mathematicians have discussed the relationship between representations of the Virasoro algebra and the geometry of

Algebraic Analysis, Volume II

ISBN 0-12-400466-0

the moduli spaces of curves, [1-7]. Such a trend of study originates from string theory in particle physics. We believe that it will bring a new development in geometry. It means to study the geometry of the moduli spaces in a framework of quantum field theory, or of representations of infinite-dimensional algebras. Among the results cited above, we are especially interested in the papers of Arbarello *et al.* [6] and Beilinson, Manin and Schechtman [4]. In [4] they showed that the Witt algebra (the Virasoro algebra without a center) acts infinitesimally on the moduli spaces of curves and that this action canonically lifts to an action of the Virasoro algebra in the Mumford sheaf λ_j with central charge $6j^2-6j+1$. Let $\pi:\mathscr{C}\to\mathscr{S}$ be a local universal deformation space of a compact Riemann surface R. Then λ_j is defined as a sheaf on $\mathscr{S}$ with the stalk

$$\lambda_{j,[R]}=\det(H^0(R,\Omega_R^{\otimes j}))\otimes\det(H^1(R,\Omega_R^{\otimes j}))^*,$$

where $[R]=\pi(R)$ and Ω_R is the sheaf of holomorphic one-forms on R. It is known ([3]) that there is an isomorphism

$$\lambda_j\simeq\lambda_1^{6j^2-6j+1}. \tag{0.1}$$

This is important in the bosonic string theory to construct the above isomorphism in a canonical way.

On the other hand, Arbarello *et al.* gave another proof in [6] of the following numerical identity on the first Chern class of the Mumford sheaves:

$$c_1(\lambda_j)=(6j^2-6j+1)c_1(\lambda_1). \tag{0.2}$$

Their method is based on the embedding of the *dressing* moduli space into Sato's universal Grassmann manifold *UGM* [8] via the Krichever map (or the Burchnal-Chaundy-Krichever theory, cf. Shiota [9] or Segal and Wilson [10]). Let $\underline{d}$ be the Witt algebra and ρ_j be its representation with spin j (ρ_j is a Lie algebra homomorphism from $\underline{d}$ to $\mathfrak{gl}(\infty)$). We denote by $\mathscr{M}_g$ the coarse moduli space of compact Riemann surfaces with genus g. They proved that there exists a homomorphism $\mu:H^2(\underline{d})\to H^2(\mathscr{M}_g)$ satisfying

$$\mu(\rho_j^*(\psi))=c_1(\lambda_j), \tag{0.3}$$

where ψ is the non-trivial two cocycle of $\mathfrak{gl}(\infty)$. Since one has

$$\rho_j^*(\psi)=(6j^2-6j+1)\rho_1^*(\psi), \tag{0.4}$$

the identity (0.2) can be deduced as a corollary of (0.3).

We formulate a super analogue Λ_j of the Mumford sheaf on a local deformation space of a superconformal curve. The super version of the isomorphism (0.1) is the following

Theorem (Deligne and Voronov, cf. [20]).

$$\Lambda_j \simeq \Lambda_1^{2j-1}. \tag{0.5}$$

Especially one deduces $\Lambda_3 \simeq \Lambda_1^5$, which will play an important role in the fermionic string theory. Although we do not know the complete proof, we will give two reasons supporting this theorem. One of them is a geometrical consideration to the restriction of Λ_j onto the base manifold of the super-moduli space. The other is a Lie theoretical discussion on the super Witt algebra and $\mathscr{gl}(\infty|\infty)$, which are the super extensions of the Witt algebra and $\mathscr{gl}(\infty)$ respectively. We will show the super version of (0.4), which holds for these Lie superalgebras.

The authors considered previously in [11, 12, 13; see also 18, 19] the super KP hierarchy, the universal super Grassmann manifold and $\mathscr{gl}(\infty|\infty)^{\tilde{}}$, which is a one-dimensional central extension of $\mathscr{gl}(\infty|\infty)$. These concepts will play an important role in the theory of the supermoduli spaces. However, in order to develop the theory more in the direction of [6], we need a deeper understanding of the super quasi-periodic solutions to the SKP hierarchy and the super Fock representations of $\mathscr{gl}(\infty|\infty)^{\tilde{}}$ (cf. [14]). We hope that the result of this note will be the first step of the theory.

The authors thank Doctor T. Shiota for stimulating discussions. He also pointed out by [20] that the isomorphism (0.5) was proved by P. Deligne and A. Voronov independently.

1.

In this section, we describe the definition of superconformal curves and state a theorem on the super analogue of the Mumford sheaves (0.1). For foundations on the theory of supermanifolds, readers should refer to Manin's article [15].

Let $C = (R, \mathcal{O}_C)$ be a complex supermanifold of superdimension $(1|1)$, and $\mathcal{T}_C$ the sheaf of holomorphic super vector fields on it. We say that C is a superconformal curve if and only if there exists a locally free $\mathcal{O}_C$-submodule $\mathcal{T}_C^{(1)}$ of $\mathcal{T}_C$ with the following property [16]. There exists an odd vector field D, a local basis of $\mathcal{T}_C^{(1)}$, such that $\{D, D^2\}$ form a local basis of $\mathcal{T}_C$. This module is of rank $(1|0)$ and is called a superconformal structure of C. We know that there exists a bijection between the isomorphism classes of the following objects [16]:

(1) Spin structures on the Riemann surface R.
(2) Superconformal curves C with the base manifold R.

We define a quotient module $\mathscr{T}_C^{(0)}$ for a superconformal curve C with the superconformal structure $\mathscr{T}_C^{(1)}$, by the following short exact sequence:

$$0 \to \mathscr{T}_C^{(1)} \to \mathscr{T}_C \to \mathscr{T}_C^{(0)} \to 0. \tag{1.1}$$

$\mathscr{T}_C^{(0)}$ is a locally free $\mathcal{O}_C$-module of rank $(1|0)$. Applying the functor $\mathscr{H}om_{\mathcal{O}_C}(\cdot, \mathcal{O}_C)$ to the above sequence, we have the following short exact sequence:

$$0 \leftarrow \Omega_C^{(1)} \leftarrow \Omega_C \leftarrow \Omega_C^{(0)} \leftarrow 0. \tag{1.2}$$

Choosing a nice coordinate system (z, ξ) on C, one can express the odd vector field D as $D = \partial/\partial\xi + \xi\,\partial/\partial z$ (cf. Shander [17]). Then the modules in (1.2) are locally represented as follows:

$$\Omega_C \simeq \mathcal{O}_C\, d\xi \oplus \mathcal{O}_C \omega, \qquad \Omega_C^{(0)} \simeq \mathcal{O}_C \omega \qquad \text{and} \qquad \Omega_C^{(1)} \simeq \mathcal{O}_C (d\xi \bmod \mathcal{O}_C \omega),$$

where $\omega = dz + \xi\, d\xi$ is an even element in Ω_C and $d\xi$ is an odd element in Ω_C.

Now let us consider a local universal deformation space $\pi : \mathscr{C} \to \mathscr{S}$ of a superconformal curve $C = (R, \mathcal{O}_C)$. The sheaf $\mathscr{T}_C^{(0)}$ is relevant to infinitesimal deformations of the curve C. In fact we have

Proposition 1 (Manin [1]).

$$T_{[C]}\mathscr{S} = H^1(R, \mathscr{T}_C^{(0)}), \qquad \text{where} \qquad [C] = \pi(C). \tag{1.3}$$

From this proposition, one can deduce that the supermoduli space $\mathscr{S}$ is of $(3g-3|2g-2)$ superdimensions for $g \geq 2$.

We want to calculate the cohomology in the right-hand side of (1.3). Put $\hat{\mathcal{O}} = \mathbf{C}[[z]]$, the ring of formal power series in z, and $\hat{\mathscr{K}} = \mathbf{C}((z))$, the field of formal Laurent series in z. The Lie superalgebra $\mathfrak{g} = \mathfrak{g}_0 \oplus \mathfrak{g}_1$ is defined by

$$\mathfrak{g}_0 = \left\{ f(z)\frac{\partial}{\partial z} + \tfrac{1}{2} f'(z)\xi\frac{\partial}{\partial \xi};\ f(z) \in \hat{\mathscr{K}} \right\},$$

$$\mathfrak{g}_1 = \left\{ f(z)\xi\frac{\partial}{\partial z} - f(z)\frac{\partial}{\partial \xi};\ f(z) \in \hat{\mathscr{K}} \right\}.$$

This Lie superalgebra is called the super Witt algebra (the Neveu–Schwarz algebra without a center). Fixing a point $p \in R$, we set

$$H^0(R, \mathcal{T}_C^{(0)}(*p)) = \{\text{meromorphic sections of } \mathcal{T}_C^{(0)}, \text{ holomorphic everywhere except at } p\}.$$

Choosing a nice local coordinate system (z, ξ) at p, such that the odd vector field D is expressed as $D = \partial/\partial\xi + \xi\,\partial/\partial z$, we make a trivialization of $\mathcal{T}_C^{(0)}$ around p by means of the local basis $\{D, D^2\}$. The space $B(C, p)$ is, by definition, the image of $H^0(R, \mathcal{T}_C^{(0)}(*p))$ in $\hat{\mathcal{K}}[\xi] = \hat{\mathcal{K}} \otimes \mathbf{C}[\xi]$ under this trivialization, which is considered as a subspace of $\mathfrak{g}$.

Proposition 2.

$$H^1(R, \mathcal{T}_C^{(0)}) \simeq \mathfrak{g}/(\mathfrak{g}_{\hat{\mathcal{O}}} + B(C, p)),$$

where $\mathfrak{g}_{\hat{\mathcal{O}}} = \mathfrak{g}/z^{-1}\mathbf{C}[z^{-1}]\mathfrak{g}$.

This proposition can be proved by making use of the Cech cohomology method. Furthermore, through the procedure of the proof, we see that $\mathfrak{g}$ acts infinitesimally on $\mathcal{S}$.

Now we introduce a super analogue of the Mumford sheaves, Λ_j, on $\mathcal{S}$. The stalk of Λ_{2j} $(j \in \frac{1}{2}\mathbf{Z})$ at $[C]$ is given by

$$\Lambda_{2j,[C]} = \mathrm{Ber}\{H^0(R, \Omega_C^{(0)\otimes j})\} \otimes \mathrm{Ber}\{H^1(R, \Omega_C^{(0)\otimes j})\}^*,$$

where "Ber" denotes the Berezinian module (cf. Manin [15]), and * is its dual module. There is the following isomorphism.

Theorem 3 (Deligne and Voronov, cf. [20]).

$$\Lambda_j \simeq \Lambda_1^{2j-1} \qquad (j \in \mathbf{Z}).$$

A ground of this theorem is as follows: If Λ_j is deduced to the base manifold $\mathcal{S}_{\mathrm{red}}$ of $\mathcal{S}$, then $\Lambda_j|_{\mathcal{S}_{\mathrm{red}}} \simeq \lambda_{j/2}(\lambda_{j/2+1/2})^{-1}$. Using (0.1), we get $\Lambda_j|_{\mathcal{S}_{\mathrm{red}}} \simeq (\Lambda_1|_{\mathcal{S}_{\mathrm{red}}})^{2j-1}$. In the next section we will give another reason for the theorem, which is Lie theoretical.

2.

First we introduce a Lie superalgebra $\mathfrak{gl}(\infty|\infty) = \mathfrak{gl}(\hat{\mathcal{K}}[\xi])$. Let $e_i^{(\alpha)} = z^{-i}\xi^\alpha$ $(i \in \mathbf{Z}, \alpha = 0, 1)$ be a basis for $\hat{\mathcal{K}}[\xi]$, and $E_{ij}^{(\alpha\beta)}$ $(i, j \in \mathbf{Z}, \alpha, \beta = 0, 1)$ be the matrix units. Namely, $E_{ij}^{(\alpha\beta)} e_k^{(\gamma)} = \delta^{\beta\gamma}\delta_{jk} e_i^{(\alpha)}$. The Lie superalgebra $\mathfrak{gl}(\infty|\infty)$

is spanned by these matrices, and their supercommutation relations are

$$[E_{ij}^{(\alpha\beta)}, E_{kl}^{(\gamma\varepsilon)}] = \delta^{\beta\gamma}\delta_{jk}E_{il}^{(\alpha\varepsilon)} - (-)^{(\alpha+\beta)(\gamma+\varepsilon)}\delta^{\alpha\varepsilon}\delta_{il}E_{kj}^{(\gamma\beta)}.$$

The second cohomology of $\mathscr{gl}(\infty|\infty)$ is of one dimension: $H^2(\mathscr{gl}(\infty|\infty)) = \mathbf{C}\psi$. The two-cocycle ψ is given by

$$\psi(E_{ij}^{(\alpha\beta)}, E_{kl}^{(\gamma\varepsilon)}) = (-)^{\alpha}\delta^{\alpha\varepsilon}\delta^{\beta\gamma}\delta_{il}\delta_{jk}(Y_+(j) - Y_+(i)),$$

where $Y_+(j) = 1 (j \geq 1)$, $=0 (j \leq 0)$.

We select a basis of $\mathscr{g}$, the super Witt algebra of NS-type (see Section 1), as follows:

$$l_m = z^{-m+1}\frac{\partial}{\partial z} + \frac{-m+1}{2} z^{-m}\xi\frac{\partial}{\partial \xi},$$

$$g_{m+1/2} = \sqrt{-1}z^{-m}\left(\xi\frac{\partial}{\partial z} - \frac{\partial}{\partial \xi}\right), \qquad (m \in \mathbf{Z}).$$

They enjoy the following supercommutation relations:

$$[l_m, l_n] = (m-n)l_{m+n},$$

$$[l_m, g_{n+1/2}] = \left(\frac{m}{2} - n - \frac{1}{2}\right)g_{m+n+1/2},$$

$$[g_{m+1/2}, g_{n+1/2}] = 2l_{m+n+1}.$$

The second cohomology of $\mathscr{g}$ is also one-dimensional. Its non-trivial two cocycle σ is defined by

$$\sigma(l_m, l_n) = \tfrac{1}{4}(m^3 - m)\delta_{m+n,0},$$

$$\sigma(g_{m+1/2}, g_{n+1/2}) = (m^2 + m)\delta_{m+n+1,0}, \qquad \sigma(l_m, g_{n+1/2}) = 0.$$

The Lie superalgebra $\mathscr{D}(1)$ of superdifferential operators of at most first order is, by definition, $\mathscr{D}(1) = \hat{\mathscr{K}}[\xi] \oplus \mathscr{g}$.

Remark. $H^2(\mathscr{D}(1))$ is of three dimensions. We give a basis α_j. Let $x_m = z^{-m}$, $y_{m-1/2} = z^{-m}\xi$ be a basis for the subalgebra $\hat{\mathscr{K}}[\xi]$:

$$\alpha_1(l_m, l_n) = \tfrac{1}{4}(m^3 - m)\delta_{m+n,0},$$

$$\alpha_1(g_{m+1/2}, g_{n+1/2}) = (m^2 + m)\delta_{m+n+1,0},$$

$$\alpha_1 = 0 \qquad \text{otherwise}.$$

$$\alpha_2(l_m, x_n) = m(m-1)\delta_{m+n,0},$$

$$\alpha_2(g_{m+1/2}, y_{n-1/2}) = -2\sqrt{-1}m\delta_{m+n,0},$$

$$\alpha_2 = 0 \qquad \text{otherwise.}$$

$$\alpha_3(x_m, x_n) = -m\delta_{m+n,0},$$

$$\alpha_3(y_{m-1/2}, y_{n-1/2}) = \delta_{m+n-1,0},$$

$$\alpha_3 = 0 \qquad \text{otherwise.}$$

Now we construct representations φ_j of $\mathscr{D}(1)$. The representation space of φ_j is $V_j = \omega^{\otimes j/2} \cdot \hat{\mathscr{K}}[\xi]$ ($j \in \mathbf{Z}$), and is said to be of spin $j/2$. For $P = f(z, \xi) + X \in \mathscr{D}(1)$ ($X \in \mathscr{g}$), we define

$$\varphi_j(P)(\omega^{\otimes j/2}h(z, \xi)) = \omega^{\otimes j/2}(f(z, \xi)h(z, \xi)) + L_X(\omega^{\otimes j/2}h(z, \xi)),$$

where L_X is the Lie derivative associated with the super vector field X. For

$$X = g(z)\frac{\partial}{\partial z} + \tfrac{1}{2}g'(z)\xi\frac{\partial}{\partial \xi},$$

we have

$$L_X(\omega^{\otimes j/2}h(z, \xi)) = \omega^{\otimes j/2}\left(X(h)(z, \xi) + \frac{j}{2}g'(z)h(z, \xi)\right),$$

and for

$$X = g(z)\left(\xi\frac{\partial}{\partial z} - \frac{\partial}{\partial \xi}\right),$$

we have

$$L_X(\omega^{\otimes j/2}h(z, \xi)) = \omega^{\otimes j/2}(X(h)(z, \xi) + j\xi g'(z)h(z, \xi)).$$

Then φ_j is actually a Lie superalgebra homomorphism from $\mathscr{D}(1)$ to $\mathscr{gl}(V_j) \simeq \mathscr{gl}(\infty|\infty)$. The restriction of φ_j on $\mathscr{g}$, which is denoted by ρ_j, provides a Lie superalgebra homomorphism from $\mathscr{g}$ to $\mathscr{gl}(\infty|\infty)$.

Proposition 4. *We have*

$$\rho_j^*\psi = (2j-1)\rho_1^*\psi \qquad (j \in \mathbf{Z}),$$

where $\rho_j^*: H^2(\mathscr{gl}(\infty|\infty)) \to H^2(\mathscr{g})$ *is the pull back by* ρ_j.

We finally remark that one can construct the representation of spin $j/2$ of the super Witt algebra of Ramond-type. In this construction we use a one form $\tilde{\omega} = d \log z + \xi\, d\xi$.

References

[1] Yu. I. Manin, Quantum strings and algebraic curves, talk at the ICM-86, Berkeley.
[2] Yu. I. Manin, *Funct. Anal. Appl.* **20** (1987) 244.
[3] A. A. Beilinson and Yu. I. Manin, *Comm. Math. Phys.* **107** (1986) 359.
[4] A. A. Beilinson, Yu. I. Manin, and V. A. Schechtman, Localization of the Virasoro and Neveu-Schwarz algebras, very preliminary draft.
[5] T. Eguchi and H. Ooguri, *Nucl. Phys.* **B282** (1987) 308.
[6] E. Arbarello, C. De Concini, V. Kac, and C. Procesi, Moduli space of curves and representation theory, talk at AMS-87.
[7] N. Kawamoto, Y. Namikawa, A. Tsuchiya, and Y. Yamada, Geometric realization of conformal field theory on Riemann surfaces, Nagoya Univ., preprint.
[8] M. Sato and Y. Sato, *Lecture Notes in Num. Appl. Anal.* **5** (1982) 259.
[9] T. Shiota, *Inventiones math.* **83** (1986) 333.
[10] G. Segal and G. Wilson, *Publ. IHES.* **61** (1985) 5.
[11] K. Ueno and H. Yamada, *RIMS-Kokyuroku* **554** (1985) 91.
[12] K. Ueno and H. Yamada, *Lett. Math. Phys.* **13** (1987) 59.
[13] K. Ueno and H. Yamada, to appear in *Advanced Studies in Pure Math.*, Kinokuniya.
[14] V. Kac and J. W. van de Leur, Super boson-fermion correspondence, MIT., preprint.
[15] Yu. I. Manin, *J. Soviet Math.* **30** (1985) 3.
[16] A. M. Baranov, Yu. I. Manin, I. V. Frolov, and A. S. Schwarz, *Comm. Math. Phys.*, **111** (1987) 373.
[17] V. N. Shander, *Funct. Anal. Appl.* **14** (1980) 91.
[18] Yu. I. Manin and A. O. Radul, *Comm. Math. Phys.* **98** (1986) 65.
[19] M. Mulase, Solvability of the super KP equation and a generalization of the Birkhoff decomposition, UCLA., preprint.
[20] T. Shiota, private communication.

New Elliptic Solitons

J.-L. Verdier
Ecole Normale Supérieure
Paris, France

In the light of the fundamental work of Sato and the Japanese school on the Kadomtsev–Petviashvili and Korteweg–De Vries equations [4, 9] we take up again the study of the rational, trigonometric and elliptic solutions of those equations, initiated by Calogero [2], Airault, McKean and Moser [1] and Krichever [7]. In this short note, we would like to formulate the problem as we see it now, mention some new results and examples obtained in collaboration with A. Treibich, and raise some questions. As for the results, the detailed proofs will be given elsewhere [12]. Following the tradition in the subject we use the abbreviations KP and KdV.

1. Solitons

Let us call a *soliton* a meromorphic function of three or two complex variables of the type

$$u(x, y, t) = 2\frac{\partial^2}{\partial x^2}(\log \tau_w(x, y, t)), \tag{1}$$

Algebraic Analysis, Volume II

ISBN 0-12-400466-0

$$u(x, y) = 2\frac{\partial^2}{\partial x^2}(\log \tau_{\mathbf{w}}(x, t)), \tag{2}$$

where in (1), $\mathbf{w}$ is a point of the infinite grassmannian Gr, and in (2) $\mathbf{w}$ is in the subgrassmannian $\mathrm{Gr}^{(2)}$. Here we use the results and notations of Segal and Wilson [10]. So the solitons make up a class of solutions of the KP equation for (1) or of the KdV equation for (2). Among the solitons we are interested in the subclass of *solitons of finite type.* They are characterized by the fact that the orbit of the point $\mathbf{w}$ under the KP flows is *finite-dimensional.* They correspond by the Krichever construction to some algebraic data that we will make precise now. We first take a curve Γ, i.e., a reduced, compact, irreducible, one-dimensional complex analytic (or complex algebraic) variety. In other words, Γ is a compact Riemann surface with a finite number of singularities. We fix a point p in the open subset Γ^0 of smooth points of Γ. Associated to Γ and p we have the Jacobian manifold $\mathrm{Jac}\,\Gamma$ and the Abel map $A_\Gamma : \Gamma^0 \to \mathrm{Jac}\,\Gamma$ which associates to a point $m \in \Gamma^0$ the divisor class of $\{m\} - \{p\}$. The manifold $\mathrm{Jac}\,\Gamma$ is a connected commutative algebraic group of dimension the arithmetic genus $g = h^1(\Gamma, \mathcal{O}_\Gamma)$. We have an exact sequence of commutative algebraic groups: $0 \to L \to \mathrm{Jac}\,\Gamma \to \mathrm{Jac}\,\tilde{\Gamma} \to 0$, where the last map comes from the normalization map $\tilde{\Gamma} \to \Gamma$. The group L, called the *local group* in [11], has a purely local description in terms of the singularities of Γ. It is a commutative and connected algebraic linear group and is therefore a canonical product $\mathbf{U} \times \mathbf{T}$ where $\mathbf{U}$ is the unipotent part, a product of additive groups and $\mathbf{T}$ is the maximal algebraic torus, a product of multiplicative groups. Any algebraic linear subgroup of $\mathrm{Jac}\,\Gamma$ is contained in L. The group $\mathrm{Jac}\,\tilde{\Gamma}$ is an abelian variety and it can be shown from [11] that the extension above $\mathrm{Jac}\,\tilde{\Gamma}$ by L is trivial if and only if Γ is a rational curve (i.e. $\mathrm{Jac}\,\tilde{\Gamma} = 0$) or a smooth curve (i.e., $L = 0$). Let $I(\Gamma)$ be complete algebraic moduli space of isomorphism classes of rank 1 torsion-free sheaves F on Γ which are minimal (i.e., $\mathcal{O}_\Gamma = \mathcal{E}nd(F)$) and such that $\chi(\Gamma, F) = h^0(\Gamma, F) - h^1(\Gamma, F) = 0$. The group $\mathrm{Jac}\,\Gamma$ acts on $I(\Gamma)$ via the tensor product of sheaves on Γ. The subset of classes $F \in I(\Gamma)$ such that $h^0(\Gamma, F) \neq 0$, is a divisor denoted by Θ. Let us take now $U = \partial/\partial u(A_\Gamma)|_{u=0}$, $V = \partial^2/\partial u^2(A_\Gamma)|_{u=0}$, $W = \partial^3/\partial u^3(A_\Gamma)|_{u=0}$, three tangent vectors at the origin of $\mathrm{Jac}\,\Gamma$ obtained as derivatives of A_Γ with respect to a local coordinate u on Γ at p. Then there exists an algebraic rational function $\mathbf{p}$ on $I(\Gamma)$, section of $\mathcal{O}_{I(\Gamma)}(2\Theta)$, such that for any $F \in I(\Gamma)$, the meromorphic function

$$u(x, y, t) = 2\mathbf{p}(\exp(xU + yV + tW) \cdot F), \tag{3}$$

is a KP soliton of finite type. Moreover, the $\mathbf{p}$ function is locally of the type $\partial^2/\partial x^2 \log \tau(\exp(xU)F)|_{x=0}$. If Γ is hyperelliptic at p, and if u is an antisymmetric coordinate, then $V=0$, and the meromorphic function

$$u(x, t) = 2\mathbf{p}(\exp(xU + tW) \cdot F), \tag{4}$$

is a KdV soliton of finite type. Moreover there is a one-to-one correspondence between solitons of finite type and the isomorphism classes of data (Γ, p, U, V, W, F) that we call *Krichever data.* In the following, we will identify solitons of finite type and Krichever data. The pointed curve (Γ, p) is called the *spectrum* of the soliton; g is the arithmetic genus of the soliton; F is its initial data.

If the curve Γ is smooth, the map $L \mapsto L((1-g)p)$ is an isomorphism between $I(\Gamma)$ and $\operatorname{Jac}\Gamma$, and the function $\mathbf{p}$ is of the type $L \mapsto 2\partial^2/\partial x^2 \log \theta(L((1-g)p) + xU + K)|_{x=0} + \mathrm{Cst}$ [10], where θ is a Riemann θ-function.

2. Real Periodic Solitons

Let $S = (\Gamma, p, U, V, W, F)$ be a soliton of finite type. We say that S is a *real soliton,* if everything in sight is equipped with a real structure. Let us denote by $G_{\mathbf{R}}$ the real one-parameter subgroup of $\operatorname{Jac}\Gamma$ generated by U. The group $G_{\mathbf{R}}$ depends only on the real pointed curve (Γ, p) and not on the choice of U, V, W, F. The soliton S is said to be *real special* if $G_{\mathbf{R}}$ is closed. So we have two cases: $G_{\mathbf{R}}$ is isomorphic to $\mathbf{R}$ or $G_{\mathbf{R}}$ is isomorphic to the circle $\mathbf{S}^1$. In the latter case we say that we have a *real periodic soliton.* The condition for a real soliton to be periodic is a strong condition. For instance, the moduli of smooth real solitons of genus $g > 1$ is a real analytic "manifold" of dimension $4g+1$ ($3g-3$ parameters for the moduli of Γ, 1 parameter for the point p, 3 parameters for U, V, W, g parameters for the initial data F). Let $S = (\Gamma, p, U, V, W, F)$ be a real periodic soliton. We then have an exact sequence $0 \to H^1(\Gamma, \mathbf{Z}) \to H^1(\Gamma, \mathcal{O}_\Gamma) \to \operatorname{Jac}\Gamma \to 0$. Let Σ be the singular locus of Γ. We have a canonical surjective map c: $H_1(\Gamma - \Sigma, \mathbf{Z}) \simeq H^1_c(\Gamma - \Sigma, \mathbf{Z}) \to H^1(\tilde{\Gamma}, \mathbf{Z})$. Since the space $\mathbf{R}U \subset H^1(\Gamma, \mathcal{O}_\Gamma)$ is tangent to a $\mathbf{S}^1$, it contains some nonzero real element $c(\alpha) \in H^1(\Gamma, \mathbf{Z})$, and we have $c(\alpha) \in \mathbf{R}U \Leftrightarrow \int_\alpha \omega = 0$, $\forall \omega \in H^0(\Gamma, \omega_\Gamma(-p))$, where ω_Γ is the sheaf of regular differentials on Γ [11]. But the latter condition is a set of $g-1$ equations that make sense for the solitons near S, if Γ is smooth, for instance, and if we allow α to vary in its locally constant system, and therefore the moduli of smooth KP real periodic solitons should be a real analytic space of dimension $3g+2$. I don't know any complete proof of this fact. A similar dimension count shows that the moduli of KdV real solitons of genus $g>0$, is a real analytic manifold of dimension $3g+1$, and the moduli of KdV real periodic solitons should be a real analytic subspace of dimension $2g+2$.

3. Special Complex and Algebraic Solitons

Let $S=(\Gamma, p, U, V, W, F)$ be a soliton of finite type. Let us denote by G the complex one-parameter subgroup of $\operatorname{Jac}\Gamma$ generated by U. The group G depends only on the pointed curve (Γ, p) and not on the choice of U, V, W, F. We will say that S is a *special complex* soliton if the group G is a closed analytic subgroup. We will say that S is an *algebraic soliton* if G is an algebraic subgroup. If G is an elliptic curve, one can show that G is algebraic and in this case we will simply call S an *elliptic soliton.* But if G is additive or multiplicative it is not necessarily closed for the analytic topology. And if it is closed, it is not necessarily algebraic. Actually if S is a special complex soliton with an additive or multiplicative G, then Γ is a singular curve; if S is additive or multiplicative algebraic then Γ is a rational singular curve, and S is additive algebraic if and only if Γ is a rational singular curve with only unibranched singular points. In particular S is a KdV additive algebraic soliton if and only if (Γ, p) is isomorphic to the curve obtained by the desingularization at infinity of the plane curve of equation $y^2=x^{2g+1}$, pointed by the point at infinity. This case has been studied in [1]. It gives all the rational solutions of the KdV equation. The multiplicative algebraic case has not been studied yet. As for the complex special soliton of additive or multiplicative type, only a few examples are understood. Let us look for instance at a very simple case. Let E be an elliptic curve, $s \in E$ a point, and Γ be the curve deduced from E by pinching the point s to the order two. Let $p \in \Gamma$ be a smooth point of Γ. It can be shown that the Jacobian of Γ is analytically (but not algebraically) isomorphic to $\mathbf{C}^{\times} \times \mathbf{C}^{\times}$ and that Γ is not the spectrum of an algebraic soliton. Let $\omega \in H^0(E, \Omega_E(2p-s))$, be the unique (up to a scalar multiple) non-zero section and α and β two linearly independent elements of $H_1(E, \mathbf{Z})$. Then, any soliton of spectrum (Γ, p) is a special complex soliton of multiplicative type if and only if the two numbers $\int_\alpha \omega$ and $\int_\beta \omega$ are linearly dependent over $\mathbf{Q}$. Any soliton of spectrum (Γ, p) is not a special complex soliton if and only if the two numbers $\int_\alpha \omega$ and $\int_\beta \omega$ are linearly dependent over $\mathbf{R}$ and independent over $\mathbf{Q}$. Any soliton of spectrum (Γ, p) is a special soliton of additive type if and only if the two numbers $\int_\alpha \omega$ and $\int_\beta \omega$ are linearly independent over $\mathbf{R}$. If we let p vary in Γ^0, we are in the third case outside of a real subcurve γ of Γ^0, we are in the first case in an everywhere dense countable subset of γ, and we are in the ugly second case in the complement.

The curve Γ is hyperelliptic at p if and only if the divisor $2(s-p)$ is rationally equivalent to zero. Then p can no longer move continuously in E. But we can let the curve E vary in the moduli space, and one can show that (Γ, p) is the spectrum of a special complex soliton of multiplicative type when the invariant $j(E)$ belongs to a specified countable subset.

4. Elliptic Solitons

Let $S = (\Gamma, p, U, V, W, F)$ be an elliptic soliton. Then any soliton of spectrum $(\tilde{\Gamma}, p)$ where $\tilde{\Gamma}$ is the normalization of Γ, is again an elliptic soliton and one can therefore first study the smooth elliptic solitons. Let us compute the dimension of the moduli of the smooth elliptic solitons of genus $g > 1$. Let α and β be two linearly independent cycles in $H_1(\Gamma, \mathbf{Z})$ which belong to $\mathbf{C}U \subset H^1(\Gamma, \mathcal{O}_\Gamma)$ and allow them to vary in their locally constant system in a neighbourhood of Γ. Then the equations $\int_\alpha \omega = \int_\beta \omega = 0$, $\forall \omega \in H^0(\Gamma, \Omega_\Gamma(-p))$ is a set of $2g-2$ local equations of the submoduli of the elliptic solitons. It should therefore be of dimension $2g+3$. We have a complete proof of this fact [12]. As for the moduli of the KdV smooth elliptic solitons of genus $g \geq 1$, a similar dimension count gives $g+3$. In particular, if we fix the j invariant of the elliptic curve G, we should find a dimension zero set of possible spectra of genus g. We also have a proof of this fact.

The singular elliptic solitons have not been studied yet. Let us take for instance the very simple case of a curve Γ obtained by adding a double point to a smooth elliptic curve E. There are two cases. If the double point is a pinch point, then, for any $p \in \Gamma^0$, (Γ, p) is not the spectrum of an elliptic soliton. If Γ is obtained by glueing a and b in E, $a \neq b$, then there exists a $p \in \Gamma^0$, such that (Γ, p) is the spectrum of an elliptic soliton if and only if the divisor $(a-b)$ is torsion. If this is so, there are in general two points p for which (Γ, p) is the spectrum of an elliptic soliton. For special values of the invariant $j(E)$, in a countable subset, there is only one point p for which (Γ, p) is the spectrum of an elliptic soliton. In this case, the curve Γ is hyperelliptic at p.

Let (Γ, p, U, V, W, F) be an elliptic soliton, for any $F \in I(\Gamma)$ and fixed y and t, the function $x \mapsto 2\mathbf{p}(\exp(xU+yV+tW) \cdot F)$ is a rational function on the elliptic curve G whose divisor at infinity is $2(G \cdot F \cap \Theta)$ and it is a second derivative of a logarithm. This function is therefore, for a generic F, of the type $2(\sum_1 \mathbf{p}_G(x-a_1(y,t)))+\mathrm{Cst}$, where $\mathbf{p}_G$ is the Weierstrass function with a pole in $1/x^2$ at the origin. The number of poles of this function is the intersection number $(G.F).\Theta$ and is called the *degree* of the elliptic soliton. For a general F, one can show by an argument of pole expansion, using the fact that this function is a solution of the KP equation, that this function in x is of the type $2(\sum_j m_j \mathbf{p}_G(x-a_j(y,t)))+\mathrm{Cst}$, where the m_j are triangular numbers and $\sum_j m_j = (G.F).\Theta$ [1]. Conversely we know that any solution of the KP equation which is, as a function in x, a rational function on E is of the type mentioned above (this is at least proved for the KdV equation in [1]), and that any solution of the type above is an elliptic soliton.

5. Tangent Covers

To study the smooth elliptic solitons, one can use the theory of tangent covers. This comes from the simple fact that the Jacobian of a smooth curve is self-dual. Let (Γ, p, U, V, W, F) be an elliptic soliton. The embedding $G \to \mathrm{Jac}\,\Gamma$ gives, by duality, a projection $\mathrm{Jac}\,\Gamma \to G$. If we compose with the Abel's embedding $\Gamma \to \mathrm{Jac}\,\Gamma$, we obtain a finite cover of a special type, which we call a *tangent cover* [12]. Strictly speaking, a tangent cover of a pointed elliptic curve (E, q), is a pointed curve (Γ, p) smooth at p, together with a finite morphism $\pi: \Gamma \to E$, such that $\pi(p) = q$ and such that the image of E in $\mathrm{Jac}\,\Gamma$ by the natural map is tangent at the origin to the image of Γ by the Abel map. A tangent cover $\pi: \Gamma \to E$ is said to be *primitive* if the natural map $E \to \mathrm{Jac}\,\Gamma$ is injective. Therefore, smooth spectra of elliptic solitons give primitive tangent covers and, conversely, primitive tangent covers provide spectra of elliptic solitons. We have three numerical invariants of a tangent cover $\pi: \Gamma \to E$: The arithmetic genus g of Γ, the geometric genus $\tilde{g}$ of Γ, and the degree of π. When the tangent cover is primitive, the degree of π is the intersection number $E.\Theta$, computed in $\mathrm{Jac}\,\Gamma$. We say that a tangent cover $\pi: \Gamma \to E$ is *minimal* if there exists no factorization of π in a nontrivial birational map $\Gamma \to \Gamma'$ and a tangent cover $\Gamma' \to E$. The main result in this section [12] is that for any n, there exists moduli of minimal tangent covers of degree n, and that these moduli are an affine linear space of dimension $n-1$. The generic elements in this moduli space are smooth tangent covers of genus n. This point of view of tangent covers allows us to recover and make precise the results of Krichever [7] on elliptic KP solitons. The exact connection between our approach and Krichever's approach can be obtained through the C.M.K. (Calogero, Moser, Krichever) theory presented in [12], and we have been inspired by the cohomological interpretation of Lax equations given in [5]. To describe this moduli space, let us introduce the rank-two vector bundle V on E which is a non-trivial extension of two rank-one trivial vector bundle, let us denote by S the corresponding ruled surface over E, and by $\pi: S \to E$, the projection. Let $C_0 \subset S$ be the unique curve of genus >0 and of self-intersection zero, $S_q \subset S$ the fiber over $q \in E$, p the unique intersection point $C_0 \cap S_q$. The automorphism group of S over E is isomorphic to $\mathbf{C}$. For any $n > 0$, the linear system $|nC_0 + S_q|$ is of dimension n and p is a fixed point of this linear system. Any irreducible $\Gamma \in |nC_0 + S_q|$ pointed at p, with the projection $\pi: \Gamma \to E$, is a minimal tangent cover of E of degree n. These irreducible elements define an open linear affine subspace $V(n, E)$ of $|nC_0 + S_q|$. The moduli of minimal tangent covers of degree n is $V(n, E)/\mathrm{Aut}(S/E)$. We can also describe the tangent covers by equations.

6. Hyperelliptic Tangent Covers

Let us fix a curve $C_1 \in |C_0 + S_q|$. There exists a unique involution τ of S preserving C_1 and S_q. An element $\Gamma \in V(n, E)$ is said to be *symmetric* if it is preserved by τ. The symmetric elements make a linear affine subspace $SV(n, E)$ of $V(n, E)$ of dimension $[n/2]$. The generic elements in $SV(n, E)$ are smooth tangent covers Γ of degree n and genus n, equipped with an involution τ over the canonical involution of (E, q), and the quotient Γ/τ is a smooth curve of genus $[n/2]$. We denote by $HV(n, E)$ the set of $\Gamma \in SV(n, E)$ such that the normalized curve of Γ is hyperelliptic at p. Any hyperelliptic tangent cover $\pi: \Gamma \to E$ factors through a nontrivial birational map $\Gamma \to \Gamma'$ where $\Gamma' \in HV(n, E)$. Therefore to study the KdV elliptic solitons we are reduced to study the closed subspace $HV(n, E)$ of $SV(n, E)$. For any integer n, set $\gamma(n) = \sup(p \in \mathbf{N} \mid p(p+1)/2 \leq n)$. The first result which we have is that the arithmetic genus g of any hyperelliptic tangent cover is

$\leq \gamma(n)$ and when n is triangular (i.e., $n = \gamma(n)(\gamma(n)+1)$, we can deduce from a result of Ince [6] that there exists a unique hyperelliptic tangent cover of E of genus $\gamma(n)$. Let us consider, for a given integer n, the set of equations among four positive integers $(\mu_0, \mu_1, \mu_2, \mu_3)$:

$$\begin{aligned} \sum_i \mu_i^2 &= 2n+1, \\ \mu_0 &= n+1 \text{ (mod. 2)}, \\ \sum_i \mu_i &= 2\gamma(n)+1. \end{aligned} \tag{5}$$

Let us denote by $\psi(n)$ the number of distinct solutions of those relations. It has been proved for us by J. Oesterlé that we have $\psi(n) = \rho_3(8n+4-a^2)/8$ where a is the biggest odd number such that $a^2 \leq 8n+4$ and $\rho_3(m)$ is the number of $(z_1, z_2, z_3) \in \mathbf{Z}^3$ such that $m = \sum_i z_i^2$. Since the function ρ_3 is a known arithmetic function, we get the inequalities $1 \leq \psi(n) \leq o(n^{1/4} \log^2 n)$. We have two conjectures: (A) For any integer n and any elliptic curve E, the set $HV(n, E)$ is finite; (B) For any integer n and elliptic curve E, there exists at least $\psi(n)$ distinct hyperelliptic tangent covers of degree n and geometric genus $\gamma(n)$*, and those covers are smooth curves. These conjectures have been checked for small n ($n \leq 11$). It should be pointed out that, since we know explicitly the equations of tangent covers, the conjecture (B) can be checked mechanically in each degree n. To motivate the conjecture (B) let us explain the geometrical meaning of the μ_i. Let $\Gamma \in SV(n, E)$. The fixed points of τ acting on Γ consist of p plus possibly four other points (r_i), $0 \leq i \leq 3$. In order to lower the genus of $\tilde{\Gamma}/\tau$, one can try to increase the number of fixed points in $\tilde{\Gamma}$ and to achieve this goal, one can try to impose multiplicities of order μ_i at the points r_i. If the μ_i satisfy the first two relations of (5), to impose the multiplicities μ_i at the points r_i, amounts to annihilating $[n/2]$ affine linear forms on $SV(n, E)$. Therefore if those forms are *linearly independent*, there exists only one curve Γ fulfilling those conditions. Furthermore, if we have a curve Γ fulfilling those conditions, and if the singularities at the points r_i are ordinary one can prove that the curve $\tilde{\Gamma}$ is hyperelliptic. Finally the third equality of (5) implies that the genus of $\tilde{\Gamma}$ is $\gamma(n)$. Therefore the conjecture (B) is just the statement that two systems of linear equations with complex coefficients, depending on the choice of an elliptic curve E, have a maximal rank.

When we have a smooth spectrum (Γ, p) of a KdV soliton, let us call the "source potential" the function $u(x) = 2\,\partial^2/\partial x^2 \log \theta(xU)$ (up to a constant).

* Footnote added in proof: those conjectures have been proven (Sept. 1988) [13].

The corresponding solitons will be deduced from the function u by adding a constant and applying the different KP flows. When (Γ, p) is the spectrum of an elliptic soliton, the source potential is of the type $u(x) = 2\sum_i m_i \mathbf{p}_E(x-a_i)$, where $\mathbf{p}_E$ is the Weierstrass function of an elliptic curve E, and the m_i are triangular numbers. The degree of the corresponding solitons is $n = \sum m_i$. Since the soliton is KdV, the genus g of the soliton is smaller than $\gamma(n)$. According to the conjecture (B), there should be, for any integer n, at least $\psi(n)$ different elliptic source potentials of degree n and genus $\gamma(n)$. For any integer g, Ince's potential $u(x) = g(g+1)\mathbf{p}_E(x)$ is a KdV elliptic source potential of degree $n = g(g+1)/2$, and genus $g = \gamma(n)$ and we have $\psi(g(g+1)/2) = 1$. The source potentials of maximal genus, for small degrees are: $n=1$, $u(x) = 2\mathbf{p}_E(x)$, $g=1$, $\psi(1)=1$; $n=2$, $u(x) = 2(\mathbf{p}_E(x) + \mathbf{p}_E(x-a))$ where a is a two-division point, $g=1$, $\psi(2)=3$; $n=3$, $u(x) = 6\mathbf{p}_E(x)$, $g=2$, $\psi(3)=1$; $n=4$, $u(x) = 2(3\mathbf{p}_E(x) + \mathbf{p}_E(x-a))$ where a is a two-division point, $g=2$, $\psi(4)=3$; $n=5$,

$$u(x) = 2(3\mathbf{p}_E(x) + \mathbf{p}_E(x-a) + \mathbf{p}_E(x-b))$$

where a and b are two distinct two-division points, $g=2$, $\psi(5)=3$, and there are six other source potentials; $n=6$, $u(x) = 12\mathbf{p}_E(x)$, $g=3$, $\psi(6)=1$. It would be interesting to know a systematic way to write down all the source potentials.

7. Solitons Versus Diffusons

The class of solitons is not the only known class of solutions of the, say KdV, equation. The classical scattering theory, solves the initial value problem for the KdV equation in a space of real functions decreasing at infinity and a theory of curves of infinite genus has been developed in [8] to solve the initial value problem for real periodic potentials. The relations between the different classes is not completely understood in my opinion. For instance, the classical scattering theory associates to any initial data: (a) a function called the *reflexion coefficient*, and (b) a finite set of eigenvalues and norming constants [4]. If the reflexion coefficient is zero, then the corresponding solution is a rational KdV soliton called the *Hirota solution.* If the reflexion coefficient is not zero, then the corresponding solution is not a soliton. Let us call *diffuson* a solution decreasing at infinity such that the set of eigenvalues is empty. A diffuson is therefore completely described by its reflexion coefficient, and the general solution of the KdV equation among decreasing functions is produced by an interaction between a real rational soliton and a diffuson as explained in [4]. We would like to ask the following question: Does there exist a solution of the KdV equation which displays an interaction between a smooth soliton of finite type and a diffuson? Of course such a solution cannot be a function decreasing at infinity but should be a bounded function in x and tend to a quasi-periodic soliton when the time goes to infinity.

Part of this work has been done at the Institute Mittag Leffler and the Forschungsbereich Institute of the University of Göttingen. The author is a member of the UA 212 in the University of Paris 7.

References

[1] H. Airault, H. P. McKean, and J. Moser, Rational and elliptic solutions of the Korteweg–De Vries equation and a related many-body problem. *Comm. Pure Appl. Math.* **30** (1977) 95–148.

[2] F. Calogero, Solution of the one-dimensional n-body problems with quadratic and/or inversely quadratic pair potentials. *J. Math. Phys.* **12** (1971) 419–436.

[3] E. Date, M. Jimbo, M. Kashiwara, and T. Miwa, Transformation groups for soliton equations. I. *Proc. Japan Acad.* **57A** (1981) 342–347; II. *Ibid.*, 387–392; III. *J. Phys. Soc. Japan* **50** (1981) 3806–3812; IV. *Physica* **4D** (1982) 343–365; V. *Publ. RIMS, Kyoto Univ.* **18** (1982) 1111–1119; VI. *J. Phys. Soc. Japan* **50** (1981) 3813–3818; VII. *Publ. RIMS, Kyoto Univ.* **18** (1982) 1077–1110.

[4] P. Deift and E. Trubowitz, Inverse scattering on the line. *Comm. Pure Appl. Math.* **32** (1979) 121–251.

[5] P. A. Griffiths, Linearizing flows and a cohological interpretation of Lax equations. *Am. J. Math.* **107**, No. 6 (1985) 1455–1484.

[6] E. L. Ince, Further investigations into the periodic Lamé function. *Proc. Roy. Soc. Edinburgh* **60** (1940) 83–99.

[7] I. M. Krichever, Elliptic solutions of the Kadomtsev-Petviashvili equation and integrable systems of particles. *Funk. Anal.* **14:4** (1980) 45–54.

[8] H. P. McKean and E. Trubowitz, Hill's operator and hyperelliptic function theory in the presence of infinitely many branched points. *Comm. Pure Appl. Math.* **29** (1976) 143–226.

[9] M. Sato and Y. Sato, Soliton equations as dynamical systems on an infinite dimensional Grassmann manifold. Proc. US-Japan seminar, *Non-linear partial differential equations in applied science.* H. Fujita, P. D. Lax, and G. Strang, ed.) Kinokuniya/North-Holland, 1982, pp. 259–271.

[10] G. Segal and G. Wilson, Loop groups and equations of KdV type. *Publ. Math.* **61** (1985) 5–65.

[11] J.-P. Serre, *Groupes algébriques et corps de classes.* Act. sc. industr. 1264. Hermann Paris, 1959.

[12] A. Treibich and J.-L. Verdier, *Varietés de IMCE.* A paraitre.

[13] A. Treibich and J.-L. Verdier, *Solitons elliptiques.* A paraitre.

On the Microlocal Smoothing Effect of Dispersive Partial Differential Equations, I: Second-Order Linear Equations

Masao Yamazaki
Mathematical Sciences Research Institute
Berkeley, California and
Department of Mathematics
Faculty of Science
University of Tokyo
Tokyo, Japan

0. Introduction

Several authors have found that a number of dispersive partial differential equations have local smoothing effect. Namely, the solutions of such equations have local regularity better than that of the Cauchy data, if the Cauchy data satisfy certain conditions. Here, by a dispersive equation, we mean a non-Kowalevskian evolution equation on $\mathbf{R}^n$ whose evolution operators form a one-parameter group.

First, Kato [5] proved the smoothness of the solutions of the KdV equation, assuming the exponential decay of the Cauchy data in a space

Algebraic Analysis, Volume II

ISBN 0-12-400466-0

direction. Then Hayashi–Nakamitsu–Tsutsumi [3], [4] showed that the regularity of the solutions of the nonlinear Schrödinger equations reflects the decay order of the Cauchy data. In these works, the decay of the Cauchy data plays an important role.

On the other hand, Constantin–Saut [1] obtained regularizing effect of general dispersive equations. Here the gain of regularity is determined by the order of the equation, and the decay of the Cauchy data was not considered. In this paper, they considered equations with principal symbols elliptic with respect to the space variables.

The purpose of this paper is to describe a microlocal version of the regularizing effect of some systems of the second-order linear partial differential equations. Namely, we show that certain decay of the Cauchy data in some direction implies microlocal regularity of solutions at positive time. Here the principal symbol is assumed to be non-characteristic or of principal type at the point. Furthermore, the decay of the Cauchy data is required only in some conic set, not in the whole space.

Our method is a modification of that of Hayashi–Nakamitsu–Tsutsumi [3], [4]. Namely, we make use of a differential operator which almost commutes with the equation, whose idea is similar to the method of geometrical optics for the hyperbolic equations. However, as opposed to the case of the hyperbolic equations, the order of the differential operator constructed above varies as time progresses, and this is why the regularizing effect appears.

Similar results for higher-order equations and nonlinear equations will be published in a forthcoming paper.

1. Notations and the Statement of the Main Results

Let n and N be positive integers, and let I_N denote the $N \times N$ identity matrix. For $\alpha \in \mathbf{N}^n$ and $x \in \mathbf{R}^n$, we put

$$\alpha ! = \alpha_1 ! \cdots \alpha_n !, \qquad |\alpha| = \alpha_1 + \cdots + \alpha_n, \qquad x^\alpha = x_1^{\alpha_1} \cdots x_n^{\alpha_n},$$

$$\partial_{x_l} = \partial/\partial x_l, \qquad \partial_x^\alpha = \partial_{x_1}^{\alpha_1} \cdots \partial_{x_n}^{\alpha_n}, \qquad D_x^\alpha = (-i)^{|\alpha|} \partial_x^\alpha \quad \text{and} \quad \langle x \rangle = (1 + |x|^2)^{1/2},$$

where $\mathbf{N}$ denotes the set of natural numbers (=nonnegative integers), $\mathbf{R}$ denotes the set of real numbers, and $i = \sqrt{-1}$.

For $m \in \mathbf{R}$, let S^m denote the set of functions $a(x, \xi)$ defined on $\mathbf{R}_x^n \times \mathbf{R}_\xi^n$

satisfying

$$|\partial_\xi^\alpha \partial_x^\beta a(x, \xi)| \leq C_{\alpha\beta} \langle \xi \rangle^{m-|\alpha|}$$

for every α, $\beta \in \mathbf{N}^n$. For $a(x, \xi) \in S^m$ and $u(x) \in \mathcal{S}'(\mathbf{R}^n)$, we put

$$a(X, D)u(x) = (2\pi)^{-n} \int_{\mathbf{R}^n} \exp(ix \cdot \xi) a(x, \xi) \hat{u}(\xi)\, d\xi,$$

where $\hat{u}(\xi)$ denotes the Fourier transform of $u(x)$ defined by

$$\hat{u}(\xi) = \int_{\mathbf{R}^n} \exp(-ix \cdot \xi) u(x)\, dx.$$

For an N-vector $p = (p_1, \ldots, p_N)$ and an $N \times N$-matrix $A = (a_{jk})$, put $|p| = (\sum_{j=1}^N |p_j|^2)^{1/2}$ and $|A| = (\sum_{j,k=1}^N |a_{jk}|^2)^{1/2}$. Further, let H^s denote the Sobolev space on $\mathbf{R}^n$, and let $\mathbf{H}^s$ denote the direct product of N copies of H^s equipped with the norm $\|f\|_{\mathbf{H}^s} = (\sum_{j=1}^N \|f_j\|_{H^s}^2)^{1/2}$. For a Sobolev space or a symbol space X, let $L^\infty(A, X)$ denote the set of bounded X-valued functions on A.

Now, let T be a positive number, and consider the following Cauchy problem for a system of second order linear partial differential equations on $[0, T] \times \mathbf{R}^n$:

$$\begin{aligned} -i\partial_t u(t, x) &= A(D) I_N u(t, x) + B(D) u(t, x) + C(t, x) u(t, x) + f(t, x), \\ u(0, x) &= u_0(x), \end{aligned}$$

where $u = (u_1, \ldots, u_N)$ are the unknown functions, $u_0 = (u_{0,1}, \ldots, u_{0,N})$ is the given Cauchy data, and $f = (f_1, \ldots, f_N)$ are the given functions.

We suppose that $A(\xi) = \sum_{j,k=1}^N a_{jk} \xi_j \xi_k / 2$, where (a_{jk}) is a real-valued symmetric matrix, $B(\xi) = \sum_{l=1}^n B_l \xi_l$, where each B_l is a real-valued diagonal matrix, $C(t, x)$ is an $N \times N$ matrix whose entries are smooth functions of $(t, x) \in [0, T] \times \mathbf{R}_x^n$ satisfying the estimate $|\partial_x^\beta C(t, x)| \leq C_\beta \langle x \rangle^{-1}$ for every $|\beta| \geq 1$.

Then, since the operator $iA(D)I_N + iB(D)$ is skew-selfadjoint on $\mathbf{H}^s$ and since the operator $C(t, X)$ is uniformly bounded on $\mathbf{H}^s$, Stone's theorem implies that the above system has a unique solution $u \in L^\infty([0, T], \mathbf{H}^s)$ for $u_0 \in \mathbf{H}^s$ and $f \in L^\infty([0, T], \mathbf{H}^s)$.

We now begin to state our results. Let $\xi_0 \in \mathbf{R}^n \backslash \{0\}$ be a point satisfying $\nabla A(\xi_0) = \sum_{k=1}^n a_{jk} \xi_{0,k} \neq 0$, and let $\Phi(x) \in C^\infty(\mathbf{R}^n)$ be a function such that $\Phi(x) \equiv 1$ holds on a conic neighborhood of $\nabla A(\xi_0)$. Then we have the following theorem.

Theorem 1. *Let s be a real number, and let k be a positive integer. Suppose that* $f\equiv 0$, $u_0\in \mathbf{H}^s$, $\Phi(x)\langle x\rangle^k\cdot u_0(x)\in \mathbf{H}^s$. *Then, for every* $\chi(x)\in C_0^\infty(\mathbf{R}^n)$, *there exists a symbol* $\psi(\xi)\in S^0$ *and a constant C satisfying the following conditions:*

$$\begin{aligned}&\psi(\xi)\equiv 1 \qquad \textit{on a conic neighborhood of } \xi_0.\\ &\|\chi(x)\{\psi(D)u\}(t,x)\|_{\mathbf{H}^{s+k}}\le Ct^{-k} \qquad \textit{for every } 0<t\le T.\end{aligned}\tag{1.1}$$

Next we state a more general result concerning the case where f is not equal to zero. For this purpose we introduce some notations. Suppose that $\phi(x)$, $\phi_0(x)\in C^\infty(\mathbf{R}^n)$, $\psi(\xi)$, $\psi_0(\xi)\in S^0$ and $q=(q_1,\ldots,q_n)\in\mathbf{R}^n$ satisfy the conditions

$$\begin{aligned}&q\cdot x\ge c|x| \quad \text{holds for } x\in\operatorname{supp}\phi_0,\\ &\text{The set } F_{\psi_0}=\{\nabla A(\xi);\ \xi\in\operatorname{supp}\psi_0\} \quad \text{does not contain } 0,\\ &\phi_0(x)\equiv 1 \quad \text{on a conic neighborhood of } \operatorname{supp}\phi,\\ &0\le\phi_0(x)\le 1,\\ &\phi(x)\equiv 1 \quad \text{on a conic neighborhood of } F_{\psi_0},\\ &0\le\phi(x)\le 1,\\ &\psi_0(\xi)\equiv 1 \text{ on a conic neighborhood of } \operatorname{supp}\psi,\\ &\psi(\xi)\equiv 1 \quad \text{on a conic neighborhood of } \xi_0,\\ &|\partial_x^\beta\phi(x)|\le C_\beta\langle x\rangle^{-|\beta|} \text{ and } |\partial_x^\beta\phi_0(x)|\le C_\beta\langle x\rangle^{-|\beta|} \text{ holds for } |\beta|\ge 1,\\ &v\cdot\nabla\phi(x)\ge C|v|\cdot|\nabla\phi(x)| \text{ and } v\cdot\nabla\phi_0(x)\ge C|v|\cdot|\nabla\phi_0(x)| \text{ holds with some}\\ &\qquad \text{positive constant } C \text{ for } v\in F\psi_0.\end{aligned}\tag{1.2}$$

Further, we put $V(t,x,\xi)=\sum_{j=1}^n(t\sum_{k=1}^n a_{jk}\xi_k+x_j)q_j$. Then we have the following

Theorem 2. *Let s and k be the same as in Theorem 1. Suppose that* $u_0\in\mathbf{H}^s$, $\phi_0(x)(q\cdot x)^k\psi_0(D)u_0\in\mathbf{H}^s$, $f\in L^\infty([0,T],\mathbf{H}^s)$ *and* $\phi_0(x)V(t,X,D)^j\psi_0(D)f\in L^\infty([0,T],\mathbf{H}^s)$ *for* $j=1,\ldots,k$. *Then we have* $\phi(x)V(t,X,D)^k\psi(D)u\in L^\infty([0,T],\mathbf{H}^s)$.

2. Proof of Theorem 1

We derive Theorem 1 from Theorem 2. We proceed by induction on k.

Given $\Phi(x)$ and $\chi(x)$ in Theorem 1, we can take $q\in\mathbf{R}^n\backslash\{0\}$, $\psi_0(\xi)\in S^0$, an open cone $U \subset \mathbf{R}^n$ with vertex x_0 satisfying the following conditions:

$\Phi(x) \equiv 1$ on $\{x; |x| > M\} \cap U$ holds for some $M > 0$,

$\operatorname{supp} \chi \subset U$,

U is a conic neighborhood of F_{ψ_0},

$q(x - x_0) \geq C|x - x_0|$ holds on U for some $C > 0$,

$\psi_0(\xi) \equiv 1$ on a conic neighborhood of ξ_0.

Now, making use of a parallel transformation, we may assume $x_0 = 0$. Further, we can choose $\phi(x), \phi_0(x) \in C^\infty(\mathbf{R}^n)$, $\psi(\xi) \in S^0$ and $\tilde{\chi}(x) \in C_0^\infty(\mathbf{R}^n)$ satisfying (1.2),

$$\phi(x) \equiv 1 \qquad \text{holds on a neighborhood of } \operatorname{supp} \tilde{\chi}, \tag{2.1}$$

and

$$\tilde{\chi}(x) \equiv 1 \qquad \text{holds on a neighborhood of } \operatorname{supp} \chi. \tag{2.2}$$

Then, from the assumption $\Phi(x)\langle x\rangle^k u_0(x) \in \mathbf{H}^s$ and the relation

$$(q \cdot x)^k \psi_0(D) = \sum_{|\alpha| \leq k} (i^{|\alpha|}/\alpha\,!)\partial_\xi^\alpha \psi_0(D) \partial_x^\alpha (q \cdot x)^k,$$

it follows that $\phi_0(x)(q \cdot x)^k \psi_0(D) u_0 \in \mathbf{H}^s$. Hence Theorem 2 yields

$$\phi(x)\, V(t, X, D)^k \psi(D) u \in L^\infty([0, T], \mathbf{H}^s).$$

From this and (2.1) we have

$$\tilde{\chi}(x)\, V(t, X, D)^k \psi(D) u \in L^\infty([0, T], \mathbf{H}^s). \tag{2.3}$$

On the other hand, we have

$$\tilde{\chi}(x)\, V(t, X, D)^k = t^k \tilde{\chi}(x) \cdot \left(\sum_{l,m=1}^{k} q_l a_{lm} D_m \right)^k + \sum_{j=0}^{k-1} t^j \tilde{\chi}(x) P_j(t, X, D), \tag{2.4}$$

where $P_j(t, X, D)$ is a differential operator of order j.

Hence, by the induction hypothesis, we have

$$t^j \tilde{\chi}(x) P_j(t, X, D) \psi(D) u \in L^\infty([0, T], \mathbf{H}^s) \tag{2.5}$$

for $j = 0, \ldots, k-1$.

It follows from (2.3), (2.4) and (2.5) that

$$t^k\tilde{\chi}(x)\cdot\left(\sum_{l,m=1}^{n} q_l a_{lm} D_m\right)^k \psi(D)u \in L^\infty([0, T], \mathbf{H}^s). \tag{2.6}$$

Now, since $|\sum_{l,m=1}^{n} q_l a_{lm}\xi_m| = |q\cdot\nabla A(\xi)| \geq C|\nabla A(\xi)|$ holds for $\xi \in \operatorname{supp}\psi_0$, there exists an elliptic symbol $P(\xi)\in S^k$ such that

$$\tilde{\chi}(X)(q\cdot\nabla A)(D)^k\psi(D) = \tilde{\chi}(X)P(D)\psi(D).$$

Hence we have

$$t^k\chi(x)P^{-1}(D)\tilde{\chi}(X)(q\cdot\nabla A)(D)^k\psi(D)u \in L^\infty([0, T], \mathbf{H}^{s+k}). \tag{2.7}$$

On the other hand, we have

$$t^k\chi(X)P^{-1}(D)\tilde{\chi}(X)(q\cdot\nabla A)(D)^k\psi(D)u - t^k\chi(x)\psi(D)u$$
$$= t^k\chi(X)[P^{-1}(D), \tilde{\chi}(X)]P(D)\psi(D)u \in L^\infty([0, T], \mathbf{H}^{s+k}), \tag{2.8}$$

since (2.2) implies that the operator $\chi(X)[P^{-1}(D), \tilde{\chi}(X)]$ is associated with a symbol in S^{-k-m}.

Now the conclusion follows from (2.7) and (2.8).

3. A Proposition

In this section we prove the following

Proposition 1. *Suppose that $\phi(x)$, $\phi_0(x)$ and $\psi_0(\xi)$ satisfy the condition (1.2), and that the family of functions $\{u_\lambda(t, x)\}_{\lambda\in\Lambda}$ and $\{f_\lambda(t, x)\}_{\lambda\in\Lambda}$ satisfy the conditions*

$$(1-\psi_0)(D)(\phi_0 u_\lambda)(t, x) \in L^\infty(\Lambda\times[0, T], \mathbf{H}^{s+1}), \tag{3.1}$$

$$\phi_0 u_\lambda(0, x) \in L^\infty(\Lambda, \mathbf{H}^s), \tag{3.2}$$

$$\phi_0 u_\lambda(t, x) \in L^\infty(\Lambda\times[0, T], \mathbf{H}^{s-1/2}), \tag{3.3}$$

$$\phi_0 u_\lambda(t, x) \in L^\infty([0, T], \mathbf{H}^s) \qquad \textit{for every} \quad \lambda\in\Lambda, \tag{3.4}$$

$$\phi_0 f_\lambda(t, x) \in L^\infty(\Lambda\times[0, T], \mathbf{H}^s) \tag{3.5}$$

$$(-i\partial_t - A(D)I_N - B(D) - C(t, X))u_\lambda(t, x) = f_\lambda(t, x). \tag{3.6}$$

Then we have $\phi u_\lambda(t, x) \in L^\infty(\Lambda\times[0, T], \mathbf{H}^s)$.

Proof. We have, from (3.4) and (3.6),

$$\begin{aligned}\tfrac{1}{2}\partial_t\|\phi u_\lambda\|^2_{\mathbf{H}^s} &= \mathrm{Re}(\phi\cdot\partial_t u_\lambda \mid \phi u_\lambda)_{\mathbf{H}^s}\\ &= -\mathrm{Im}(\phi(X)\{A(D)I_N + B(D)\}u_\lambda \mid \phi u_\lambda)_{\mathbf{H}^s}\\ &\quad -\mathrm{Im}(\phi(x)C(t,x)u_\lambda|\phi u_\lambda)_{\mathbf{H}^s} - \mathrm{Im}(\phi f_\lambda|\phi u_\lambda)_{\mathbf{H}^s}\\ &= J_1 + J_2 + J_3.\end{aligned} \tag{3.7}$$

First, it is easy to see that

$$|J_2| \le C\|\phi u_\lambda\|^2_{\mathbf{H}^s}, \tag{3.8}$$

$$|J_3| \le \|\phi_0 f_\lambda\|_{\mathbf{H}^s}\cdot\|\phi u_\lambda\|_{\mathbf{H}^s} \le C\|\phi u_\lambda\|_{\mathbf{H}^s} \tag{3.9}$$

from (3.5).

Next we have

$$\begin{aligned} J_1 &= -\mathrm{Im}(\{A(D)I_N + B(D)\}(\phi u_\lambda)|\phi u_\lambda)_{\mathbf{H}^s}\\ &\quad +\mathrm{Im}\left\{(-i)\left(\sum_{j=1}^n (\partial_{x_j}\phi)(X)\left(\sum_{k=1}^n a_{jk}I_N D_k + B_j\right)u_\lambda \mid \phi u_\lambda\right)_{\mathbf{H}^s}\right.\\ &\quad \left. -\tfrac{1}{2}\sum_{j,k=1}^n (\partial_{x_j}\partial_{x_k}\phi(x)a_{jk}u_\lambda|\phi u_\lambda)_{\mathbf{H}^s}\right\}\\ &= -\mathrm{Re}(P(X,D)u_\lambda|\phi u_\lambda)_{\mathbf{H}^s},\end{aligned} \tag{3.10}$$

where

$$P(x,\xi) = \sum_{j=1}^n \partial_{x_j}\phi(x)\left(\sum_{k=1}^n a_{jk}\xi_k + B_j\right).$$

Then, since $\phi_0 \equiv 1$ holds on supp ϕ, we have

$$\begin{aligned} P(X,D) &= P(X,D)\phi_0(X)\\ &= P(X,D)\psi(D)\cdot\phi_0(X) + P(X,D)(1-\psi)(D)\phi_0(X).\end{aligned} \tag{3.11}$$

Now (3.1) implies

$$\begin{aligned}&|(P(X,D)(1-\psi)(D)(\phi_0 u_\lambda)|\phi u_\lambda)_{\mathbf{H}^s}|\\ &\quad \le C\|(1-\psi)(D)(\phi_0 u_\lambda)\|_{\mathbf{H}^{s+1}}\|\phi u_\lambda\|_{\mathbf{H}^s} \le C\|\phi u_\lambda\|_{\mathbf{H}^s}.\end{aligned} \tag{3.12}$$

On the other hand, we have

$$\begin{aligned}&(P(X,D)\psi(D)\phi_0 u_\lambda|\phi u_\lambda)_{\mathbf{H}^s}\\ &\quad = (\phi(X)\langle D\rangle^{2s}P(X,D)\psi(D)\phi_0 u_\lambda|\phi_0 u_\lambda)_{\mathbf{H}^0}\\ &\quad = (\langle D\rangle^{-s}\phi(X)\langle D\rangle^{2s}P(X,D)\psi(D)\langle D\rangle^{-s}\cdot\langle D\rangle^s\phi_0 u_\lambda|\langle D\rangle^s\phi_0 u_\lambda)_{\mathbf{H}^0}.\end{aligned} \tag{3.13}$$

Then there exist symbols $Q_j(x, \xi) \in S^j$ for $j = -1, 0, 1$ such that $Q_1(x, \xi) = \phi(x)P(x, \xi)\psi(\xi)$, $Q_0(x, \xi)$ is pure imaginary scalar-valued, and

$$\langle D\rangle^{-s}\phi(X)\langle D\rangle^{2s}P(X, D)\psi(D)\langle D\rangle^{-s} = \sum_{j=-1}^{1} Q_j(X, D). \tag{3.14}$$

Using this notation, we have

$$\begin{aligned} |(Q_{-1}(X, D)\langle D\rangle^s\phi_0 u_\lambda|\langle D\rangle^s\phi_0 u_\lambda)_{\mathbf{H}^0}| &\le C\|\langle D\rangle^s\phi_0 u_\lambda\|^2_{\mathbf{H}^{-1/2}} \\ &= C\|\phi_0 u_\lambda\|^2_{\mathbf{H}^{s-1/2}} \le C \end{aligned} \tag{3.15}$$

and

$$-\mathrm{Re}(Q_0(X, D)\langle D\rangle^s\phi_0 u_\lambda|\langle D\rangle^s\phi_0 u_\lambda)_{\mathbf{H}^0} \le C\|\langle D\rangle^s\phi_0 u_\lambda\|^2_{\mathbf{H}^{-1/2}} \le C \tag{3.16}$$

from (3.3).

Finally, we have

$$\begin{aligned} &-\mathrm{Re}(Q_0(X, D)\langle D\rangle^s\phi_0 u_\lambda|\langle D\rangle^s\phi_0 u_\lambda)_{\mathbf{H}^0} \\ &= -\sum_{l=1}^{n}\left(\phi(X)\left\{\sum_{j=1}^{n} \partial_{x_j}\phi(X)\cdot\{(\nabla A)_j(D) + b_{jl}\}\right\}\psi(D)\right. \\ &\qquad\qquad \left.\times\langle D\rangle^s\phi_0 u_{\lambda,l}|\langle D\rangle^s\phi_0 u_{\lambda,l}\right)_{\mathbf{H}^0}, \end{aligned}$$

where $B_l = (b_{jl}\delta_{jk})$. Moreover, there exists a constant $C_0 > 0$ such that we have

$$\phi(x)\left(\sum_{j=1}^{n}\{\partial_{x_j}\phi(x)\cdot(\nabla A)_j(\xi) + b_{jl}\} + C_0\langle\xi\rangle^{-1}\right)\psi(\xi) \ge 0$$

for every $l = 1, \ldots, n$.

Now, applying the sharp Gårding inequality of Fefferman–Phong [2] and making use of the assumption (3.3), we conclude

$$-\mathrm{Re}(Q_1(X, D)\langle D\rangle^s\phi_0 u_\lambda\,|\langle D\rangle^s\phi_0 u_\lambda)_{\mathbf{H}^0} \le C\|\langle D\rangle^s\phi_0 u_\lambda\|^2_{\mathbf{H}^{-1/2}} \le C. \tag{3.17}$$

From (3.13)–(3.17) we obtain

$$-\mathrm{Re}(P(X, D)\psi(D)\phi_0 u_\lambda\,|\,\phi u_\lambda)_{\mathbf{H}^s} \le C. \tag{3.18}$$

It follows from (3.10), (3.11), (3.12) and (3.18) that

$$J_1 \le C(1 + \|\phi u_\lambda\|_{\mathbf{H}^s}). \tag{3.19}$$

Substituting (3.8), (3.9) and (3.19) into (3.7), we conclude $\partial_t\|\phi u_\lambda\|_{\mathbf{H}^s} \le C$. Now the conclusion follows from this, (3.2) and the Gronwall inequality.

4. Proof of Theorem 2

First we choose functions $\phi_j(x) \in C^\infty(\mathbf{R}^n)$ and symbols $\psi_j(\xi) \in S^0$ for $j = 1, \ldots, 3k$ satisfying the following conditions:

$$0 \le \phi_j(x) \le 1,$$

$$\phi_{j-1}(x) \equiv 1 \quad \text{holds on a conic neighborhood of supp } \phi_j,$$

$$\psi_{j-1}(\xi) \equiv 1 \quad \text{holds on a conic neighborhood of supp } \psi_j,$$

$$\phi_{3k}(x) \equiv \phi(x), \qquad \psi_{3k}(\xi) \equiv \psi(\xi),$$

$$|\partial_x^\beta \phi_j(x)| \le C_\beta \langle x \rangle^{-|\beta|},$$

$$v \cdot \nabla \phi_j(x) \ge C|v| \cdot |\nabla \phi_j(x)| \text{ holds for } v \in F\psi_0.$$

Further, we put $S_\varepsilon(\xi) = \langle \varepsilon \xi \rangle^{-1/2}$ for $\varepsilon \in]0, 1[$, and

$$v^{(j,h,\varepsilon)}(t, x) = V(t, X, D)^k S_\varepsilon(D) \psi_h(D) u(t, x)$$

for $j = 0, 1, \ldots, k$, $h = 0, 1, \ldots, 3k$ and $0 < \varepsilon \le 1$.

Then, since the function $\phi(x) V(t, X, D)^k \psi(D) u(t, x)$ is the limit of the sequence $\{\phi_{3k}(x) v^{(k,3k,\varepsilon(m))}(t, x)\}_{m=1}^\infty$ in the sense of distributions as $\varepsilon(m)$ tends to 0, the conclusion is an immediate consequence of the following

Proposition 2. *The family* $\{\phi_{3j} v^{(j,3j,\varepsilon)}\}$ *belongs to the space* $L^\infty(]0, 1] \times [0, T], \mathbf{H}^s)$.

In the sequel we prove this proposition by induction on j. The case $j = 0$ is trivial.

First we put

$$g^{(m,h,\varepsilon)}(t, x) = (-i\partial_t - A(D) I_N - B(D) - C(t, X)) v^{(m,h,\varepsilon)} \tag{4.1}$$

for $m = 0, 1, \ldots, j$ and $h = 0, 1, \ldots, 3j$. Then we have

$$g^{(m,h,\varepsilon)} = -i(q \cdot \nabla A)(D) v^{(m-1,h,\varepsilon)} + V(t, X, D) g^{(m-1,h,\varepsilon)}$$
$$+ [V(t, x, D), A(D) I_N + B(D) + C(t, X)] v^{(m-1,h,\varepsilon)}$$

by straightforward calculation. Hence, observing the equality $[V(t, X, D), A(D)] = i(q \cdot \nabla A)(D)$, we conclude

$$g^{(m,h,\varepsilon)} = V(t, X, D) g^{(m-1,h,\varepsilon)} + [V(t, X, D), B(D) + C(t, X)] v^{(m-1,h,\varepsilon)}.$$

From this formula we can easily see

$$g^{(m,h,\varepsilon)}=V(t,X,D)^m g^{(0,h,\varepsilon)}+[V(t,X,D)^m,B(D)+C(t,X)]v^{(0,h,\varepsilon)}$$

by induction on k.

On the other hand, choosing $R_h(t,x,\xi)\in L^\infty([0,T],S^{-1})$ satisfying $R_h(t,X,D)=[\psi_h(D),C(t,X)]$, we can write

$$g^{(0,h,\varepsilon)}=S_\varepsilon(D)\psi_h(D)f+R_h(t,X,D)v^{(0,h-1,\varepsilon)}.$$

Combining these formulae we have

$$g^{(j,h,\varepsilon)}=V(t,X,D)^j S_\varepsilon(D)\psi_h(D)f+V(t,X,D)^j R_h(t,X,D)v^{(0,h-1,\varepsilon)}$$
$$+[V(t,X,D)^j,B(D)+C(t,X)]v^{(0,h,\varepsilon)}.$$

Next, we remark the following lemma, which can be easily verified by induction on k.

Lemma 1. *For every operators P and Q we put $Q^{(0)}=Q$ and $Q^{(k+1)}=[P,Q^{(k)}]$ for $k=0,1,\dots$. Then we have*

$$P^kQ=\sum_{j=0}^{k}\binom{k}{j}Q^{(j)}P^{k-j}$$

for every $k=0,1,\dots$.

In view of this lemma, we can write

$$g^{(j,h,\varepsilon)}=V(t,X,D)^j S_\varepsilon(D)\psi_h(D)f$$
$$+\sum_{m=0}^{j}\binom{j}{m}R_h^{(m)}(t,X,D)V(t,X,D)^{j-m}v^{(0,h-1,\varepsilon)}$$
$$+\sum_{m=1}^{j}\binom{j}{m}(B^{(m)}(D)+C^{(m)}(t,X))V(t,X,D)^{j-m}v^{(0,h,\varepsilon)},$$

where $B^{(m)}(D)$, $C^{(m)}(t,X)$ and $R_h^{(m)}(t,X,D)$ are defined by

$$B^{(0)}(D)=B(D),\quad B^{(m+1)}(D)=[V(t,X,D),B^{(m)}(D)],$$
$$C^{(0)}(t,X)=C(t,X),\quad C^{(m+1)}(t,X)=[V(t,X,D),C^{(m)}(t,X)],$$
$$R_h^{(0)}(t,X,D)=R_h(t,X,D)$$

and

$$R_h^{(m+1)}(t,X,D)=[V(t,X,D),R_h^{(m)}(t,X,D)].$$

Then we have $B^{(1)}(D)=\sum_{l=1}^{n} q_l B_l$, $B^{(2)}=B^{(3)}=\cdots=0$, $C^{(m)}(t,x)=(-it\sum_{l,h=1}^{n} q_h a_{hl}\partial_{x_l})^m C(t,x)$. Furthermore, it follows from the assumption on $C(t,x)$ and the definition of $R_h(t,x,\xi)$ that $\langle x\rangle R_h^{(0)}(t,x,\xi)\in L^\infty([0,T],S^{-1})$ for every h.

Hence, putting

$$T_h^{(j,1)}(t,x,\xi)=jR_h^{(1)}(t,x,\xi)+jB^{(1)}\psi_h(\xi)+jC^{(1)}(t,x)\psi_h(\xi)$$
$$+R_h^{(0)}(t,x,\xi)V(t,x,\xi)-i\sum_{l=1}^{n} q_l\frac{\partial R_h^{(0)}}{\partial \xi_l}(t,x,\xi)$$

and

$$T_h^{(j,m)}(t,x,\xi)=\binom{j}{m}R_h^{(m)}(t,x,\xi)+C^{(m)}(t,x)\psi_h(\xi)\qquad(m=2,\ldots,j),$$

we conclude $T_h^{(j,m)}(t,x,\xi)\in L^\infty([0,T],S^0)$ and

$$g^{(j,h,\varepsilon)}=V(t,X,D)^jS_\varepsilon(D)\psi_h(D)f+\sum_{m=0}^{j-1}T_h^{(j,j-m)}(t,X,D)v^{(m,h-1,\varepsilon)}.\tag{4.2}$$

On the other hand, we have the following

Lemma 2. *Let σ be a real number, and let k be a natural number. Suppose that $\phi(x)$, $\tilde\phi(x)\in C^\infty(\mathbf{R}^n)$ satisfy the conditions*

$$\tilde\phi(x)\equiv 1\qquad\text{on a neighborhood of }\operatorname{supp}\phi,\tag{4.3}$$

$$|\partial_x^\beta\phi(x)|\le C_\beta\langle x\rangle^{-|\beta|}\qquad\text{for every }\beta\in\mathbf{N}^n.\tag{4.4}$$

If a family of functions $\{u_\lambda(t,x)\}_{\lambda\in\Lambda}$ satisfies

$$\{u_\lambda\}\in L^\infty(\Lambda\times[0,T],\mathbf{H}^t)\qquad\text{for some }t\in\mathbf{R}\tag{4.5}$$

and

$$\tilde\phi(x)V(t,X,D)^ju_\lambda\in L^\infty(\Lambda\times[0,T],\mathbf{H}^\sigma)\qquad\text{for all }j=0,1,\ldots,k\tag{4.6}$$

and if $\{S_\lambda(t,x,\xi)\}_{\lambda\in\Lambda}$ and $\{T_\lambda(t,x,\xi)\}_{\lambda\in\Lambda}$ are families of symbols such that $\{S_\lambda(t,x,\xi)\}\in L^\infty(\Lambda\times[0,T],S^m)$ and $\{T_\lambda(t,x,\xi)\}\in L^\infty(\Lambda\times[0,T],S^l)$, we have

$$\tilde\phi(x)T_\lambda(t,X,D)V(t,X,D)^kS_\lambda(t,X,D)u_\lambda\in L^\infty(\Lambda\times[0,T],\mathbf{H}^{\sigma-m-l}).$$

Proof. For $j=0,1,\ldots,k$, define the symbols

$$S_\lambda^{(j)}(t,x,\xi)\in L^\infty(\Lambda\times[0,T],S^m)$$

inductively by the formulae $S_\lambda^{(0)} = S_\lambda$ and

$$S_\lambda^{(j+1)}(t, x, \xi) = i \sum_{h=1}^{n} \left(q_h \frac{\partial S_\lambda^{(j)}}{\partial \xi_h}(t, x, \xi) - \sum_{\mu=1}^{n} a_{h\mu} \frac{\partial S_\lambda^{(j)}}{\partial x_\mu}(t, x, \xi) \right).$$

Then we have

$$S_\lambda^{(j+1)}(t, X, D) = [V(t, X, D), S_\lambda^{(j)}(t, X, D)].$$

Hence, Condition (4.3) and Lemma 1 yield

$$\begin{aligned} &\phi(x) T_\lambda(t, X, D) V(t, X, D)^k S_\lambda(t, X, D) u_\lambda \\ &\quad = \phi(x)\tilde{\phi}(x) \sum_{j=0}^{k} \binom{k}{j} T_\lambda(t, X, D) S_\lambda^{(j)}(t, X, D) V(t, X, D)^{k-j} u_\lambda \\ &\quad = \phi(x) \sum_{j=0}^{k} \binom{k}{j} T_\lambda(t, X, D) S_\lambda^{(j)}(t, X, D) \tilde{\phi}(X) V(t, X, D)^{k-j} u_\lambda \\ &\qquad + \sum_{j=0}^{k} \binom{k}{j} \phi(x) [\tilde{\phi}(x), T_\lambda(t, X, D) S_\lambda^{(j)}(t, X, D)] V(t, X, D)^{k-j} u_\lambda . \end{aligned} \tag{4.7}$$

It follows immediately from Condition (4.6) that the first summation in the right-hand side of (4.7) belongs to $L^\infty(\Lambda \times [0, T], \mathbf{H}^{\sigma-m-l})$.

On the other hand, the asymptotic expansion formula and Condition (4.4) ensures the existence of symbols $P_{\lambda,j}(t, x, \xi) \in L^\infty(\Lambda \times [0, T], S^{t-\sigma+m+l-k})$ for every $j = 0, 1, \ldots, k$ such that $\langle x \rangle^h P_{\lambda,j}(t, x, \xi) \in L^\infty(\Lambda \times [0, T], S^{t-\sigma+m+l-k})$ holds for every $h = 0, 1, \ldots, k$.

Then, for every polynomial $A(x)$ of order at most k, we have

$$P_{\lambda,j}(t, X, D) A(X) = \sum_{|\alpha| \le k} (-i)^{|\alpha|} \left(\frac{\partial^\alpha A}{\partial x^\alpha} \cdot \frac{\partial^\alpha P_{\lambda,j}}{\partial \xi^\alpha} \right)(t, X, D), \tag{4.8}$$

which corresponds to a family of symbols in $L^\infty(\Lambda \times [0, T], S^{t-\sigma+m+l-k})$.

On the other hand, the function $V(t, X, D)^{k-j} u_\lambda$ can be represented as a sum of terms of the form $t^l A(X) B(D) u_\lambda$, where A and B are polynomials of order at most k.

From this fact and (4.8), we conclude that the second summation in the right-hand side of (4.7) belongs to $L^\infty(\Lambda \times [0, T], \mathbf{H}^{\sigma-m-l})$. This completes the proof.

We return to the proof of Proposition 2. First, for every positive integers μ and h, we put $\Lambda =]0, 1]$, $T_\lambda(t, x, \xi) = 1$, $S_\lambda(t, x, \xi) = S_\varepsilon(\xi)\psi_h(\xi)$, $u_\lambda =$

$\psi_0(D)f$, $\tilde{\phi}(x)=\phi_0(x)$ and $\phi(x)=\phi_\mu(x)$. Then Lemma 2 yields

$$\phi_\mu(X)V(t,X,D)^jS_\varepsilon(D)\psi_h(D)f\in L^\infty(]0,1]\times[0,T],\mathbf{H}^s). \qquad (4.9)$$

On the other hand, we have the following

Lemma 3. *For $m=0,1,\dots,j-1$ and $\mu=0,1,\dots,m$, we have*

$$\phi_{3m}(x)v^{(\mu,3m,\varepsilon)}\in L^\infty(]0,1]\times[0,T],\mathbf{H}^s).$$

Proof. We proceed by induction on m. The case $m=0$ is trivial.

Assume that $\phi_{3m-3}(x)v^{(\mu,3m-3,\varepsilon)}\in L^\infty(]0,1]\times[0,T],\mathbf{H}^s)$ holds for every $\mu=0,1,\dots,m-1$. Then, putting $\Lambda=]0,1]$, $T_\lambda(t,x,\xi)=1$, $S_\lambda(t,x,\xi)=\psi_{3m}(\xi)$, $\tilde{\phi}(x)=\phi_{3m-3}(x)$, $\phi(x)=\phi_{3m}(x)$ and $u_\lambda(x)=v^{(0,3m-3,\varepsilon)}\in L^\infty(]0,1]\times[0,T],\mathbf{H}^s)$, and applying Lemma 2, we obtain $\phi_{3m}(x)v^{(\mu,3m,\varepsilon)}\in L^\infty(]0,1]\times[0,T],\mathbf{H}^s)$ for every $\mu=0,1,\dots,m-1$.

The case $\mu=m$ is one of the induction hypotheses. This completes the proof.

Now, for $m=0,1,\dots,j-1$ and $h\geq 3m+2$, put

$$T_\lambda(t,x,\xi)=T_h^{(j,j-m)}(t,x,\xi)$$
$$S_\lambda(t,x,\xi)=\psi_{h-1}(\xi),$$
$$\tilde{\phi}(x)=\phi_{3m}(x),$$
$$\phi(x)=\phi_{h-1}(x),$$
$$\theta_\lambda(t,x)=v^{(0,3m,\varepsilon)}(t,x).$$

Then, in view of Lemma 3, we can apply Lemma 2 to obtain

$$\phi_{h-1}(x)T_h^{(j,j-m)}(t,X,D)v^{(m,h-1,\varepsilon)}\in L^\infty(]0,1]\times[0,T],\mathbf{H}^s). \qquad (4.10)$$

Now (4.2), (4.9) and (4.10) imply

$$\phi_{h-1}(x)g^{(j,h,\varepsilon)}\in L^\infty(]0,1]\times[0,T],\mathbf{H}^s) \qquad (4.11)$$

if $h\geq 3j-1$.

Further, for every polynomials $A(x)$ and $B(x)$, the asymptotic expansion formula implies that the operator $(1-\psi_0)(D)\phi_{h-1}(X)A(X)B(D)\psi_h(D)$ can be represented as $P(t,X,D)$ with a symbol $P(t,x,\xi)\in L^\infty([0,T],S^\sigma)$ for every $\sigma\in\mathbf{R}$. This fact and the fact $S_\varepsilon(D)u\in L^\infty(]0,1]\times[0,T],\mathbf{H}^s)$ imply

$$(1-\psi_0)(D)\phi_{h-1}(X)v^{(j,h,\varepsilon)}\in L^\infty(]0,1]\times[0,T],\mathbf{H}^{s+1}) \qquad (4.12)$$

for every $h\geq 1$.

Hence, if the condition

$$\phi_{3j-2}(x)v^{(j,3j-2,\varepsilon)} \in L^\infty(]0,1]\times[0,T],\mathbf{H}^{s-1}) \tag{4.13}$$

holds, we obtain

$$\phi_{3j-1}(x)v^{(j,3j-1,\varepsilon)} \in L^\infty(]0,1]\times[0,T],\mathbf{H}^{s-1/2})$$

by putting $h=3j-1$ in (4.11) and (4.12) and applying Proposition 1.

Applying Proposition 1 again with $h=3j$ in (4.11) and (4.12), we obtain the conclusion

$$\phi_{3j}v^{(j,3j,\varepsilon)} \in L^\infty(]0,1]\times[0,T],\mathbf{H}^{s}).$$

Now it remains only to prove (4.13) from the induction hypothesis. First we remark

$$v^{(j,3j-2,\varepsilon)} = (q\cdot x)v^{(j-1,3j-2,\varepsilon)} + t(q\cdot\nabla A)(D)v^{(j-1,3j-2,\varepsilon)}. \tag{4.14}$$

For the second term of this right-hand side, we apply Lemma 2 by putting $T_\lambda(t,x,\xi)=(q\cdot\nabla A)(\xi)$, $S_\lambda(t,x,\xi)=\psi_{3j-2}(\xi)$ and $u_\lambda = v^{(0,3j-3,\varepsilon)}$. Then, in view of Lemma 3 and the fact $u_\lambda \in L^\infty(]0,1]\times[0,T],\mathbf{H}^s)$, we conclude

$$\phi_{3j-2}(x)(q\cdot\nabla A)(D)v^{(j-1,3j-2,\varepsilon)} \in L^\infty(]0,1]\times[0,T],\mathbf{H}^{s-1}). \tag{4.15}$$

Next we treat the first term. Fix a function $\Phi(x)\in C^\infty(\mathbf{R}^n)$ satisfying $\Phi(x)\equiv 1$ on a neighborhood of $\operatorname{supp}\phi_{3j-2}(x)$ and $\phi_{3j-3}(x)\equiv 1$ on a neighborhood of $\operatorname{supp}\Phi(x)$. Next we put $P(X,D)= -[A(D)I_N+B(D),(q\cdot X)\phi_{3j-2}(X)]$. Then $P(X,D)$ is a first-order differential operator whose coefficients have bounded derivatives.

On the other hand, (4.1) and (4.2) imply

$$\begin{aligned}
&(-i\partial_t - A(D)I_N - B(D) - C(t,X))\phi_{3j-2}(X)(q\cdot X)v^{(j-1,3j-2,\varepsilon)}\\
&\quad = -[A(D)I_N+B(D),(q\cdot X)\phi_{3j-2}(X)]v^{(j-1,3j-2,\varepsilon)}\\
&\qquad + \phi_{3j-2}(x)(q\cdot x)g^{(j-1,3j-2,\varepsilon)}\\
&\quad = P(X,D)\Phi(X)v^{(j-1,3j-2,\varepsilon)} + \phi_{3j-2}(X)V(t,X,D)^jS_\varepsilon(D)\psi_{3j-2}(D)f\\
&\qquad - t\phi_{3j-2}(X)(q\cdot\nabla A)(D)\psi_{3j-3}(X)V(t,X,D)^{j-1}S_\varepsilon(D)\psi_{3j-2}(D)f\\
&\qquad + \sum_{m=0}^{j-2}\Big\{\phi_{3j-2}(x)\sum_{l=1}^{n} iq_l\partial_{\xi_l}T_{3j-2}^{(j-1,j-1-m)}(t,X,D)v^{(m,3j-3,\varepsilon)}\\
&\qquad + \phi_{3j-2}(x)T_{3j-2}^{(j-1,j-1-m)}(t,X,D)v^{(m+1,3j-3,\varepsilon)}\\
&\qquad - t\phi_{3j-2}(x)T_{3j-2}^{(j-1,j-1-m)}(t,X,D)(q\cdot\nabla A)(D)v^{(m,3j-3,\varepsilon)}\Big\}\\
&\quad = I_1+I_2+I_3+\sum_{m=0}^{j-2}(I_4^{(m)}+I_5^{(m)}+I_6^{(m)}).
\end{aligned} \tag{4.16}$$

We now study each term of the right-hand side.

First, (4.9) implies

$$I_2 \in L^\infty(]0,1]\times[0,T], \mathbf{H}^s). \tag{4.17}$$

Next we have

$$\phi_{3j-3}(x)V(t,X,D)^{j-1}S_\varepsilon(D)\psi_{3j-2}(D)f \in L^\infty(]0,1]\times[0,T], \mathbf{H}^s).$$

In fact, this is easy to see if $j=1$, and this follows from (4.9) if $j\geq 2$. It follows that

$$I_3 \in L^\infty(]0,1]\times[0,T], \mathbf{H}^{s-1}). \tag{4.18}$$

Next, put $T_\lambda(t,x,\xi)=1$, $S_\lambda(t,x,\xi)=\psi_{3j-2}(\xi)$ and $u_\lambda = v^{(0,3j-3,\varepsilon)}$. Then Lemma 3 ensures that we can apply Lemma 2 to obtain $\Phi(x)v^{(j-1,3j-2,\varepsilon)} \in L^\infty(]0,1]\times[0,T], \mathbf{H}^s)$. This implies

$$I_1 \in L^\infty(]0,1]\times[0,T], \mathbf{H}^{s-1}). \tag{4.19}$$

Finally, put $T_\lambda(t,x,\xi)=T_{3j-2}^{(j-1,j-1-m)}(t,x,\xi)$, $S_\lambda(t,x,\xi)=\psi_{3j-2}(\xi)$ and $u_\lambda = v^{(0,3j-3,\varepsilon)}$. Then, applying Lemma 2 again, we conclude

$$I_5^{(m)} \in L^\infty(]0,1]\times[0,T], \mathbf{H}^s) \qquad \text{for every } m=0,1,\ldots,j-2. \tag{4.20}$$

In the same way we have

$$I_4^{(m)} \in L^\infty(]0,1]\times[0,T], \mathbf{H}^s) \qquad \text{for every } m=0,1,\ldots,j-2 \tag{4.21}$$

and

$$I_6^{(m)} \in L^\infty(]0,1]\times[0,T], \mathbf{H}^{s-1}) \qquad \text{for every } m=0,1,\ldots,j-2. \tag{4.22}$$

From (4.16) to (4.22) we conclude

$$\begin{aligned}(-i\partial_t - A(D)I_N - B(D) - C(t,X))&\\ \times \phi_{3j-2}(x)(q\cdot x)v^{(j-1,3j-2,\varepsilon)} &\in L^\infty(]0,1]\times[0,T], \mathbf{H}^{s-1}).\end{aligned} \tag{4.23}$$

On the other hand, we have

$$\begin{aligned}\phi_{3j-2}(q\cdot x)v^{(j-1,3j-2,\varepsilon)}(0,x)&\\ &= \phi_{3j-2}(x)\phi_0(x)(q\cdot x)^j S_\varepsilon(D)\psi_{3j-2}(D)\psi_0(D)u_0(x)\\ &= \phi_{3j-2}(x)S_\varepsilon(D)\psi_{3j-2}(D)\cdot\phi_0(X)(q\cdot X)^j\psi_0(D)u_0(x)\\ &\quad + \phi_{3j-2}(x)[\phi_0(X)(q\cdot X)^j, S_\varepsilon(D)\psi_{3j-2}(D)]u_0(x)\\ &= J_1 + J_2.\end{aligned} \tag{4.24}$$

It follows from the assumption that

$$J_1 \in L^\infty(]0,1], \mathbf{H}^s). \tag{4.25}$$

On the other hand, the asymptotic expansion formula and the condition on $\phi_0(x)$ guarantees that the operator

$$\phi_{3j-2}(X)[\phi_0(X)(q\cdot X)^j, S_\varepsilon(D)\psi_{3j-2}(D)]$$

is a pseudodifferential operator associated with a symbol in $L^\infty(]0,1], S^{-j})$. Hence we have

$$J_2 \in L^\infty(]0,1], \mathbf{H}^{s+j}). \tag{4.26}$$

From (4.23), (4.24), (4.25) and (4.26) we conclude

$$\phi_{3j-2}(x)(q\cdot x)v^{(j-1,3j-2,\varepsilon)} \in L^\infty(]0,1]\times[0,T], \mathbf{H}^{s-1}). \tag{4.27}$$

Now (4.13) follows from (4.14), (4.15) and (4.27). This completes the proof of Proposition 2.

Acknowledgements

The author expresses his sincere gratitude to Professor Yoshikazu Giga for valuable suggestions, and to Professor Nakao Hayashi for helpful discussions.

References

[1] P. Constantin and J.-C. Saut, Effets régularisants pour des équations dispersives générales, *C. R. Acad. Sc. Paris* **304** (1987) 407–410.

[2] C. Fefferman and D. H. Phong, On positivity of pseudo-differential operators, *Proc. Natl. Acad. Sci. USA* **75** (1978) 4673–4674.

[3] N. Hayashi, K. Nakamitsu, and M. Tsutsumi, On solutions of the initial value problem for the nonlinear Schrödinger equations in one space dimension, *Math. Z.* **192** (1986) 637–650.

[4] N. Hayashi, K. Nakamitsu, and M. Tsutsumi, On solutions of the initial value problem for the non-linear Schrödinger equations, *J. Funct. Anal.* **71** (1987) 218–245.

[5] T. Kato, The Cauchy problem for the Korteweg-de Vries equations, in *Nonlinear partial differential equations and their applications*, College de France Seminar, I, 293–306, Pitman, Boston, 1979.

Locally Prehomogeneous Spaces and Their Transverse Localizations

Tamaki Yano
Department of Mathematics
Faculty of Science
Saitama University
Urawa, Japan

The purpose of this paper is to introduce the notion of a locally prehomogeneous space and its transverse localization along an orbit. The main result is Theorem 4.1 which shows that the localization procedure preserves several properties, for example, the local prehomogeneity. This gives a unified approach to the theory of prehomogeneous vector spaces and the theory of logarithmic vector fields. A special case was treated in [7].

1. Introduction

Let S be a paracompact connected complex manifold of dimension n, and let $\mathcal{O}$, $\mathcal{X}$ be the sheaf of holomorphic functions, the sheaf of holomorphic vector fields on S, respectively.

Let $\mathcal{G}$ be an $\mathcal{O}$-coherent Lie subalgebra of $\mathcal{X}$. We denote by $\exp(\mathcal{G})$ the pseudogroup of local transformations generated by local 1-parameter groups

Algebraic Analysis, Volume II

ISBN 0-12-400466-0

$\exp(tX)$, $X \in \mathscr{G}$, $t \in \mathbf{C}$, $|t| \ll 1$. A pair $(\mathscr{G}, S)$ is called *locally prehomogeneous* (abbrev. LPH), if there is a closed analytic proper subset D of S, such that $S - D$ is a single $\exp(\mathscr{G})$ orbit. (In Section 2, we will prove local criteria for a pair to be LPH.) Let U be an open set in S. A non-zero holomorphic function $f \in \mathcal{O}(U)$ is called a *relative invariant* (abbrev. RI) of $(\mathscr{G}, S)$ on U if the following formula holds.

$$Xf \in \mathcal{O}f, \qquad \forall X \in \mathscr{G}. \tag{1.1}$$

We now give two typical examples of locally prehomogeneous pairs.

Let (G, ρ, V) be a prehomogeneous vector space (abbrev. PV). For any $X \in \mathrm{Lie}\, G$, $d\rho(X)$ defines a vector field on V in a natural manner. If $\mathscr{G}$ is a $\mathcal{O}$-coherent Lie subalgebra of $\mathscr{X}$ generated by $\{d\rho(X) \mid X \in \mathrm{Lie}\, G\}$, then $(\mathscr{G}, S)$ is LPH (see [3], [4] for prehomogeneous vector spaces).

Let D be a reduced hypersurface of S. A vector field X is called *logarithmic along D*, if X is tangent to D. Let $\mathrm{Der}_S(\log D)$ be the coherent sheaf of logarithmic vector fields on S. Then, $(\mathrm{Der}_S(\log D), S)$ is LPH, and a local defining function of D is a RI. (See [1] Lemma 3.4).

Return to our situation. We call a pair $(\mathscr{G}, S)$ *logarithmic* (abbrev. LG) if there is a reduced hypersurface D, such that $\mathscr{G} = \mathrm{Der}_S(\log D)$. We call a pair $(\mathscr{G}, S)$ *logarithmically free* (abbrev. LGF) if it is LG and $\mathscr{G}$ is locally free of rank n. Such D is called a *Saito divisor* (d'après Cartier).

Let $(\mathscr{G}, S)$ be a pair, not necessarily LPH. Let O be an orbit of $\exp(\mathscr{G})$, and $p \in O$. Let H be a smooth submanifold of S, which is transverse to O at p. In Section 3, we will define a pair $(\mathscr{G}_H, H)$ in a neighborhood of p, where $\mathscr{G}_H$ is a Lie algebra of vector fields on H, which is defined as a restriction of an appropriately chosen $\mathcal{O}$-coherent Lie subalgebra $\mathscr{H}$ of $\mathscr{G}$ that stabilizes O (see Section 3). In Sections 2, 4 we will show the following:

Theorem 1.1. (1) *A pair $(\mathscr{G}, S)$ is LPH if and only if* $\mathrm{rank}_{\mathcal{O}}\, \mathscr{G} = n$.

(2) *Let $(\mathscr{G}, S)$ be a pair. Let $(\mathscr{G}_H, H)$ be the transverse localization of $(\mathscr{G}, S)$ on H. If $(\mathscr{G}, S)$ is LPH (or LG, LGF, having a RI), so is $(\mathscr{G}_H, H)$.*

I will briefly explain the motivation to this paper. Let [PV], [LG], [LGF], [LPH] denote the theory of the corresponding notion. Then, [PV] or [LG] is not the subset of the other, but [PV]$\cap$[LG]$\subset$[PV having a RI]. There are a few but important classes in [regular PV]$\cap$[LGF], and when we treat them, we must cut the zero-locus of a RI by an affine plane transverse to an orbit, as noted by M. Sato. In general, the result is not PV. Such a slice was already treated by K. Saito implicitly in [LG], but that procedure is important also for [PV]\[LG]. The present paper gives a setting of the theory [LPH] $\supset$ [PV] $\cup$ [LG], and shows that the class LPH behaves well for the transverse localization.

I want to express my hearty gratitude to Professor M. Sato and Professor K. Saito, who made [PV] and [LG].

2. Locally Prehomogeneous Space

We retain the notations in Section 1. We denote by TS, T_pS the tangent bundle of S, the tangent space of S at p, respectively. For an element X of $\mathscr{X}$, $X_p \in T_pS$ denotes the value of X at p. For a sheaf $\mathscr{F}$, $\mathscr{F}_p$ means its stalk at p, and $\mathrm{Supp}(\mathscr{F})$ is its support. For a coherent sheaf of ideals $\mathscr{I}$, $\sqrt{\mathscr{I}}$ and $V(\mathscr{I})$ denote its radical and the analytic set defined by $\mathscr{I}$.

Definition 2.1. We define

$$\mathscr{G}(p) = \{X_p \in T_pS \mid X \in \mathscr{G}\}, \tag{2.1}$$

$$\mathscr{I} = \Omega_S^n \bigotimes_{\mathscr{O}_S} \bigwedge^n \mathscr{G}. \tag{2.2}$$

Lemma 2.2.

(1) *$\mathscr{I}$ is a coherent ideal of $\mathscr{O}$.*
(2) *If $(\mathscr{G}, S)$ is LPH, then* $D = \mathrm{Supp}(\mathscr{O}/\mathscr{I}) = V(\sqrt{\mathscr{I}})$.
(3) *Let O be an orbit and $p \in O$. Then* $\dim_p O = \dim_{\mathbf{C}} \mathscr{G}(p)$.

Proof. The ideal $\mathscr{I}$ is obviously coherent by (2.2). We note that if we take local coordinates (x) at p, and express a vector field as $X = \sum a_i(x)\partial/\partial x_i$, the stalk $\mathscr{I}_p$ is generated by the determinant of the coefficients of all n-tuples of vector fields in $\mathscr{G}_p$. Therefore, $S - \mathrm{Supp}(\mathscr{O}/\mathscr{I})$ becomes the open dense orbit if the pair is LPH and hence $D = \mathrm{Supp}(\mathscr{O}/\mathscr{I})$. (3) is obvious.

The following theorem gives local criteria for the prehomogeneity.

Theorem 2.3. *Let $(\mathscr{G}, S)$ be a pair. Then the following conditions are equivalent.*

(1) *$(\mathscr{G}, S)$ is LPH.*
(2) $\dim_{\mathbf{C}} \mathscr{G}(p) = n$ *for some $p \in S$.*
(3) $\mathrm{rank}_{\mathscr{O}_p} \mathscr{G}_p = n$ *for any $p \in S$. (Theorem* 1 (1)).
(4) $\mathrm{rank}_{\mathscr{O}_p} \mathscr{G}_p = n$ *for some $p \in S$.*

(5) $\mathcal{I}_p \neq (0)$ *for any* $p \in S$.
(6) $\mathcal{I}_p \neq (0)$ *for some* $p \in S$.
(7) $\mathcal{I}_p = \mathcal{O}_p$ *for some* $p \in S$.

Proof. (1)⇒(2) is obvious by definition and Lemma 2.2 (3). (3)⇒(4) is trivial. (4)⇒(3) because $\mathcal{G}$ is coherent and S is connected. (2)⇔(7) can be seen from the proof of Lemma 2.2 (2). (5)⇒(6) and (7)⇒(6) are trivial.

(1)⇒(3). Assume $(\mathcal{G}, S)$ is LPH and suppose there is a point $p \in S$, such that $\operatorname{rank}_{\mathcal{O}_p} \mathcal{G}_p = m < n$. Let k_p be the field of quotients of $\mathcal{O}_p$. We can take a neighborhood U of p in S and $f_i \in \mathcal{O}(U)$, $X_i \in \mathcal{G}(U)$, $1 \leq i \leq m$ such that the image of $\{(1/f_1)X_1, \ldots, (1/f_m)X_m\}$ in $k_p \otimes_{\mathcal{O}_p} \mathcal{G}_p$ is a k_p basis. Since $\mathcal{G}$ is coherent, there is a neighborhood $V \subset U$ and $Y_1, \ldots, Y_h \in \mathcal{G}(V)$ such that Y_j's generate $\mathcal{G}_q$ for any $q \in V$. We can find a neighborhood W of p, $W \subset V$, and meromorphic functions g_{ij}, $1 \leq i \leq h$, $1 \leq j \leq m$, in W, such that $Y_i = \sum_j g_{ij}(1/f_j)X_j$, $1 \leq i \leq h$. Set $Y = \bigcup_{i,j}$ {the set of poles of g_{ij}/f_j}. Then, for any $q \in W \setminus Y$, $\mathcal{G}_q$ is generated by $\{X_1, \ldots, X_m\}$. Hence $\dim \mathcal{G}(q) \leq m < n$, and therefore $(W \setminus Y) \cap (S - D) = \varnothing$, which implies $D \supset W \setminus Y$, a contradiction. Therefore, $\operatorname{rank}_{\mathcal{O}_p} \mathcal{G}_p \geq n$. Since $\mathcal{X}$ is locally free of rank n, we must have $\operatorname{rank}_{\mathcal{O}_p} \mathcal{G}_p = n$.

(3)⇒(5). Assume $\mathcal{I}_p = (0)$ for some p. By the definition of $\mathcal{I}$, this assumption implies that any n elements in $\mathcal{G}_p$ are linearly dependent over $\mathcal{O}_p$, hence any n elements in $k_p \otimes \mathcal{G}_p$ are linearly dependent over k_p. Therefore $\operatorname{rank}_{\mathcal{O}_p} \mathcal{G}_p < n$.

(6)⇒(1). Since $\mathcal{I}$ is a coherent non-zero ideal, $D = V(\sqrt{\mathcal{I}})$ is a closed analytic proper subset of S. Let us consider $q \in S \setminus D$. Then, there are $X_1, \ldots, X_n \in \mathcal{G}$ such that $(X_1)_q, \ldots, (X_n)_q \in T_q S$ are linearly independent. This implies $\dim \mathcal{G}(q) = n$ for all $q \in S \setminus D$. We can take a neighborhood U_q of q such that $\forall r \in U_q$ is of the form $e^X q$ for some $X \in \mathcal{G}$. Since $S - D$ is paracompact, there exists a locally finite subcovering of $S - D = \bigcup_{q \in S} U_q$. Now, $S - D$ is a connected paracompact complex manifold, and hence metrizable and arcwise connected. Therefore, for $\forall p$, $\forall p' \in S - D$, if we take a path $L_{pp'}$ connecting p and p', $L_{pp'}$ is covered by a union of a finite number of neighborhoods $\bigcup_j U_{q_j}$, and hence $p' = e^{X_h} e^{X_{h-1}} \cdots e^{X_1} p$ for some $X_j \in \mathcal{G}$. Therefore, $S - D$ is a single $\exp(\mathcal{G})$ orbit.

We have proved

$$\begin{array}{ccccccc} (2) & \Leftarrow & (1) & \Rightarrow & (3) & \Leftrightarrow & (4). \\ \Updownarrow & & \Uparrow & & \Downarrow & & \\ (7) & \Rightarrow & (6) & \Leftarrow & (5) & & \end{array} \tag{2.3}$$

Hence the Theorem

Let $f \in \mathcal{O}$ be a RI of $(\mathcal{G}, S)$. Then, f is relatively invariant under the action of $\exp(\mathcal{G})$. More precisely, we have

Lemma 2.4. *Let $f \in \mathcal{O}(U)$, and $X \in \mathcal{X}(U)$, $U \subset S$. Suppose,*

$$Xf(p) = a(p)f(p), \qquad \forall p \in U. \tag{2.4}$$

Then we have, for small $t \in \mathbf{C}$,

$$f(e^{tX}p) = A(t, p)f(p), \tag{2.5}$$

$$A(t, p) = \exp\left(\int_0^t a(e^{tX}p)\, dt\right). \tag{2.6}$$

Proof. In general, for $g \in \mathcal{O}$, we have $g(e^{tX}p) = (e^{tX}g)(p)$ by Taylor's formula. Set $F(t, p) = f(e^{tX}p)$. Then

$$\begin{aligned} \frac{d}{dt}F(t,p) &= \frac{d}{dt}((e^{tX}f)(p)) \\ &= (e^{tX}Xf)(p) \\ &= (e^{tX}(af))(p) \\ &= a(e^{tX}p)F(t,p). \end{aligned} \tag{2.7}$$

By integrating (2.7) we get (2.5) and (2.6).

Remark 2.5. We can expand (2.6) explicitly:

$$A(t, p) = \sum_{n \geq 0} \frac{t^n}{n!} e_n(a, Xa, \ldots, X^{k-1}a, \ldots), \tag{2.8}$$

where $e_n(u_1, u_2, \ldots, u_k, \ldots)$ is the universal polynomial defined as follows. Let $s_k = (k-1)! \sum_i \xi_i^k$ be the k-th power sum in indeterminates (ξ). Then e_n makes the following formula an identity.

$$e_n(s_1, s_2, \ldots, s_k, \ldots) = n! \sum_{(i)} \xi_{i_1}\xi_{i_2} \cdots \xi_{i_n}, \qquad i_1 \leq i_2 \leq \cdots \leq i_n. \tag{2.9}$$

For example, $e_0 = 1$, $e_1 = u_1$, $e_2 = u_2 + u_1^2$, $e_3 = u_3 + 3u_1u_2 + u_1^3$.

Proposition 2.6. *We assume that $(\mathcal{G}, S)$ is LPH. Let $p \in S$ and let us take a decomposition into irreducible components of D at p. Suppose that $D_1, \ldots, D_m$ are components of codimension one of D. Let f_i $1 \leq i \leq m$ be a local defining function of D_i. Then each f_i is a RI.*

Conversely, if f is a RI and $f(p) = 0$, then the hypersurface $\{q \mid f(q) = 0\}$ is a subset of D. Each irreducible factor of f is a RI.

Proof. Let $g_t = e^{tX}$ be a local transformation at p. Since D is $\exp(\mathcal{G})$-invariant, $f = f_1 \dots f_m$ is relatively invariant by g_t. Then each f_i must be relatively invariant because $\mathcal{O}_p$ is a unique factorization domain. The last assertion is clear.

When $(\mathcal{G}, S)$ is LPH, we can stratify D in several manners. One is the dimensional stratification, starting from D_{reg}, the regular part of D, $(D - D_{\mathrm{reg}})_{\mathrm{reg}}$, and so on. Another one can be given by the stratification of the ideal $\mathcal{I}$. In our case, D consists of $\exp(\mathcal{G})$ orbits. Each orbit is a maximal integral submanifold of $\mathcal{G}$, thus it is smooth connected quasi-regularly imbedded submanifold of S. Hence we can define the following closed analytic sets A_r (due to K. Saito in the case LG):

$$A_r = \{p \in S \mid \dim \mathcal{G}(p) \leq r\}, \qquad r = 0, 1, \dots, n. \tag{2.10}$$

Then by the definition of $\mathcal{G}(p)$, it is clear that

$$A_r = \bigcup O_\alpha, \qquad \text{where } O_\alpha\text{'s are all orbits of } \dim O_\alpha \leq r. \tag{2.11}$$

Since there may be a continuous family of orbits, the following relation *does not hold* in general (see Example 5.2).

$$\text{For } \forall p \in A_r - A_{r-1}, \qquad \dim_p(A_r - A_{r-1}) = r. \tag{2.12}$$

The following gives a simple example of LPH, which is neither LG nor PV.

Example 2.7. $S = \mathbf{C}^3$, $\mathcal{G}$ is generated by $x^2\,\partial/\partial x$, $xy\,\partial/\partial x$, $y^2\,\partial/\partial x$, $x^2\,\partial/\partial y$, $xy\,\partial/\partial y$, $y^2\,\partial/\partial y$, $z\,\partial/\partial z$. Then D and $\mathcal{I}$ are given by $D = \{(0, 0, z)\} \cup \{(x, y, 0)\}$, $\mathcal{I} = \mathfrak{m}_{x,y}^4 \mathcal{O}_S z$, $\mathfrak{m}_{x,y} = \mathcal{O}_S x + \mathcal{O}_S y$.

3. Localization

Let $(\mathcal{G}, S)$ be a pair, O an $\exp(\mathcal{G})$ orbit of codimension m, $p \in O$. Let H be an m-dimensional submanifold of S being transverse to O at p. From now on, we consider locally near p. The next proposition is the key to the subsequent consideration.

Proposition 3.1. *There exist* (i) *a neighborhood* U *of* p, (ii) *a system of local coordinates* $\{\varphi, (z_1, \dots, z_m, y_1, \dots, y_{n-m})\}$ *of* U *at* p, (iii) *a family of local biholomorphic mappings* $\{\tau_s\}_s$ *being parametrized by* (s) *in a neighborhood of* 0 *of* $\mathbf{C}^{n-m}$, *and* (iv) *a family of local holomorphic automorphisms* $\{\pi_{s,u}\}_{s,u}$ *of* H *fixing* p *and being parametrized by* (s, u) *in a neighborhood of*

$(0, 0)$ *of* $\mathbf{C}^{n-m} \times \mathbf{C}^{n-m}$, *satisfying the following property.*

(1) $(\varphi^{-1}(0, y)) = O$.

(2) $(\varphi^{-1}(z, 0)) = H$.

(3) Set $H_{y_0} = \{\varphi^{-1}(z, y_0)\}$ *for a fixed* y_0, *and* $O_{z_0} = \{\varphi^{-1}(z_0, y)\}$ *for a fixed* z_0. *Then* $\tau_s(H) = H_s$, $\tau_s\varphi^{-1}(z, 0) = \varphi^{-1}(z, s)$.

(4) $\tau_s\tau_u\varphi^{-1}(z, 0) = \tau_{s+u}\pi_{s,u}\varphi^{-1}(z, 0)$ *when* $\tau_s\tau_u\varphi^{-1}$ *is defined.*

(5) *At a point* $q = \varphi^{-1}(z, y)$, $T_qS = T_qH_y \oplus T_qO_z$ *gives a decomposition into a direct sum of the tangent space.*

Proof. *Step 1.* Let us take $X_1, \ldots, X_{n-m} \in \mathscr{G}$, whose images in T_pS span $\mathscr{G}(p)$. Take a system of local coordinates $\{\rho, (u_1, \ldots, u_m)\}$ of H with $\rho^{-1}(0) = p$. Consider the holomorphic mapping defined in a neighborhood W of $0 \in \mathbf{C}^m \times \mathbf{C}^{n-m}$:

$$\psi' : (u, t) \mapsto \exp\left(\sum_{i=1}^{n-m} t_i X_i\right)\rho^{-1}(u). \tag{3.1}$$

If W is sufficiently small, ψ' is biholomorphic because $(X_1)_p, \ldots, (X_{n-m})_p$ span $\mathscr{G}(p)$. Thus the inverse of ψ' defines a system of local coordinates. In these coordinates X_i is written in the form

$$X_i = \sum a_{j_1 i}\frac{\partial}{\partial t_{j_1}} + \sum b_{j_2 i}\frac{\partial}{\partial u_{j_2}}, \qquad a_{ji}(0) = \delta_{ji} \text{ (Kronecker's delta)}. \tag{3.2}$$

Now take X'_k, $k = 1, \ldots, h$ such that $\{X_1, \ldots, X_{n-m}, X'_1, \ldots, X'_h\}$ generates $\mathscr{G}$ in a neighborhood U' of p. Set

$$X'_k = \sum a'_{j_1 k}\frac{\partial}{\partial t_{j_1}} + \sum b'_{j_2 k}\frac{\partial}{\partial u_{j_2}}. \tag{3.3}$$

Define new vector fields $Z_1, \ldots, Z_h$, $Y_1, \ldots, Y_m$ by

$$\begin{aligned}(Z, Y) &= \left(\frac{\partial}{\partial u}, \frac{\partial}{\partial t}\right)\left(\begin{array}{c|c}(b'_{j_2k}) - (b_{j_2i})(a_{j_1i})^{-1}(a'_{j_1k}) & (b_{j_2i})(a_{j_1i})^{-1} \\ \hline O_{n-m,h} & I_{n-m}\end{array}\right). \\ &= (X', X)\left(\begin{array}{c|c} I_h & 0 \\ \hline -(a_{j_1i})^{-1}(a'_{j_1k}) & (a_{j_1i})^{-1}\end{array}\right).\end{aligned} \tag{3.4}$$

Clearly, (Z, Y) also generates $\mathscr{G}$. We finally define the local coordinates $\{\varphi, (z, y)\}$ by the inverse of the following local biholomorphism.

$$\psi : (z, y) \mapsto \exp\left(\sum_{i=1}^{n-m} y_i Y_i\right)\rho^{-1}(z). \tag{3.5}$$

We define

$$\mathcal{H} = \sum_{k=1}^{h} \mathcal{O}_S Z_k. \tag{3.6}$$

We remark that if we set

$$\mathcal{H}(q) = \{Z_q \mid \forall Z \in \mathcal{H}\} \subset T_q S \qquad \text{for } q \in U. \tag{3.7}$$

then, by the definition of $\mathcal{H}$, $\mathcal{H}(q) \subset T_q H_y$ for $\forall q \in U$.

Step 2. We will study φ and $\mathcal{H}$. First we check the following:

Lemma 3.2.
(1) *$\mathcal{H}$ is a coherent Lie subalgebra of $\mathcal{G}$.*
(2) $[Y_i, Y_j] \in \mathcal{H}, \forall i, j.$
(3) $[Y_i, Z] \in \mathcal{H}, \forall i, \forall Z \in \mathcal{H}.$
(4) *$\exp(tZ)q \in H$ for $\forall q \in H$, $\forall Z \in \mathcal{H}$ and small $t \in \mathbf{C}$.*

Proof. The module $\mathcal{H}$ is easily seen to be characterized by the following.

$$\mathcal{H} = \{Z \in \mathcal{G} \mid Zt_j = 0, \forall j\}. \tag{3.8}$$

Therefore $\mathcal{H}$ is a coherent $\mathcal{O}$ module and is a Lie algebra.
Since $[Y_i, Y_j]t_k = 0$, $[Y_i, Z]t_k = 0$, (2) and (3) are consequences of (3.8). (4) is also a consequence of (3.8) because $H = \{t_j = 0 \mid \forall j\}$.

Next, we check that (Z, Y) has the similar form as (3.4) in the coordinate φ:

$$(Z, Y) = \left(\frac{\partial}{\partial z}, \frac{\partial}{\partial y}\right)\left(\begin{array}{c|c} C & B \\ \hline O_{n-m,h} & I_{n-m} \end{array}\right). \tag{3.9}$$

That is, we will show

$$Z_k y_j = 0, \tag{3.10}$$

$$Y_i y_j = \delta_{ij}, \qquad \text{for } 1 \le j \le m, \forall i, k. \tag{3.11}$$

The function y_j is characterized by $y_j(\psi(z, c)) = c_j$. Therefore, if we set $Z = Z_k$ and $q = \psi(z, c)$, $Z_q \in T_q S$ is determined by the following.

$$Z_q y_j = \left(\frac{d}{dt} y_j(e^{tZ}\psi(z, c))\right)\bigg|_{t=0}. \tag{3.12}$$

We need the following formula (abbrev. BCH).

Lemma 3.3 (Baker–Campbell–Hausdorff formula. See [6]). *Let X, $Y \in \mathfrak{X}$. Then, the following formal identity holds.*

$$\exp(X)\exp(Y) = \exp(Z), \tag{3.13}$$

$$Z = \sum_{n\geq 1} c_n(X, Y), \qquad c_1 = X + Y, \tag{3.14}$$

$$(n+1)c_{n+1}(X, Y) = \tfrac{1}{2}[X - Y, c_n] + \sum_{\substack{p\geq 1\\ 2p\leq n}} \frac{(-1)^{p+1}B_{2p}}{(2p)!} \sum_{|k|=n} [c_{k_1}, [\cdots[c_{k_{2p}}, X+Y]\cdots]], \quad n \geq 1. \tag{3.15}$$

Here, B_{2p}, $p \geq 1$ are Bernoulli's numbers, $k = (k_1, \ldots, k_{2p})$, $k_j \in \mathbf{N}$, $|k| = \sum k_j$. If X and Y are small enough (in a topology of $\mathfrak{X}$), the series (3.14) converges absolutely and (3.13) is an identity in $\exp(\mathfrak{X})$.

There is a formula of c_n due to Dynkin.

$$nc_n = \sum_{m=1}^{n} \frac{(-1)^{m-1}}{m} \sum_{\substack{p_i+q_i\geq 1\\ |p|+|q|=n}} \frac{\beta(X^{p_1}Y^{q_1}\cdots X^{p_m}Y^{q_m})}{p_1!q_1!\cdots p_m!q_m!}, \tag{3.16}$$

$$\beta(X) = X, \quad \beta(Y) = Y, \quad \beta(Z_1\cdots Z_k) = [Z_1, [Z_2\cdots[Z_{k-1}, Z_k]]\cdots],$$
$$Z_j \in \{X, Y\}, k \geq 2.$$

Using BCH twice, we can find $Z(t) \in \mathscr{H}$ depending holomorphically on t as a TS-valued function such that

$$e^{tZ}\exp(\sum c_i Y_i) = \exp(\sum c_i Y_i)\, e^{tZ(t)}. \tag{3.17}$$

Here, we used Lemma 3.2 (2), (3). Therefore, by (3.12), (3.17) and Lemma 3.2 (4), we have, for some $\mathbf{C}^m$-valued holomorphic function $z(t)$,

$$\begin{aligned} y_j(e^{tZ}\psi(z, c)) &= y_j(\exp(\sum c_i Y_i)\, e^{tZ(t)}\rho^{-1}(z)) \\ &= y_j(\psi(z(t), c)) \\ &= c_j. \end{aligned} \tag{3.18}$$

Hence we get (3.10). Similarly, we have

$$\begin{aligned} y_j(e^{tY_k}\psi(z, c)) &= y_j\left(\exp\left((t + c_k) + \sum_{i\neq k} c_i Y_i\right) e^{tZ_k(t)}\rho^{-1}(z)\right) \\ &= \delta_{jk}t + c_j. \end{aligned} \tag{3.19}$$

Thus we proved (3.11) and much more. Let the notation be the same as in Proposition 3.1. Consider the fibering

$$\bigcup H_y \to O, \qquad \psi(z, y) \mapsto \psi(0, y). \tag{3.20}$$

Then, by (3.18) $\exp(\mathscr{H})$ preserves the fibre structure (3.20), and by using (3.19) repeatedly, for $b = (b_1, \ldots, b_{n-m}) \in \mathbf{C}^{n-m}$ with $\|b\|$ being sufficiently small, $\tau_b := \exp(\sum b_i Y_i)$ induces a translation $(y) \mapsto (y+b)$. The $\pi_{s,u}$ can be constructed similarly.

Thus we have proved Proposition 3.1.

We remark that $e^{tZ(t)} = \exp(\mathrm{ad}(-\sum c_i Y_i))\, e^{tZ}$.

Definition 3.4. (1) Let $\iota : H \to S$ be the imbedding and let ι^* denote the inverse image. Define

$$\mathscr{G}_H = \mathscr{O}_H \underset{\iota^*\mathscr{O}_S}{\otimes} \iota^*\mathscr{H}. \tag{3.21}$$

Then $\mathscr{G}_H$ is clearly a coherent $\mathscr{O}_H$ module having a Lie algebra structure as vector fields on H.

(2) The pair $(\mathscr{G}_H, H)$ is called the *localization* of $(\mathscr{G}, S)$ on H.

Our construction of the localization depends on the system of local coordinates ρ on H, and the selection of $\{Y_1, \ldots, Y_{n-m}\}$ that depends on the selection of $\{X, X'\}$. The characterization of $\mathscr{H}$ in (3.8) depends on functions t_j's and is not an intrinsic one related only to the orbit O. The $\mathscr{G}_H$ is, however, well-defined. To make precise, we have the following

Proposition 3.5. *Let $(Z_1, \ldots, Z_h, Y_1, \ldots, Y_{n-m})$ be the system of generators of $\mathscr{G}$ defined in the proof of Proposition 3.1 with respect to φ. Let $\{\varphi', (z', y')\}$ be another system of coordinates associated with $(Y'_1, \ldots, Y'_{n-m})$, and $(Z'_1, \ldots, Z'_{h'})$ be a corresponding system of generators of $\mathscr{H}'$. Then, there is a system of generators $(Y''_1, \ldots, Y''_{n-m}, Z''_1, \ldots, Z''_{h'})$ of $\mathscr{G}$ where $(Z''_1, \ldots, Z''_{h'})$ is a system of generators of $\mathscr{H}$ and $Y''_i \equiv Y_i \bmod \mathscr{H}\,\forall i$, such that*

$$(Z') = (Z'') + (Y'')\left(\frac{\partial y}{\partial z'}\right)[Z'_j z'_i]. \tag{3.22}$$

Here $[X_j x_i]$ denotes a matrix whose (i, j) entry is $X_j x_i$.

Proof. By direct calculation.

We omit the proof but note that since H is fixed as $\{\varphi^{-1}(z,0)\}$ and $\{\varphi^{-1}(z',0)\}$ in both coordinates, $(\partial y/\partial z')$ vanishes on H, which implies that $\mathscr{G}_H$ does not depend on the choice of $\{X, X'\}$.

Owing to Proposition 3.1, we get the following.

Proposition 3.6. *Set* $\mathscr{G}_{H_s} = (\tau_s)_* \mathscr{G}_H$ $(=\{\tau_s Z \tau_{-s} \mid Z \in \mathscr{G}_H\})$. *Then we get a family of pairs,*

$$\bigcup_s (\mathscr{G}_{H_s}, H_s) \to O. \tag{3.23}$$

We finally remark that in Lemma 3.2 (3), a slightly stronger result holds:

$$[a(y)Y_i, \mathscr{H}] \in \mathscr{H}, \text{ for a holomorphic function } a \text{ in } y. \tag{3.24}$$

If we consider the $\mathscr{O}_S$ module $\mathscr{G}/\mathscr{H}$, it is not a Lie algebra. If, however, we take the $\mathscr{O}_O$ module $\overline{\mathscr{G}/\mathscr{H}} := \mathscr{O}_O \otimes_{\mathscr{O}_S} \mathscr{G}/\mathscr{H}$, then by (3.24), $\overline{\mathscr{G}/\mathscr{H}}$ inherits a Lie algebra structure. In $\overline{\mathscr{G}/\mathscr{H}}$, the $\bar{Y}_i$'s (the images of Y_i) satisfy the relation $[\bar{Y}_i, \bar{Y}_j] = 0$, and $\bar{Y}_i = \partial/\partial t_i$. Thus we can consider the pair $(\overline{\mathscr{G}/\mathscr{H}}, O)$ on O. In Section 4, we will see some cases where we can really take $Y_i = \partial/\partial y_i$ on S.

4.

The localization procedure preserves several properties of $(\mathscr{G}, S)$.

Theorem 4.1. *Let* $(\mathscr{G}, S)$ *be a pair,* $(\mathscr{G}_H, H)$ *be the localization on* H *along an orbit* O *at* p. *Then the following holds.*

(1) *If* $(\mathscr{G}, S)$ *is LPH, (LG, LGF, respectively), then* $(\mathscr{G}_H, H)$ *is LPH (LG, LGF, resp.).*

(2) *Suppose a pair* $(\mathscr{G}, S)$ *has a RI* f *near* p. *Then* $f_H = f|_H$, *the restriction of* f *on* H, *is a RI of* $(\mathscr{G}_H, H)$. *Moreover, there is a system of local coordinates* (z, y) *at* p, *not depending on* f, *such that*

$$O = \{(z, y) \mid z = (0)\}, \tag{4.1}$$

$$f(z, y) = g(z, y) f_H(z), \qquad g \text{ is a unit in } \mathscr{O}_p. \tag{4.2}$$

Proof. (2) Take a coordinate system φ constructed in the proof of the Proposition 3.1. Then (4.1) is satisfied. Now, set

$$Y_i f = b_i f, \qquad i = 1, \ldots, n-m, \qquad e(y) = (\exp \textstyle\sum c_i Y_i)|_{c=y}. \tag{4.3}$$

Then, by Lemma 2.4,

$$f(e(y)\rho^{-1}(z)) = \exp\left(\sum y_i \int_0^1 e(ty)b_i\,dt\right) f(\rho^{-1}(z)). \tag{4.4}$$

Thus we get (4.2).

Let $Z \in \mathscr{G}_H$ and Z' be the corresponding element in $\mathscr{H}$. Then, $Z'f = a(z, y)f$. Since $Z'y_i = 0$, we have $Zf_H = a(z, 0)f_H$.

(1) Assume $(\mathscr{G}, S)$ to be LPH. Then there is $q \in H$ in a neighborhood of p such that $\dim \mathscr{G}(q) = n$. If not, H is included in D and then p cannot belong to O. It is clear by the forms of $Y_1, \ldots, Y_{n-m}$ in the proof of Proposition 3.1 that $\dim \mathscr{H}(q) = m$. Then $\dim \mathscr{G}_H(q) = m$ by Definition 3.4, which implies $(\mathscr{G}_H, H)$ is LPH.

Assume $(\mathscr{G}, S)$ to be LG. That is, $(\mathscr{G}, S)$ is LPH, $\mathscr{G} = \mathrm{Der}_S(\log D)$ with a reduced hypersurface D. As we have seen, $(\mathscr{G}_H, H)$ is LPH. We must show that $D_H := D \cap H$ is reduced and that $\mathscr{G}_H = \mathrm{Der}_H(\log D_H)$. The first assertion is clear by the trivialization given by Proposition 3.1. The inclusion $\mathscr{G}_H \subset \mathrm{Der}_H(\log D_H)$ is clear. Hence we show $\mathscr{G}_H \supset \mathrm{Der}_H(\log D_H)$. Let $Z \in \mathrm{Der}_H(\log D_H)$. Then, using Proposition 3.1, we can construct a family of vector fields $Z_y \in \mathrm{Der}_{H_y}(\log D_{H_y})$ by $Z_y = (\tau_y)_* Z$. Therefore, the correspondence $\psi(z, y) \mapsto Z_y$ gives a holomorphic vector field X on S, and $X \in \mathrm{Der}_S(\log D) = \mathscr{G}$. Therefore, $X \in \mathscr{H}$ and hence $Z \in \mathscr{G}_H$.

Assume $(\mathscr{G}, S)$ to be LGF. We have only to show that $\mathscr{G}_H$ is free. Saito's criterion says that $\mathscr{G} = \mathrm{Der}_S(\log D)$ is free if and only if $\Omega_S^n \otimes_{\mathscr{O}} \bigwedge^n \mathscr{G} = \mathscr{O}f$, $D = V(f)$. By the construction of $\mathscr{H}$, $\Omega_S^m \otimes_{\mathscr{O}} \bigwedge^m \mathscr{H} \simeq \mathscr{O}f$. This implies $\Omega_H^m \otimes_{\mathscr{O}_H} \bigwedge^m \mathscr{G}_H = \mathscr{O}_H f_H$, where $f_H = f|_H$, $D_H = V(f_H)$. We have already seen that $\mathscr{G}_H = \mathrm{Der}_H(\log D_H)$. Hence by Saito's criterion, $\mathscr{G}_H$ is free.

As a corollary, we get the following assertion, which is in [1], (3.6).

Corollary 4.2. *Let $(\mathscr{G}, S)$ and $(\mathscr{G}_H, H)$ be as in Theorem 4.1. Assume $(\mathscr{G}, S)$ be LG. Then, $(\mathscr{G}_H, H)$ is LG and there exists a system of local coordinates (z, y) at p, and a reduced* RI *f_H of $(\mathscr{G}_H, H)$, such that*

$$D = \{(z, y) \mid f_H(z) = 0\}, \qquad D_H = \{(z) \mid f_H(z) = 0\}, \qquad O = \{(0, y)\}. \tag{4.5}$$

Let the situation be the same as in Theorem 4.1 (2). In general, it happens that all RI's of $(\mathscr{G}, S)$ at p are units in $\mathscr{O}_p$. Assume that there exists a RI in $\mathfrak{m}_p$ (the maximal ideal at p), and let $f_1, \ldots, f_k$ be the totality of representatives (up to unit) of irreducible RI's in m_p. Recall that f_i's correspond to one-codimensional irreducible components $D_1, \ldots, D_k$ of D (Proposition 2.6). Set $D' = D_1 \cup \cdots \cup D_k = \{f_1 f_2 \ldots f_k = 0\}$, and consider $\mathscr{G}' = \mathrm{Der}_S(\log D')$ in a neighborhood of p. Then the following relation can be proved similarly as in [5, Prop. 2.4]:

$$\text{hom dim}_{\mathcal{O}_p} \mathscr{G}'_p + 2 \geq \text{codim}_p \operatorname{Sing}(D'). \tag{4.6}$$

When we study the pair $(\mathscr{G}, S)$, it is important to study the system of differential equations satisfied by

$$f^s = f_1^{s_1} f_2^{s_2} \cdots f_k^{s_k}, \qquad s = (s_1, \ldots, s_k) \in \mathbf{C}^k.$$

Let $\mathscr{D}_S$ be the sheaf of holomorphic linear differential operators on S. Consider the system of differential equations $\mathscr{M}_s$ and $\mathscr{M}(\mathscr{G})_s$:

$$\mathscr{M}_s = \mathscr{D}_S f^s, \tag{4.7}$$

$$\mathscr{M}(\mathscr{G})_s = \mathscr{D}_S \Big/ \Big(\sum_{X \in \mathscr{G}} \mathscr{D}_S (X - s \cdot a(X)) \Big), \quad \text{where}$$
$$a(X) = (a_1(X), \ldots, a_k(X)),\ X f_i = a_i(X) f_i,\ 1 \leq i \leq k. \tag{4.8}$$

Since the denominator of (4.8) annulates f^s, $\mathscr{M}_s$ is a quotient of $\mathscr{M}(\mathscr{G})_s$. Therefore, if we consider them on T^*S, the cotangent bundle of S, $\operatorname{Supp}(\mathscr{E} \otimes_{\mathscr{D}} \mathscr{M}(\mathscr{G})_s) \supset \operatorname{Supp}(\mathscr{E} \otimes_{\mathscr{D}} \mathscr{M}_s)$, where $\mathscr{E}$ is the sheaf of micro-differential operators on T^*S. The condition for the coincidence of those two systems will be treated in a continuing article.

Let $f_{H,1}, \ldots, f_{H,m}$ be RI of $(\mathscr{G}_H, H)$ defined by $f_{H,i} = f_i|_H$. Let us define $\mathscr{M}_{H,s}$ and $\mathscr{M}(\mathscr{G}_H)_s$ similarly as in (4.7) and (4.8). Then, the following is also a consequence of Theorem 4.1.

Corollary 4.3. *The following isomorphisms hold*:

$$\mathscr{M}_s \simeq \mathscr{D}_S \underset{\mathscr{D}_H}{\otimes} \mathscr{M}_{H,s}, \qquad \mathscr{M}(\mathscr{G})_s \simeq \mathscr{D}_S \underset{\mathscr{D}_H}{\otimes} \mathscr{M}(\mathscr{G}_H)_s. \tag{4.9}$$

Proof. Using the system of local coordinates (z, y) stated in Theorem 4.1 (2), and changing $\{f_i\}_i$ into $\{f_{H,i}\}_i$, the denominator of (4.8) equals $\sum \mathscr{D}_S\, \partial/\partial y_i + \sum_{X \in \mathscr{G}_H} \mathscr{D}_S (X - s \cdot a(X))$. Here, we identify $\mathscr{G}_H$ with $\iota_* \mathscr{G}_H$, $\iota: H \to S$ is the imbedding. Hence the second isomorphism of (4.9) holds. The first can be treated similarly.

5. Examples

Example 5.1. Let $\Delta_n(y_0, y_1, \ldots, y_n)$ be the discriminant of the binary form $y_0u^n + y_1u^{n-1}v + \cdots + y_nv^n$. Then in the following four cases, $(\mathrm{Der}_S(\log D), S)$ is LGF.

(1) $D = \{\Delta_n(x_0, \ldots, x_n) = 0\}$, $S = \mathbf{C}^{n+1}$, $n \geq 3$.
(2) $D = \{\Delta_{n+1}(1, 0, x_2, \ldots, x_{n+1}) = 0\}$, $S = \mathbf{C}^n$.
(3) $D = \{x_n\Delta_n(1, x_1, x_2, \ldots, x_n) = 0\}$, $S = \mathbf{C}^n$.
(4) $D = \{\Delta_n(1, x_1, x_2, \ldots, x_{n-1}, x_n^2) = 0\}$, $S = \mathbf{C}^n$.

Cases (2), (3), (4) are discriminants of finite reflection groups of types A_n, B_n, D_n and were already treated in [1], [2], [8]. Case (1) can be derived from the case for B_n by using the result in [8].

Example 5.2. Let us consider a pair $(\mathscr{G}, S)$ where $S = \mathbf{C}^4$ and $\mathscr{G} = \sum_{i=1}^5 \mathscr{O}X_i$,

$$(X_1, X_2, X_3, X_4, X_5) = (\partial_x, \partial_y, \partial_z, \partial_t)\begin{pmatrix} 2x & 0 & 0 & 0 & 0 \\ y & y & x & 0 & 0 \\ 0 & -2z & 2yz^2 & x+zy^2 & 0 \\ 0 & -2t & 2yt^2 & 0 & x+ty^2 \end{pmatrix}. \tag{5.1}$$

Then $X_1f = 6f$, $X_2f = 0$, $X_3f = 2y(t+z)f$, $X_4f = y^2f$, $X_5f = y^2f$, where $f = x(x+zy^2)(x+ty^2)$.

We have $[X_3, X_4] = -2yzX_4$, $[X_3, X_5] = -2ytX_5$, and all the others commute. There is one relation, $xX_2 - yX_3 + 2zX_4 + 2tX_5 = 0$. We have

$$\mathscr{I} = m \cdot f, \qquad m = x\mathscr{O} + y\mathscr{O} + z\mathscr{O} + t\mathscr{O}.$$

This pair is LPH and has relative invariants x, $x+zy^2$, $x+ty^2$. It is also LG with $D = \{f = 0\}$. Orbits and the decomposition mentioned in Section 2 are given by

$$O_1 = \{x = 0, yzt \neq 0\}, \qquad O_2 = \{x + zy^2 = 0, yzt \neq 0, z \neq t\},$$

$$O_3 = \{x + ty^2 = 0, yzt \neq 0, z \neq t\},$$

$$O_4 = \{(0, y, z, 0) \mid yz \neq 0\}, \qquad O_5 = \{(0, y, 0, t) \mid yt \neq 0\},$$

$$O_6 = \{(-y^2z, y, z, z) \mid yz \neq 0\}, \qquad O_7 = \{(0, y, 0, 0) \mid y \neq 0\},$$

$$O_{c,d} = \{(0, 0, z, t) \mid cz - dt = 0\} \backslash \{(0)\}, \qquad \forall c,d \in \mathbf{C},$$

$$O_8 = \{(0)\}, \qquad \text{and the open dense orbit } O_0 = S - D.$$

$$A_4 = S, \qquad A_3 = D, \qquad A_2 = \overline{O_4 \cup O_5 \cup O_6},$$

$$A_1 = \overline{O_7} \cup \{(0, 0, z, t)\}, \qquad A_0 = O_8.$$

We note that $K = \{(0, 0, z, t)\}$ is a singularity of D but consists of infinite orbits $O_{c,d}$. If we take into account the singularities and the Whitney condition (b), it is preferable to take the decomposition as

$$B_4 = S, \qquad B_3 = D, \qquad B_2 = A_2 \cup \{(0, 0, z, t)\},$$

$$B_1 = \overline{O_7} \cup O_{1,0} \cup O_{1,1} \cup O_{0,1}, \qquad B_0 = O_8,$$

and to give the stratification by $B_r \backslash B_{r-1}$, $r = 0, \ldots, 4$.

(1) Localization along $O_{1,0}$. $H = \{(x, y, z, 1)\}$.

$$(Z_1, Z_2, Z_3) = (\partial_x, \partial_y, \partial_z) \begin{pmatrix} 2x & 0 & 0 \\ y & x + y^2 & 0 \\ 0 & -2(1-z)zy & x + zy^2 \end{pmatrix}. \tag{5.2}$$

Then $Z_1 f_H = 6 f_H$, $Z_2 f_H = 2y(1+z) f_H$, $Z_3 f_H = y^2 f_H$, $f_H = x(x+y^2)(x+zy^2)$.

The localization is LGF with $D_H = \{f_H = 0\}$. Again, the point $\{(0, 0, c)\}$ forms a single orbit for each $c \in \mathbf{C}$.

(2) Localization along O_7. (LGF). $H = \{(x, 1, z, t)\}$.

$$(Z_1, Z_2, Z_3) = (\partial_x, \partial_z, \partial_t) \begin{pmatrix} x & 0 & 0 \\ z & x + z & 0 \\ t & 0 & x + t \end{pmatrix}. \tag{5.3}$$

Then $Z_1 f_H = 3 f_H$, $Z_2 f_H = f_H$, $Z_3 f_H = f_H$, $f_H = x(x+z)(x+t)$.

(3) Localization along $O_{1,1}$. $H = \{(x, y, 1+u, 1-u)\}$.

$$(Z_1, Z_2, Z_3, Z_4)$$

$$= (\partial_x, \partial_y, \partial_u) \begin{pmatrix} 2x & 0 & 0 & 0 \\ y & x + (1+u^2)y^2 & y(x+y^2) & uy^3 \\ 0 & 2(1-u^2)uy & -2ux & 2(x + y^2 - u^2 y^2) \end{pmatrix}. \tag{5.4}$$

Then

$$Z_1 f_H = 6f_H,\ Z_2 f_H = 4y f_H,\ Z_3 f_H = 4y^2 f_H,\ Z_4 f_H = 0,$$
$$f_H = x(x + y^2 + uy^2)(x + y^2 - uy^2).$$

The localization is LG with $D_H = \{f_H = 0\}$. Each point $\{(0, 0, u)\}$ is a single orbit.

We remark that $\Lambda = T^*_{\{0\}}S$ is a simple holonomic variety

$$\{(z-t)\partial_z\partial_t + (\partial_z - \partial_t)s\}f^s = 0. \tag{5.5}$$

We remark that

$$\partial_t(X_4 - y^2 s) - \partial_z(X_5 - y^2 s) = y^2\{(z-t)\partial_z\partial_t + (\partial_z - \partial_t)s\}.$$

References

[1] Saito, K., Theory of logarithmic differential forms and logarithmic vector fields, *J. Fac. Sci. Univ. Tokyo, Sect. IA* **27** (1980) 265–291.

[2] Saito, K., Yano, T., and Sekiguchi, J., On a certain generator system of the ring of invariants of a finite reflection group, *Comm. in Algebra* **8** (1980) 373–408.

[3] Sato, M. and Kimura, T., A classification of irreducible prehomogeneous vector spaces and their relative invariants, *Nagoya Math. J.* **65** (1977) 1–155.

[4] Sato, M., Kashiwara, M., Kimura, T., and Oshima, T., Micro-local analysis of prehomogeneous vector spaces, *Inventiones Math.* **62** (1980) 117–179.

[5] Terao, H., Arrangements of hyperplanes and their freeness I, *J. Fac. Sci. Univ. Tokyo, Sect. IA* **27** (1980) 293–312.

[6] Varadarajan, V. S., *Lie groups, Lie algebras, and their representations*, Prentice-Hall, 1974.

[7] Yano, T., Complete localization of prehomogeneous vector spaces, *Saitama Math. J.* **5** (1987) 27–33.

[8] Yano, T. and Sekiguchi, J., The microlocal structure of weighted homogeneous polynomials associated with Coxeter systems I, II, *Tokyo J. Math.* **2** (1979) 193–220; **4** (1981) 1–34.

On Carlson's Theorem for Holomorphic Functions

Kunio Yoshino
Department of Mathematics
Jochi University
Tokyo, Japan

1. Brief History of Carlson's Theorem

In 1914, F. Carlson proved the following theorem.

Theorem 1. *Let $F(z)$ be holomorphic and of the form $O(\exp(k|z|))$ for* $\operatorname{Re} z<0$ *and let $F(z)=O(\exp(-a|z|))$, where $a>0$, on the imaginary axis. Then $F(z)$ vanishes identically.*

As a corollary of Theorem 1, he also proved the following Theorem 2.

Theorem 2. *If $F(z)$ is holomorphic and of exponential type k $(<\pi)$ in the left half plane* $\operatorname{Re} z<0$ *and $F(-n)=0$ for $n=1,2,3,\ldots$, then $F(z)$ vanishes identically.*

ISBN 0-12-400466-0

There are several ways to prove Carlson's theorem. For the details, we refer to [3], [7] and [13]. In 1975, V. Avanissian and R. Gay generalized Carlson's theorem to the n-dimensional case by using the theory of analytic functionals. They consider Carlson's theorem for entire functions of exponential type. Inspired by their work, M. Morimoto and the author proved Carlson's theorem by using the theory of analytic functionals with non-compact carrier, so-called Fourier-ultrahyperfunctions in the 1-dimensional case. After this work, P. Sargos and M. Morimoto proved the following Theorem 3.

Theorem 3. *Let $F(z_1, z_2, \ldots, z_n)$ be holomorphic in the direct product of left half planes $\{(z_1, z_2, \ldots, z_n) \in \mathbf{C}^n;\ \mathrm{Re}\, z_i < 0\}$ and satisfy the following estimate: for every positive numbers ε and ε', there exists $C_{\varepsilon,\varepsilon'}$ such that*

$$|F(z_1, z_2, \ldots, z_n)| \leq C_{\varepsilon,\varepsilon'} \exp\left(\sum_{i=1}^{n} a_i x_i + k_i |y_i| + \varepsilon |z| \right)$$

$(x_i = \mathrm{Re}\, z_i \leq -\varepsilon')$, where $k_i < \pi$ and $F(-m_1, -m_2, \ldots, -m_n) = 0$ $((m_1, m_2, \ldots, m_n) \in \mathbf{N}^n)$. Then $F(z_1, z_2, \ldots, z_n)$ vanishes identically.

Here we give a sketch of the proof of Carlson's theorem by using the theory of analytic functionals with non-compact carrier. First we consider the 1-dimensional case. By virtue of a Paley–Wiener type theorem for analytic functionals with non-compact carrier, there exists an analytic functional T such that $F(z) = \tilde{T}(z) = \langle T_\zeta, \exp(\zeta z) \rangle$, where $\tilde{T}(z)$ denotes the Fourier–Borel transformation of the analytic functional T. The Avanissian-Gay transform of T is defined by

$$G_T(w) = \langle T_\zeta, (1 - w \exp(\zeta))^{-1} \rangle.$$

$G_T(w)$ has following properties:

(1) $G_T(w)$ is holomorphic in $\mathbf{C} \backslash \exp(-L)$, where $L = [a, \infty) + i[-k, +k]$.

(2) $G_T(w) = -\sum_{n=1}^{\infty} T(-n) w^{-n}$ $(|w| > \exp(-a))$

(3) (inversion formula)

$$\tilde{T}(z) = (2\pi i)^{-1} \int_{\partial L} G_T(\exp(-\zeta)) \exp(z\zeta)\, d\zeta.$$

From the assumption for $F(z)$, $\tilde{T}(-n) = F(-n)$ is equal to zero for all natural numbers n. So $G_T(w)$ vanishes identically because of (2). Hence we obtain the desired result by virtue of the inversion formula (3). Now we consider the 2-dimensional case. We fix the second variable $z_2 = -m_2$, where m_2 is a natural number. We apply Carlson's theorem to $F(z_1, -m_2)$. Immediately, we have $F(z_1, -m_2) = 0$. Next we consider $F(z_1, z_2)$ fixing z_1. Then we obtain the 2-dimensional Carlson's theorem. For the details of the proofs by means of analytic functionals, we refer to [2], [3], and [12].

2. Some Applications of Carlson's Theorems

In this section, we give some applications of Carlson's theorems, namely, Theorem 1, Theorem 2 and Theorem 3. We use special terminologies in quantum physics without explanations.

2.1. Unique Analytic Continuation of Scattering Amplitude in the Angular Momentum Plane

The scattering amplitude $f(E, \cos\theta)$ has the following partial wave expansion:

$$f(E, \cos\theta) = (2ik)^{-1} \sum_{l=0}^{\infty} (2l+1)(S_l(k)-1)P_l(\cos\theta),$$

where $S_l(k)$ is the l-th partial wave scattering matrix and $P_l(\cos\theta)$ denotes the Legendre polynomial of degree l. If $S_l(k)$ is continued analytically to the complex l-plane (the angular momentum plane), then we obtain

$$f(E, \cos\theta) = (ik)^{-1} \int_\Gamma (l+\tfrac{1}{2})(S_l(k)-1)P_l(\cos\theta)(-1)^l \cos(\pi l)^{-1}\, dl$$

where $P_l(\cos\theta)$ is the Legendre function of the first kind and Γ is the contour shown in Fig. 1.

By the residue theorem, we have

$$f(E, \cos\theta) = -(2k)^{-1} \int_{-i\infty}^{+i\infty} (S_l(k)-1)P_l(-\cos\theta)(\cos\pi l)^{-1}\, dl$$

$$-\pi k^{-1} \sum_{j=0}^{M} \sigma_j \lambda_j(k)(\cos\pi\lambda_j)P_{\lambda_j - 1/2}(-\cos\theta),$$

where the $\lambda_j(k)$ are known as Regge poles. This is the so-called *Watson-Sommerfeld transformation.* Carlson's theorem invokes the uniqueness of

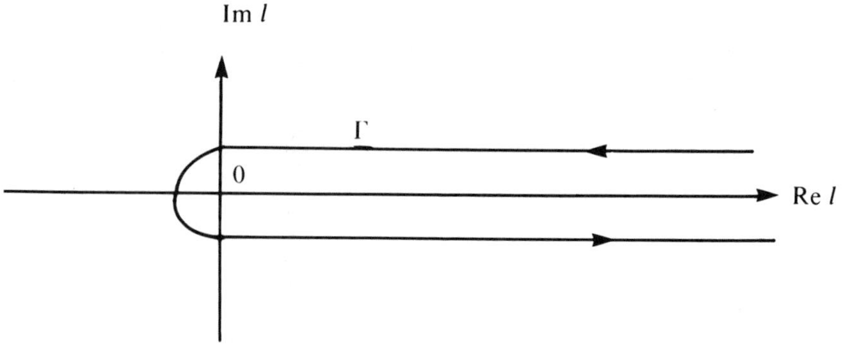

FIGURE 1

analytic continuation of $S_l(k)$ in the l-plane (the complex angular momentum plane). For the details of these discussions, we refer to [4] and [9].

2.2.

Carlson's theorem also invokes the uniqueness of analytic continuation from the temperature Green function to the 2-time Green function in the theory of quantum statistical mechanics. For the details of these discussions, we refer to [1].

2.3.

Carlson's theorem (Theorem 1) is also useful in proving the absence of positive eigenvalues of N-body Schrödinger operators. For the details of this argument, we refer to [13].

2.4.

We can make use of Carlson's theorem (Theorem 1) to prove the extension problem of real analytic solutions of partial differential equation $P(D)u = 0$, where $P(D)$ is not an elliptic operator. For the details, we refer to [5].

3. Existence of Interpolating Function

$\mathrm{Exp}(\mathbf{C}^n; K)$ denotes the space of entire functions of exponential type of $H_K(z)$, where $H_K(z)$ is a supporting function of the compact convex set

K in $\mathbf{C}^n$. As shown in Section 1, the mapping

$$\begin{array}{ccc} \operatorname{Exp}(\mathbf{C}^n; K) & \rightarrow & \text{sequence space} \\ \Psi & & \Psi \\ F(z) & \mapsto & \{F(-m_1, -m_2, \ldots, -m_n)\} \end{array}$$

is injective under suitable conditions. In this section we investigate the surjectivity of the above mapping.

Theorem 4. *Following conditions are equivalent.*

(4) $a_{m_1,m_2,\ldots,m_n} = F(-m_1, -m_2, \ldots, -m_n)$ *where F belongs to* $\operatorname{Exp}(\mathbf{C}^n; K)$ *and* K_i *(i-th projection K) is contained in* $\{z_i \in \mathbf{C};\ |\operatorname{Im} z_i| < \pi\}$.

(5) *The power series* $\sum a_{m_1,m_2,\ldots,m_n} w_1^{m_1} w_2^{m_2} \ldots, w_n^{m_n}$ *is analytically continued to* $\prod_{i=1}^{n} (\mathbf{C} \backslash \exp(-K_i))$ *and vanishes at infinity.*

(6) *For any positive number* ε, *there exists a constant* C_ε *such that*

$$|\Sigma\, a_m c_m| \leq C_\varepsilon \sup_{\zeta \in K_\varepsilon} |\Sigma\, c_m \exp(m\zeta)|,$$

where c_m *is a sequence such that the* c_m*'s are all 0 except for finitely many terms, and* K_ε *denotes the* ε*-neighbourhood of K.*

In the case of 1-dimension, the equivalence between (4) and (5) is due to Le Roy and Lindelöf. The equivalence between (4) and (6) is proved by G. Rauzy. See [11] and [12].

Example 1 ([15]). Let a_n be $P_n(\cos\theta)$, the Legendre polynomial of degree n. The following generating expansion is well known.

$$\sum_{n=0}^{\infty} P_n(\cos\theta) w^n = (1 - 2w\cos\theta + w^2)^{-1/2}$$

Note that $(1 - 2w\cos\theta + w^2)^{-1/2}$ is holomorphic in $\mathbf{C}\backslash\exp(-K)$, where $K = [-i\theta, i\theta]$, and vanishes at infinity. In this case the interpolating function is given by the Schlöfli integral

$$P_z(\cos\theta) = (2\pi i)^{-1} \int_C w^z (1 - 2w\cos\theta + w^2)^{-1/2}\, dw,$$

where C is the contour shown in Fig. 2.

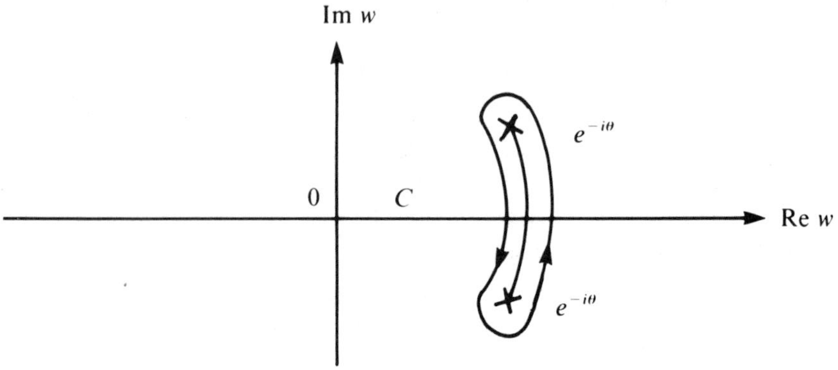

FIGURE 2

4. Carlson's Theorem for Holomorphic Functions with Proximate Order $1+\log(\log r)/\log r$

In this section we consider Carlson's theorem for holomorphic functions with proximate order $1+\log(\log r)/\log r$. First we define the Mellin transform of such functions. Let $F(z)$ be holomorphic in the right half plane $\{z;\ \mathrm{Re}\ z>0\}$ and satisfy the following estimate: For any $\varepsilon>0$ and $\varepsilon'>0$, there exists $C_{\varepsilon,\varepsilon'}$ such that

$$|F(z)|\leq C_{\varepsilon,\varepsilon'}\exp(x\log x+k|y|+\varepsilon|z|) \qquad (*)$$

for $x=\mathrm{Re}\ z\geq\varepsilon'$.

Now we define the Mellin transform $MF(w)$ of $F(z)$ as follows:

$$MF(w)=(2i)^{-1}\int_{c-i\infty}^{c+i\infty}F(z)(-w)^z(\sin\pi z)^{-1}\,dz.$$

$MF(w)$ has the following properties.

(*i*) $MF(w)$ is holomorphic in $\{w\in\mathbf{C};\ k<|\arg w|\leq\pi\}$

(*ii*) $MF(w)$ has the asymptotic expansion

$$MF(w)\sim\sum_{n=1}^{\infty}F(n)w^n.$$

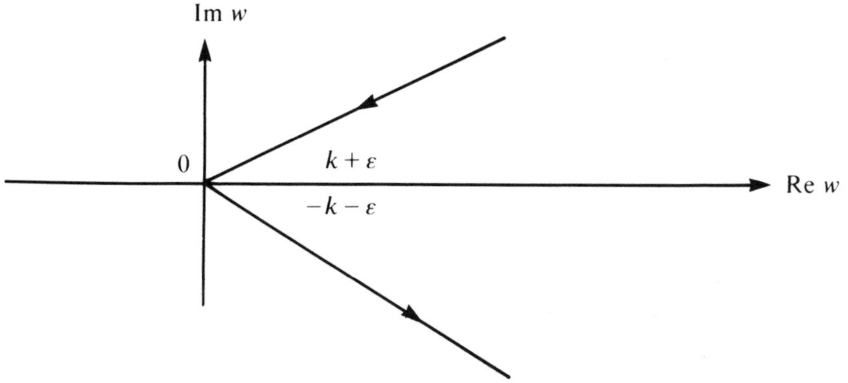

FIGURE 3

If k is less than $\pi/2$, this is a strong asymptotic expansion. For the details of (i) and (ii), we refer to [6] and [16]. Furthermore, we have the following integral formula:

$$F(z)=(2\pi i)^{-1}\int_{\Gamma_1} MF(w)w^{-z-1}\,dw,$$

where Γ_1 is the contour shown in Fig. 3.

Now we will prove the following Theorem 5.

Theorem 5. *Let $F(z)$ be a holomorphic function defined on the right half plane $\{z\in\mathbf{C}\colon \operatorname{Re} z>0\}$, and satisfying the estimate $(*)$. If $F(n)=0$ valid for all $n\in\mathbf{N}$ and* k is less than *$\pi/2$, then $F(z)$ vanishes identically.*

Proof. First we consider the Mellin transform $MF(w)$ of $F(z)$. From the property (ii) and the assumption $F(n)=0$, we see that $MF(w)$ has a 0-asymptotic expansion. Note that this is a strong asymptotic expansion. So $MF(w)$ vanishes identically (see [13]). By virtue of (i), we see that $F(z)$ vanishes identically.

Remark 1. We can generalize Theorem 5 to the n-dimensional case by the same argument as in Theorem 3.

Remark 2. The assumption $k<\pi/2$ is crucial in Theorem 5. For example $F(z)=1/\Gamma(1-z)$ (where Γ denotes the Gamma function) satisfies all assumptions in Theorem 5 with $k=\pi/2$. But in this case $F(z)$ does not vanish identically.

References

[1] A. A. Abrikosov, L. P. Gorokov, and I. E. Dzyalonski, *Methods of Quantum Field Theory in Statistical Physics.* Prentice Hall, 1963.

[2] V. Avanissian and R. Gay, Sur une transformation des fonctionnelles analytiques et ses applications aux fonctions entieres des plusieurs variables, *Bull. Soc. Math. France* **103** (1975) 341–384.

[3] R. Boas, *Entire Function.* Academic Press, New York, 1954.

[4] V. De Alfaro and T. Regge, *Potential Scattering.* North Holland, Amsterdam, 1965.

[5] A. Kaneko and T. Kawai, Hyperfunction and Partial differential equation, *Sugaku* **25**, No. 3 (1973) 47–61 (in Japanese).

[6] Yu. A. Kubyshin, Sommerfeld–Watson summability method of perturbation series, *Theoretical and Mathematical Physics*, **58** No 1 (1984) 91–96.

[7] B. Ja. Levin, *Distribution of Zeros of Entire Functions*, Translation of Mathematical Monograph. American Mathematical Society, 1964.

[8] M. Morimoto and K. Yoshino, A uniqueness theorem for holomorphic functions of exponential type, *Hokkaido Math. J.* **7** (1978) 259–270.

[9] H. M. Nussenzveig, *Causality and Dispersion Relations.* Academic Press, New York, 1972.

[10] V. P. Palamodov, From hyperfunction to analytic functionals, *Soviet Math. Dokl.* **18** (1977) 975–979.

[11] G. Rauzy, Les zeros entieres des fonctions entieres de type exponentielle, *Seminaire de Theorie des Nombres*, Annee 1976–1977.

[12] P. Sargos and M. Morimoto, Transformations des fonctionnelles analytiques a porteurs non-compacts, *Tokyo J. Math.* **4** (1981) 457–492.

[13] M. Reed and B. Simon, *Analysis of Operators* (Method of Modern Mathematical Physics Vol 4). Academic Press, New York, London, 1978.

[14] E. C. Titshmarsh, *Theory of Functions.* Oxford Univ. Press, London, New York, 1958.

[15] K. Yoshino, Some examples of analytic functionals and their transformations, *Tokyo J. Math.* **5** (1982) 479–490.

[16] K. Yoshino, Lerch's theorem for analytic functionals with non-compact carrier and its applications to entire functions, *Complex Variables* **2** (1984) 303–318.